ESTATES GAZETTE LAW REPORTS

Estates Gazette Law Reports

1977

Volume 1

Series edited by
Barry Denyer-Green LLM PhD FRICS
of the Middle Temple, barrister

Cases edited by
J Muir Watt OBE MA
of the Inner Temple, barrister

Audrey Boyle BA
Publisher's editor

Estates Gazette

Routledge
Taylor & Francis Group
LONDON AND NEW YORK

EG Books is an imprint of Routledge
2 Park Square, Milton Park, Abingdon, Oxon OX14 4RN
711 Third Avenue, New York, NY 10017, USA
Routledge is an imprint of the Taylor & Francis Group, an informa business

ISBN for complete set of 2 volumes: 0 7282 0221 2
ISBN for this volume 0 7282 0222 0
ISSN 0951-9289

For information on all EG Books publications
visit our website www.routledge.com

CONTENTS

TABLE OF CASES

Titles of cases shown in bold type
Names of parties reversed shown in ordinary type

INDEX OF SUBJECT-MATTER

AGRICULTURE

A

COURT OF APPEAL

February 15 1977

(Before Lord Justice MEGAW, Lord Justice LAWTON and Lord Justice Geoffrey LANE)

JOHNSON AND ANOTHER v MORETON

Estates Gazette March 12 1977

(1977) 241 EG 759

Agricultural Holdings Act—Security of tenure—Whether "Contracting-out" clause in agreement valid and effective —Fixed term—Agreement not to serve counternotice under section 24(1)—Policy of Act—Clause of no effect—

B **Argument based on section 24 (2) (e) self-contradictory— Tenant's appeal from county court decision allowed**

This was an appeal by the tenant from the decision of His Honour Judge Harrison-Hall at Warwick County Court, holding in favour of the landlords, in a case stated by the arbitrator, Edward Righton FRICS, that an agreement to exclude the operation of section 24 (1) of the Agricultural Holdings Act 1948 was not contrary to the Act nor against public policy and that a notice by the landlords under section 24 (2) (e), based on the tenant's breach of this agreement, was effective to render section 24 (1) inapplicable.

C G B H Dillon and Derek Wood (instructed by Ellis & Fairbairn, agents for Wright, Hassall & Co, of Leamington Spa) appeared for the appellant, and P G Langdon-Davies (instructed by Rotherham & Co, of Coventry) represented the respondents.

Giving the judgment of the court, MEGAW LJ said: This appeal raises an important question as to the security of tenure which is provided for tenants of agricultural holdings by the Agricultural Holdings Act 1948, which we shall call "the Act." One of the purposes of that Act is to provide generally that a tenancy of an agricultural holding, even though by the agreement creating the tenancy it has been

D agreed that it shall last only for a definite number of years, shall nevertheless continue indefinitely after the expiration of that period, unless an independent body of experts, the agricultural land tribunal, decides, on one or other of a number of specific grounds set out in the Act, that a notice to quit, given by the landlord, should be effective. The question raised by this appeal is, in effect, whether it is permissible for the parties to agree that that security of tenure shall not apply, so that the tenancy can, at the wish of the landlord, be made to terminate on the expiration of the period stated in the tenancy agreement. In other words, is the Act to be read as permitting its general purpose of overriding a

E contractual provision for a fixed term itself to be overridden by another contractual provision? We say at once that we think the answer is clearly "No."

The present landlords of Village Farm, Offchurch, Warwickshire, are the plaintiffs in this action, Henry Leslie Johnson and Miss Caroline Anne Johnson. Village Farm is a farm of some 236 acres. It is agricultural land within the meaning of that phrase in section 1 (2) of the Act. By a lease dated August 11 1967 the landlords' predecessors in title granted to the tenant, Samuel Moreton, a tenancy of the farm for 10 years, the term commencing on January 1 F

1967. Clause 27 of the lease reads:

> The tenant agrees to give possession of the whole of the farm to the landlords immediately upon the determination of the term hereby granted and not in any event to serve a counternotice under section 24 (1) of the Agricultural Holdings Act 1948 or to take any steps to claim the benefit of any statutory provision granting security of tenure which may be in force at the time of the determination thereof.

The intention of the clause is to deprive the tenant, so far as agreement between the parties can do so, of the benefit of any statutory provision intended to give an G agricultural tenant security of tenure. The reference to section 24 (1) of the Act is because that is the vital provision for this purpose in the Act.

In order that the facts of this case—few and undisputed—may be understood, it is necessary first to refer to the relevant statutory provisions. We read first section 3 subsections (1), (2) and (4) of the Act:

> (1) A tenancy of an agricultural holding for a term of two years or upwards shall, instead of terminating on the expiration of the term for which it was granted, continue (as from the expiration of that term) as a tenancy from year to year, but otherwise on the terms of the original tenancy so far as applicable, H unless, not less than one year nor more than two years before the date fixed for the expiration of the term, a written notice has been given by either party to the other of his intention to terminate the tenancy.
>
> (2) A notice given under the foregoing subsection shall be deemed, for the purposes of this Act, to be a notice to quit. . . .
>
> (4) This section shall have effect notwithstanding any agreement to the contrary.

So far, then, the Act provides, with reference to a tenancy such as that with which we are concerned, that, although the parties have by their contract agreed that it shall terminate at the end of a specified number of years, nevertheless the contract shall not have that effect unless a notice to quit J shall have been given, complying with the timetable in subsection (1). Subsection (4) provides that this statutory requirement may not be overridden by contract. If this were all, the effect, as regards security of tenure, would be small. But it is not all. It is the foundation on which the structure of security of tenure is built. The Act, having by this section specified the requirement of a notice to quit, then by section 24 (1) goes on to provide what shall be the effect, if and when such notice is given. The landlords' contention in the present case is that, although the parties cannot contract out of the requirements as to the giving of notice laid down by the Act, they are nevertheless free K to contract out of the consequences of the giving of that notice, as laid down by the Act.

Before we come to section 24 we should refer briefly to section 23, although it is of no direct relevance to the present case. The landlords rely on the fact that it, in conformity with section 3 but in contrast with section 24, contains express words overriding inconsistent contractual provisions. The opening words of section 23 (1) are: "A notice to quit an agricultural holding or part of an agricultural holding

A shall (notwithstanding any provision to the contrary in the contract of tenancy of the holding) be invalid if it purports to terminate the tenancy before the expiration of 12 months from the end of the then current year of tenancy." There follows a proviso which sets out circumstances in which the general provision as to the expiry date of the notice shall not apply. They include, among others, cases where a receiving order has been made against the tenant and cases where a public utility authority or a government department or a local authority require possession of the land for use for certain purposes.

Then comes the vital section, section 24, which is marginally B noted "Restrictions on operation of notice to quit." This, as we have said, is the effective enactment in the scheme of the Act for providing security of tenure. Section 24 (1), amended, reads:

> (1) Where notice to quit an agricultural holding or part of an agricultural holding is given to the tenant thereof, and not later than one month from the giving of the notice to quit the tenant serves on the landlord a counternotice in writing requiring that this subsection shall apply to the notice to quit, then, subject to the provisions of the next following subsection, the notice to quit shall not have effect unless the agricultural land tribunal consents to the operation thereof.

C It will be remembered that clause 27 of the lease in the present case, in addition to its opening words, which are plainly and admittedly of no effect because of the Act, and in addition to its closing words, which purport to override *any* statutory provision granting security of tenure, specifically refers to section 24 (1) in these words: "The tenant agrees . . . not in any event to serve a counternotice under section 24 (1). . . ."

Section 24 (2) then specifies in subparagraphs (a) to (g) certain circumstances in which section 24 (1) shall not apply. We propose to quote only subparagraph (e):

> D (2) The foregoing subsection shall not apply where . . . (e) at the date of the giving of the notice to quit the interest of the landlord in the agricultural holding to which the notice relates had been materially prejudiced by the commission by the tenant of a breach, which was not capable of being remedied, of any term or condition of the tenancy that was not inconsistent with the fulfilment by the tenant of his responsibilities to farm in accordance with the rules of good husbandry, and it is stated in the notice that it is given by reason of the matter aforesaid.

Section 25 then sets out the specific and limited circumstances in which the agricultural land tribunal may allow the notice to quit to be effective.

E Against this background of the relevant provisions of the Act we can summarise briefly the undisputed facts and the proceedings which have led up to this appeal by the tenants. We have already at the outset of this judgment mentioned the relevant terms of the lease of August 11 1967. The 10 years specified in it would expire on December 31 1976. The landlords desired to recover possession on that date. For that purpose they initiated what we would describe as a ritual dance. The landlords' first step was to serve a notice (in fact, three notices, but that complication may be disregarded) on November 27 1975. These were notices properly given under section 3 (1) of the Act, notifying the F tenant of the landlords' intention to terminate the tenancy on December 31 1976. So section 24 (1) of the Act came into operation, unless, at any rate, its operation was effectively excluded by the terms of clause 27 of the lease. The tenant then, acting or purporting to act under section 24 (1), on December 16 1975 served the counternotice contemplated by that subsection, invoking its operation. On December 22 the landlords took their next step in the ritual by serving another notice on the tenant. This second notice purported to be given under section 24 (2) (e) of the Act. We quote from it (page 33 of the bundle of documents):

> . . . for the following reasons: That at the date of the giving G of this notice the interest of the landlords in the abovementioned holding has been materially prejudiced by the commission by you of a breach, which is not capable of being remedied, of a term or condition of the said lease being a term or condition which is not inconsistent with the fulfilment of your responsibilities to farm in accordance with the rules of good husbandry, namely serving a counternotice contrary to clause 27 of your said lease.

So the landlords were saying: "You, the tenant, have bound yourself by contract not to serve a counternotice under section 24 (1). Your action in serving the counternotice is a breach of that term of the tenancy, and it is a H breach which has all the characteristics, positive and negative, specified in section 24 (2) (e). Therefore section 24 (1) does not apply. Therefore your purported counternotice is ineffective." This reasoning is not sound unless (i) the relevant part of clause 27 is an effective, legally binding, contractual term, despite the provisions of the Act; and (ii) the counternotice is a breach of the terms of the tenancy such as is contemplated by, and covered by, section 24 (2) (e). We think neither of those conditions is fulfilled.

The tenant then, on December 23 1975, served a further notice on the landlords, stating that, without prejudice to the question whether the landlords' second notice is bad in law, J he, the tenant, wishes to contest the reason given in the notice, and he requires the question to be determined by arbitration under the Act. So issue was joined. An arbitrator was appointed. His task was an easy one. The parties agreed that he should state a special case for the opinion of the court in accordance with paragraph 24 of the 6th Schedule to the Act, without hearing evidence. The terms of the special case and the question for the court were agreed between the parties or their advisers. No material facts are stated in the special case beyond those which we have mentioned. The contentions of the parties were set out. K The question of law was stated as: " . . . whether on the agreed facts of this case the said reason" (that is, "the reasons"—so described in the plural—in the landlords' second notice, which we have already quoted) "is good or not."

The special case was argued before Judge Harrison-Hall. On October 13 1976 he gave his decision in a full and careful reserved judgment. He held in favour of the landlords, taking the view that it was not contrary to the express or implied words of the Act, nor against public policy, that the parties should agree to exclude the operation of section 24 (1); and that the provisions of section 24 (2) (e) were applicable, so that the service by the tenant of the counter- L notice operated as a breach of contract under that paragraph, so that section 24 (1) did not apply.

The tenant appeals. Broadly, his submissions are, first, that it is clear from the terms of the Act that any contractual term purporting to exclude the operation of section 24 (1) is of no effect; and, secondly, that, even if that were not so, section 24 (2) (e), on which the landlords rely as an essential part of their case, can on its own wording have no application so as to exclude the operation of section 24 (1).

On the first issue the landlords contend, as they contended successfully in the county court, that there is nothing in the M Act, whether by its express terms or by implication, or on any ground related to the policy of the Act or public policy, why the parties' express contractual agreement, that section 24 (1) should be excluded, is ineffective. They rely upon the principle, stated in various authorities and textbooks, that it is open to a party to a contract to exclude by a contractual term a statutory provision which has been enacted for his benefit, provided that the intended exclusion does not prejudice the rights of other persons and provided that it is not inconsistent with the policy of the Act or public policy. The landlords also rely, as an indication of the true meaning and effect of the Act, in relation to this question of

A contractual exclusion, upon the fact that section 24 contains no stipulation such as "notwithstanding any agreement to the contrary"; whereas such stipulations are expressly included in section 3 by subsection (4) thereof, in section 11 (1), in section 15, section 23 (1) and in section 65 (1). The landlords rely on such cases as *Premier Dairies Ltd v Garlick* [1920] 2 Ch 17 and *Kennedy v Johnstone* 1956 SC 39, in which, in relation to other provisions of the Act, it was held that they could be excluded, validly, by contractual terms. In the latter case it was a provision, referable to Scots law, in the corresponding Agricultural Holdings (Scotland) Act 1949. The landlords also rely on *Gladstone v Bower* [1960] B 2 QB 384. In that case this court held that it was unable to fill in a gap in the Act, even though it was satisfied that the gap had been left inadvertently by the legislature, which, if it had noticed the gap, would have filled it. We are told that the legislature has not, in fact, done anything since then about the supposed gap: and it is suggested that this may show that the court was wrong in its supposition regarding the policy of the Act as to security of tenure.

In answer to these submissions, the tenant, in addition to maintaining his own primary submission that it would make nonsense of the security of tenure provisions of this Act if the parties were able by their agreement to exclude section C 24 (1), relied upon certain other authorities, such as *Salford Guardians v Dewhurst* [1926] AC 619, as defining the general principle. The tenant also submitted that the express inclusion in other sections of the Act of "notwithstanding any agreement to the contrary" carried no inference that the legislature intended that section 24 (1) could effectively be rendered inoperative by agreement between the parties. The tenant submitted that the true test was as stated by Streatfeild J in *Coates v Diment* [1951] 1 All ER 890 at p 896/C. The tenant also relied on what he claimed was the analogy of the Rent Restrictions Acts, where, he asserted, contracting out of D the provisions of the various Rent Acts plainly was not permitted, even though there were no express stipulations to that effect in the relevant Acts. To this argument, the landlords replied that the Rent Restrictions Acts provide no true analogy; for the right to contract out of the security of tenure provisions of these Acts is, indeed, expressly precluded by the wording of the Acts, even though it is not done in the form of a stipulation saying "notwithstanding any agreement to the contrary." The tenant for his part further submits that *Gladstone v Bower* is not relevant: there is here no gap in the legislation.

We think that the tenant is right on the first issue, and this E appears without the necessity of reference to authorities or analogies. It is simply a matter of the plain, commonsense meaning and effect of the Act, apparent on its own terms. The point is essentially a simple one. Sections 3 and 24 must be looked at together. Their purpose is, we should have thought, beyond dispute. They apply to cases in which the landlord and the tenant have, by their private contract, purported to agree that the tenancy which they are agreeing to create shall end after a specified number of years. For simplicity, let us take the present case: after 10 years. The whole purpose of section 3 and section 24, read together, is to provide that, notwithstanding the purported agreement of F the parties, the agreement shall *not* necessarily and automatically end after 10 years, whatever notice may have been given. It would make nonsense of the legislation if it were to be read as providing that: (i) although the parties may have agreed that the tenancy shall end after 10 years, that agreement *shall not* have effect; nevertheless (ii) the parties may, validly and effectively, agree that the provision that the tenancy shall end after 10 years *shall* have effect.

Whatever subtleties may be canvassed, the clear and simple fact remains: if the landlords are right in their submission, there was no real purpose in the enactment of sections 3 and 24. Parliament might just as well have left matters alone. On that basis the parties can agree that the tenancy shall end G after 10 years, and that term of the contract shall be binding and effective. Why go through the elaborate farce of legislation which appears in the first instance to prevent the parties from making a binding and effective agreement limiting the tenancy to 10 years, if, at the same time, it is open to the parties, by another clause in the self-same agreement, to achieve the very thing which they are supposed to be prevented from achieving? The answer, surely, must be that the operation of section 24 (1) cannot be excluded by agreement, and the provision of clause 27 of the lease on which the landlords rely is, indeed, of no effect.

H Having reached that conclusion on that simple ground, which is in no way affected by any of the authorities cited, either as establishing general propositions of construction or in relation to this particular Act, we do not propose to discuss at any great length the other detailed and elaborate submissions which were put before us. It will, we hope, be clear that, if we are right in our view on the main issue, there was no reason why section 24 should have included any express words, even if as a matter of drafting it would have been practicable, such as "notwithstanding any agreement to the contrary between the parties."

However, we should add that in our opinion it would be J impossible to hold—the second proposition which the landlords have to establish in order to succeed—that section 24 (2) (e) is applicable so as to exclude the operation of section 24 (1). For the landlords, as we understand it, it is admitted —indeed, we believe it is put forward as an essential part of their submissions—that, even though (as they contend) clause 27 is a perfectly valid and effective contractual provision which would be recognised by the courts as binding on the tenant, nevertheless the tenant cannot be stopped from serving his counternotice under section 24 (1). The counternotice is in breach of a valid contractual provision; but, since the Act says that the counternotice may be given, it K may be given. If that part of the landlords' submission is right, we should have thought that it was the end of their case. For if section 24 (1) applies, all its consequences must follow. It cannot both apply and at the same time not apply. Yet this is what the landlords' argument appears to involve. Assume, however, in favour of the landlords, that we are wrong in that view: section 24 (1) having applied, the landlords then treat the tenant's counternotice as a breach of contract. They then serve a further notice under section 24 (2) (e). If this is a valid, proper and effective notice under section 24 (2) (e), the effect of it is that, by reason of the opening words of section 24 (2), section 24 (1) "shall not L apply." Having applied, it shall, so it appears to be submitted, as a result of its application, "not apply." But section 24 (2) (e) includes as one of its requirements that the breach committed by the tenant (in this case, it is said, the service of the counternotice) "was not capable of being remedied." If the landlords' argument were right, so that they were entitled to succeed in this action, the breach committed by the tenant was, indeed, "capable of being remedied." It was capable of being remedied by the simple procedure adopted by the landlords: the service of a second notice, which, if the landlords were right in all other respects, would M produce the result that section 24 (1) "shall not apply." The argument goes round in a circle.

There are various other difficulties which we think would face the landlords, if they reached the stage of establishing "a breach of a term of the tenancy" (namely, of the provision of clause 27, purporting to prevent the tenant from serving a counternotice under section 24 (1)). But we do not propose to lengthen this judgment by pursuing them. We allow the appeal and answer the question stated in the special case in these terms:

"The reason stated in paragraph 11 of the special case is not a good reason."

A The appeal was allowed with costs in the Court of Appeal and below. Leave to appeal to the House of Lords was refused.

COURT OF APPEAL

January 20 1977

(Before Lord Justice CAIRNS, Lord Justice ROSKILL and Lord Justice BRIDGE)

AVON COUNTY COUNCIL v CLOTHIER

B

Estates Gazette June 25 1977

(1977) 242 EG 1048

Agricultural Holdings Act 1948, proviso to section 2(1)—" Use of the land only for grazing or mowing during some specified period of the year "—Licence for grazing—Effect of later agreement for stabling ponies—Gratuitous permission or ancillary use—Proviso applied—Appeal from county court dismissed and possession order upheld

This was an appeal from a decision of Judge Ewart James in favour of the respondents, Avon County Council, who had claimed possession of certain grazing land occupied by the C appellant under agreements with the respondents. The respondents contended that the agreements were covered by the proviso to section 2(1) of the Agricultural Holdings Act 1948. The appellant contended that he had become entitled under section 2(1) to a tenancy from year to year. The facts are stated in the judgment of Bridge LJ.

C F Sara (instructed by Riders, agents for Bobbetts, Harvey & Grove, of Bristol) appeared on behalf of the appellant; B K Levy (instructed by the Solicitor, Avon County Council) represented the respondents. The court did not call on Mr Levy.

D Giving the first judgment at the invitation of Cairns LJ, BRIDGE LJ said: This is an appeal from a judgment and order of His Honour Judge Ewart James, dated June 4 of last year, sitting at Bristol whereby he ordered that the plaintiffs, the Council of the County of Avon, should recover possession of land, about half an acre in extent, forming part of a property known as Beech House, Stover Road, Yate, in the suburbs of Bristol.

The facts of the matter are in a very short compass. The plaintiff county council bought Beech House for the purpose of a future road-widening scheme. Early in 1975 an officer of the county council discovered that the defendant, without E authority, had *de facto* occupied the land. The property comprised the house and curtilage known as Beech House and a separate paddock on which stand certain outbuildings. Finding the defendant in occupation of the paddock, Mr Bartlett had a conversation with him, and the result of that was that the county council then wrote to the defendant offering to grant him a licence to occupy the paddock. That offer is contained in a letter dated April 7 1975, which reads, so far as material, as follows: " Dear Sir, Grazing Land at Beech House, Stover Road, Yate. The land shown edged red on the attached plan is available for grazing cattle, sheep and horses from April 8 1975 to December 30 1975, and if F you wish to take a licence on the undermentioned terms for this period, please sign the enclosed additional copy of this letter where indicated, and return it to me together with the consideration of £10.00." The undermentioned terms are set out in 11 numbered paragraphs, of which I think I need only refer to a few. The licensee undertakes the responsibility for the erection and maintenance of stock-proof fencing, and for the maintenance of fences, gates and ditches. He is to be responsible for keeping the land clear of thistles and weeds. Then no 7 is important: " This licence shall at all times be construed as a personal agreement with the licensee for the grazing of animals owned by him and any other use of the said land or any part thereof shall be a breach of these conditions and the licensee shall not assign or sublet any of his rights thereunder." G And for good measure, by paragraph 10 it is provided: " No agricultural tenancy will be created under the terms of section 2(1) of the Agricultural Holdings Act 1948." A few days after the defendant had replied, signing his agreement to accept the licence on the terms offered in the letter, Mr Bartlett returned to the property and had a further conversation with the defendant, as to which the learned judge in giving judgment made the following finding: " Mr Bartlett had a conversation with the defendant concerning use of the outbuildings and thought that there was no harm in the defendant using the outbuildings." H As a result of that conversation the county council wrote to the defendant on April 18 1975 a letter signed by the Director of Estates Services saying: " I understand that arrangements have now been made for you to be granted occupation of the outbuildings at Beech House in connection with the stabling of ponies."

The short point raised in this appeal by the defendant is whether upon those facts the defendant became entitled to a tenancy from year to year of agricultural land under the terms of section 2(1) of the Agricultural Holdings Act 1948. That subsection provides, so far as material: J

> . . . where under an agreement made on or after the first day of March, nineteen hundred and forty-eight, any land is let to a person for use as agricultural land for an interest less than a tenancy from year to year, or a person is granted a licence to occupy land for use as agricultural land, and the circumstances are such that if his interest were a tenancy from year to year he would in respect of that land be the tenant of an agricultural holding, then, unless the letting or grant was approved by the Minister before the agreement was entered into, the agreement shall take effect, with the necessary modifications, as if it were an agreement for the letting of the land for a tenancy from year to year: Provided that this subsection shall not have effect in relation to an agreement for the letting of land, or the granting K of a licence to occupy land, made (whether or not the agreement expressly so provides) in contemplation of the use of the land only for grazing or mowing during some specified period of the year. . . . "

Mr Sara's submission on behalf of the appellant/defendant is that the language of the first half of the subsection applies to the licence which was granted to the defendant because it was a licence to occupy the land for use as agricultural land, and he submits that the proviso is inapplicable because, by virtue of the permission granted to the defendant to stable ponies in the outbuildings, the licence cannot be said to have been made " in contemplation of the use of the land only for L grazing or mowing " within the meaning of that phrase in the proviso.

In my judgment, there are two answers to that submission. First, if one looks at the licence granted by the letter dated April 7 1975, while it is true that the land edged red on the plan does include the outbuildings in question, the terms of the licence being limited to the use of the land for grazing and being subject to the express restriction imposed by paragraph 7, which I have read, can have no application to those outbuildings, because animals cannot, of course, graze within outbuildings. Accordingly, the subsequent agreement, using the word in a non-technical sense, reached following the M conversation between Mr Bartlett and the defendant, and recorded in the county council's letter of April 18, granting permission to the defendant to occupy the outbuildings for stabling ponies, was a purely gratuitous licence, because no additional consideration other than the £10, which had been required and paid pursuant to the licence of April 7, was required or paid for the further permission. The licence to occupy the outbuildings is, therefore, unaffected by section 2 of the Act because the phrase in subsection (1), " where . . . a person is granted a licence to occupy land for use as agricultural land," must clearly be construed as applying only to a contractual licence, and a gratuitous licence is not a contractual licence. But if, contrary to this view, one assumes, as I will for the purposes of the remainder of this judgment, that there was a single licence granted to the defendant on the terms of the two letters of April 7 and April 18 1975, and that that single licence was a contractual licence to which, subject to the proviso, subsection (1) of section 2 applied, nevertheless, this is clearly a case where the licence is taken out of the operation of the section by the terms of the proviso.

Mr Sara accepts, and in my judgment rightly accepts, that a use of buildings permitted by the terms of a licence to occupy land in contemplation of the use of the land for grazing only, which was purely ancillary and incidental to the use of the land for grazing, would not take the licence out of the ambit of the proviso. The question therefore is whether it was ever in the contemplation of the parties to this licence that the defendant should have the use of the outbuildings for purposes other than such as were incidental and ancillary to the use of the land for grazing. The onus on that issue, it seems to me, lay upon the defendant and was not discharged. Even if the onus had been the other way, the evidence suggests that what was in the contemplation of the parties was the use of the outbuildings for purposes incidental to the grazing of the land and nothing more.

Mr Sara advances two arguments to seek to controvert that conclusion. He says, first, that the outbuildings were extensive. The evidence does not disclose their nature or size, but Mr Sara relies on what one can see from the plan attached to the letter of April 7 1975 and he says that they would have provided very much more stabling accommodation than was necessary for one pony, and the evidence showed that one pony was all the land was sufficient to support. He relies, in particular, upon the use of the word "ponies" in the plural, as opposed to "pony" in the singular, in the letter of April 18 1975. I am quite unable to accept that argument. There seems to me to be nothing in the evidence here to indicate that the county council had in contemplation anything other than a reasonable permission granted to the defendant to use the buildings in connection with the grazing of any animal for the time being on the land. That, I think, is the finding which the learned judge intended to make as expressed in his judgment when he said: "The fact that the use of the outbuildings was subsequently granted does not in my view prevent it being a licence (meaning thereby a grazing licence) because the use of the buildings was in connection with use as a licence." Secondly, Mr Sara says that it is not open to the court to take that view because before occupation of buildings by animals could properly be regarded as ancillary to the use of land for grazing it would be necessary for the evidence to show that that occupation of the buildings was essential or required as a matter of practical farming to support of that proposition he relied upon a dictum of the late Ungoed-Thomas J in the case of *Lory v Brent London Borough Council* [1971] 1 WLR 823, at p 829 to this effect: "The proviso, however"—and he is referring to the proviso to section 2(1) of the Agricultural Holdings Act 1948—"is directed to contemplation of 'the use' of the land. Such matters as fencing and shelter for the grazing animals may be essential or required as a matter of practical farming to enable the animals to be grazing on the land; and such fencing and shelter might then be 'use of the land for grazing only.'" I am sure that the late Ungoed-Thomas J would have been horrified if he had thought that his, if I may say so respectfully, perfectly sensible and proper observation in the course of giving judgment in that case were to be taken as having the force of statute and providing an exhaustive definition of what can or cannot be regarded as use of buildings incidental or ancillary to the use of land for grazing only, when the question of the application of the proviso to section 2(1) is in dispute.

For those reasons, I think the learned judge came to the correct conclusion on the evidence, and I would dismiss the appeal.

Agreeing, ROSKILL LJ said: I agree that this appeal should be dismissed. Manifestly this case does not fall within section 2 of the Agricultural Holdings Act 1948.

Also agreeing, CAIRNS LJ said: For my part, I am clearly of opinion that the letter of April 18 1975 granted no more than a gratuitous permission for the use of the outbuildings and was not a contractual licence or a variation of the original licence, there being no consideration for any such variation. But if I am wrong about that, then I agree that the contemplated use of the outbuildings under the letter of April 18 1975 was purely ancillary to the use for grazing. Stress was laid by Mr Sara on the use of the plural word "ponies" in the letter of April 18. It may be noticed that the original licence of April 17 granted a licence "for the grazing of cattle, sheep and horses." In those circumstances, it was natural to use the word "ponies" in the later letter. Accordingly, I agree that the appeal should be dismissed.

The appeal was dismissed, the respondents to have an order for costs against the appellant, such order not to be enforced without further order of the court. The court referred to the registrar of the county court the investigation of the appellant's means under Regulation 20(2)(b) of the Legal Aid (General) Regulations 1971.

STILES v FARROW

Severance of the reversion—No apportionment of rent agreed by tenants—Still only one contract of tenancy and one holding—Separate notices by reversioners demanding arbitration as to rents of parts not valid

Estates Gazette February 26 1977
(1976) 241 EG 623

This was a case stated by an arbitrator, J E Pallett FRICS, for an opinion on a question of law concerning an arbitration under section 8 of the Agricultural Holdings Act 1948 between R A G B Stiles, J M R Stiles, J B Stiles and F J B Stiles, the landlords of Manor Farm, Wellow, near Yarmouth, Isle of Wight, and J W Farrow and C G Farrow, the tenants. The case was heard by His Honour Judge R Stock in Newport, Isle of Wight, County Court on December 10 1976.

Hubert Dunn (instructed by Day, Whately & Gilbert White, of Godalming) appeared for the landlords. H A C Densham, of Burges, Salmon & Co, of Bristol, represented the tenants.

The facts were that Manor Farm, an agricultural holding of some 860 acres, was let to the tenants under an agreement dated October 29 1962 entered into with the then landlords, Magdalen College, Oxford. In August 1972 Magdalen College sold part of the holding, comprising 364 acres, to Mr and Mrs R A G B Stiles and the remaining part, of 496 acres, to Mr and Mrs J B Stiles, and notice of the change in identity of the landlords was given to the tenants by the college's solicitors. Notice of severance of the freehold reversion was given by the new landlords' agents in September 1972, and on completion of the severance, the tenants were informed of the apportionment of the rent, at that time £3,550 per annum, agreed between themselves by the new owners, but the tenants never agreed to nor disagreed with that apportionment. The rent was paid half-yearly in a single sum. In July 1973 the terms and conditions of the tenancy were varied by a supplemental agreement and the rent was increased to £8,550 as a single unapportioned sum.

In 1975 two notices dated September 25 were given to the tenants by the landlords' agent, D A W Brown FRICS, demanding arbitration as to rent under section 8 of the

A Agricultural Holdings Act 1948. One notice was in respect of "the rent to be paid for the holding which you hold as tenants of" Mr and Mrs J B Stiles "being part of the property known as Manor Farm. . . ." and stated that the other part was vested in Mr and Mrs R A G B Stiles. The second was given of behalf of Mr and Mrs R A G B Stiles in respect of their part of the holding and was *mutatis mutandis* in similar terms, referring to the other part being owned by Mr and Mrs J B Stiles. No notice under section 8 in respect of the whole holding was served either by the landlords or by the tenants.

B The questions of law submitted by the arbitrator for the opinion of the court were whether the two notices dated September 25 1975 were valid and effective notices to implement the provisions of section 8; whether the arbitrator had jurisdiction to determine the rent either (i) for the whole holding, or (ii) for the severed parts thereof; and, if the answer to (ii) was in the affirmative, whether the arbitrator should make separate awards in respect of each of the severed parts.

Giving judgment, JUDGE STOCK said that Mr Dunn for the landlords had agreed, indeed could hardly have contended otherwise, that the whole property was comprised C in one tenancy and constituted one agricultural holding. It was clear from the judgment of Stamp LJ in *Jelley v Buckman* [1974] QB 488 that severance of the freeholder's reversionary interest did not create a new tenancy, and bearing in mind that provision and looking at the agreement between the parties dated July 2 1973 it was quite clear that the whole farm, comprising over 800 acres, constituted one holding and was comprised in one contract of tenancy.

Mr Dunn had submitted that the two notices given under section 8 should be read together; but in his (Judge Stock's) view that was not possible. One only had to look at each of the demands for reference to arbitration. Both specifically D included part of the holding and specifically excluded the remaining part. It was not possible to read the two notices together. It had then been argued on behalf of the landlords that the combined effect of section 140 and section 141 (2) of the Law of Property Act 1925 was that to each of the severed parts of the freehold interest in the land there was attached the right to give notice pursuant to section 8 of the 1948 Act in respect of that part. The obligations of the tenant in the original tenancy agreement included an obligation "To pay the said rent hereinafter reserved or any rent substituted therefor by agreement or by arbitration under the provisions of section 8 of the Agricultural Holdings Act 1948. . . ." E Mr Dunn argued that there was thus an obligation in the tenancy agreement whereby the tenant was under a contractual duty to pay the rent as fixed by arbitration or agreed; therefore section 141 (2) of the Law of Property Act 1925 operated as there was a covenant or provision which the owner of a severed part was entitled to enforce by taking proceedings in respect of his part of the reversion. On the other hand, Mr Densham, for the tenants, submitted that sections 140 and 141 related only to contractual rights and had no bearing on statutory rights, and that seemed to him (his Honour) to be correct. The right to appoint an F arbitrator and to proceed under the 1948 Act was a statutory right; all the lease did was to impose an obligation to pay rent as fixed following the exercise of the statutory right. It was implicit that the landlord could recover the increased rent even without the express contractual provision. That would exist in any event. It did not affect the particular manner in which the statutory right was to be exercised. That was contained in the statutory provisions and in particular the statutory provision which stated that "the landlord may give notice. . . ." G

Finally it was said on behalf of the landlord that one should look at the statute as a whole, and in particular sections 60 and 61 and the compensation provisions contained in sections 32 to 59, and that it was clear from those provisions that the term "holding" was used so as to include "part of a holding." It seemed to him (Judge Stock) clear from those provisions, without looking at them in detail, that when the Act intended the provisions to apply to part of a holding it said so. So where, as in section 8, there was no provision for notice to be given in respect of part of a holding, one could read into that section such a provision on the basis that that fell within the general scheme of the Act. It H followed, therefore, that the answer to the first question posed in the statement of special case was in the negative; the two notices were not valid and effective notices because they should have been given by the landlord (the singular including the plural) and in respect of the holding and not by the landlords of the severed parts in respect of the portions of the holding which they respectively owned.

When one looked at the other questions one saw the great difficulties that would be presented if any other answer were given. If, for example, the arbitration had been pursued in respect of one part only, what would be the rent payable for J the other part and what would happen to it? It had not been apportioned, but supposing the arbitrator were entitled to, and did in fact, assess the rent payable for the part in respect of which the arbitration had been pursued at, say, £6,000 per annum, what would be the rent for the other part? There would be no guide at all if the owner of one part were to apply for arbitration and the owner of the other part were not. There was considerable force in the argument that when there was only one holding the rent could only be revised in respect of that whole holding. It was true, as Mr Densham said, that the sum of the parts (in rental terms) did not necessarily equal the whole. It could readily be seen, as he K submitted, that if the freehold interest in the holding were severed so that one part comprised all the buildings and not much land and the other part comprised most of the land but no buildings an anomalous situation would arise. So far as the land with the buildings were concerned, in view of perhaps onerous repairing obligations on the tenant those buildings would in large part prove to be a liability, since the excess buildings would be useless for the small acreage. Then, if one looked at the larger acreage without the buildings, the rent would be depressed in respect of that part because of the large acreage not provided with any buildings. It was clear from section 2 of the Agriculture Act 1958, which L amended the valuation provisions contained in section 8 (1) of the 1948 Act, that the arbitrator must look at the rent of the holding referred to him and it was not possible for him to take into account other property. There was also force in another illustration given by Mr Densham where the ownership of a small piece of land with very high amenity value because of the demands, for example, for the use of the land as grazing for ponies, might be severed from the remainder of the holding though remaining subject to the tenancy, and then the reversionary owner of that small part might go to arbitration and obtain a rental wholly disproportionate for that part when viewed as part of a whole holding. The whole scheme M of the 1948 Act said that the landlord of the holding might demand arbitration and therefore where there was only one holding, only one valid notice could be given and that should be given by the landlord(s), ie all the four members of the Stiles family.

The landlords were ordered to pay the costs of the case stated, to be taxed on scale 4.

ARBITRATION

A

QUEEN'S BENCH DIVISION

April 1 1976

(Before Mr Justice FORBES)

PURSER & CO (HILLINGDON) LTD v JACKSON AND ANOTHER

Estates Gazette May 28 1977

(1976) 242 EG 689

Defects in house—Requirements of National House-Builders' Registration Council—Whether second arbitration is excluded by rule of res judicata—Whether house-owners are estopped from claiming for new defects after original B ones have been the subject of an arbitration—Declaration in favour of house-owners

This was a special case by an arbitrator in an arbitration in which the builders of a house, Purser & Co (Hillingdon) Ltd, claimed that the purchasers, Mr and Mrs George Henry Jackson, of 36 West Park, The Parks Estate, Minehead, Somerset, were estopped from seeking compensation for newly-discovered defects in the house after they had been awarded £105 for original defects in an arbitration award of 1972. The builders maintained that all the defects had been, or should have been, before the original arbitration and that the matter was now *res judicata* on the ground that C where a competent tribunal had decided on an issue between two parties neither of them could raise it again. The purchasers, the builders contended, should not be allowed a second arbitration.

John Dyson (instructed by Gordon Dadds & Co) appeared for the builders; Alastair R MacGregor (instructed by Lewis Silkin & Partners, agents for Bowmans, of Minehead) represented the purchasers.

Giving judgment, FORBES J said that the National House-Builders' Registration Council scheme applied to the contract. If there were defects in the construction of the building they were to be made good by the builder, subject to certain D conditions. Any dispute could be referred to arbitration. Defects did occur and the purchasers were awarded £105 in an arbitration. After that, roof defects were claimed and the claimants applied for the appointment of an arbitrator to decide whether they were the result of non-compliance with National Council standards. He claimed that they had not been considered in the previous arbitration.

Mr Dyson, for the builders, had cited *Conquer v Boot* [1928] 2 KB 336. The facts there were not dissimilar. The Divisional Court ruled that the cause of action, breach of contract, was the same in two actions. As the issue had been decided in the first action it was then *res judicata* and the E plaintiff could not claim damages in a second action. This was a binding decision so far as it related to causes of action. There appeared, however, to be no direct authority that this rule applied to arbitrations as well as causes of action. *Conquer v Boot* was said by Mr Dyson to be based on the maxim *inter est republicae ut sit finis litium* (it concerns the state that lawsuits should not be protracted). But arbitration was a private matter and there was not the same pressing need to ensure that arbitrators should not be troubled more than once as there was in the case of judges.

He (his Lordship) was not willing to extend the draconian F rule of *Conquer v Boot* and there appeared to be no authority to make him do so.

It was clear from the terms of the contract that the parties intended serial arbitrations. Defects in the building might arise from time to time and this meant that there could be arbitration from time to time. There would be no estoppel per *rem judicatam* to the raising on a subsequent occasion of a defect which was not covered by the first arbitration. The purchasers were not to be debarred from claiming relief in respect of breaches not included in the terms of reference of the first arbitration. There would be a declaration to that effect and the respondents would pay the claimants' costs. G

QUEEN'S BENCH DIVISION

March 2 1977

(Before Mr Justice FORBES)

SI PENSION TRUSTEES LTD v WILLIAM HUDSON LTD

Estates Gazette May 28 1977

(1977) 242 EG 689

H

Arbitration Act 1950, section 27—Power of court to extend time for commencing arbitration proceedings—Late application to president of RICS to appoint arbitrator—Rent review provision in underleases—Considerations to be taken into account by court in exercising discretion—Relief granted and time extended

The applicants in the present proceedings, SI Pension Trustees Ltd, who were also plaintiffs in proceedings which had been started in the Chancery Division, asked the court to exercise the power given by section 27 of the Arbitration Act 1950 to extend the time allowed under the underleases of property in Maltravers Street, Westminster, to apply to J the president of the Royal Institution of Chartered Surveyors for the appointment of an arbitrator in pursuance of a rent review clause in the underleases. The underlessees were the defendants, William Hudson Ltd.

D P Friedman (instructed by J E Kennedy & Co, of Harrow) appeared for the plaintiffs; Gavin Lightman (instructed by Herbert Oppenheimer, Nathan & Vandyk) represented the defendants.

Giving judgment, FORBES J said: In this matter the plaintiffs apply under section 27 of the Arbitration Act 1950 for extension of time within which to make application for the appointment of an arbitrator pursuant to clauses in certain K underleases. I heard argument on these matters in chambers but both counsel have told me that there is no authority directly in point, that the matter is of some general importance, and that it would therefore be convenient if I adjourned judgment into open court.

The two underleases are dated February 18 1971. The property is in Maltravers Street, Westminster. The periods of the leases are for 10 years from June 24 1970. There is no material difference between the terms of the two leases.

A The lessors were BPC Investments Ltd and the lessees were the defendants. The reversions expected on the underleases were assigned to the plaintiffs in June of 1971. The properties are, in the first lease an office block, and in the second lease car-parking space in an underground car-park. The rents reserved are £73,000 per annum for the first lease and £2,500 per annum for the second. These rents are payable until what is called "the review date," that is December 25 1975. There are provisions for rent reviews thereafter.

I shall deal only with one of the leases. As I have said, they are in identical terms. The reddendum is in these terms:

> B Yielding and paying therefore yearly during the said term and so in proportion for any less time than a year the respective rents following that is to say (a) until the 25th day of December 1975 (hereinafter called "the review date") the rent of £73,000 and (b) from "the review date" and until the expiration hereof the greater of a yearly rent of £73,000 or such amount as may be agreed upon in writing between the Lessor and the Lessee before the review date or in the absence of such agreement as may be determined by an arbitrator to be nominated by the President for the time being of the Royal Institution of Chartered Surveyors on the application of the Lessor made before but not more than two quarters before the review date. . . .

C and there then follows a provision setting out the criteria which the arbitrator is to use.

On June 19 1975 the plaintiffs, mindful of the approach of "the review date," wrote to the defendants a letter in these terms: "Under provisions of the underlease dated February 18 1971"—and they set out the parties—"the rent reserved by such underlease and payable in respect of the above mentioned property is due for review on December 25 1975. I shall be glad therefore if you will let me know the name and address of the person with whom our surveyor should communicate with a view to reaching agreement on the amount of the new rent to be payable with effect from D December 25 next." The defendants replied on July 2 1975: "We would inform you that the person dealing with the review is the writer, C J Robeson, and your surveyor should make contact accordingly." From then to November 1975 there was correspondence between the plaintiffs and the defendants about the review of rent and on October 28 the plaintiffs suggested a rent of £135,000, to which the defendants replied that that was excessive but they were looking into the matter. On November 20 1975 the defendants wrote an important letter in these terms: "We write further to our letter concerning the rent review relating to the above premises and we would inform you that we have taken advice E from our professional advisers, and in the light of their advice we have come to the conclusion that there should be no increase in the rent payable under our underlease as the rent reserved under this lease is an adequate reflection"—I think it should be—"of current market values. We have now given full instructions on this matter to Messrs Savills, where a Peter Oswald is dealing with the same on our behalf." On November 26 there was a letter from Messrs Savills confirming that: "We are arranging a detailed inspection of the premises now we have received formal instruction to advise in this matter, and will be in touch with you in the near future."

F Thereafter I think the history is best set out in paragraph 8 of the affidavit of S M Dutnall, who dealt with the matter on behalf of the plaintiffs. He says this:

> From November 26 1975, when Savills were first instructed by the defendants, to the period commencing with the Christmas break I telephoned them [that is Savills] on numerous occasions but I was not always successful in speaking to Mr Oswald. I actually succeeded in speaking to him on three or four occasions and mainly these conversations consisted of inquiries by me as to whether the promised inspection had taken place and whether he had formulated his opinion of the proposed rental value. Mr Oswald in each conversation said that he had been precluded from giving the matter his immediate attention because of pressure of G work, but he nevertheless promised to give the matter his early attention. At no time did I gain the impression that Savills' instructions were to be confined to a formal arbitration but that they had authority to proceed with full negotiations to a final stage.

Pausing there, it seems quite clear to me that the plaintiffs took the letters of November 20 and 26 as indicating that negotiations were to continue with the defendants with a view to possible agreement about the new rent, but with Savills instead of with the defendants direct. It is clear also that this was how Savills understood their instructions. In my view it is almost inconceivable that the defendants should H have intended their letter to be taken in any other sense. At all events, believing that the defendants did intend that letter to be read in that sense, the plaintiffs allowed the review date to pass without applying to the president of the Royal Institution of Chartered Surveyors for the appointment of an arbitrator. They were more than surprised, indeed, to receive a letter on January 9 1976 which was in these terms:

> We write further to our letter to you dated November 20 and we have checked with the Royal Institution of Chartered Surveyors, who have informed us that you have not made an application to them for the appointment of an arbitrator, and, accordingly, you are no longer entitled to a review in the rent of the J above premises. Under the circumstances our instructions to Savills have been withdrawn.

Further correspondence ensued after that.

The attitude adopted at this stage by the defendants can be judged from a paragraph in their letter of January 20 in these terms:

> It was abundantly clear that there was a very substantial difference in our opinions regarding the current rental value of the above property and the chance of us negotiating a rental in those circumstances was remote, and that the matter would thus go to arbitration. Because of this we thought it necessary to have independent surveyors who could in due course present our case K at arbitration.

Despite reasoned remonstrance from the plaintiffs in subsequent correspondence, this is the attitude which the defendants, with some brusqueness, maintained.

It will be plain from what I have said earlier as to my view of the clear meaning of the defendants' letter of November 20 of 1975 that the attitude they assumed in the correspondence subsequent to the review date was not, as I see it, justified. It is in my opinion clear that the defendants certainly allowed, if not actually encouraged, the plaintiffs to believe that they were expected to engage in genuine L negotiations with Savills after November 20, with a view to agreement on a revised rent. The suggestion made by the defendants in the letter of January 20—that their earlier letter was merely indicating that Savills would present their case at arbitration—does not bear close examination. If the letter of November 20 had really meant to convey this impression, then, in my view, its equivocal nature is the result of deliberate, rather than accidental, draftsmanship.

In this state of affairs, the rival contentions of Mr Friedman and Mr Lightman can shortly be indicated. Shorn of all embellishments, what they say is this: Mr Friedman says that this is a case which falls squarely within section 27 of M the Arbitration Act 1950, which gives the court a discretion to extend time in certain circumstances, and that that discretion should be exercised in the plaintiffs' favour in this case. Mr Lightman says that section 27 has no application to a case of this kind, and that, even if it had, the court's discretion should not be exercised.

It is necessary to look at section 27. It is a short section:

> Where the terms of an agreement to refer future disputes to arbitration provide that any claims to which the agreement applies shall be barred unless notice to appoint an arbitrator is given or an arbitrator is appointed or some other step to commence

arbitration proceedings is taken within a time fixed by the agreement, and a dispute arises to which the agreement applies, the High Court, if it is of opinion that in the circumstances of the case undue hardship would otherwise be caused, and notwithstanding that the time so fixed has expired, may, on such terms, if any, as the justice of the case may require, but without prejudice to the provisions of any enactment limiting the time for the commencement of arbitration proceedings, extend the time for such period as it thinks proper.

The main contention between the parties is thus that Mr Friedman says that section applies to this case. Mr Lightman says "No," that there is a distinction between time limits laid down for the bringing into operation of rights and limitation periods, and section 27 applies to limitation periods and not to such other time limits. These time limits, he says, in rent review clauses are conditions precedent and it is not right to regard them as barring claims. For these propositions Mr Lightman seeks to rely on *Samuel Properties (Developments) Ltd v Hayek* [1972] 1 WLR 1296, *United Scientific Holdings Ltd v Burnley Borough Council* [1976] Ch 128 [reversed by House of Lords on March 23 1977] and *Nea Agrex SA v Baltic Shipping Company Ltd* [1976] 1 QB 933. It is true that in the *Samuel Properties* case Russell LJ, as he then was, referred to the rent review clause in that case as an option, and there have been cases in which a distinction has been made between rent review clauses which were options and those which were obligations. But this distinction was seen by all the members of the Court of Appeal in the *United Scientific Holdings* case to be an artificial one. If this distinction were a matter of importance, I would take the view that the rent review clause with which I am concerned is not an option subject to time limits; it is machinery whereby both sides may either negotiate a revised rent, or, if that proves impossible, may present their respective evidence and arguments to an arbitrator in order that a revised rent may be determined.

Mr Lightman urges that the review clauses should be construed strictly against the landlord, because, he says, they are inserted for his benefit. True it is that it is to the advantage of the landlord to obtain an upward revision and to the advantage of the tenant to resist this, but once the landlord has set the machinery in motion (as this landlord did by his letter of June 19) that machinery exists for the benefit of both. I cannot see that such a situation gives rise to a necessity to construe this clause as if it excluded section 27 of the Arbitration Act. It may be necessary at this point to emphasise what it seems to me I am *not* called upon to decide. There is a world of difference between deciding whether or not section 27 of the Arbitration Act can be invoked in these proceedings and whether or not time is of the essence of the contract embodied in the rent review clause. The latter point falls properly to be argued in the proceedings in the Chancery Division which the plaintiffs have already started. No doubt, if it be there held that time is of the essence of the contract, that division will be urged to grant relief to the plaintiffs on the ground that, by reason of the unconscionable conduct of the defendants, the plaintiffs had been led to believe that strict adherence to the terms of the contract would not be insisted upon. But the criteria by which it must be decided whether section 27 applies are not the criteria by which to judge whether time is of the essence of the contract. The cases on which Mr Lightman relies are dealing with this latter question and not with section 27 at all. Indeed I understand both counsel to say that there is no authority directly in point as to whether section 27 applies to rent review clauses of this kind. Similarly, if it be decided that this court has jurisdiction to extend time under section 27, I am not thereafter concerned with any question of whether the defendants' conduct can be said to be unconscionable. That, again, is a matter for Chancery proceedings. This court would, in those circumstances, be solely concerned with the question whether time should be extended because "in the circumstances of the case undue hardship would otherwise be caused." I therefore reject Mr Lightman's argument based on options and conditions precedent. The right approach, it seems to me, is to consider, first, whether, on a true construction of the clause, it can be said to bar any claims to which the agreement applies if notice is not given by the time allowed. If the answer is that this clause does have that effect, then the clause must be read as being subject to the court's power to extend that time. For the answer to that point I turn first to the judgment of Lord Denning in the *Nea Agrex* case [1976] 1 QB 933, and the passage to which I refer is in the Master of the Rolls' judgment at p 946. He says this:

> I know that section 27 only applies "where the terms of an agreement to refer future disputes to arbitration provides," etc, but I do not think this should be confined to express terms. It should include also implied terms. When a charterparty incorporates a clause paramount, which in turn incorporates article III, rule 6, there is to be implied (into the agreement to refer future disputes to arbitration) a provision that any claims shall be barred unless arbitration is commenced within one year. On such an implication being made, the way is clear for the introduction of section 27. If the court is of opinion that undue hardship would be caused by insisting on the one year limitation the court can extend the time on the principle stated in the *Pegasus* case.

The effect of that passage, it seems to me, is that it is unnecessary for the arbitration clause to say in precise terms that claims are barred unless notice is given in time. If that is the true effect of the clause, that is enough. Here the one point of agreement between both sides is that, subject to any argument in the Chancery proceedings, failure to apply for the appointment before the review date effectively prevents an arbitrator determining the revised rent and therefore prevents the landlord from claiming one. I have no doubt that the effect of this clause is to bar the landlord's claims and that, therefore, it is a clause to which section 27 applies. In *The International Tank and Pipe SAK v Kuwait Aviation Fuelling Co KSC* [1975] QB 224 Lord Denning MR, at p 233, said: "I look upon section 27 as being, in effect, an additional statutory term written into the arbitration clause." I am therefore justified in reading this clause as if the time limit contained in it was subject to the court's power to extend time "if it is of the opinion that in the circumstances of the case undue hardship would otherwise be caused."

When considering whether undue hardship would otherwise be caused, I derive considerable assistance from the decision of the Court of Appeal in *Liberian Shipping Corporation "Pegasus" v A King & Sons Ltd* [1967] 2 QB 86, and there are passages in the judgments in that case to which I should refer. The first is that of Lord Denning MR at p 98 E:

> It does appear that in the past the courts have been inclined to use this argument: The words used in the statute are "undue hardship." A "hardship" by itself is something severe which is hard to endure. It is always a "hardship" to be barred by a time limit. If a man does not read the contract and is a day or two late, it is a "hardship," but it is not an "undue hardship": because it is his own fault. I cannot accept this narrow interpretation of the statute. These time-limit clauses used to operate most unjustly. Claimants used to find their claims barred, when, by some oversight, they were only a day or two late. In order to avoid that injustice, the legislature intervened so as to enable the courts to extend the time whenever "in the circumstances of the case undue hardship would otherwise be caused." "Undue" there simply means excessive. It means greater hardship than the circumstances warrant. Even though a claimant has been at fault himself, it is an undue hardship on him if the consequences are out of proportion to his fault. Applying this test, it seems to me that if a claimant makes a mistake which is excusable, and is in consequence a few days out of time, then if there is no prejudice to the other side, it would be altogether too harsh to deprive him of all chance for ever of coming and making his claim. All the more so if the mistake is contributed to or shared by the other side. That indeed is this very case. I am quite

A prepared to accept that the charterers, when they went to the meeting of June 27 did not intend to mislead the owners. They were both under a misapprehension. Neither of them realised that the time had already expired. But it is pretty plain that the conduct of the charterers put the owners off their guard. The owners would not contemplate that they would be barred while negotiations were still going on. As soon as they realised that the negotiations were not going to be fruitful, they at once took the necessary steps. They were only nine days late. In these circumstances it seems to me it would be undue hardship to hold that the owners are barred absolutely. The case has a resemblance to *Hughes v Metropolitan Railway Co* (1877) where there were negotiations for a settlement and it was held that, pending those
B negotiations, the strict rights of the parties did not apply.

I should also, I think, refer to passages from the judgment of Salmon LJ (as he then was) starting at p 107 D:

> I have no doubt at all that if two ordinary businessmen entering into this contract had been asked if it would cause undue hardship to refuse to extend the time should circumstances such as the present occur, they would both unhesitatingly have answered "yes." I am not prepared to hold that the court's powers under the section should be very rarely exercised. Still less that they should be exercised freely. The question as to whether or not those powers should be exercised must turn exclusively on the particular
> C facts of each case in which the question arises. In considering this question the court must take all the relevant circumstances of the case into account, the degree of blameworthiness of the claimants in failing to appoint an arbitrator within the time, the amount at stake, the length of the delay, whether the claimants have been misled, whether through some circumstances beyond their control it was impossible for them to appoint an arbitrator in time. In the last two circumstances I have mentioned, which do not arise here, it is obvious that normally the power would be exercised [and I mentally underline that sentence: Forbes J] but those are not the only circumstances and they are not to my mind necessary circumstances for the exercise of the power to extend time. I do not intend to catalogue the circumstances to be taken into
> D account, but one very important circumstance is whether there is any possibility of the other side having been prejudiced by the delay. Of course, if there is such a possibility, it might be said that it is no undue hardship upon the claimants to refuse an extension of time because, if the hardship is lifted from their shoulders, some hardship will fall on the shoulders of the respondents and, after all, the delay is the claimants' fault. But there is no such possibility in this case. This is a simple case of a few days' delay, caused by an excusable piece of inadvertence which has done the respondents no conceivable harm and which, unless we exercise the powers which in my view we undoubtedly have, will deprive these claimants of what may turn out to be a valid claim for £33,000 and lift the liability off the shoulders of
> E the respondents. I have no doubt but that this would cause the claimants undue hardship.

It should be noted that the fact that the exercise of the court's discretion would, or might, mean that the other party would no longer be relieved of the liability which insistence on the time limit would lift from his shoulders is not a matter which appears to have troubled either Lord Denning or Lord Salmon. Similarly, I do not think here that I should be influenced by the fact that a decision to extend time may possibly result in hardship to the defendants in the sense that they may find themselves liable to pay an increased rent, a
F liability from which insistence on the time provision would relieve them. In any event it is the defendants' contention that arbitration would result in no rent increase. If one looks at the second paragraph of the affidavit of Mr Robeson, sworn on behalf of the defendants, where he is challenging paragraph 15 of the plaintiffs' affidavit—it is headed "Paragraph 14" but that is clearly a mistake—he says this: "I challenge that an assessment of the rent would result in a figure in excess of the unreviewed rent and that is the contention which the defendants have at all times maintained. They have never to my knowledge suggested a rental of £8 per sq ft. The figure of 12,000-odd sq ft is a gross error and not net usable space and is therefore not the appropriate figure for the purposes of G
a rent review. The office area, including reception area, is only so many square feet and so on." He sets out the reasons for his contention that, if arbitration proceedings did go on, they would result in no increase in the rent payable, using the criteria set out in the rent review clause. So that even if one were to take into account the fact that there might, in a sense, be hardship on the defendants, in that they might have to pay an increased rent, it is clearly the view of the defendants that they would be successful in any arbitration proceedings and would not in fact have to pay such an increased rent. This is a matter, it seems to me, of some importance when one is considering the whole question of undue hardship in this H
case.

Mr Lightman also suggests that there would be no hardship to the plaintiffs here, because they might have a claim in negligence against either their surveyors or their solicitors. It seems plain from the affidavit of Mr Dutnall that he, and he alone, was concerned with the negotiations, and also (I might add) with the failure to apply for an appointment in time. Whether there would be a claim against the solicitors therefore looks to me a matter for mere speculation. I am in no better position with regard to any liability of Mr Dutnall or of his employers. Mr Dutnall describes himself as a
chartered surveyor employed by Smiths Industries Ltd to J
manage the investment properties of the plaintiffs to these proceedings. I am left to speculate as to the position between Smiths Industries—his employers—and the plaintiffs. The existence of the initials "SI" in the plaintiffs' title might lead one to assume that the plaintiffs are a subsidiary company of Smiths Industries formed to hold the landed properties on which the pension fund of the parent company in part relies, but that is no more than a guess. And whether, if it proved to be right, the plaintiffs would have a claim in negligence against its parent company is a field in which I decline to forage. Nor have I any evidence to say one way or the other
whether, if Mr Dutnall was negligent, he has the means to K
discharge any judgment on his own. I am left therefore with the plaintiffs' claim that they might succeed in persuading an arbitrator that the proper rent was at least £110,000 per annum in place of the present £75,000. To be deprived of the opportunity of making this claim seems to me to amount to undue hardship when I take into account the contemporary correspondence, the effect of which I have already referred to. I am left in no doubt that the plaintiffs were misled here into overlooking the strict time limit imposed in the clause, a matter which Salmon LJ thought would "obviously"—the word is his—result in an extension of time; that the amount L
at stake is considerable, and that the plaintiffs are trustees of a pension fund, but, nevertheless, are sensitive to the needs of their tenants (see paragraph 10 of Mr Dutnall's affidavit) and, looking at these matters in the light of the guidance given in the *Pegasus* case, it seems to me proper to come to the conclusion that there would be undue hardship if I did not make the order asked for.

There remains only one consideration, namely the length of time which elapsed before this application was made. The skeleton calendar is of this order: on December 25 1975 the period expired; on January 9 1976 the defendants indicated
that they were relying on the expiry of the period to defeat M
any claim for a revised rent. This application was not made until May 7 1976. Mr Lightman says this is a delay of a character to put the plaintiffs out of court. But such a calendar is misleading. In fact there was correspondence between the parties up to February 10 1976, received on February 12, which amounted to an attempt by the plaintiffs to get the defendants to change their minds—not unreasonable, in my view, in the circumstances. Thereafter, on March 12 1976 the plaintiffs' solicitors did apply to the president of the Royal Institution of Chartered Surveyors to appoint an arbitrator, but the defendants' solicitors had forestalled them.

A They had already written to the president on February 25 objecting in advance. On the same day as that application the plaintiffs' solicitors issued an originating summons in the Chancery Division followed by interlocutory proceedings in that division. While it is true there was some delay in taking out *this* originating summons in this division, there was obviously considerable activity and the defendants could have been under no misapprehension about the determination of the plaintiffs to have the question of their right to a review of the rent litigated. In the circumstances, I do not think there was undue delay which should disentitle the plaintiffs to the relief sought.

B I think the proper order—and I am open to argument about this—is that the time should be extended to March 15 1976, C thus making the application to the president of March 12 one which falls within the time prescribed.

I ought to add, I think, that counsel have been good enough to refer me, after I had reserved judgment, to the case of *C H Bailey Ltd v Memorial Enterprises Ltd* [1974] 1 WLR 728, and that I have in fact read it, but I came to the conclusion that it did not add anything, in effect, to the view I had formed on the application of section 27.

The plaintiffs were granted an order that the time for applying for the appointment of an arbitrator should be treated as extended to March 15 1976. The costs of the present application to be costs in the arbitration. D

COMPULSORY PURCHASE AND COMPENSATION

A **QUEEN'S BENCH DIVISION**

February 9 1976

(Before Douglas FRANK QC, sitting as a deputy judge of the division)

BRINKLOW AND ANOTHER v SECRETARY OF STATE FOR THE ENVIRONMENT AND ANOTHER

Estates Gazette February 12 1977

(1976) 241 EG 461

Compulsory acquisition of small site with non-conforming uses—If housing need exists, secondary motive to extinguish uses irrelevant—Existence of other powers to
B **extinguish uses also irrelevant—But Secretary of State erred in law in not taking into account effects of absence of alternative sites for relocation—Order quashed**

This was an application by C D Brinklow and Croft Bros (London) Ltd for an order to quash the London Borough of Haringey (Land and Buildings adjoining 41 Lynmouth Road N2) Compulsory Purchase Order 1973, made by the second respondents, the London Borough of Haringey, and confirmed by the first respondent, the Secretary of State for the Environment, on June 23 1975.

C J Lockhart-Mummery (instructed by Stevensons) appeared for the applicant Croft Bros (London) Ltd; G R G
C Roots (instructed by C N Frere-Smith) for the applicant Mr Brinklow; and H K Woolf (instructed by the Treasury Solicitor) represented the respondents.

Giving judgment, DOUGLAS FRANK said: The order was made under Part V of the Housing Act 1957 and relates to an area of 0.33-acre on which stands a single-storeyed warehouse with access from Lynmouth Road and an area of land with access from Fortis Green. The warehouse is used for the storage of flooring and wall tiles and the open land is used for the storage of crashed motor vehicles awaiting repair at a garage situated about 90 ft to the west.
D Mr Brinklow holds both areas on a 21-year lease from September 30 1975 but has sublet the warehouse to Croft Brothers Ltd. The applicants objected to the compulsory purchase order on the grounds that:

(a) the council's principal object was to liquidate two alleged non-conforming uses; the provision of housing accommodation was a secondary motive and this order, made under Part V of the Housing Act 1957 was, therefore, *ultra vires;*

(b) the cost, inclusive of compensation, for the acquisition of this site for the provision of housing accommodation would be excessively high whereas the proposed housing gain would be minimal; and

E (c) there would be no alternative accommodation for these two displaced uses.

The Secretary of State caused a local inquiry to be held and the inspector in his report submitted on September 6 1974 found as follows:

4. The council state as their reasons for making the order that the present uses are non-conforming to the zoning in the initial development plan and have given rise to complaints from local residents. There was an overwhelming housing need in the borough and the seven 4-person houses that could be built on this site would contribute a valuable housing gain.

5. The objectors claim that acquisition of the site would result F
in total extinguishment of these two uses. The total site cost of such acquisition would be about £5,500 per habitable room of the proposed redevelopment, exclusive of building cost. This was about five times the normal rate for site acquisition in this area; the council did not contradict this estimate.

6. The objectors made the following points on the making of the order:

a. The council's case had been principally made by the planning witness who had stated in cross-examination that the action of the council in making the Part V compulsory purchase order had been prompted by complaints from local residents about noise, obstruction and loss of visual amenity. G

b. In the report to the Development Control Panel of July 4 1973 it was evident that the motivation behind the making of the Part V order was on planning grounds (document no 5).

c. Evidence of complaints mostly turned on a bundle of letters, mainly from one person.

d. The weight of the council's planning case for this Part V order rested largely on the use of the front part of the site for the storage of damaged vehicles, to which the letters of complaint mainly related.

e. The real purpose of the council should have been approached through the provisions of section 51 of the Town and Country Planning Act 1971. This Part V order was *ultra vires.* H

The inspector concluded:

1. The council have a pressing housing need and even a small site providing 100 per cent housing gain would make a valuable contribution towards the borough's housing requirements.

2. It is reasonable to expect an environmental gain from housing redevelopment on Part V compulsory purchase order lands. Such environmental improvement is, however, secondary to the primary purpose of the acquisition of land and buildings under Part V of the Housing Act 1957. This purpose is primarily the provision of housing accommodation.

3. The primary motivation for the making of this Part V J
compulsory purchase order seems to have been for the extinguishment of the two non-conforming uses on this site because of complaints received by letters from several local residents.

4. No action has been taken by the council under the provisions of section 51 of the Town and Country Planning Act 1971 by the making of a discontinuance order. Under these provisions objections might well have been heard at a joint inquiry within the complete framework of applicable housing and planning legislation. Local residents who complained could have appeared as witnesses and in turn been cross-examined by the objectors.

5. In my opinion this inquiry was incomplete. The true purpose of Part V of the Housing Act 1957 is being used as a means to achieve an end not primarily intended under the K
provisions of Part V of the Act. Confirmation of this order, if unchallenged, would set a dangerous precedent.

The inspector recommended that the order should be not confirmed.

The Secretary of State's letter of decision dated June 23 1975 stated that he accepted the findings of facts but did not accept the conclusion that the purpose of the order was one not primarily intended under the provisions of Part V of the Housing Act 1957. The letter then went on:

A 9. In the view of the Secretary of State the question for decision is whether the council have established a need to acquire the order lands for housing purposes and considerations relating to the extinguishment of non-conforming uses are not relevant to this decision. The overwhelming housing need in the borough has been clearly established and the Secretary of State accepts that the acquisition of even a small site providing maximum housing gain would make a valuable contribution to the borough's housing requirements. He is therefore satisfied that the council's need to acquire the order lands for housing purposes has been established.

10. The estimates of the cost of acquiring the order lands, including compensation, presented in support of one of the B objections to the order have been noted, but it is not considered that they demonstrate that the acquisition cost would be unacceptably high for housing development in the borough. With regard to the absence of alternative sites for the relocation of the dispossessed uses, the view is held that this is not a factor which is relevant to the consideration of the confirmation of the order.

It was then stated that the Secretary of State had decided to confirm the order subject only to an unimportant modification.

The applicants' grounds are that the confirmation of the order is not within the powers conferred by the Acts under C which it was made and confirmed and further or in the alternative that the interests of the applicants have been substantially prejudiced by failure to comply with the Compulsory Purchase by Local Authorities (Inquiries Procedure) Rules 1962. I will deal first with the allegation that there was a breach of the rules. The submission here, put succinctly, was that the inspector found as a fact that the purpose in making the compulsory purchase order was to extinguish the non-conforming uses, that the Secretary of State differed from the inspector in those findings of fact, that he came to his decision without affording the applicants an opportunity of making representations in writing and, D accordingly, that he was in breach of rule 9 of the 1962 rules. If in truth the Secretary of State differed from the inspector on a finding of fact there can be no doubt as to a breach of the rules, and Mr Woolf conceded that the order would have to be quashed.

It had been an important part of the applicants' case that the council's motive for the making of the order was not for housing but to terminate the non-conforming uses and was therefore *ultra vires*. It was submitted for the applicants that in paragraph 9 of his decision letter the Secretary of State had differed from his inspector on the question of E motivation. However, as I construe that paragraph, the Secretary of State in effect agreed with the submission that an order made under the Housing Act 1957 must be made for housing purposes, and in so far as there was any other motive it could not be relevant; further, the Secretary of State, having disregarded the non-conforming uses, applied his mind to the question of housing and found that there was an overwhelming need, and he was satisfied that the council's need to acquire the order lands for housing purposes had been established. The fact that the council may have had two motives for the acquisition of the land, one for extinguishing the uses and the other for housing, does not F prevent them acquiring the land under the Housing Act provided that one of the purposes is for housing; that there were other powers to extinguish the uses is not relevant: see *Westminster Bank Ltd v Minister of Housing and Local Government* [1971] AC 508. Counsel for the applicants conceded that, if para 10 of the decision letter is to be construed in the way that I have just explained, his point is a false one. In my judgment motivation is irrelevant if the circumstances for confirmation of a compulsory purchase order exist. In my view the Secretary of State in para 9 put his finger upon the logical error into which the inspector fell, disregarded motivation and satisfied himself that the circumstances for confirmation, namely housing needs, exist. G Accordingly, the applicants' submission fails in this respect. I should add that, had the council no intention of using the land for housing, different considerations would arise. The compulsory purchase order would have been made in bad faith and might then have been void *ab initio*.

The next ground of complaint against the Secretary of State's decision is that he reached the conclusion which, on the evidence and as a reasonable person, he could not have come to. Mr Roots submitted that the court can interfere if the Secretary of State has come to a conclusion to which, on the evidence, he could not reasonably come, and referred H to *Ashbridge Investments Ltd v Ministry of Housing and Local Government* [1965] 3 All ER 371 at 374. He again stressed, by reference to the inspector's report, that the motivation for the making of the order was to terminate the existing uses and that housing was only introduced as a method of dealing with the land after the termination of those uses, and he referred to para 5 of the inspector's conclusions. However, in my judgment, the same considerations apply here as in considering the previous submission, namely, whether there was evidence before the Secretary of State on which he could conclude that there was a need for the land for housing purposes. J In concluding that there was such need he had the evidence set out before him in the body of the report and in the first paragraph of the inspector's conclusions. Accordingly, the appeal fails on this ground also.

Finally, it is contended that the Secretary of State erred in law in deciding that the absence of alternative sites for the relocation of the dispossessed uses was not relevant to the consideration of the confirmation of the order. Mr Woolf submits that the Secretary of State can disregard considerations other than housing, that is to say, that he can attach no weight to them on the ground that housing considerations are paramount. K He refers to and adopts the judgment at p 374 D in the case of *Ashbridge Investments Ltd v Ministry of Housing and Local Government* (*supra*) and submits that the Secretary of State can decide whether to attach relevance to a matter or not and that in the context no relevance is synonymous with no weight. He referred to *Anisminic Ltd v Foreign Compensation Commission* [1969] 2 AC 147, in which Lord Pearce at p 195 said that the court's jurisdiction over inferior tribunals is supervision and not review and if the tribunal directs itself to the right inquiry, asking the right questions, the court will not intervene merely because the tribunal has or may have L come to the wrong answer, provided that it is an answer which lies within the jurisdiction.

It is necessary, I think, first to examine some of the statutory provisions relating to this matter. Section 96 (a) of the Housing Act 1957 empowers a local authority to acquire any land as a site for the erection of houses. Under section 97 (1) land for that purpose may be acquired by agreement or the local authority may be authorised to purchase land compulsorily and the Seventh Schedule to the Act then applies. By para 1 of the Seventh Schedule, the Acquisition of Land (Authorisation Procedure) Act 1946 M applies to a compulsory purchase order under Part V of the Act as if the 1957 Act had been in force immediately before the commencement of the 1946 Act. Under paras 1, 2 and 3 of the First Schedule to the 1946 Act the local authority are required in this case to make a compulsory purchase order for confirmation by the Secretary of State, but before doing so they are required to take the specified steps to ensure that any owner, lessee or occupier is aware of the making of the order and of the right of objection. If no objection is made the confirming authority may confirm the order, but if there is objection, as in this case, then subparagraphs (2) to (4) of paragraph 4 apply, and they provide:

(2) If any objection duly made as aforesaid is not withdrawn, the confirming authority shall, before confirming the order, either cause a public local inquiry to be held or afford to any person by whom any objection has been duly made as aforesaid and not withdrawn an opportunity of appearing before and being heard by a person appointed by the confirming authority for the purpose, and, after considering the objection and the report of the person who held the inquiry or the person appointed as aforesaid, may confirm the order either with or without modifications.

(3) If any person by whom an objection has been made avails himself of the opportunity of being heard, the confirming authority shall afford to the acquiring authority, and to any other persons to whom it appears to the confirming authority expedient to afford it, an opportunity of being heard on the same occasion.

(4) Notwithstanding anything in the two last foregoing sub-paragraphs, the confirming authority may require any person who has made an objection to state in writing the grounds thereof, and may disregard the objection for the purposes of this paragraph if the confirming authority is satisfied that the objection relates exclusively to matters which can be dealt with by the tribunal by whom the compensation is to be assessed.

Mr Roots submits that the proper conclusion to draw from those provisions is that any matter raised by an objector and which is a direct consequence of the compulsory purchase order is relevant, except the assessment of compensation, which falls to be dealt with by the Lands Tribunal, and that the purpose of the hearing was to enable the Secretary of State to decide whether it was expedient to confirm the order and that in doing so he must make a balance, that is to say, he must weigh the need for housing against the effect on the objectors. He referred to the well-known case of *Associated Provincial Picture Houses Ltd v Wednesbury Corporation* [1948] 1 KB 223, where at p 228 Lord Greene MR said:

If, in the statute conferring the discretion, there is to be found expressly or by implication matters which the authority exercising the discretion ought to have regard to, then in exercising the discretion it must have regard to those matters.

So, in the case of *Hanks v Ministry of Housing and Local Government* [1963] 1 QB 999 the learned judge said:

If the authority fails to take into account as a relevant factor something which is relevant, and which is or ought to be known to it, and which it ought to have taken into account, the exercise of the power is normally bad. I say "normally" because I can conceive that there may be cases where the factor wrongly taken into account, or omitted, is insignificant, or where the wrong taking into account, or omission, actually operated in favour of the person who later claims to be aggrieved by the decision.

In my judgment, thus far Mr Roots' submission is amply supported by the express terms of subpara (2) of para 4 of the First Schedule to the Act of 1946 and by the authorities. Thus, where a person entitled to object raises a matter relevant to the effect of the proposed compulsory purchase upon him, he is entitled to have that matter considered and taken into account by the confirming authority. I should only add that how much weight is given to that matter is entirely within the discretion of the confirming authority, provided that that discretion is not exercised in such a way as no reasonable man would exercise it.

However, that is not the end of the matter. I now have to consider whether the Secretary of State took into account that there were no alternative sites for the relocation of the uses carried out by the objectors. This matter was not dealt with by the inspector either in his findings of fact or in his conclusion, but in the decision letter it was dealt with in one sentence, which I repeat:

With regard to the absence of alternative sites for the relocation of the dispossessed uses, the view is held that this is not a factor which is relevant to the consideration of the confirmation of the order.

Thus the issue resolves itself into the question of the correct construction of that sentence. On the one hand, the applicants say that in effect the Secretary of State took no account of the relocation question, whereas Mr Woolf says that he took it into account and then said it carried no weight. My approach to this question is this: if, notwithstanding the words used, I am confident that the Secretary of State took the relocation matter into account then I should find in his favour. On the other hand, if there is reasonable doubt, this being a compulsory purchase case, the order should be quashed so that he should have a further opportunity of considering that matter. I find it difficult to equate the words "no relevance" with the words "no weight." I prefer to adopt the words of Lord Greene MR in *Tomkins v Tomkins* [1948] P 170 at 175, namely:

When I say relevant I mean this, so nearly touching the matter in issue as to be such that a judicial mind ought to regard it as a proper thing to be taken into consideration.

Ex hypothesi a factor which is not relevant is one which ought not to be regarded as a proper thing to be taken into consideration.

It follows, by applying that test in the instant case, that the Secretary of State regarded the relocation issue as irrelevant, as a question which he ought not to take into account, and not one which he took into account and then dismissed as having no weight. I feel fortified in that conclusion by a reference, first, to para 9 of the decision letter, in which he used very similar language and in particular the word "relevant," but there showed that in effect he was agreeing with the objectors that he considered that the question was not one which ought to be taken into account at all rather than that he had taken it into account but attached no weight to it.

I further think that the dispossession of two businesses, taking into account the consequent effect upon those who rely upon them for employment and their families, and those who rely upon them for the services they provide, could not be considered by a reasonable man as a matter of no weight. Accordingly, I have reached the conclusion that, on the balance of probabilities, the Secretary of State used the word "relevant" in the sense that I have described and did not take into account and put into the balance the absence of alternative sites for the relocation of the dispossessed uses.

Accordingly, in my judgment, the applicants have established a ground on which they are entitled to succeed and the compulsory purchase order must be quashed.

ESTATE AGENTS

A QUEEN'S BENCH DIVISION

February 16 1976

(Before His Honour Judge B GIBBENS, sitting as a deputy judge of the division)

LUCK v DAVENPORT-SMITH

Estates Gazette May 7 1977

(1976) 242 EG 455

**Various disputes between estate agent and former employee —Commission payable on sales negotiated before, but completed after, leaving principal's service—Covenant restricting employee from setting up in practice or accept-
B ing employment in estate agency within a certain radius— Principles governing such restrictive covenants—Restriction in present case held to be unreasonable in scope and therefore void as contrary to public policy**

This was an action brought by Keith Luck against his former employer, Michael Gordon Davenport-Smith FSVA for relief on various grounds, including a claim for commission on sales negotiated by the plaintiff while still in the defendant's employment. The defendant, who carried on the business of an estate agent from offices at Peacehaven, Saltdean and Patcham in Sussex, raised a number of matters by way of defence and counterclaimed to enforce a covenant in
C restraint of trade against the plaintiff.

M D Kennedy (instructed by David M Laing, Brighton) appeared for the plaintiff, and B Galpin (instructed by Weeks, Legg & Dean, Brighton) represented the defendant.

Giving judgment GIBBENS J said: The outline of the history of the relationship of the parties is this. When he was 18 years of age Mr Luck, having had some small experience with a one-man firm, applied for employment by Mr M G Davenport-Smith, who was carrying on the business of an estate agent in Peacehaven, Saltdean and Patcham in Sussex. His letter of application of March 6 1965 was successful, because on March 19 1965 Mr Davenport-Smith wrote to
D give Mr Luck employment on the terms set out in that letter, the employment being subject to an initial period of three months' probation, after which, subject to approval of both parties, it says, "You will then become a permanent member of our staff." On January 1 1969 Mr Luck was moved from the office in Peacehaven to Patcham, where he became branch manager; it was a small office but it entitled him to that appellation. Again in the same year, on August 30 1969, he was transferred to the Saltdean office, this time with the title of senior negotiator which, although it sounds less, I think was not in fact in any way a demotion, because he was
E proving to be a valuable servant and assistant to Mr Davenport-Smith and it was indeed a more responsible job. But Mr Luck was, quite understandably, ambitious and wished to improve his position, and, as a result, in May 1970 he accepted employment with another firm of estate agents in the Sussex field, Dutton Briant & Watts, who, I think, have their main offices in Brighton. But, having formed a good opinion of Mr Luck, Mr Davenport-Smith sought to dissuade him from leaving and proposed a partnership between them. Negotiations went on for a partnership, during which time Mr Luck was really treated as a partner; but eventually,
on March 17 1972, the negotiations were abruptly broken F
off by a letter which indicated that as several drafts had been prepared without reaching agreement the negotiations would now be at an end. At that time, owing to the change in prospects, Mr Luck and Mr Davenport-Smith had a discussion as to the terms which should now govern Mr Luck's employment for continued service with the defendant; and, on p 33 of the bundle of agreed correspondence, we find Mr Luck's own holograph note of what took place in that discussion and what he says was agreed. Mr Davenport-Smith's version of the matter is to be found in his own note made on the cover of Mr Luck's employment file, which has been
produced in this case, exhibit D2. On July 4 1973 Mr Luck G
wrote to the defendant, saying, "After much consideration, I have decided to relinquish my position in your Saltdean Park Road office," and giving one calendar month's notice to expire on Saturday August 4 of that year. He explained his departure by saying that he trusted his new job with Fox & Sons (another firm of estate agents principally situated in Brighton but having wide connections over the whole of the South Coast down, I understand, as far as Cornwall) could give him improved prospects of becoming a manager in years to come. Although the notice was expressed to expire on Saturday August 4 Mr Luck did in fact stay on to help out,
because of a temporary staff shortage experienced by the H
defendant, and he helped out by staying in the office at Peacehaven as assistant manager on the terms set out in a letter at p 36 of the bundle of correspondence. Again, on p 37, he indicates (in a letter to Mr Davenport-Smith) that there was further discussion between them, but he still felt that it would be in his best interest to leave as "whilst you have made a most generous offer in financial terms the prospects of actual management seem very remote." Those letters indicate how highly Mr Davenport-Smith valued the services of the plaintiff. On September 25 1973 the plaintiff finally
left the defendant's employment and began his service with J
Fox & Sons. During all the time that he had been with the defendant firm the plaintiff's remuneration had been varied from time to time in different ways, and the chronology is set out in paragraph 3(a) of the amended statement of claim, which I do not feel I need repeat.

That is the outline of the plaintiff's service with the defendant. I begin my examination of the evidence by saying this, that Mr Davenport-Smith has paid a tribute to the honesty of Mr Luck, saying that he is a completely honest as well as efficient servant. I think that the same tribute should really be offered to Mr Davenport-Smith, too. I think
that perhaps, as happens even with the most honest of people, K
the pressures of cogitation in the litigation that has transpired have caused Mr Davenport-Smith to make mistakes in recollection which enabled counsel for Mr Luck to impeach Mr Davenport-Smith's evidence in some respects; but I do not think that he was in any way dishonest intentionally, and being faced with the contrary recollections of two people trying to be absolutely honest makes it none the easier for a judge to sort out where the actual truth lies.

The issues in this matter may be summarised in this way. Firstly, there is an allegation by the defendant of a most trivial matter, namely that Mr Luck was overpaid to the extent of £5 in car allowance. That has been conceded by the plaintiff as being unworthy of argument, and, therefore, lest I forget it, the defendant is entitled to that sum. Secondly, there was a sum mentioned in the reamended defence and counterclaim, paragraph 5 (particulars of sums overpaid to or retained by the plaintiff or paid out without authority) subparagraph (b), where it is alleged that the plaintiff, acting for the defendant, collected rent from the tenants of 9 Cissbury Crescent, Saltdean, which the defendant as managing agent for that property had transferred to the landlord Mr Poultney. The amount is £80. It is quite clear that a mistake did arise—Mr Luck has freely admitted it—and that it was his fault. The reason for the mistake was that the tenants had to pay for the period ending, as the correspondence shows, on January 12 1973, the sum of £80; and they paid that on January 13 1973 and at the same time were presented with a demand for the next and final period of their tenancy, namely, from January 13 to February 10 1973, exactly the same amount. It seems that, having received that demand, the tenant—assuming good faith on his part—thought that it was the same bill which had come back again and he remonstrated with Mr Luck, who fell into the same confusion and wrote a letter apologising for the mistake, as he thought it was, and gave a receipt for that £80 which he had in fact never received. That sum of money, less commission, was paid to the landlord out of the defendant's funds. That was a mistake on Mr Luck's part. But it is said by the defendant that the result is that the rent so overacknowledged to the landlord is irrecoverable. This is very unsatisfactory, because in fact the defendant knows the present address of the then tenant, Mr Coulter (it appears indeed in his reamended defence, in paragraph 14) and no effort at all has been made to recover that sum from Mr Coulter. It is true that Mr Coulter might, if he was less than honest or if he still laboured under the same mistake, have said, "I have a receipt for it." That would occasion a good deal of trouble to Mr Davenport-Smith in trying to enforce the claim, but it is not very impressive to find out that he has not even troubled to ask Mr Coulter whether he realises the mistake that was made and would he pay up. I think this item, therefore, constitutes an undoubted error and inefficiency on the part of Mr Luck, but I do not think that it is recoverable in the circumstances by Mr Davenport-Smith, simply because he has not taken the trouble to establish that part of his claim that he could not have got it back willingly from Mr Coulter. I would therefore find in that respect in favour of Mr Luck.

The next item (but it comes in subparagraph (a) of the same particulars) is the question of a sum of £100 that was lost at the time when the plaintiff Mr Luck was serving at the Saltdean office. It represents a deposit in cash received by Mr Luck in respect of a house called 34 Chichester Drive. The money apparently was paid in cash, and he gave a receipt for it (which is to be found on p 14 of bundle B) on January 4 1973. Though that came in, as far as I understand it, during ordinary business hours, Mr Luck was not able to pay the money into the bank, probably because the bank had closed, though he does not really recall very accurately one way or another. So Mr Luck put the money into the only safe in the office. He did not put it into his drawer, which would be locked with a key, or anywhere else. He put it into the safe, which was an article that belied its name, because the lock was defective and the key had been lost; so it was used only for storing documents and there was no receptacle for money that required to be kept safely. Mr Luck had been told by Mr Davenport-Smith not to leave money in that safe overnight but rather to take it home if he could not do anything else with it; bank it preferably, but to take it home otherwise, or indeed, if he liked, to take it over to the police station office and leave it there. On this occasion, having put it in the safe, Mr Luck left it there overnight, and the next morning when he went to pay the money into the bank he found that it was gone. He says—and I accept it—that he told Mr Davenport-Smith at once that the money had been lost and asked whether he wanted to notify the police or the insurance company. Mr Davenport-Smith was annoyed but decided that nothing else should be done. I think that that is to be accepted as the fact, because Mr Davenport-Smith acknowledged that at that time he did not realise his insurance policy would cover the transport of money to an employee's home and nor did he feel, as I am sure was the case, that the insurance company would pay up if the money had been left unguarded in a safe which could not be locked; so he did nothing about it. But later on he discovered that his insurance policy would cover money taken home by the employee up to the sum of £100, so he then made a claim, and it is obvious from the letter that came back from the insurance company that he made the claim on the basis that Mr Luck had taken the money home, and he was unwilling to admit that the loss had really occurred in that safe. The plaintiff was responsible for that money; he was the manager or the senior person in that office. However unreasonable it was—and I think it was unreasonable—to be told to take the money home, he had been so instructed and, either negligently or deliberately, he failed to take it home; he left it in the safe. Of course, the office being locked, one would have thought that it was pretty safe anyway. No forcible breaking and entering occurred as far as could be seen—no burglary—and the money must have been taken by someone in the office when the office was open, or who had gained access with a key. So Mr Luck's failure to take the money home, or leaving it overnight in the safe, seems to have occurred at an unfortunate time. Nonetheless, I think that Mr Luck has failed to show that he took reasonable care of that money and, by disregarding the instructions, that he must be responsible for that amount. But Mr Davenport-Smith is not by any means to be exculpated for what happened. I think it was a great pity that he should have put forward his claim to the insurance company on what I consider to be less than a frank basis. But leaving that aside, he had known for a long time that the safe was useless for security. He had chosen not to get a key—easily enough done—and to put the safe in order. He had chosen instead to tell employees to take the money home, which exposed them to possible suspicion if the money were lost, however accidentally, and imposed upon them responsibilities which were not a part of their employment. I think he was very much to blame, and I will put his responsibility for this loss at 75 per cent, which means that Mr Luck will need to pay him no more than £25.

The fourth issue raised in the case is the question of Mr Luck's claim to commission after leaving Mr Davenport-Smith's employment. In paragraph 6 of the amended statement of claim, the plaintiff says that he is entitled to receive the sums calculated on the net profits of the defendant's practice in the way that was clearly agreed during his service, and he also asks for commission due in respect of properties the sale of which he negotiated but for which the sales were not completed until after he left. The defendant does not deny that he would be liable to an account if commission is due or a share of profits is due. But, as regards the commission, the plaintiff claims only in respect of sales negotiated by him while he was at the Saltdean office between June 1972 and July 1973, and at the Peacehaven office from the beginning of August 1973 to September 1973, the time when he was helping out, so to speak, as I have already mentioned. This is a matter which of course depends on the interpretation of his terms of employment. It is quite clear that upon the authorities he is entitled to commission, notwithstanding that he left the defendant's service, if, before

A leaving, he had done something to earn the commission and if the contract allowed that he should be paid notwithstanding that the entitlement was concluded after his service ceased. I need not refer to all the authorities that have been cited to me, but I take that to be the result of the various cases that I have studied.

The commission situation, if I may call it that, in the defendant's service may be illustrated by letters in the course of the plaintiff's service. In the letter of March 19 1965 paragraph 3 reads: "Commission of 7½ per cent on all sales effected by you," and then, in paragraph 12, there is reference to a special situation, giving a Mr Cross, manager B of the Peacehaven office, priority on all applicants; but it is perhaps relevant to note that, in the circumstances there which would entitle the plaintiff to commission, the last sentence reads, "Commissions are paid on the first pay day after the completion of a sale or letting." On March 24 1966, when the plaintiff received another letter containing what was said to be new terms of employment from March 21 1966, "Commission is 2½ per cent on all sales negotiated at the Peacehaven office, also to include introductory fees for insurance," etc, and it goes on, "In fact this applies to all business negotiated from March 23 1966 until you have been requested by the Saltdean office to accompany one of C their applicants to newer properties and you effect a sale immediately to that applicant you will still be paid 2½ per cent commission on the sale." In April 1967 the commission was increased to 3½ per cent, and to quote the letter, "to be paid on all sales completed from June 29 1967." Likewise, on November 13 1968, the letter appointing him senior negotiator says "4 per cent commission on all sales negotiated from January 1 1969," and so on; I need not read all of the letter. Except in the letter of March 23 the word "sale" is always qualified. Coming nearer to the present time, on June 1 1970 Mr Luck received a letter saying: "Your commission will be reduced to 3½ per cent for sales D negotiated *which proceed to completion* from June 1 to June 28 inclusive of this year." When the negotiations for a partnership were terminated Mr Luck's note reads that he was promised commission at 5 per cent as well as a share of the profits. Mr Davenport-Smith's note on the file, exhibit D2, reads: "From April 1"—this was in 1972—"commission of 4½ per cent on all sales, Saltdean and Roedean," and later on, "From April 1 1972, sales negotiated *which are completed*, commission 4½ per cent." Of course it is well known that the courts have decided, in claims by estate agents against their clients or customers for commission, that when E the word "sale" is used, just as when the word "purchaser" is used, that means a concluded sale where someone has in fact bought the property. But in this case, where, as I have shown, the expression used is "sales negotiated," and distinguished from "completion," I think the meaning is that when the plaintiff has negotiated the proposed sale of property to such an extent that the purchaser or intending purchaser has agreed subject to contract, and the matter has passed from the estate agent's hands to the hands of solicitors for final completion, he has then earned his commission, though he will not be paid unless and until the intending purchaser finally completes the sale and the commission is received by F Mr Davenport-Smith. An example of that is the letter of June 1 1970, which I have already read, where the words are "for sales negotiated which proceed to completion." I think it indicates that there are two parts to the matter: first of all, the earning of the commission by Mr Luck; and, secondly, the condition precedent to his receiving the money. His own aim is to get the matter to such a head that nothing then remains but for the conveyance to be completed.

I take the view that that is how the two parties to this agreement understood the matter. Indeed the course of conduct shows that it was so understood. Mr Luck received payments after leaving the service of the defendant in many instances. G So did other people, like Mr Ost, who left the defendant's employment and who were paid without question for matters that they had negotiated during their service, but which had not been completed, and for which, therefore, Mr Davenport-Smith had not received the commission until after their service had ended. Mr M C Ayling, who was asked about this matter, said that when he left the defendant he was senior negotiator at Saltdean and he always understood that he was entitled to commission on the introduction of a prospective purchaser to a property, and he said, "There never was a suggestion that I would not be paid commission after I had left. I expected to be paid, and I would have insisted on being paid." H Mr W J Hannington, who was a man of great experience and has been for, I think he said, 50 years in the estate agency business, told the court that the negotiator would always be entitled to commission after leaving the service of a particular agent on sales negotiated before leaving; the matter would have to be brought to the stage of "subject to contract," he said, and, having done that, the negotiator would merely have to wait for the completion to get his commission. In his evidence Mr Davenport-Smith said that the negotiator is entitled to commission on completion of the sale but his entitlement after leaving the service of a particular estate agent is only moral entitlement or a moral obligation to pay, J and he said "I paid Mr Ost and Mr Ayling after they left, because I had no reason not to pay them. Indeed I paid the plaintiff after leaving because I had no reason not to pay him." "But," he said, "I did not know that, having gone to Fox & Sons he was about to be planted in the Saltdean area, which I raised with the plaintiff subsequently, and which I say is in breach of his restrictive covenant. Consequently I withheld payment." I think it is perfectly clear from the evidence and on the interpretation of the letters, to which I have referred, that the negotiator is entitled to commission on matters negotiated by him to the point where the papers pass from the estate K agent to the solicitor; and, as I say, "negotiated by him" means, of course, when he was with the defendant. He cannot have done it otherwise. In those circumstances, I do not think that the defendant is entitled to withhold payment because he disapproves of the plaintiff's subsequent conduct or because he thinks the plaintiff has broken a restrictive covenant. Matters of set-off may arise, but there is no doubt that the plaintiff is entitled to insist on his commission being brought into account and, unless there is a set-off, to be paid that commission. Consequently, finding in his favour, I would direct that an account should be taken of the amount of commission and profits payable to him. L

Now I come to the matter which is perhaps of major importance in this case, that is the defendant's claim to enforce against the plaintiff a covenant restraining him from setting up a practice or from being employed by another estate agent within a certain area. The matter arose in this way. In the original letter of employment on March 19 1965, when, of course, the plaintiff was only 18 years of age, the defendant Mr Davenport-Smith said that one of the terms of employment was that the plaintiff would sign an agreement "not to join or open an estate office within a one-mile radius of our present offices." That is void, for many reasons, M among which, of course, are that, although the plaintiff signed that letter to indicate his acceptance of all those terms, he had not in fact signed an agreement at that time including that restraint, but it is also void for being unreasonable in point of time and having regard to the plaintiff's age and circumstances which I need not go into more fully. But when he became 21 the plaintiff was presented with a formal agreement, dated November 13 1967, which reads in its material parts as follows. "Whereas the employee on March 22 1965 entered into the employment of the employer upon the terms and conditions set out in the letter of March 19 1965 and whereas it was prior to the date

A hereof agreed as part of the terms of employment of the employee that the parties would enter into this agreement now it is hereby agreed as follows: "Employer will employ"—the plaintiff—"in his aforesaid business and the employee expressly agrees as follows," and the subparagraph of that latter agreement reads:

> Not at any time during a period of three years after the determination of his employment hereunder (whether the same shall be determined by notice or otherwise) to undertake or carry on or be employed either directly or indirectly concerned or interested as employer, journeyman, manager, clerk, canvasser or otherwise howsoever (except only as a servant of the employer) in any B business of a house, land and estate agent . . . at any place within a radius of 1 mile from the said business premises in Saltdean, Peacehaven, and Patcham aforesaid or within the like radius from the premises of the particular business of the employer in which he shall be employed at the time of such determination.

The breach in this case alleged is "within 1 mile of the Saltdean offices," because in July 1973 he joined Fox & Sons, who have a large number of offices over the South Coast, and in April 1974 they moved him to be manager of their offices in Saltdean. They took the view that the covenant which I have referred to was not enforceable and C they say that they wished to move the plaintiff to Saltdean simply because he was continuing to live there (no one could prevent him from living there), and so it was more handy, so to speak.

The question I have to determine is, first of all, whether that agreement has contractual effect, that is to say, whether there is consideration for it. What did the defendant give as a *quid pro quo* to the plaintiff for entering into that restrictive agreement? Did he give anything new to the plaintiff? The plaintiff had, as the agreement recites, been employed under the terms of the letter of March 19 1965, which provided that the employment should be determined D at one month's notice on either side. It is to be noted that the agreement in 1967 merely says, "The employer will employ the employee in his aforesaid business": there is no provision whatever for a period of notice or other method of determination. The only indication of a notice is in the body of the covenant at issue, where in parenthesis it says "whether the same"—that is, whether the employment—"shall be determined by notice or otherwise." Under that agreement, it would appear to me that less is offered than was previously offered. I see no reason why it should not be terminated fairly swiftly, even on reasonable notice, landing the plaintiff in a situation where he had got no job and is E prevented by the covenant which he had now entered into from practising elsewhere. However, he had been employed on the basis that he would enter into that covenant, and he knew—and has acknowledged—that if he refused to enter into that covenant then, at any rate at one month's notice, his employment would be terminated. So he had got an agreement which was really putting into effect past consideration, on the basis that he would continue in the old employment. So with some hesitation, particularly after reading the case of *Coleborne v Kearns* (1912) 46 ILT 305, a case decided in 1912, I think that there was consideration for that new agreement. I do not think it was superseded, F as alleged by the plaintiff, by any subsequent change in the terms of his employment. It is noteworthy that, although that covenant was adumbrated (that new agreement was foreseen in the letter in 1965) it had never again been mentioned until it was entered into. I do not think I need go into the dispute between the witnesses as to whether Mr Davenport-Smith spoke to the plaintiff about it before putting it forward for his signature. There is a direct conflict of evidence on that, and I do not think it helps one way or the other to decide that issue. That agreement was intended to be the continuing and basic contract of service, as it were, but the various incidental details could be separately agreed from time to time; that is why it was put in a separate G document.

That is not the end of the matter, because an agreement in restraint of trade is *prima facie* void as being contrary to public policy, unless the defendant who maintains it shows that it is both reasonable in the particular circumstances of the case and reasonable having regard to what is called public policy, namely, the public interest. The evidence that I have heard may be thought to indicate that there is no goodwill (I quote Mr Hannington) in an estate agent's business, at any rate so far as buying and selling of houses is concerned. There may be in the management of property—I do not know—but as regards the buying and selling of houses H there is no goodwill, he says. Therefore a restrictive covenant is never required, never reasonable, between estate agents and employees. Mr Hannington strongly rejects any suggestion that there should be such covenants in his business, except of course where principals are partners and wish to restrain each other from breaking off and setting up as principals elsewhere, because, he said, in such circumstances obviously the personality and connections of one man may be much more important than in the case of an employee. That view of the matter cannot be right, as demonstrated by the fact that there are many cases in the reports where covenants of that nature between estate agents and their employees have J been enforced by the courts. Consequently such covenants do exist and have been thought in some instances to be reasonable and in the public interest. In the case of *Scorer v Seymour Jones* [1966] 1 WLR 1419 an estate agent who had a long-established business in Dartmouth, Devonshire, opened a new office in Kingsbridge in the same county and engaged the defendant, a Mr Jones, to work in that office as a clerk and negotiator and to conduct the business of the office with the assistance of a typist. At the time when the defendant entered into that employment he understood that there would be some restriction imposed by the plaintiff on his business activities should he leave the plaintiff's employ- K ment, but no formal agreement as to the restrictions to be imposed was reached at that time. In June of the same year, 1964, the Contracts of Employment Act made it necessary for the terms of employment to be set out in writing, and the parties entered into a written agreement setting out the terms of the defendant's employment, and clause 8 of that agreement contained a restriction that during a period of three years after leaving the plaintiff's service the defendant, Mr Jones, would not be employed or interested directly or indirectly in any capacity whatsoever in the business of an auctioneer, surveyor or estate agent or in any ancillary L business carried on within a radius of 5 miles of the plaintiff's premises nor use or disclose any information concerning the plaintiff's business. That contract was upheld though it was severed in respect of part of it. Salmon LJ, in that case, referred to the case of *Bowler v Lovegrove* [1921] 1 Ch 642, which was a case which decided that a covenant intended for no other purpose than to restrict competition was void, and Salmon LJ said that that case went to the verge of getting the defendant home. But he goes on, at p 1425, as follows:

> The county court judge in his judgment says (rightly, I now think) that the circumstances of the instant case are quite different M from the circumstances of *Bowler v Lovegrove*. He did not, however, state the respects in which the circumstances were different. Our attention has been drawn to these very important matters by Mr Crawford on behalf of the plaintiff. P O Lawrence J said: "It is true that the defendant came into personal contact with the plaintiffs' customers. But this fact loses its significance when the nature of the business carried on by the plaintiffs and the duties of the defendant in connection with such business are considered. The plaintiffs' customers with whom the defendant came into personal contact were not the ordinary recurring customers such as exist in most other businesses." He was saying that the fact that a servant comes into personal contact with

A customers in the normal case might justify such a covenant, but that it lost its significance in the case he was considering because the customers were not recurring customers. In the instant case, the judge has accepted the evidence which was called by the plaintiff; and there was cogent evidence on behalf of the plaintiff that he had many recurring customers. This is the fact which to my mind distinguishes the instant case from *Bowler v Lovegrove*. When the defendant was in the Kingsbridge office he *was* in effect the Kingsbridge office. Every customer who came into that office dealt with him. He was in a position in which he would have every opportunity of gaining knowledge of the customers' business and influence over the customers.

I think that the whole of the law on this subject is set out in
B two short passages in the speech of Lord Parker of Waddington in *Morris (Herbert) Ltd v Saxelby* [1916] 1 AC 688. He said, speaking of covenants such as these: " Wherever such covenants have been upheld it has been on the ground, not that the servant or apprentice would, by reason of his employment or training, obtain the skill and knowledge necessary to equip him as a possible competitor in the trade, but that he might obtain such personal knowledge of and influence over the customers of his employer . . . as would enable him, if competition were allowed, to take advantage of his employer's trade connection. . . . "

If one remembers that in the instant case there were recurring customers who formed a large part of the plaintiff's trade connection, this seems to me to be an obvious illustration of a case
C where a man, because of his employment, is enabled to gain personal knowledge of and influence over the customers which will allow him to take advantage of his employer's trade connection unless there is a covenant such as this in his service agreement."

As I say, and need not repeat by reading it, the case of *Bowler v Lovegrove* was a case where, although the employee came into contact with the customers, they were not recurring customers. That, as Salmon LJ said, distinguished the case from *Scorer's* case. The evidence in this case directed to the reasonableness of requiring a covenant of any kind restraining
D competition, seems to amount to this. The plaintiff continued living in Saltdean, a relatively small community, in which, up till that time, namely, July 1973, the defendant was, so to speak, *the* estate agent. [*At this point in the judgment the defendant's counsel, Mr Galpin, interrupted to say that the plaintiff lived in Peacehaven, not Saltdean.*] The plaintiff lived in Peacehaven, a small community, and to explain the error of what I was saying, fairly close to Saltdean, especially having regard to present communications. In that area, the defendant was really *the* estate agent, and the plaintiff became a very successful manager of the Saltdean office. So that one may assume that his personality and efficiency impressed
E themselves upon customers, and played some part in his success. The success, however, had not been achieved by 1967, and at that time the agreement was entered into by the parties merely because of Mr Davenport-Smith's wish to protect himself against a man who might in the future be successful and who might in the future be able, either on his own or as someone else's employee, to take away some of his customers. But there is absolutely no evidence that the plaintiff in fact took any customer away from the defendant. In fact the evidence is to the contrary that he ever made any attempt to seduce clients of the defendant or that the defendant's clients came to him because they
F found out that he was in a rival office. In paragraph 14 of the amended defence, the defendant has done the best he could to set out the instances of people he believed were lost to him because the plaintiff is now with Fox & Sons at the Saltdean office, and really that allegation in paragraph 14 has proved disastrous for the defendant because in about eight out of ten cases it seems that the property was not sold by or under the influence of the plaintiff, and, indeed, in at least two instances the property was sold by the defendant himself. So I find that the evidence is, so far as it goes (and Mr Galpin has made the point that further discovery and minute examination might prove otherwise) to the contrary
that any customer has followed the plaintiff to the rival firm, G
and I accept that. The reason is (according to Mr Hannington's experience—and I accept his evidence) that people just do not go back to one particular estate agent. The tendency is to spread their custom among a number of estate agents; just as people advertise in more than one newspaper to attract greater attention, so they spread their tentacles to different estate agents. According to Mr Hannington, " sole agencies are not now general practice for ordinary people, except perhaps banks and that sort of organisation, but even where sole agencies are given they are for a restricted time, and as far as I know "—and I accept his evidence on this point
—" the plaintiff has never asked anyone for a sole agency." H
One sole agency was, I think, virtually forced on him by one of the customers—I think it was Mr Venus or Mr Back, but the name does not matter: but that was the only instance. Mr Hannington maintains that an estate agency has no goodwill, so you have nothing to protect; and he has never had a negotiator coming to him, seeking employment, who has come labouring under the burden of a restrictive covenant from his previous employer, except Mr Ost, who came from the defendant. His firm does not seek to impose any restrictive covenant on employees, because they have found that such covenants are unnecessary, and, in so far as statistics
go, recurring trade is minimal. He said that over 10 years J
he has found that they have resold for only 2.8 per cent of their customers in Saltdean; in Rottingdean it is rather more, 7 per cent; and Peacehaven, 6.8 per cent; and it is agreed now that on the defendant's figures the highest at which recurring trade could be put is 8 per cent. Another witness, M H de Silva, was called on behalf of the defendant. He is a chartered surveyor and a partner in the firm of Parsons, Son & Basley, estate agents and surveyors, who practise in Brighton and Hove, and he is speaking from experience in the business since 1948. He said that there are about 20 firms in the Brighton and Hove area, with, between
them, about 43 offices, and he was told at those offices there K
is a restrictive covenant demanded of some kind or other; the more senior the employee, the more need, he said, for a restrictive covenant, and the more isolated the community the more usual it is to have it. It is less usual for negotiators in the town centre to be put under a restrictive covenant. But he said, in any case, one should introduce the covenant at an early stage so as to make sure that the employee is not going to leave on the crest of sudden success without restraint upon him.

This, as I have said, was a small community, in which the
plaintiff was a near neighbour (I am talking now about L
Saltdean). His personality must have been known, but in 1967 it must have been almost totally unknown and there could have been little danger, at the age of 21, of any immediate threat of competition from Mr Luck. I think it was not reasonable to have this restrictive covenant in 1967. But if I am wrong about that, was it reasonable in its scope? The expert witnesses are agreed about that. In my judgment, it would be unreasonable to have a limitation extending for three years, because, as I have said, the recurring trade is small and is likely to be confined, if it occurs at all, to the immediate community, not very far afield. In this instance,
the plaintiff is bound in a radius of one mile from the M
business at Saltdean and Patcham, and, though one mile is reasonable, according to Mr Hannington and to Mr de Silva, I think that, combined with the requirement that it should be for a period of three years, the whole covenant becomes unreasonable.

That being so, there being no question of any severance of the covenant, I must hold that the covenant is void and unenforceable. Therefore the counterclaim on that issue fails. I perhaps ought to add that I am not saying that in no case is an estate agent's restrictive covenant enforceable. Mr Davenport-Smith has other covenants with all his employees,

A so it seems. I do not, of course, pronounce upon them, because every covenant depends upon its own facts, and there may be those which are enforceable and those which are not.

Judgment was given for the plaintiff for an account of the commission and profits due to him and judgment for the defendant for £30 on his counterclaim. There was no order as to costs except liberty to apply failing agreement. B

HIGHWAYS

A **QUEEN'S BENCH DIVISION**

November 3 1976

(Before Mr Justice SLYNN)

WATERS v SECRETARY OF STATE FOR THE ENVIRONMENT

Estates Gazette January 1 1977

(1976) 241 EG 30

Highways—Dual-lane road to be constructed to relieve section of A38 near Plymouth used primarily for local traffic—Secretary of State has power to build such a road
B **and is not limited to roads to be used by through traffic, alone or primarily—No force in further points taken against confirmation of line order for the new road—Motion to quash order dismissed**

This was an application by John Duncan Waters ARICS to quash the London to Penzance Trunk Road (Marsh Mills to, Tamar Bridge) line order made by the respondent, the Secretary of State for the Environment, on March 12 1976.

The applicant appeared in person. H K Woolf (instructed by the Treasury Solicitor) represented the respondent.

Giving judgment, SLYNN J said that on December 15 1972
C the respondent Secretary of State published a draft order, "London to Penzance Trunk Road (Marsh Mills to Tamar Bridge)." That order was to be made under sections 7 and 44 of the Highways Act 1959 and section 8 of the Highways Act 1971. It referred to certain highways that the Secretary of State proposed to construct along routes described, and to certain other highways already in existence, and the purpose of the order was that these routes should become trunk roads from the dates specified in the order. A public inquiry was held between March and June 1974 which, together with site inspections, lasted some 37 days. By a letter dated March 25 1976 the Secretary of State said that having considered repre-
D sentations, objections, alternative routes and counter-proposals and the report of the inspector, he had decided to accept all the inspector's recommendations and had made the main order and a slip road order as published and two side road orders with modifications. The main order was in fact made on March 12 1976 and came into operation on March 26 1976.

Mr Waters now applied under para 2, schedule 2 to the Highways Act 1959 to quash the main order. He did so as a person who had appeared at the public inquiry while a resident of Plymouth, although he now lived in London. In his notice of motion he set out his main grounds supplemented by an affidavit. This had been most carefully and ably
E prepared and had been of great assistance. First he said that the order was *ultra vires* of the Secretary of State under section 7 of the Highways Act 1959. Secondly he said that if the order was *intra vires*, false and misleading statements had been made by witnesses called to support the making of the order, either because they knowingly made false statements or because they were misled by the Secretary of State; and it followed that there had been a breach of section 279 of the Highways Act 1959, read with section 240 of the Local Government Act 1933. Thirdly, Mr Waters said that the

Secretary of State had changed his case so radically at or F
towards the end of the public inquiry that objectors were not able to deal with the true case and that there should have been, but was not, an adjournment; accordingly he (Mr Waters) had been prejudiced, and on that ground the order should be quashed. Mr Waters had made it clear that of his three grounds it was the second, which, if any, he preferred over the others.

At the inquiry, the Department of the Environment presented a statement of their case which described in paragraph 6 their proposals for improving the A38 at Plymouth by the construction of a new road. The draft section 7 order published in 1972 would, if made, authorise the Secretary of G
State to construct about 5½ miles of new trunk road. It was made plain that the existing stretch of road was simply not coping with the amount of traffic which it was having to take, and that for the future more traffic was expected. A distinction was made between through and local traffic. The history of the matter was recounted, and showed that from the 1930s proposals had been made for dealing with the problem. By 1961 a dual-lane highway was suggested from Marsh Mills to Tamar Bridge. In June 1971 the Department of the Environment accepted responsibility for preparation and construction of this, including procedures to establish the H
route as a trunk road. In his report, the inspector remarked that there had been considerable doubt in his mind about the purpose of the road. He said that he found himself puzzled as to the circumstances in which the Department had decided to pursue responsibility, having regard to their claim that the road was for the solution of Plymouth's own traffic problems. It was clear from later paragraphs of the report that some at least of those appearing at the inquiry had fully appreciated that the user of the road would be primarily by local traffic. In his decision letter of March 25 1976 the Secretary of State said that before the inquiry some 34 proposals had been considered. He was satisfied that the new J
road would be open for the purposes of through traffic, although he appreciated that it would be used by local traffic too; and he had accepted the inspector's recommendations and had made the order.

It was clear from all this material that there was a need to deal with the volume of traffic on the A38. The first question for him (his Lordship) was whether the order was *intra vires.* Mr Waters submitted that section 7 entitled the Secretary of State to make a road a trunk road only where the purpose of the road was for through traffic alone. In his (Slynn J's) judgment, section 7 required the Secretary of State to keep under review the national system of routes for through K
traffic in England and Wales. The question was therefore whether the Secretary of State could reasonably have concluded that to make this a trunk road was expedient for expanding or reorganising the system of routes for through traffic. He (his Lordship) did not consider that Parliament intended by this section to limit the Secretary of State to the sort of case where the road was to be used exclusively for through traffic. In his judgment it was possible under this section for a road to be made a trunk road even if it was

A mainly to be used by local traffic. In his judgment, therefore, the Secretary of State was entitled to make the order as he did, and the order was *intra vires*.

Mr Water's second point was the allegation that some of the witnesses had given false and misleading evidence. As to this, he (his Lordship) considered that in dealing with the user of the road the witnesses concerned were making it clear that the principal user of the road would be by local traffic. He did not accept that the witnesses were giving false evidence, either deliberately or because they had been duped by the Secretary of State. Looking at the matter in a general way, it might be that the ratepayers of Plymouth were in a
B sense lucky, in that the Department were to take over 100 per cent rather than, as was usually the case, 75 per cent of the cost of the road concerned. Nor did he (Slynn J) accept that the inspector had been fooled as to what the Secretary of State's intention was, even though he had expressed puzzle- C
ment as to the financial aspect of the matter. Lastly there was the argument that there should have been an adjournment. It seemed to him (his Lordship) that from the beginning the problem had been made clear. He saw no change at any stage in the nature of the user of the road which was being talked about by various witnesses. Mr Waters had not asked for an adjournment at the hearing, and seemed to be presenting his legal arguments and proposals for an alternative route with skill and care. He (Slynn J) was not satisfied that Mr Waters had been prejudiced. He would therefore not accept that there were grounds upon which he D
could quash the line order, and Mr Waters, despite the skill and care of his arguments, had not satisfied him that it would be right for him to do so.

HOUSING ACTS

A **COURT OF APPEAL**

November 24 1976

(Before Lord Justice MEGAW, Lord Justice SCARMAN and Lord Justice BROWNE)

CAMPDEN HILL TOWERS LTD AND ANOTHER v GARDNER AND ANOTHER

Estates Gazette April 30 1977

(1976) 242 EG 375

Housing Act 1961 — Important decision on interpretation of landlord's statutory covenants under section 32 — Covenant to pay a share of increased cost of repairs, including repairs
B **to block in which flat was situated — "Structure and exterior" held to extend to parts not included in demise, such as the outside wall of a flat, but "installations" confined to what is within the demised dwelling**

This was an appeal by Campden Hill Towers Ltd and Davstone (Holdings) Ltd, lessors successively of two blocks of flats—Gate Hill Court and Campden Hill Towers—and other property in Notting Hill Gate, London, from the refusal of Judge Leslie at Bloomsbury and Marylebone County Court to award them the whole sum they had claimed from S W Gardner and his wife, lessees of 20 Gate Hill Court, in respect of the lessees' covenant to contribute towards the cost of
C repairs. Camden Hill Towers Ltd, the first plaintiffs, had assigned their reversion to Davstone (Holdings) Ltd, the second plaintiffs, during the period covered by the claim.

J S Colyer QC and Robert Pryor (instructed by Thornton, Lynne & Lawson) represented the appellants, and Derek Wood (instructed by Allen & Son) appeared for the respondents.

Giving the judgment of the court, MEGAW LJ said that by an underlease, Campden Hill Towers Ltd let to a Miss Lawson and a Miss Field, a third-floor flat, 20 Gate Hill Court. During the currency of the underlease Davstone (Holdings) Ltd took over Campden's interests in Gate Hill Court and an adjacent block. Miss Lawson married Mr
D Gardner and they became lessees of flat no 20. By proceedings in the county court, the lessors claimed against the lessees the recovery of so-called "service charges" of £197.22 for 1973 and 1974. The claim was based on a covenant in the underlease. The lessees, relying on section 32 of the Housing Act 1961, contended that the lessors were not entitled to recover the whole of the sum claimed. The lessors now agreed that the underlease was a lease to which section 32(1) applied.

Section 32 provided:

(1) In any lease of a dwelling-house, being a lease to which
E this section applies, there shall be implied a covenant by the lessor—(a) to keep in repair the structure and exterior of the dwelling-house (including drains, gutters and external pipes); and (b) to keep in repair and proper working order the installations in the dwelling-house—(i) for the supply of water, gas and electricity, and for sanitation (including basins, sinks, baths and sanitary conveniences but not, except as aforesaid, fixtures, fittings and appliances for making use of the supply of water, gas or electricity), and (ii) for space heating or heating water, and any covenant by the lessee for the repair of the premises (including any covenant to put in repair or deliver up in repair, to paint, point or render or to pay money in lieu of repairs by the lessee F
or on account of repairs by the lessor) shall be of no effect so far as it relates to the matters mentioned in paragraphs (a) and (b) of this subsection.

His Lordship said that by virtue of the definition of "dwelling-house" in section 32(5) the relevant "dwelling-house" was flat no 20, not the whole block, for the flat was the relevant "part of a building" let as a private dwelling. Section 32(1) provided that there should be implied in the underlease a covenant by the lessors in the terms of paragraphs (a) and (b) of section 32(1), whatever they meant, and that any covenant inconsistent therewith, purporting to remove from the landlords, or to reduce, the obligations G
comprised in paragraphs (a) and (b) should be ineffective. The subsection did not, however, purport to render ineffective a covenant putting on the lessees any obligations as to repairs outside the scope of paragraphs (a) and (b). Section Section 32(1) provided that there should be implied in "any covenant by the lessee for the repair of the premises (including any covenant . . . to pay money . . . on account of repairs by the lessor) . . . so far as it relates to the matters mentioned in sub-paragraphs (a) and (b) of this subsection."

The intention and effect of these provisions, particularly as regards covenants to pay money were, in some respects, obscure. For example, what did "the premises" mean? H
However, it did not matter for the decision in this case how wide or how narrow a meaning ought to be given to "the premises" because paragraphs (a) and (b) referred to "the dwelling-house," not to "the premises." It was common ground that clause 4(2) of the underlease, incorporating the third schedule thereto, was a covenant affected by section 32(1). It related in some degree to the matters mentioned in paragraphs (a) and (b). The question was—in what degree? That would involve an examination of the meaning of these paragraphs.

By the definition in the underlease, "the premises" included not only the whole block of flats in which the J
relevant flat, "the dwelling-house," was situated, but another block of flats and other property. This was the definition of "premises" for the purpose of the underlease, not, of course, for the purpose of the statute. There were 140 flats altogether. Clause 2 of the underlease demised "all that flat . . . numbered 20 on the third floor of Gate Hill Court," but expressly excluded its outside walls from the demise. The underlease also provided for the yearly rent and the service charge. Both parties accepted that, to the extent that the "service charge" involved payment by the lessees in respect of matters which fell within paragraphs (a) and (b), K
the lessors could not require payment by the lessees. To that extent the covenant for the "service charges" was unenforceable. But to the extent that the service charges relate to matters outside the scope of paragraphs (a) and (b), the covenant was enforceable and the lessees must pay.

The third schedule listed the undertakings by the lessors towards the costs in respect of which the lessees were to pay the service charge. They included keeping in repair gas and water pipes, gutters, sewers, drains and electric cables "in, under and upon the premises"; keeping in repair the common

A parts; and redecorating and repairing the exterior of the premises and the main timbers and party walls thereof. That schedule throughout was concerned with "the premises" as defined in the underlease. However widely the words of paragraphs (a) and (b) of section 32(1) were to be interpreted, no one could sensibly suggest that they should be treated as applying to repairs to buildings which were quite separate from the block of flats in which "the dwelling-house" in question was situated. Therefore the lessees' obligation under clause 4 (2) of the underlease to pay one 140th share of the extra cost of such part of any repairs as related to buildings other than Gate Hill Court was not abrogated by section
B 32(1). But that negative proposition did not help much in deciding what the positive proposition should be. It merely emphasised an apparent oddity: the less directly the repairs in a covenant appeared to be for the lessees' immediate benefit, the more likely it was that they would have to pay for them. Whether that was Parliament's intention, their Lordships would not speculate.

The judge had rejected the lessors' argument that the Act only imposed an obligation to keep in repair the structure and exterior of the particular flat demised. He expressed the view that the intention of the legislature must have been to
C regard the whole building in which the dwelling-house was comprised as "the exterior of the dwelling-house." Their Lordships did not think that that was the correct interpretation of paragraph (a) even if "the whole building" for that purpose was to be treated as meaning only Gate Hill Court. In their Lordships' opinion anything which, in the ordinary use of words, would be regarded as a part of the structure, or of the exterior, of the particular "dwelling-house," regarded as a separate part of the building, would be within the scope of paragraph (a), even if not part of what was demised. That paragraph therefore applied to the outside wall or walls of the flat; the outside of inner party walls of the flat; the
D outer sides of horizontal divisions between flat no 20 and flats above and below; and the structural framework and beams supporting the floors, ceilings and walls of the flat. It went no further than that, however. It was not concerned with other obligations on the lessors, whether by virtue of obligations in leases or under other statutory provisions.

Paragraph (2), in their Lordships' judgment, meant that the installations within the physical confines of the flat must be kept in repair and capable, so far as their own structural and mechanical condition was concerned, of working properly. But no more than that. "Installations" in paragraph (b) did not, for example, cover a boiler outside the
E flat. The lessors might be under additional obligations, but if so they did not arise from the statute.

[*Note:* As the parties had previously reached an agreement as to the amount which would be payable by the lessees to the lessors in the circumstances, the appeal was dismissed by consent, no order being made as to the costs of the appeal, but the order as to costs below to stand. Megaw LJ said that, so far as the decision had a bearing on other cases, it should be noted that the terms of the decision, and the reasons why the appeal had been dismissed, were peculiar to the unusual procedure adopted in the case.]

F

QUEEN'S BENCH DIVISIONAL COURT

March 30 1977

(Before Lord WIDGERY CJ, Mr Justice EVELEIGH and Mr Justice WIEN)

FAWCETT v NEWCASTLE UPON TYNE CITY COUNCIL

Estates Gazette May 28 1977

(1977) 242 EG 695

Housing Act 1974, section 47—Council's appeal against Crown Court's decision quashing conviction—Landlord's failure to notify local authority of service on tenant of notice to quit premises in housing action area—Invalidity of notice to quit no defence to prosecution for failure to notify—Landlord properly convicted by magistrates—Appeal allowed G

This was an appeal, by way of case stated by Newcastle upon Tyne Crown Court, by the City Council of Newcastle upon Tyne against the Crown Court's decision on February 3 1976 allowing an appeal by Albert Fawcett against his conviction by Newcastle upon Tyne justices for failing, without reasonable excuse, to notify the local authority that he had served a notice to quit on the tenant of premises in Harrison Place Housing Action Area contrary to section 47(1) of the Housing Act 1974. H

Robert Stewart (instructed by T Brown, chief clerk) appeared for Newcastle City Council; the respondent, Albert Fawcett, appeared in person.

Giving the first judgment, WIEN J said that the respondent owned property in the Harrison Place Housing Action Area in Newcastle upon Tyne and in May 1975 he served a notice to quit upon a tenant. The notice was defective in that it did not give the tenant the proper length of time, but was served in order that the tenant could be rehoused. It was, however, irrelevant that the notice was given in order to provide the tenant with a document to show he had to be rehoused. J

Section 47(1) of the Housing Act 1974 provided that:

> Within the period of seven days beginning with the day on which a notice to quit is served:
>
> (a) in respect of land consisting of or including housing accommodation in a housing action area, and
>
> (b) on a tenant who occupies as a dwelling the whole or any part of that land,
>
> the landlord by or on whose behalf the notice is served shall notify the local authority in accordance with Schedule 4 to this Act, that the notice has been served.

By section 47(5)(b) of the Act: K

> A local authority who receive a notification given in compliance with any provision of subsections (1) to (3) about shall . . .
>
> (b) within the period of four weeks beginning with the day on which the notification was received, inform the person by whom it was furnished what action, if any, they propose to take as a result of the notification with respect to the land to which the notification relates.

The effect of notification of the service of any notice purporting to be a notice to quit was to impose on a local authority, within four weeks of receipt of the notice, a duty to inform the person on whose behalf the notice was served L what action, if any, they proposed to take.

The short point was what did the phrase "notice to quit" mean when used in section 47(1)? Did it mean a valid notice to quit or did it mean any notice purporting to be a notice to quit? The only purpose in informing a local authority of the service of a notice to quit was to enable the local authority to take action. The local authority could not possibly inquire into the validity of a notice to quit. Provided the document purported to be a notice to quit that was enough. There was nothing wrong with the original conviction and no justification for the Crown Court's decision. The appeal should be allowed. M

LORD WIDGERY and EVELEIGH J agreed.

The appeal was allowed and the original conviction restored. No order was made as to costs.

A **COURT OF APPEAL**

July 23 1976

(Before Lord Justice MEGAW, Lord Justice ORR and Lord Justice BROWNE)

R v KERRIER DISTRICT COUNCIL, EX PARTE GUPPYS (BRIDPORT) LTD

Estates Gazette June 18 1977

(1976) 242 EG 955

Houses unfit for human habitation—Word "shall" in sections 9 and 16 of Housing Act 1957 imperative—No discretion to proceed instead under Public Health Act 1936—Appeal by district council dismissed

B This was an appeal by Kerrier District Council from a judgment of the Queen's Bench Divisional Court in favour of Guppys (Bridport) Ltd on an application by the company for an order of mandamus directed to the council. The Divisional Court had held that "shall" in sections 9 and 16 of the Housing Act 1957, which deal with the duty of a local authority when satisfied that a house is unfit for human habitation, was imperative, and that a local authority was in consequence bound to take the appropriate action under those sections once satisfied of the matters set out therein. The Divisional Court's decision was reported at (1975) 236 EG 645.

C Raymond Sears QC and Anthony Dinkin (instructed by Sharpe, Pritchard & Co, agents for F J Pearson, Council Offices, Camborne) appeared for the council; Robert Pryor (instructed by Turner, Peacock, agents for Hollowell & Bollam, of Weymouth) represented the respondents.

Giving the first judgment, ORR LJ said that Guppys owned two back-to-back houses, which, it was common ground, were unfit for human habitation for the purposes of section 4 of the Housing Act 1957. It had not been judicially determined whether they were capable of being made fit at reasonable expense. Both were let to protected tenants, who had refused an offer of other accommodation by Guppys. Guppys had D been willing at all times to convert the houses into a single house, but they could not do so without obtaining vacant possession. The council claimed that because of intense pressure they were unable to find other accommodation for the tenants.

On March 19 1974 the council served notice on Guppys under section 93 of the Public Health Act 1936, requiring them to repair the roofs of the houses which were alleged to constitute a nuisance under the Act. That was not done, a complaint was made by the council and a summons was issued under section 94 of the Act. On July 1 1974 Guppys' solicitors wrote to the council: "... the council is in default E in exercise of its mandatory duties under the Housing Act 1957, section 9 or 16 and section 69. Our clients have provided information and the council has no doubt also had official representations that these houses are unfit for human habitation, so that the duty arises to consider this information F and make the appropriate order, which in our submission should be under section 16 rather than section 9." The council replied: "The council accepts that the above properties are unfit for human habitation, but it is contended that neither 9 nor ... 16 ... is mandatory, ... and that the council has the discretion as to the use of Housing Act powers or section 93 of the Public Health Act 1936. As you know, in this instance my council chose the latter procedure after taking into account all the circumstances of the case."

The repairs not having been carried out, the magistrates' court, on the council's summonses, made an order that the work should be done within two months, and an appeal by G Guppys against that order was dismissed. Meanwhile, on October 15 1974 Guppys applied for mandamus. An affidavit on behalf of the council stated that the houses had been repaired but were still unfit within sections 4 and 18 of the Housing Act 1957.

The issue before the Divisional Court was whether the provisions of sections 9 and 16 were mandatory or whether the council could proceed under one or other of the sections or under the provisions of the 1936 Act. Looking at the 1957 Act itself, and the word "shall" in the context of that Act, it was difficult to see how it could be treated as other than imperative. The Act used "shall" or "may" very frequently. H It plainly drew a distinction between them.

As to the wider aspects of the council's argument, he (his Lordship) had not been satisfied that the provisions in the Housing Act 1974 as to improvement grants would be stultified, but it was in any event impossible to treat the 1974 Act as being in any way a guide to the construction of words used in an Act passed 17 years earlier. The fact that a meaning given to "shall" in the earlier Act might have an untoward effect in the later Act might be of some relevance, but it was a very minor matter in relation to the canons of construction. He (his Lordship) had not been led by the J council's argument based on other sections to conclude that the two "shalls" in sections 8 and 16 should be construed as other than imperative. He did not propose to refer to all the statutory provisions on which the council had relied.

As to the council's other main general argument that there would be a great danger of a vast number of applications for mandamus, he was not satisfied that that was likely to happen, having regard *inter alia* to the fact that the issue of mandamus was discretionary. Be that as it might, however, it was a matter for Parliament to deal with, and the council's argument did not persuade him (his Lordship) to doubt the conclusion which he would otherwise reach that "shall" in K sections 9 and 16 was to be construed as imperative.

MEGAW LJ and BROWNE LJ agreed.

The appeal was dismissed with costs.

Leave to appeal to the House of Lords was refused.

LANDLORD AND TENANT

GENERAL

A **COURT OF APPEAL**

June 30 1976

(Before Lord Justice MEGAW, Lord Justice BROWNE and Lord Justice Geoffrey LANE)

NORFOLK CAPITAL GROUP LTD v KITWAY LTD

Estates Gazette February 5 1977

(1976) 241 EG 383

Consent to assignment by limited company—Landlords' refusal held reasonable in a case where the assignee would after five years' occupation be in a position to enfranchise —Rent Acts cases held distinguishable if tending against B this result

This was an appeal by Norfolk Capital Group Ltd, assignees of a lease of properties in Princes Square, London W2, from a judgment of Judge Leslie in the Bloomsbury and Marylebone County Court on November 7 1975 refusing to declare that the respondents, Kitway Ltd, assignees of the reversion, had unreasonably withheld their consent to the assignment of the balance of the leasehold interest in one of the demised properties, 9 Princes Mews, to a Mr Freeman.

D M W Barnes (instructed by Ashurst, Morris Crisp & Co) appeared for the appellants, and Miss S Morgan (instructed by Clintons) represented the respondents.

C Giving judgment, MEGAW LJ said that the lease in question had been granted by the Prudential Insurance Co to Eccleston Hotel Co Ltd for 60 years from June 24 1950, so that in 1975 some 25 years of the term had expired and there were some 35 years still to run. The demise was of property in Princes Square, London W2. It comprised a number of buildings, houses and mews properties: the houses faced north on to Princes Square and were numbered consecutively 8 to 16, while the mews properties were at the back, facing on to Princes Mews and numbered consecutively 6 to 14. Each of the mews properties consisted of a garage or garages D below, with a residential flat or maisonette above. The rent for the whole of the demise was £2,300 a year. The lease contained a covenant as to user and a prohibition against assignment without consent of the whole or any part of the demised premises, "such consent not to be unreasonably withheld in the case of a person shown . . . as being respectable and responsible and suitable for carrying on the business or user of the demised premises permitted by [the covenant as to user already referred to]." In 1973 the appellants, Norfolk Capital Group Ltd, took an assignment of the whole of the demise, their title being registered in November of that year. In 1975 they were minded to assign the leasehold E interest in 9 Princes Mews—that was to say, a single property consisting of a garage with flat or maisonette above—to a Mr Freeman, and they made application to the respondent reversioners, Kitway Ltd, for the necessary consent. The respondents refused, giving as their only reason the fact that if Mr Freeman became assignee of no 9 he would, after residing there continuously himself for five years, become entitled to acquire the freehold of his property under the provisions of the Leasehold Reform Act 1967. The appellants took the matter to the county court, but their application was dismissed. They now came to the Court of Appeal.

By way of preliminary observation, he (his Lordship) F would say that if one were asked, without having been taken into any legal authorities relating to the matter, whether or not, in the circumstances just outlined, it was unreasonable for the respondents to refuse their consent when the consequences of giving that consent were likely to be that they would be deprived of their freehold interest in five years' time, he found it very difficult to think that anyone would find it possible to say that the refusal was unreasonable. It was however said that on the authorities that was the only proper conclusion in law. Counsel for the appellants said that there was a principle evolved in comparable Rent Acts cases which was of general application and governed the G present matter. The cases in question were *Lee v K Carter Ltd* [1949] 1 KB 85, CA; *Swanson v Forton* [1949] Ch 143, CA; *Dollar v Winston* [1950] Ch 236, Roxburgh J; and *Thomas Bookman Ltd v Nathan* [1955] 1 WLR 815, CA. The principle which it was claimed these authorities established was that where the effect of a desired assignment was likely to be that the proposed assignee would thereafter gain the protection of the Rent Acts, the landlord was not entitled to refuse consent provided the assignment was to be regarded as a "normal" assignment; he could refuse only if there were some feature of the assignment which made it an "abnormal" assignment. An assignment was "abnormal" H where the sole intention (evidently on the part of the intended assignor and the intended assignee) was to give the intended assignee statutory protection. Where the effect of the proposed assignment would be to give such protection, but that was not within the intention of the parties to the assignment, the assignment was "normal" and consent could not be refused. He (his Lordship) thought that that argument involved many difficulties both of law and of practical application, but he would not go into the authorities cited, because he was content for the purposes of the present appeal to assume that counsel had correctly stated the position in J Rent Acts cases. Even so, there was no authority to the effect that, and he (Megaw LJ) saw no reason why, the supposed principle of "normal" and "abnormal" assignments should be imported into cases concerned with the application of the Leasehold Reform Act. That principle, if indeed it was a principle, was one which he for his part thought should not be extended beyond the sphere in which it had been applied—if in fact it had. Cases under the Act of 1967 were distinguishable in that the proposed assignee would in course of time obtain the right to deprive the landlord of the ownership of the property: and he (his Lordship) saw no reason why the landlord should be said K to be acting unreasonably in refusing to consent to a transaction which would, or would be likely to, bring about for him that potential serious disadvantage, a disadvantage different in kind, in his (Megaw LJ's) judgment, from that which the courts had had to consider in relation to the Rent Acts. There appeared to be nothing contrary to this view in *Welch v Birrane* (1975) 235 EG 501 or *In re Cooper's Lease* (1968) 19 P & CR 541. The appeal should be dismissed.

BROWNE and GEOFFREY LANE LJJ delivered brief concurring judgments, and the appeal was dismissed with costs.

A **COURT OF APPEAL**

July 8 1976

(Before Lord DENNING MR, Lord Justice ORR and Lord Justice WALLER)

BICKEL AND OTHERS v DUKE OF WESTMINSTER AND OTHERS

Estates Gazette February 5 1977

(1976) 241 EG 385

Consent to assignment by trustees—Landlords' refusal held reasonable in a case where the assignee would after five years' occupation be in a position to enfranchise—Different B **approaches in the Court of Appeal to the question of the applicability of comparable cases involving Rent Act protection**

This was an appeal by the trustees of the Westminster Estate, owners of the freehold of the Grosvenor Belgravia Estate, from a judgment of Judge Ruttle at Westminster County Court holding that they had unreasonably withheld their consent to the assignment by the respondents the trustees of the Ancient Order of Foresters Friendly Society, of the residue of the terms of seven leases.

M Essayan QC (instructed by Boodle Hatfield & Co) appeared for the appellants, and J R Reid (instructed by C Whittington, Tilling & Knight) represented the respondents.

Giving judgment, LORD DENNING said that the case arose out of a difference of opinion between two highly respectable bodies of trustees about nine houses in Burton Mews, off Chester Square. Typical was no 5a, owned by the Grosvenor Estate and let to the Foresters on a long lease for 37½ years expiring at Christmas 1983. The rent was only £12 a year, the rateable value £388, so the tenancy was a long one at a low rent within the Leasehold Reform Act. Foresters could not claim to enfranchise it, because it was not occupied by them as their residence. They held it only as an investment, and had sublet it to a lady on an D underlease for 35½ years, due to expire at Christmas 1983 less three days. This subtenant had occupied the house concerned for some years as her residence, but as she did not pay a low rent she could not claim to enfranchise her underlease. She and Foresters had put their heads together, and their plan was for Foresters to assign the head lease to her. If they assigned lawfully to her, she would have the right after five years to buy the freehold from Grosvenor, because she would then be the tenant of a leasehold house on a long tenancy at a low rent and would have occupied the house as her residence in right of that tenancy for the previous E five years. Foresters could not assign the head lease to her without Grosvenor's written consent, but there was an express covenant in the head lease providing that such consent should not be unreasonably withheld. Accordingly Foresters applied to Grosvenor in 1974 for consent to an assignment to the tenant in question, stating that they wished "to dispose of the leases by sale in order to re-invest the funds in other investments." Grosvenor refused consent since they believed that "by doing so this will prevent the purchasers from enfranchising under the provisions of the Leasehold Reform Act." They added that they were willing to negotiate a fair price for the surrender. In those circumstances the question F arose whether the trustees were unreasonably withholding consent. The county court judge held that they were, and Grosvenor appealed.

Two propositions had been canvassed as if they were propositions of law. First it was said that in order to be reasonable, the landlord's refusal must be based either on the personality of the assignee or on the user or occupation of the premises: and it was said that *Houlder Bros & Co Ltd v Gibbs* [1925] Ch 198, 575 bound the court despite the doubt expressed in *Tredegar v Harwood* [1929] AC 72, 82. If such were the law, it followed that Grosvenor could not reasonably refuse consent. Secondly it was submitted that G where a house was subject to the Rent Acts, the landlord could not reasonably refuse his consent to a "normal" assignment during the contractual term, even though it meant that the assignee would be able to stay on afterwards as a statutory tenant: *Thomas Bookman Ltd v Nathan* [1955] 1 WLR 815. It was true that the landlord could reasonably refuse consent in the case of an "abnormal" assignment of the fag-end of the contractual term made for the purpose of giving the assignee the benefit of the Acts: see *Lee v K Carter Ltd* [1949] 1 KB 85; *Swanson v Forton* [1949] Ch 143; and *Dollar v Winston* [1950] Ch 236. But here, the respondents argued, the assignment was an absolutely normal H assignment of the lease for the last seven years of the term. Once again, if the cases cited could properly be regarded as laying down a proposition of law, Grosvenor's refusal of consent must be treated as unreasonable.

He (his Lordship) did not think that the authorities referred to established propositions of law with regard to refusal of consent. The words of the covenant before the court were perfectly clear English words: "such licence shall not be unreasonably withheld." When those words came to be applied in any particular case, the court neither could nor should determine by strict rules the grounds on which the landlord might or might not reasonably refuse J consent. The landlord had to exercise his judgment in all sorts of circumstances, and it was impossible for him, or for the courts, to envisage them all. The utmost that the courts could do was to give guidance to those who had to consider the problem. As one decision followed another people would get to know the likely result in any given set of circumstances, but no one decision could be a binding precedent as a strict rule of law. The reasons given by the judges were to be treated as propositions of good sense in relation to the particular case, rather than propositions of law applicable to all cases.

Turning to the present case, the authorities afforded little K guidance with regard to the problem thrown up, a new situation consequent upon the Leasehold Reform Act and one which was never envisaged before. He (Lord Denning) would test it by considering first the position of the landlords. They held a large estate which they desired to keep in their own hands so as to develop it in the best possible way. That would be impeded if one house after another was bought up by sitting tenants. Further, if they were compelled to sell under the Leasehold Reform Act they would suffer much financial loss, because the price was much less than the value of the house. Next, the position of the L tenants. They held the premises as an investment and wanted to sell it, and the landlords had said they would be willing to negotiate a fair price for it. Then the position of the subtenant. When she took her sublease she had no possible claim to enfranchisement. She was quite well protected by the Rent Acts so far as her own occupation was concerned, and the only result of a refusal would be that she would not be able to buy up the freehold for a very low figure. Taking all the circumstances into consideration, he (his Lordship) did not think that Grosvenor were withholding their consent unreasonably, and he would allow the appeal.

M Agreeing, ORR LJ said that the *Houlder Brothers* case had been distinguished by the court in subsequent appeals, in particular in *Lee's* case. Tucker LJ's judgment in the latter dispute contained the well-known allusion to the contractual relation which would have existed had consent been given to the assignment as being "pregnant with future possibilities," a phrase subsequently quoted with approval by Lord Greene in the *Swanson* case and by Danckwerts LJ in *Pimms Ltd v Tallow Chandlers Company* [1964] 2 QB 547. On the authorities, the withholding of consent in the present case was reasonable, because it related first, to the personality of the proposed assignee, in that she would

A be eligible in due course to acquire the freehold by virtue of the Leasehold Reform Act; secondly, to the effect of the proposed assignment on the user and occupation of the premises; and thirdly, to the relationship of landlord and tenant in regard to the subject-matter of the demise. There was also the point that on the evidence the object of the refusal was based on views which a reasonable man could well entertain as to the proper management of the lessors' estate, of which the premises in question formed part. He (his Lordship) thought that the judge was in error in holding that the proposed assignment was for present purposes a "normal" and not an "abnormal" one. He (the county B court judge) had wrongly attached importance in that connection to the fact that the request for consent was not made at the tail-end of the tenancy, which could not be a factor of any importance where the object of the assignment was that rights should be acquired under the Leasehold Reform Act, a statute which required five years' occupancy for that purpose.

Also agreeing, WALLER LJ said that the proposed assignment would indeed be pregnant with future possibilities, because if it were permitted, the assignee would be in a position to enfranchise his leasehold interest. That clearly C affected both the property which was the subject-matter of the lease and the relationship between the landlord and the tenant, and accordingly the landlord was entitled to refuse consent. The refusal was wholly within the principles of the *Houlder* case, *Lee* and *Pimms*.

The appeal was allowed.

CHANCERY DIVISION

April 13 1976

(Before Mr Justice FOX)

D MOORE PROPERTIES (ILFORD) LTD v McKEON AND OTHERS

Estates Gazette February 5 1977

(1976) 241 EG 387

Possession—Summary procedure under order 113 available against defendant claiming to be "holding over," but only as an unlawful subtenant of a tenant whose interest had ended

This was a summons by Moore Properties (Ilford) Ltd under order 113 of the Rules of the Supreme Court against E J P McKeon, Evelyn Beury, John Hudson, B Celleher and other persons unknown for an order for possession of a furnished self-contained flat at Alyn Bank, 21 Crescent Road, Crouch End, London N8.

J S Trenhaile (instructed by Shavin & Co) appeared for the plaintiffs, and E H Bailey (instructed by Windsor & Co) represented Miss Beury. The other defendants took no part in the proceedings.

Giving a reserved judgment, Fox J said that the flat in question had been let by the plaintiff landlords by a written agreement of December 8 1971 to Nicholas Gaffney and another person for one year from January 1 1972 and there- F after from month to month, with a proviso that the tenancy could be terminated by either party giving one month's notice in writing expiring at the end of the term or of any subsequent rental period. From November 1 1974 Mr Gaffney was accepted by the plaintiffs as the sole tenant on the terms of the December 1971 agreement. Those terms included an absolute prohibition against assigning, underletting or parting with possession of the flat or any part of it. On November 22 1975 Mr Gaffney wrote to the landlords saying that he had vacated the flat on September 22 and apologising for not having given adequate notice. He (his Lordship) had come to the conclusion that the landlords were correct in treating that letter as a notice to terminate the tenancy rather than as a surrender, and G that as a result the tenancy came to an end on December 31 1975, being the end of the relevant rental period, and any subtenancy came to an end with it.

After receiving Mr Gaffney's letter, an agent for the landlords inspected the premises in December 1975 and found that the four named defendants were in occupation of various rooms. Three said that they had been in occupation before Mr Gaffney left and had paid him rent. The fourth had by mid-December been in occupation for about six weeks, had paid no rent to Mr Gaffney and had not seen him. All had entered into possession without the landlords' knowledge or consent. From inquiries in January 1976 it appeared that H further persons, whose names could not be ascertained, were living in the flat. The second defendant, Miss Beury, who had appeared before the court, said that she had been given a rent book by Mr Gaffney. In the circumstances, he (Fox J) would assume she was a subtenant of Mr Gaffney and not merely a licensee. It was submitted on her behalf that order 113 had no application to the case because she, at least, was "a tenant ... holding over after the termination of the tenancy." Order 113, rule 1 read:

> Where a person claims possession of land which he alleges is occupied solely by a person or persons (not being a tenant or tenants holding over after the termination of the tenancy) who J entered into or remained in occupation without his licence or consent or that of any predecessor in title of his, the proceedings may be brought by originating summons in accordance with the provisions of this order.

He (his Lordship) had come to the conclusion that Miss Beury's point was bad. It seemed to him that the words "not being a tenant or tenants holding over after the termination of the tenancy" were directed only to persons who, as against the person claiming possession, could establish that they were holding over under tenancies binding upon him. There seemed no reason of principle why unlawful subtenants should be K treated differently from other unauthorised occupiers for the purpose of order 113 in circumstances such as the present. If the words "not being a tenant or tenants" etc were treated as extending to any subtenancy, whether binding on the person seeking possession or not, the scope of Order 113 could be seriously narrowed. If, for example, a tenant near the end of his term chose to grant, in breach of covenant against subletting, subtenancies to would-be squatters, Order 113 could not be applied. Further, quite apart from the deliberate creation of subtenancies in that way, it would be at least exceedingly inconvenient that where, for example, premises were occupied by squatters, and unlawful subtenants whose L subtenancy had come to an end, the whole claim for possession could not be dealt with under Order 113.

The construction which he (his Lordship) had adopted did not prejudice any right which an unlawful subtenant might have under the general law or the Rent Acts. Section 18 of the Rent Act 1968 had been referred to, but the protection given by that section was in terms applicable only to lawful subtenancies. It was clear that the subtenancy now in question was granted in breach of an absolute prohibition against subletting. There was no suggestion at all of any consent by the landlords or of any waiver. The landlords' evidence, which was not disputed, was that the defendants entered M into possession without the landlords' knowledge or consent. Any subtenancy which any of the defendants might have had was not and never had been binding on the landlords. There would be an order for possession accordingly.

A **QUEEN'S BENCH DIVISION**

November 29 1976

(Before Mr Justice O'CONNOR)

CO-OPERATIVE INSURANCE SOCIETY v FREEMAN FOX & PARTNERS

Estates Gazette February 5 1977

(1976) 241 EG 391

Rent for business premises which by November 5 1972 was still only subject to contract held the relevant rent for purposes of the counter-inflation legislation

This was a claim by the Co-operative Insurance Society
B against their tenants, Freeman Fox & Partners, for arrears of rent amounting to £66,000 odd in respect of rooms and offices on the first floor of Abford House, Wilton Road, Westminster, London SW1.

B Galpin (instructed by Allen & Son) appeared for the plaintiffs, and G Godfrey (instructed by Richards, Butler & Co) represented the defendants.

Giving judgment O'CONNOR J said that the defendants were in occupation of business premises in London under a lease from the plaintiffs that expired on May 9 1972. The rent under that old lease was £6,500 a year. In the
C early part of 1972 the two parties were negotiating a new lease. It was agreed that the defendants were going to remain in occupation, and everyone knew that the new rent would be a great deal more than the £6,500 a year, which was fixed in 1959 or thereabouts. On June 7 1972 the parties reached agreement, subject to contract, for a new lease from May 10 1972 to June 1976, the rent as from September 1 1972 to be £28,000 a year. There could be no doubt that both parties had used the technical and effective expressions necessary to keep all negotiations subject to contract, and that in law there was no binding agreement by November 5 1972 for the new lease. Indeed,
D the new lease was not finally executed and exchanged until April 16 1974, and in the meantime the plaintiffs' agents went on asking for rent at £6,500 a year, this being all that was paid throughout 1973 and into 1974. As soon as the formalities were completed, the plaintiffs requested payment of the arrears of rent due to make up £28,000 a year from September 1 1972, only to be met by a claim by the defendants that the increase was prevented as from November 5 1972 by the counter-inflation legislation.

The validity of that argument turned on the rent payable for the premises in question on November 5 1972. If one asked today what was the rent of the premises on that
E day, there would be not a shadow of doubt as to the answer, because the lease of April 16 1974 made clear that the rent then was payable at the rate of £28,000 a year. If one had asked the same question on November 5 1972 itself, again there was not a shadow of doubt as to the answer: each party would have stated that only a few weeks before, it had been determined that the rent from September 1 was to be £28,000 a year. That was the business sense of the matter, and he (his Lordship) was satisfied it was not putting any hole through the counter-inflation legislation to hold that in the present case the rent of the premises within the meaning of the order at November 5 1972 was £28,000
F a year.

The plaintiffs were given judgment for £66,149.32, with agreed interest of £11,024 and their costs.

QUEEN'S BENCH DIVISION G

April 12 1976

(Before Mr Justice MACKENNA)

WEST CENTRAL INVESTMENTS LTD v BOROVIK AND OTHERS

Estates Gazette February 26 1977

(1976) 241 EG 609

Lessors' claim for arrears of service charges in respect of maintenance of flats—Failure by lessors to prepare annual accounts under the terms of the lease—Time not of the essence and lessees liable—Defence of waiver of forfeiture rejected as service of writ operated as final election to H **determine lease**

This was a claim by West Central Investments Ltd, of Hanover Square, London W1, against Mark Borovik and his wift, Audrey Borovik, of flat 8, Cranfield House, Southampton Row, London WC1, and against the tenants of five other flats in Cranfield House, for money and for forfeiture of the respective leases on the ground of non-payment of maintenance charges.

N Irvine QC and C McKay (instructed by A Oldschool & Co) appeared for the plaintiffs, and J Mulcahy (instructed by J
J L Williams) represented the defendants.

Giving judgment, MACKENNA J said that the plaintiffs relied on a covenant in common form which appeared in the lease of each of the six flats and obliged the tenants to pay service or maintenance charges. The defendants claimed that they were not liable for the payments demanded, because the plaintiffs had failed to prepare accounts and serve notices within the times laid down by the covenant's terms. The main question for him (his Lordship) was whether time was of the essence of the provisions contained in the covenant. He would express his judgment by reference to the covenant in the lease of flat 8, occupied by Mr and Mrs Borovik, that K
being typical of the covenants in the other leases.

By the provisions in their covenant, for the first three years of a 75-year term from February 1967, Mr and Mrs Borovik were to pay service charges of £150 a year by quarterly instalments in advance. After that time they were to pay an amount, to be certified by the plaintiffs' accountants, equal to 3 per cent of the costs incurred by the lessors in managing the property. This amount, or the balance thereof over any amounts previously paid, was to be paid within 21 days of the plaintiffs' serving a notice in writing on the defendants of the sum due. The plaintiffs were obliged under the lease to arrange for the preparation and audit each year of accounts L
relating to the costs, charges and expenses incurred in maintaining the flats and to serve on the lessees within two months of the date of the accounts a notice of the proportionate sum due. In fact the plaintiffs did not have accounts prepared for the years 1970, 1971, 1972 or 1973. Accounts were not prepared until 1974, when the defendants were notified of the amounts they owed for the four preceding years. The defendants did not pay within 21 days of receiving the notices and further notices were then sent to them alleging breach of covenant. The defendants still did not pay and writs were issued before the end of 1974. The defendants had continued to pay the £150 annually in advance. The plaintiffs prepared M
accounts again at the end of 1974, and served the required notices on the defendants requiring them to pay that year's additional service charge. A further point was taken on this by the defendants, who pleaded that as these notices demanded payments in respect of the period after the issue of the writs, the plaintiffs had treated the leases as subsisting and so waived forfeiture.

Mr Mulcahy, for the defendants, had cited *United Scientific Holdings Ltd v Burnley Borough Council* [1976] 2 All ER 220. On the basis of that case he had argued that the parties in the present case had expressly stipulated that the accounts should

A be prepared and the notices served within a specified time. More than that, however, was needed to make time of the essence. It was also required that it should be expressly provided that the stipulation should be exactly complied with. This further express provision was missing in the present case. The facts in the present case were also distinguishable. There was no question here of substituting a higher figure for a lower one fixed by the lease and which would be payable if the higher figure were not substituted. In the present case there was, after the first three years, the fixed obligation to pay 3 per cent of the figure certified by the accountants and no other sum. He (his Lordship) also thought that, as in the B example given by Lord Denning MR in *C H Bailey v Memorial Enterprises Ltd* [1974] 1 WLR 728 at 732, the defendants in the present case had suffered no actual prejudice. He therefore found for the plaintiffs on the main issue.

On the issue of waiver, it was well established that the service of a writ claiming possession operated as a final election to determine the lease and no act done by the plaintiffs thereafter could operate as a waiver of the forfeiture. There would therefore be judgment for the amounts owing and for possession.

On an application by the defendants' counsel for relief against C forfeiture, his Lordship said that the terms of relief would be that the certified amounts due, together with interest at 1 per cent above bank rate from the service of the writ, and the sums due by way of mesne profits from the date when the rent had not been paid, were to be paid within 12 months. In view of the time taken by argument on matters eventually conceded by the plaintiffs, they should have two-thirds only of their costs.

COURT OF APPEAL

December 10 1976

(Before Lord DENNING MR, Lord Justice SCARMAN and Lord Justice BRIDGE)

D

GILMARTIN v WEST SUSSEX COUNTY COUNCIL

Estates Gazette April 16 1977

(1976) 242 EG 203

Construction of lease—Extent to which under terms of lease county council could deposit refuse on lessee's farm—Lord Denning's views on principles of construction not accepted by other judges—Court unanimous, however, in favour of county council's construction—Appeal allowed

This was an appeal by West Sussex County Council from a decision of Thesiger J in the Queen's Bench Division on E July 9 1976, on a preliminary issue, that West Sussex County Council was not entitled under the terms of a lease of Baythorne Farm, Itchingfield, Horsham, granted by them to Mrs Yvonne Gilmartin (there having been a sale and lease-back), to tip or deposit or cause or allow to be tipped or deposited, refuse of any kind into or over or on the whole of three areas of the 70-acre farm shown on a plan annexed to the lease.

Anthony Scrivener QC and Christopher Cochran (instructed by Sharpe, Pritchard & Co) represented the appellants; Derrick Turriff (instructed by Gregory, Rowcliffe & Co, agents for Cotching & Son, of Horsham) appeared for the respondent.

F Giving judgment, LORD DENNING MR said that the case concerned an interesting point on the modern industry of the dispersal of rubbish arising on the interpretation of a document. Nowadays the Department of the Environment had plans for the industry. Areas with ditches, holes and cuttings were found and filled up with rubbish from the district. Later, top soil was put over them to make the area suitable for agriculture. It was an excellent and comparatively modern way of dealing with refuse.

In 1964 Horsham Rural District Council made an arrangement with Mrs Gilmartin, who had a pedigree herd of Guernsey cows on some 70 acres of farmland. Under it she sold the farm to the council for some £9,000 and the council G immediately leased back the land to her under a 21-year lease at a rent. The recital to the lease said that the council had purchased the land "for the purpose of carrying out thereon a controlled refuse tipping scheme." Special clauses gave the council the right to resume possession separately of three particular areas, each about 8 acres and shown on an attached plan as "edged purple, edged blue and edged green for the purpose of the scheme."

The council was also given the right to enter on any part of the farm to carry out works specified in a schedule to the lease. The details of the works indicated an overall plan for the area to be better drained into the River Arun. There was H also a provision that at the end, when the council had made the land fit for agricultural purposes, they would give Mrs Gilmartin notice that she could have the land back.

The schedule provided for three phases. For some time everything went according to plan. Possession of the land edged purple was resumed in 1964, the land edged blue in 1970, and the works were carried out. From time to time Mrs Gilmartin complained of nuisance—the lease had a special covenant that the works should not be a nuisance—and the council apologised. The time came when her advisers looked at the lease and advised her that the tipping J area was limited by the line of a ditch shown on the plan which cut across the areas which had been repossessed. She brought an action claiming damages for nuisance and an injunction, and also claiming that there was no right to tip on parts of the land which had not been repossessed by the council. She also applied for an interlocutory injunction. Cusack J felt that the true interpretation of the lease in relation to the tipping area should be tried as a preliminary issue. Thesiger J held that Mrs Gilmartin was right.

The council, now West Sussex County Council, appealed. He (his Lordship) thought that it was unfortunate to have a trial of a preliminary issue like this, limited to the wording K of the lease, without the surrounding facts and background being known. But as the parties and the judge had agreed to deal with the preliminary point the court had to determine it on that basis. The words which Mrs Gilmartin relied on, and which the judge felt were decisive, were in para 7 of the schedule: "The tipping area as shown edged in purple . . . will be limited on the north and west sides by the new open cut ditch. Strip all topsoil from the tipping area and stockpile adjoining the access road where indicated on the plan." These words, it was said, defined the tipping area in the first phase. There were similar specific words for the second and third phases. If the words were taken by themselves without L being qualified by the rest of the document he (his Lordship) would agree with the judge that it did appear that the tipping area was limited to the area of which possession had been resumed. But the sentences must be read not only in the context of the agreement as a whole but also against the background in which it was made—to use Lord Wilberforce's phrase in *Reardon Smith Line Ltd v Hansen-Tangen* [1976] 1 WLR 989 at 997, "in the same factual matrix as that in which the parties were." One did not look only at the words but at the correspondence, the background, and everything which formed the matrix in which the parties negotiated. One important factor in this case was the actual line on the M land itself. If Mrs Gilmartin were right it would produce a hopeless state of affairs for the proper carrying out of "a controlled refuse tipping scheme," because at the end there would not be a good level field but a tipping area going down into an old ditch quite steeply and up the other side of a crevasse, half of which would be a tipping area with refuse on it and the other half not. And it might be difficult to drain, so that the last stage would be worse than the first.

In his (Lord Denning's) view, the key operative provision of the agreement was clause I (ii), which gave the council "the right to resume possession separately of each of the parts of the farm edged purple . . . blue . . . and green for the purpose of the scheme." That meant that the council was regaining possession of its own property. Clause 1 (iii) gave the council the right to carry out works specified in the schedule "with such variations as the council may consider necessary." That was a right to carry out works with a provision for variation; and it showed that there was nothing binding in law, so that when the schedule said that the tipping area should be limited as defined, even that could be varied. It was only a provisional plan and, in law and having regard to all the background circumstances, the schedule was a schedule of proposed works which could be varied in almost every respect. The governing provision was the right of the council to resume possession as it might desire. Looking at the whole background he interpreted the document differently from the judge and would allow the appeal.

Agreeing, SCARMAN LJ said that on the preliminary point the facts were not investigated at all. It was clear that there was a great area of dispute which, if Mrs Gilmartin pursued her claim for damages, would inevitably have to be investigated. He would deal with the case as it came before the court, in the framework of a construction summons as to the terms of a written agreement. After examining the lease and the agreement for sale in detail, he considered that the contractual documents did not impose on the council the inflexible duty of carrying out a particular controlled refuse tipping scheme, but left it to the council's good judgment and discretion to devise whatever scheme it thought appropriate. He had reached his conclusion for reasons deriving from analysis of the documents. One did not have to examine the matrix of the facts, which were largely uninvestigated and undetermined.

BRIDGE LJ, also agreeing, said that before the case reached the Court of Appeal it probably had not occurred to anyone that it was either necessary or helpful to look beyond the confines of the contractual documents in order to determine the disputed issue of what they meant. Of course, there were cases in which it was vital to apply the principle expressed by Lord Wilberforce in *Reardon Smith*; but he (Bridge LJ) could not see what factual matrix in 1964 could assist in the construction of the elaborate formal documents before the court. He would respectfully but vigorously express his dissent from the view that it was proper for the court to consider the pattern of land—*a fortiori* to look at photographs of what the land looked like at the present stage of the ditching operation—in order to ask itself what, from the public health, agricultural or landscape point of view, would have been the most sensible agreement for the parties to have made. On the interpretation of lamentably drafted documents, he had felt greater difficulty in reaching a conclusion than had his brethren, but he was not prepared to carry his doubts to the point of formal disagreement.

The appeal was allowed with costs.

COURT OF APPEAL

January 20 1977

(Before Lord Justice MEGAW, Lord Justice Geoffrey LANE and Sir John PENNYCUICK)

MOORE AND ANOTHER v ASSIGNMENT COURIER LTD

Estates Gazette April 23 1977

(1977) 242 EG 287

Landlords' summons claiming interim payment in pending action against tenant for forfeiture of lease—No jurisdiction in court to order that tenant pay an interim sum on basis of minimum amount for which he would be liable, either as mesne profits or as rent, whether or not landlord ultimately succeeded in action—No rule of court authorising such payment

This was an appeal by plaintiff landlords, Thomas Cyril Moore and Florence Maud Moore, from an order of Kenneth Jones J, who had dismissed their appeal from an order of Master Warren. The judge had ruled that there was no jurisdiction in the court, on a summons by a landlord in a pending action against his tenant for forfeiture of a lease on the ground of breach of covenant, and for mesne profits, to order that the tenant pay an interim sum on the basis of the minimum amount for which he would be liable, either as mesne profits or as rent, whether or not the landlord ultimately succeeded in the action.

Gavin Lightman (instructed by Macdonald, Stacey & Co) represented the appellants; the tenants were not represented.

Giving judgment, SIR JOHN PENNYCUICK said there were six consolidated actions between the parties. All raised the same issue. It was a short and interesting one, namely, whether, where a landlord purported to forfeit a lease and the tenant remained in occupation, the landlord was entitled to be paid, pending the determination of the forfeiture action, a periodic interim sum representing compensation under one head or another for the use by the tenant of the land during the period between the purported forfeiture and the determination of the action.

The landlords' claim, as appeared from their statement of claim, was that there had been breaches of the tenants' covenants. The tenants contended that either what had been done had been done with the landlords' consent or that the landlords had waived the alleged breaches. Mr Lightman's main contention on the appeal was based on the inherent jurisdiction of the court. He submitted that, where a landlord brought proceedings for forfeiture against a tenant and the tenant denied that forfeiture had occurred, the court could in the course of the proceedings make an order in respect of the tenant's continuing possession of the demised premises. Mr Lightman had submitted that the landlord must, by one road or another, be entitled to compensation for the occupation by the tenant of his property. If his claim for forfeiture was good, the compensation would be in the form of mesne profits. If it failed the relationship of landlord and tenant would continue and the landlord could get it in the form of rent under the lease. That was correct. By one road or another the landlord would be entitled to receive or be credited with compensation.

The question was, however, whether the court had any jurisdiction to order an interim payment of the minimum amount which in one way or another the landlord could recover by way of compensation. It would be impossible to formulate the circumstances in which the court had inherent jurisdiction. All that Mr Lightman had cited by way of authority for his proposition was one sentence in the judgment of Lord Denning MR in *Tiverton Estates Ltd v Wearwell Ltd* [1975] Ch 146: "These courts are master of their own procedure and can do what is right even though it is not contained in the rules." That in its context had plainly been addressed to matters of procedure and had not been intended to say that the court in matters of substance could do whatever it thought right apart from the legal principles applicable under the general law or the Rules of the Supreme Court. The present claim clearly raised an issue of substance. Apart from any provision of the rules, the answer must be that there was no jurisdiction in the court to make such an order. He (his Lordship) did not see how the court, with only a claim by the landlord for forfeiture and mesne profits before it, could make an interim order based, not on that claim, but on what the position would be if the action failed. It might be attractive to do so, but he did not see any ground on which the court had any such jurisdiction. That conclusion was reinforced by more than one consideration. First, though the

A situation must be of everyday occurrence, no authority had been cited in support of the contention that the court had such jurisdiction or showing that the courts had ever made such an order.

Secondly, section 20 of the Administration of Justice Act 1969 gave power for rules of court to be made enabling interim payments to be ordered in such circumstances as might be specified in the rules. One rule had been made under that section, Order 29, Rule 9, authorising orders for interim payments on account of damages in respect of personal injuries. No other rule had been made authorising interim payments; more especially, none had been made authorising
B interim payments in cases such as this. That afforded a considerable indication that, apart from express provisions authorised by section 20 of the 1969 Act, the court had no general jurisdiction to make interim orders, however much it might be fair to do so.

Further, considerable practical difficulties would arise in this class of case on applications for such orders. The court would have to go into the facts of each particular case, which might indicate all sorts of difficulties in calculating the minimum amount which it was fair to order. The procedure would work only in very simple cases. The answer must be that the courts did not possess the power contended for.

C Agreeing, GEOFFREY LANE LJ said that there was something to be said for the introduction of some rule, such as Mr Lightman proposed, to avoid injustice to the plaintiff, but it was not for the court to manufacture such a rule.

MEGAW LJ also agreed, and the appeal was dismissed.

COURT OF APPEAL

July 2 1976

D (Before Lord Justice MEGAW, Lord Justice BROWNE and Lord Justice GEOFFREY LANE)

POLLOCK v KUMAR

Estates Gazette April 30 1977

(1976) 242 EG 371

Landlord and tenant — Tenant evicted under possession order — Order set aside on appeal — Restitution of possession or payment of compensation — Compensation awarded

This was an appeal by Lingadass Kumar which raised a question of compensation following the setting aside of a possession order granted to his landlord, Richard Pollock,
E in respect of a top-floor flat at 20 Porchester Terrace, Paddington, London W2, by Judge Rowland at Bloomsbury and Marylebone County Court on September 5 1975.

Stephen Nathan (instructed by H Davis & Co) represented the appellant; Paul de la Piquerie (instructed by Macfarlanes) appeared for the respondent.

Giving judgment, BROWNE LJ said that so far as the order for possession was concerned it was conceded on behalf of the respondent that in the light of a decision of the Court of Appeal in *Buchmann v May* (1976) 120 SJ 384 [240 EG 49] the appeal against the possession order must succeed. The court, having read the papers, were quite satisfied
F that if that matter had been fully argued, the appeal would have succeeded, and the concession was absolutely right. Accordingly the court would allow the appeal against the order for possession.

The problem in the case arose from what had happened since then, and what should be done about it. The judge's order was made on September 5 1975 for possession on October 3. No application was made for a stay of execution at the end of the hearing, nor at any time before October 3. On October 6 a warrant for possession was issued and notice of appeal was given on the same day. At the time the appellant was still in possession of the premises. On October 8 his solicitor made an application to the county court to G
stay the warrant. That was heard by Judge Leslie on that date, and he granted a stay on the terms that the appellant should pay £27 a week into court, so long as he remained in the property. On November 4, the condition not having been complied with, the bailiffs entered on the premises and the appellant was evicted. On January 22 1976 he filed a notice of appeal seeking an order that he might be restored to possession of the top-floor flat. The question was — what should the court do in view of the fact that the appellant was no longer in possession and the respondent had relet the premises?

The court plainly had a discretion whether or not to H
order restitution of possession to the appellant. Their power was covered under Order 59, rule 10 (4) of the Rules of the Supreme Court by the last words "and the Court of Appeal may make any order, on such terms as the court thinks just, to ensure the determination on the merits of the real question is controversy between the parties." The court had ordered the payment of compensation in somewhat similar circumstances in *Tanner v Tanner* [1975] 1 WLR 1346.

The respondent had relet the premises to a Mr Burgess for a period of five years from November 11 1975 and the court had no reason to suppose that it was not a bona fide J
letting. At the time that letting was made, the respondent had been put back into possession of the premises under what was, at that time, a valid warrant for possession made under a valid and subsisting county court judgment. In these circumstances it seemed to his Lordship that it would be wrong to order the respondent to give possession back to the appellant. On the other hand, he had no doubt that an order should be made that the appellant should receive appropriate compensation. In the circumstances of the case the basis of that compensation should be the financial loss shown to have been suffered by the appellant by reason of his being evicted from the flat on November 4 last year. K
That compensation should not include any element of mental suffering or anything of that sort. It should be a purely financial calculation. It seemed to his Lordship that the Court of Appeal could not satisfactorily decide what the figure should be and the matter would have to be remitted to the county court to decide what compensation should be payable. Accordingly he would allow the appeal, set aside the order for possession and remit the matter to the county court to assess compensation on the basis he had stated.

GEOFFREY LANE LJ agreed.

Also concurring, MEGAW LJ said that in agreeing that L
the scope of the inquiry into the proper compensation payable should be limited to the financial loss suffered by the appellant by reason of the eviction, it was not to be taken that this was the only basis that would ever be appropriate.

The appeal was allowed, the possession order set aside and the matter remitted to the county court for an inquiry into the quantum of compensation.

Note: On March 24 1977 the county court judge to whom the question of the amount of compensation was remitted, Judge Llewellyn, at Bloomsbury and Marylebone County Court, assessed the compensation as follows:

	£
Net rental loss	566.87 M
Removal expenses	25.30
Travelling expenses	25.00
Bailiff's fee	5.00
Loss of amenity	100.00
	722.17

The net rental loss was calculated as the difference between the total of the rents paid by the appellant at two addresses to which he had moved after his eviction (having secured a protected tenancy in each case) and an assessment of the rent he would have paid if he had remained at 20 Porchester Terrace. The judge disregarded an address at which he had stayed for five days only and also, as too remote, the rent charge for an unprotected tenancy at the fourth address to which he had moved. The judge rejected a claim for additional examination fees and loss of earnings, holding that the appellant would probably have failed his accountancy examination in any case, apart from the eviction disturbance, if he had taken it at the time he was evicted, and that it would not have been sensible to take a full-time job while still trying to pass his examination. The sum awarded for loss of amenity included disruption, inconvenience, expenses and loss of time in searching for accommodation, and packing for his various moves.

CHANCERY DIVISION

February 2 1977

(Before Mr Justice SLADE)

LONDON GOLDHAWK BUILDING SOCIETY v EMINER AND ANOTHER

Estates Gazette May 7 1977

(1977) 242 EG 462

Possession order sought by mortgagees against mortgagor's tenants — Tenancy granted without mortgagees' consent — Tenants claimed to have purchased freehold and wished for time to redeem mortgage — Order sought under wrong procedure — Appropriate procedure under Order 88, not Order 113

This was an application by London Goldhawk Building Society, of Chiswick High Road, London W4, against Newzat Eminer and his wife, Nedila Eminer, of St John's Villas, Friern Barnet Road, London N11, for a possession order under Order 113 of the Rules of the Supreme Court.

D Hands (instructed by Warmingtons & Hasties) appeared for the plaintiff society; and P C L Clark (instructed by Tyler, Roxburgh & Dawson) represented the defendants.

Giving judgment, SLADE J said that the building society sought possession of the premises in Friern Barnet Road under Order 113 on the ground that the defendants were in occupation without their consent.

On January 10 1975 the then owner of the property, Umit Suleyman, charged it in favour of the building society. When he became in default of the mortgage, proceedings for possession were brought against him and an order for possession obtained on March 23 1976. When the society sought to execute the order three persons were caused to leave the premises with their belongings. Later two of them, the defendants, returned and remained in occupation. In their evidence the defendants claimed that in late 1975 or early 1976 a Mr Soubhi, purporting to act for Mr Suleyman as freehold owner, offered to sell them the freehold. An agreement had eventually been reached, evidenced by a memorandum dated June 19 1976, and the defendants had paid the purchase price of £3,800. On October 3 1976 Mr Soubhi died and the defendants discovered that Suleyman was an assumed name and Suleyman and Soubhi were one and the same person.

This was not an ordinary application by a landowner against a squatter. In the circumstances, it plainly fell within the provisions of Order 88, rule 1, while applied to any action by a mortgagee or mortgagor, including an action for delivery of possession to the mortgagee. Special rules as to evidence were laid down by rule 6 and the procedure differed in a number of ways from that under Order 113. He (his Lordship) doubted whether Order 113 had any application to a mortgagee's action for possession. Accepting the defendants' evidence that they had an equitable right to redeem the mortgage, the Order 88 procedure was clearly the proper one to follow.

The application would be remitted to chambers to proceed under Order 88 and an affidavit should be sworn by the plaintiff society supplementing their evidence to the extent necessary to comply with Order 88. Any further appropriate directions would be given by the master, who would also deal with the question of costs.

HOUSE OF LORDS

March 2 1977

(Before Lord DIPLOCK, Lord MORRIS of BORTH-Y-GEST, Lord SIMON OF GLAISDALE, Lord KILBRANDON and Lord EDMUND-DAVIES)

TOWN INVESTMENTS LTD AND OTHERS v DEPARTMENT OF THE ENVIRONMENT

Estates Gazette May 14 1977

(1977) 242 EG 539

Counter-Inflation (Business Rents) Orders 1972 and 1973—Leases to Minister of Works and subsequently to Secretary of State for the Environment—Whether occupation by Crown as tenant for the purpose of a business—Whether Crown or Secretary of State tenant—Premises occupied by government servants—"Crown" a term of art in constitutional law—Tenant the Crown in the sense of the government—Order applicable and rents frozen during standstill period—Court of Appeal's decision reversed

This was an appeal by the Department of the Environment from a decision of the Court of Appeal who had held in favour of respondent landlords, London County Freehold & Leasehold Properties Ltd and MEPC (Mayfair Properties) Ltd, that the Counter-Inflation (Business Rents) Orders 1972 and 1973 did not apply to tenancies of Keysign House, Oxford Street, and 17 North Audley Street, London W1. The Department did not proceed with an appeal relating to an action by Town Investments Ltd in respect of 11 and 12 Buckingham Gate, London SW1. The actions had all been commenced by originating summonses seeking declarations that the orders did not apply.

N C H Browne-Wilkinson QC and Peter Gibson (instructed by the Treasury Solicitor) appeared for the appellant department; S A Stamler QC and Christopher Priday (instructed by Beachcroft, Hyman Isaacs) represented the respondent landlords.

In a speech allowing the appeal, LORD DIPLOCK said: On November 6 1972, and at all material times thereafter, two buildings known as Keysign House, Oxford Street, and 17 North Audley Street, in London, were occupied as government offices. November 6 1972 was the starting date of the rent freeze imposed by the Counter-Inflation (Business Rents) Order 1972, made under section 2(4) of the Counter-Inflation (Temporary Provisions) Act 1972, and continued by the Counter-Inflation (Business Rents) Order 1973, made under Part II of the Counter-Inflation Act 1973, which replaced the earlier statute. These orders, as their names indicate, dealt with rents payable under "business tenancies" and their effect was to prohibit a landlord, during a standstill period which ultimately expired in March 1975, from recovering rent at a rate in excess of the rent payable on November 6 1972. Each of the orders contained a definition of "business tenancy" and a statement of what was included in the expression "business" in terms that are identical so far as they are relevant to the facts of these appeals:

> "business tenancy" means any tenancy where the property comprised in the tenancy is or includes premises which are occupied by the tenant and are so occupied for the purposes of a business carried on by him or for those and other purposes. . . .
>
> "business" includes a trade, profession or employment and includes any activity carried on by a body of persons, whether corporate or unincorporate.

The only issue before this House has been whether during the standstill period the premises to which these appeals relate were the subject of "business tenancies" within the meaning

A of the order. This in turn depends upon the answer to three questions: (1) Who was the tenant of the premises? (2) Were the premises or any part of them occupied by the tenant? And, if so (3) was the tenant's occupation for the purposes of a business carried on by him or for those and other purposes?

The leasehold title under which Keysign House was held on November 6 1972 was an underlease entered into in 1952. The landlord was the predecessor in title of the respondents, London County Freehold & Leasehold Properties Ltd; the tenant was described as follows:

> The Minister of Works (hereinafter called "the lessee" which
> B expression where the context so admits includes his successors and assigns) for and on behalf of Her Majesty.

In the case of 17 North Audley Street, the landlord was the predecessor in title of the respondent, MEPC (Mayfair Properties) Ltd. The tenant was described as follows:

> The Minister of Works (hereinafter called "the tenant" which expression shall where the context so admits include his assigns) for and on behalf of Her Majesty.

Both of these leases expired upon December 25 1972. Fresh leases for a term of five years from December 25 1972 were entered into in substantially the same terms as those which they replaced. The description of the lessee/tenant in the new
C leases was the same as in the former leases of 1952 except that the description "Secretary of State for the Environment" was substituted for that of "Minister of Works." This change in title was a consequence of the Secretary of State for the Environment Order 1970, made under the Ministers of the Crown (Transfer of Functions) Act 1946. By this order the former Ministry of Works, whose title had in 1962 been changed to that of Ministry of Public Building and Works, was dissolved and all the functions of the Minister of Public Building and Works were transferred to the Secretary of State for the Environment, to whom were also transferred all property, rights and liabilities to which that Minister was
D entitled or subject before the coming into operation of the order. Included in the functions of the Minister of Public Building and Works transferred to the Secretary of State for the Environment was the provision of accommodation for civil servants employed in departments of government other than the Ministry of which that Minister was head. At all material times the buildings that are the subject of these appeals were used to accommodate civil servants who were not engaged in the work of the Ministry of Public Building and Works or the Department of the Environment, but were engaged in the work of other government departments.

E The rents reserved under the 1952 leases and current on November 6 1972 had been in the case of Keysign House £39,690, and in the case of 17 North Audley Street £17,500. Under the new leases the rents reserved from December 25 1972 were in the case of Keysign House £369,500, and in the case of 17 North Audley Street £190,000, viz more than 10 times as much as previously. If the Counter-Inflation Orders of 1972 and 1973 applied to the premises, only the former rents were payable in respect of the standstill period. The landlords claim that these orders did not apply to either of the premises. They took out an originating summons for a declaration to this effect and for a declaration that the rents
F payable after December 25 1972 were the rents payable under the new leases and not the lower rents payable under the 1952 leases. The summonses were heard together before Foster J. The main ground on which he found in favour of the landlords was that the Crown was not entitled to avail itself of the provisions of either of the Acts or Orders in Council made thereunder, because it is expressly stated in each of the Acts that it does not bind the Crown. This ground of decision was not accepted by the Court of Appeal as correct in law, and has not been relied upon in your Lordships' House. In view of the provisions of section 31 (1) of the Crown Proceedings Act 1947, I need say no more about it.

The appeal by the Department of the Environment from G
the judgment of Foster J was dismissed by the Court of Appeal upon another ground. They held that neither of the premises was the subject of a "business tenancy" within the meaning of the Counter-Inflation Orders. Their answers to the three questions that are posed by this appeal were (1) the tenant is the Secretary of State for the Environment and not the Crown; (2) no part of the premises was occupied by the tenant, and (3) no part of the premises was occupied for the purposes of a "business."

Accordingly, I turn to these three questions:

1. *Who was the tenant of the premises?*

In the Court of Appeal this was treated as a pure question H
of construction of the leases themselves, as if Her Majesty, the Minister of Works, and the Secretary of State for the Environment were all persons to whose relationships to one another and to third parties the ordinary principles and concepts of private law applied. If this were right, it would involve a conflict between, on the one hand, the statement in the lease that the Minister or the Secretary of State is party thereto "for and on behalf of Her Majesty" and, on the other hand, the inclusion in the definition of "the lessee" in the case of Keysign House, of "his successors and assigns" and, in the definition of "the tenant" in the case of 17 North
Audley Street, of "his assigns." This conflict was pointed J
out by Buckley LJ and Sir John Pennycuick and resolved by them by holding that the tenant was the Minister or Secretary of State and not the Crown, and that the words "for and on behalf of Her Majesty" were either a mere indication that he was acting in his corporate capacity as Minister or Secretary of State (Buckley LJ and Sir John Pennycuick) or that he took the leasehold interest as trustee for the Crown (Buckley LJ and Lawton LJ).

My Lords, the fallacy in this argument is that it is not private law but public law that governs the relationships
between Her Majesty acting in her political capacity, the K
government departments among which the work of Her Majesty's government is distributed, the Ministers of the Crown in charge of the various departments and civil servants of all grades who are employed in those departments. These relationships have in the course of centuries been transformed with the continuous evolution of the constitution of this country from that of personal rule by a feudal land-owning monarch to the constitutional monarchy of today; but the vocabulary used by lawyers in the field of public law has not kept pace with this evolution and remains more apt to the constitutional realities of the Tudor or even the Norman
monarchy than to the constitutional realities of the 20th cen- L
tury. To use as a metaphor the symbol of royalty, "the Crown," was no doubt a convenient way of denoting and distinguishing the monarch when doing acts of government in his political capacity from the monarch when doing private acts in his personal capacity, at a period when legislative and executive powers were exercised by him in accordance with his own will. But to continue nowadays to speak of "the Crown" as doing legislative or executive acts of government, which, in reality as distinct from legal fiction, are decided on and done by human beings other than the Queen herself involves risk of confusion. We very sensibly speak today of
legislation being made by Act of Parliament—though the pre- M
amble to every statute still maintains the fiction that the maker was Her Majesty and that the participation of the members of the two Houses of Parliament had been restricted to advice and acquiescence. Where, as in the instant case, we are concerned with the legal nature of the exercise of executive powers of government, I believe that some of the more Athanasian-like features of the debate in your Lordships' House could have been eliminated if instead of speaking of "the Crown" we were to speak of "the government"—a term appropriate to embrace both collectively and individually all of the ministers of the Crown and parliamentary secretaries under whose

A direction the administrative work of government is carried on by the civil servants employed in the various government departments. It is through them that the executive powers of Her Majesty's government in the United Kingdom are exercised, sometimes in the more important administrative matters in Her Majesty's name, but most often under their own official designation. Executive acts of government that are done by any of them are acts done by "the Crown" in the fictional sense in which that expression is now used in English public law.

The executive acts of government with which the instant case is concerned are the acceptance of grants from lessors B who are private subjects of the Queen of leasehold interests in premises for use as government offices and the occupation of the premises by civil servants employed in the work of various government departments. The leases were executed under his official designation by the minister of the Crown in charge of the government department to which, for administrative and accounting purposes, there is entrusted the responsibility for acquiring and managing accommodation for civil servants employed in other government departments as well as that of which the minister himself is the official head. In my opinion, the tenant was the government acting through its appropriate member or, expressed in the term of art in C public law, the tenant was the Crown. I do not find it necessary for disposing of the instant appeals to trace the history of the practice of taking grants of freehold or leasehold interests of land required for government purposes in the names of commissioners or heads of government departments whether corporate or unincorporated. With the increase of government business in the early part of the 19th century the technicalities of conveyancing made it inconvenient for the King himself to be an executing party to conveyances and his direct acceptance of a leasehold tenure with its implication of service due to a feudal superior was regarded as incompatible D with his supremacy. Accordingly, one finds from the beginning of the 19th century onwards a whole variety of statutes providing for the acquisition of land required for the purposes of a particular government department or of the government generally, in the name and under the seal of designated commissioners or ministerial heads of departments. In these 19th century statutes there is no consistency in the description of the capacity in which the persons so designated are to be executing parties to conveyances of the freehold and leasehold interests which they are authorised to acquire. Sometimes the interest is referred to as being held "on behalf of His/Her Majesty," occasionally E as being held "in trust for His/Her Majesty." The use of the expression "in trust," though it does not appear in any of the statutes under which the Secretary of State for the Environment, the Minister of Works, or any of their predecessors obtained the power to acquire land for government purposes, has nevertheless been relied upon by the respondents as showing that whenever an interest in land to be used for government purposes is conveyed to an officer of state in his official capacity, the interest so conveyed becomes subject to all the incidents of a trust in private law; the legal estate is vested in the officer of state who executes the conveyance, only an equitable interest is vested in the Crown and the relationship F between him and the Crown is subject to the equitable jurisdiction of the courts.

My Lords, I would not exclude the possibility that an officer of state, even though acting in his official capacity, may in some circumstances hold property subject to a trust in private law for the benefit of a subject; but clear words would be required to do this and, even where the person to be benefited is a subject, the use of the expression "in trust" to describe the capacity in which the property is granted to an officer of state is not conclusive that a trust in private law was intended; for "trust" is not a term of art in public law and when used in relation to matters which lie within the field of public law the words "in trust" may do no more than indicate the G existence of a duty owed to the Crown by the officer of state, as servant of the Crown, to deal with the property for the benefit of the subject for whom it is expressed to be held in trust, such duty being enforceable administratively by disciplinary sanctions and not otherwise: *Kinloch v Secretary of State for India* (1882) 7 App Cas 619 per Lord Selborne LC at pp 625-626. But even if the legal relationship of trustee and cestui que trust under a trust in private law is capable of existing between an officer of state in his official capacity and a subject, the concept of such relationship being capable of existing between him as trustee and the Crown as cestui que trust is in my view wholly irreconcilable with the legal nature H in public law of the relationship between the Crown and its servants or, in more modern parlance, the government and the ministers who form part of it.

In leases such as those that are the subject of the instant appeal where the person designated as grantee is a minister of the Crown in charge of a government department, the references to "successors and assigns" or "assigns" simpliciter as being included in the definition of "lessee" or "tenant" are explicable as referring to any other minister to whom the functions of the designated minister may be transferred under the Ministers of the Crown (Transfer of Functions) Act 1946, or otherwise, and, in the case of "assigns," also to subjects to J whom the leasehold interest may be assigned. Even if they were not, I would not treat their presence in the parentheses as sufficient to displace the ordinary concepts of public law as to the relationship between the government and the ministers who form part of it.

2. *Were the premises or any part of them occupied by the tenant?*

At all relevant times the persons physically present on the premises were government servants. Expressed in the terms of art used in public law their status was that of "servants of K the Crown" and of no one else. The relationship of master and servant does not exist between a minister in charge of a government department and any other government servant employed in that department in whatever grade of the civil service he may be. They are both fellow servants of the Crown. The use of the premises by government servants for government purposes thus constituted occupation of the premises by the Crown. Holding as I do that the Crown was the tenant of the premises, I would accordingly answer this question "Yes."

3. *Was the tenant's occupation for the purposes of a business carried on by him or for those and other purposes?* L

The answer to this question depends upon how broad a meaning is to be ascribed to the word "business" in the definition of "business tenancy" in the two Counter-Inflation Orders. The word "business" is an etymological chameleon; it suits its meaning to the context in which it is found. It is not a term of legal art and its dictionary meanings as Lindley LJ pointed out in *Rolls v Miller* (1884) 27 ChD 71 at p 88 embrace:

> almost anything which is an occupation as distinguished from a pleasure—anything which is an occupation or a duty which requires attention is a business.

M That was said by the Lord Justice in connection with the construction of a covenant in a lease against the carrying on of any trade or business on the demised premises; and ever since there has been a consistent line of cases in which this broad meaning has been ascribed to the word "business" in the context of covenants in leases restricting the permitted user of the demised premises. It appears to me to be clear beyond argument that the use made of the premises that are the subject of the instant appeals to accommodate civil servants engaged in what consistently with common usage could be quite properly described as "Government business," would constitute a breach of a covenant in a lease against permitting

A any business to be carried on upon the premises. The Crown or government through its servants is carrying out there a duty which requires attention.

The wide interpretation to be put upon the word "business" in restrictive covenants of this kind is dictated by the evident object of the covenants. The evident object for which powers were conferred by the Counter-Inflation Acts 1972 and 1973 to make orders restricting increases of prices, dividends and rents, was to curb inflation by preventing more money coming into circulation without any corresponding increase in production of goods or services. Separate Counter-Inflation Orders were made under the Acts, dealing respectively
B with business rents, agricultural rents and the rents of dwelling-houses. The mischief against which these orders were directed did not depend upon who the tenant was or the use made of the premises by him but upon the receipt by his landlord of a greater sum of money in return for what produced no greater contribution to the national wealth than it had produced before. This would suggest that the evident object of the two Counter-Inflation (Business Rents) orders called for a broad construction of the word "business" in the definition of the tenancies to which they applied, so as to embrace all tenancies save those which fell within the scope of one of the two other
C orders dealing respectively with the rents of agricultural and residential tenancies or were excluded from the definition of "business tenancies" by express words—as were "building leases" under the 1973 order. I would therefore hold it to be legitimate to give to the word "business" in the definition of "business tenancy" in the two orders a meaning no less wide than that which it has been interpreted as having in covenants in leases restricting the user of demised premises. This meaning is, in my view, wide enough to include the purposes for which Keysign House and 17 North Audley Street were occupied by the Crown.

I do not therefore find it necessary to rely upon any
D extension of the ordinary meaning of the word that may be the result of the express inclusion of "a trade, profession or employment" or of "any activity carried on by a body of persons, whether corporate or unincorporate." Nevertheless these express inclusions serve not only to underline the breadth of meaning to be given to the word "business" in the orders but also to involve anomalies which cannot have been intended, if that word is not wide enough with or without the words of inclusion to embrace the exercise of functions of government by servants of the Crown. Functions of government in the United Kingdom are distributed, on no very
E consistent pattern, between the central government and local governmental and other statutory authorities composed of bodies of persons, either corporate or unincorporate. The second part of the words of inclusion clearly embrace the activities of these authorities but not those of the Crown which is in law a corporation sole. It would be remarkable were it intended in the order to exclude from the expression "business" an activity which would fall within it when carried on by servants of a corporation aggregate but not if carried on by servants of the Crown simply because it is a corporation sole. It may be that the reconciliation of the different phraseology used in the two parts of the inclusion clause to describe
F things done by or on behalf of bodies of persons and things done by or on behalf of one, is to be found in the presence in the first part of the clause of the expression "employment" —a word hardly slightly less protean than "business" itself. When used in relation to a natural person "employment" seems to me to be broad enough to include anything which, in Lord Justice Lindley's phrase, is an occupation as distinguished from a pleasure, and to be understood in the same meaning as the word "activities" used in relation to bodies of persons in the second part of the inclusion clause.

It remains to mention two further arguments which have been advanced from excluding the business of government from the meaning of the word "business" as it is used in the G
Counter-Inflation Order. The first can be disposed of shortly for only minor importance was attached to it by the Court of Appeal. It is that the definition of "business tenancy" and the inclusion clause referring to "business" were lifted by the draftsmen of the orders from section 23(1) and (2) of the Landlord and Tenant Act 1954; and that section 56(3) of the same Act contains specific provisions applying it to tenancies "held by or on behalf of a government department" and "occupied for any purposes of a government department." In an Act which contained provisions for dealing with security of tenure of tenants of business premises by the grant of new leases and rights to compensation when a new lease could not H
be granted, the fact that the draftsman of the Act has thought it prudent to make specific provision as to its application to premises occupied for government purposes is a frail support for the contention that but for those specific provisions the Act would not have applied to such premises at all and therefore tenancies of them cannot have fallen within the definition of "business tenancy" in the Act, even though in the ordinary meaning of the words of the definition, they would have been apt to include such tenancies. But any force that there might be in this contention is wholly lost when it is sought to use it to cut down the ordinary meaning of those words when used in other legislation passed with an entirely different object. J

However, a matter to which all three members of the Court of Appeal appear to have attached importance was the presence of the indefinite article "a" in the reference in the definition of business tenancy to "*a* business carried on by him" [sic the tenant]. Both Buckley and Lawton LJJ were I think influenced by the conclusion they had already reached that the relevant tenant was the Secretary of State and not the Crown. They appear to have accepted that the activities of government through the various government departments might properly be described as "business," but they considered that neither that part of those activities that was carried on K
through a particular government department nor, it would seem, those activities as a whole could properly be described as "a business."

My Lords, it has been said that Roger Casement was hanged by a comma and (as Mr Browne-Wilkinson has reminded us) that John Keats's mind was "snuffed out by an article." I think that in exercising the functions of government the civil servants of the Crown are all engaged in carrying on a single business on behalf of the Crown, ie Her Majesty's government in the United Kingdom. I do not see why the presence of an indefinite article affects the matter. I do not see why this is not L
a business carried on by the Crown on whatever premises may be used by it to accommodate civil servants employed in the various government departments, or why it is not *a* business for the purposes of which each of these premises is occupied by the Crown.

For these reasons I would allow these appeals.

LORD MORRIS OF BORTH-Y-GEST, in a dissenting speech, held that the Secretary of State for the Environment, not the Crown, was the tenant; that the Secretary of State did not occupy the premises; and that there was no business tenancy so as to make the counter-inflation orders applicable. He would have dismissed the appeal. M

LORD SIMON OF GLAISDALE, LORD KILBRANDON and LORD EDMUND-DAVIES agreed with Lord Diplock that the appeal should be allowed.

The appeal was accordingly allowed and the order of the Court of Appeal reversed (except as to costs). Each of the respondents' originating summonses was dismissed. No order was made as to costs in the House of Lords.

A **CHANCERY DIVISION**

March 9 1977

(Before Mr Justice BRIGHTMAN)

CHEF & BREWER LTD v CLARKE AND ANOTHER

Estates Gazette May 28 1977

(1977) 242 EG 687

Application for possession order under RSC Order 113—Summary proceedings for possession of land—Company claiming against former manager of one of their public houses—Manager's attempts to find alternative housing for
B **his family had failed—Eviction order needed to enable local authority to rehouse him—Possession order made—Judge's comments about the effect of his order in relation to housing priorities**

In this short cause, Chef & Brewer Ltd applied under Rules of the Supreme Court, Order 113, for possession of living quarters over one of their public houses in High Street, Milton, Notts. The former manager, Laurence Clarke, had been given two months' notice terminating his employment, but he and his family had been unable to find other accommodation and had not moved out. The premises were urgently needed to house the new manager, who was living
C in a caravanette in the public house's car park. The local authority, Daventry Council, had said they could not give the family priority on their housing list unless they were made homeless by a formal court order.

Alastair Norris (instructed by Bower Cotton & Bower, agents for Beck & Phipps, of Northampton) represented the plaintiffs; Mr and Mrs Laurence Clarke appeared in person.

Giving judgment, BRIGHTMAN J said he would make the usual possession order but wished to make it clear that he was not seeking to give any priority to the Clarkes in their local authority housing queue.

The judge asked the plaintiffs' instructing solicitors to write
D a letter to the local authority to the effect that the order for possession which he had made was intended to be without prejudice to the question whether the effect of his order was or was not to invoke any statutory or other obligation on the local authority to rehouse the defendants and that the question of rehousing which had been expressly raised in the evidence remained, so far as he was concerned, wholly in the discretion of the local authority.

E

CHANCERY DIVISION

April 22 1977

(Before MEGARRY V-C)

TOWNSENDS CARRIERS LTD v PFIZER LTD

Estates Gazette June 4 1977

(1977) 242 EG 813

Notice to determine term under option in lease—Validity of notice disputed by lessors—Notice given by lessees' subsidiary to lessors' subsidiary—Subsidiaries acting as agents and notice therefore valid—Alternatively, lessors estopped
F **from denying validity of notice—Lessors' claim for declaration fails**

This was an originating summons by Townsends Carriers Ltd, lessors of warehouse premises in Northamptonshire, claiming that provisions for determination of the lease of the premises held by Pfizer Ltd had not been complied with by the lessees. The premises had been occupied by a subsidiary company of the lessees, Unicliffe Ltd, who had paid rent and corresponded in connection with the tenancy with a subsidiary of the lessors, Wilkinson Transport Ltd. The notice terminating the lease had been sent by Unicliffe Ltd to Wilkinson Transport.

J Speed (instructed by A Barone) appeared for the plaintiffs; G
W A Blackburne (instructed by Allen & Overy) represented the defendants.

Giving judgment, MEGARRY V-C said that this originating summons raised a short point on the construction of a provision for determination contained in a lease and also of a notice purporting to exercise the power of the determination. It also raised questions of the use of the right and wrong notepaper in groups of associated companies. The lease dated August 30 1970 demised certain warehouse premises in Northamptonshire for a term of seven years from September 1 1970. The lessor was Braybrook Townsends Ltd, and the
lessee, Pfizer Ltd. The provision for determination was con- H
tained in clause 4(c) of the lease. This read: "If either the landlord or the tenant shall desire to determine the term at any time after August 31 1973 they shall give to the other party 12 months' previous notice in writing." On September 1 1972 Braybrook Townsends Ltd changed its name to Townsends Carriers Ltd, and for convenience he (his Lordship) would call the company Townsends throughout. The reversion and the term had remained vested in the original parties to the lease but a letter dated January 15 1974 recorded certain variations in the lease that were said to have been agreed.

The notice in dispute was contained in a letter dated J
August 7 1975. That letter was from a company called Unicliffe Ltd and signed by one of its directors. Unicliffe Ltd and Pfizer Ltd were both wholly-owned subsidiaries of the Pfizer Group Ltd. The letter was addressed to "Wilkinson Transport." A company called Wilkinson Transport Ltd was, like Townsends, a wholly-owned subsidiary of BG Nominees Ltd and Lex Service Group Ltd. The letter expressed the wish "formally to give notice under clause 4(c) of our desire to terminate the arrangements as they presently stand as at September 1 1976."

In these circumstances Townsends had taken out an origin-
ating summons which sought a declaration that the letter of K
August 7 1975 did not amount to a valid notice to exercise the option to determine the lease. Townsends' case was simplicity itself. The landlord was, and always had been, Townsends; the tenant was, and always had been, Pfizer Ltd. The power of determination on the tenant's part could only be exercised by Pfizer Ltd giving the appropriate notice to Townsends. This had not been done. The notice given by Unicliffe to Wilkinson Transport was thus *nihil ad rem*. If Townsends' counsel were right, the case would provide an awful warning to those concerned with associated companies to make sure they used the right notepaper. Apart from the
point mentioned no attack had been made on the validity L
of the notice.

On behalf of the defendants, Mr Blackburne had said that, although the demise was to Pfizer Ltd, the premises were used and operated by Unicliffe Ltd. From early days Townsends corresponded with Unicliffe about the premises and from early 1974 all demands for rent were addressed to Unicliffe Ltd. By the middle of 1974 Wilkinson Transport Ltd had come on the scene and thereafter rent demands and correspondence were sent to Unicliffe Ltd, not by Townsends, but by Wilkinson Transport Ltd. Throughout subsequent correspondence, Wilkinson Transport and Unicliffe Ltd dealt with
the demised premises as if they were landlord and tenant M
respectively. In those circumstances, Mr Blackburne had submitted, the maxim *qui facit per alium facit per se* had a just and beneficial operation. Townsends had allowed Wilkinson Transport to deal with the demised premises as if they belonged to them; similarly, Pfizer Ltd had allowed Unicliffe to act as if they were the tenants. In those circumstances, Wilkinson Transport had been clothed with full authority by Townsends to act as Townsends' agents, and Unicliffe Ltd were clothed with full authority by Pfizer Ltd to act as their agents in relation to the tenancy of the demised premises. There was no question of Wilkinson Transport having only

A the limited authority of a rent collector. The letter of August 7 1975 was thus sent by an authorised agent of the tenant to an authorised agent of the landlord, and on simple principles of agency was therefore valid. Alternatively, it was said that the landlord would be estopped from denying the validity of representations which he had allowed another person to make and on which the tenant had acted.

Mr Speed for the plaintiffs had said that the facts did not support this argument, but it seemed to his Lordship that the factual basis for Mr Blackburne's arguments did exist, and as for the arguments themselves, he accepted them in their essentials. Mr Speed had been constrained to agree that his
B was a strict and technical contention and from first to last he (his Lordship) had said he could see no trace of merit in it. That, of course, did not conclude the case. A claim, though devoid of merit, might be soundly based in law. But it seemed to him that Mr Blackburne's arguments provided grounds that were both sound in law and just in result. No authority cited had been directly in point but such authority as there was seemed to provide at least some support for the defence. He could see no reason why an agent who had the power to give an effective notice should not have the power to receive a notice.

If a landlord and a tenant had expressly or by implication
C each respectively consigned the management of the reversion and the tenancy to agents on their behalf, what was done in relation to the tenancy as between the landlord's and tenant's agents would be as validly done as if it had been done between the landlord and the tenant themselves. And in a case where one party had acted to its detriment, the other party would be precluded from asserting that it was not bound by what had been done.

When those principles were applied to the facts of this case, it followed that the plaintiffs' claim failed and would have to be dismissed.

D

COURT OF APPEAL

February 2 1977

(Before Lord DENNING MR, Lord Justice ROSKILL and Lord Justice LAWTON)

INDUSTRIAL PROPERTIES (BARTON HILL) LTD AND OTHERS v ASSOCIATED ELECTRICAL INDUSTRIES LTD

Estates Gazette June 18 1977
E (1977) 242 EG 955

Tenants after giving up possession dispute landlords' title on ground that it was equitable only—Liability for breach of covenant to repair accordingly repudiated—Tenants estopped from denying landlords' title in respect of their period of possession unless in peril from an adverse claim by title paramount—Authorities reviewed—Contrary statements in Harrison v Wells [1967] 1 QB 263 overruled—Tenants' appeal from county court order dismissed

This was an appeal by defendants, Associated Electrical Industries Ltd, of Stanhope Gate, Westminster, from an order
F of Judge Edgar Fay QC on the trial of preliminary issues in April 1976 holding that the plaintiff company, Industrial Properties (Barton Hill) Ltd, of Stone Park Road, Bristol, were not debarred by the facts relating to their title from suing the defendants, and that the plaintiff company's title was sufficient to support their claim for damages for breach of repairing covenants in a document purporting to be a lease dated October 24 1966. The action concerned buildings on the Barton Hill trading estate in Bristol. Philip Albert Ashley Parker and his wife, Anne Monica Parker, in whom the freehold title of the property was vested, were the second plaintiffs.

Martin Nourse QC and Gavin Lightman (instructed by G
Lewis, Lewis & Co) appeared for the appellants, and Ronald Bernstein QC, Christopher Priday and Kirk Reynolds (instructed by Wilde, Sapte & Co) represented the respondents.

Giving judgment, Lord Denning MR said that some of the buildings on the Barton Hill Estate had been occupied for years by AEI and their predecessors. That company covenanted to repair them and to yield them up in repair. The lease came to an end and AEI left the premises. They were then in a very dilapidated condition. It might cost £200,000 to put them in a proper state of repair. But AEI said they were not liable. They said that their lessors, the H
people who let the premises to them, were not the legal owners and for that reason AEI were not bound to repair the premises when they left.

The freeholders had at all times been members of a family called Parker and they had held as trustees on a trust for sale. The freehold title had been vested in members of the family at all times, from father to son, until it was now in Mr and Mrs Philip Parker, the Parker trustees. In 1959 they were advised that there were tax advantages to be gained by transferring the properties to bodies corporate, so three companies were formed. The directors and shareholders were members of the Parker family. The estate was J
divided into three portions. Each company had its portion. This case concerned one of them—Industrial Properties (Barton Hill) Ltd. By an agreement of July 7 1959 the Parker trustees agreed to sell to Industrial Properties (Barton Hill) Ltd a portion of the Barton Hill estate for £48,000. It contained a provision that the purchase was to be completed on September 29 1959. It was sold subject to a lease made on January 3 1946 by the Parker trustees to the General Electric Company Ltd, the predecessors of AEI. Under the purchase shares were allotted to members of the Parker family in respect of the price of £48,000 to pay the trustees K
for the premises. Although the purchase price was paid, nevertheless the property was never conveyed to the plaintiff company. This was not by inadvertence. It was a deliberate decision made to avoid stamp duties. The legal title remained in the Parker trustees. The plaintiff company were only the equitable owners and had remained so ever since. The agreement for sale was registered on March 9 1960 as a land charge, but no conveyance was ever executed. The solicitors —who acted for both the trustees and the companies—took the opinion of counsel, who advised that that was perfectly in order.

So the plaintiff company remained the equitable owners. L
They received the rents from the tenants and acted as landlords. In 1966, the plaintiff company negotiated a new lease to AEI. It showed the lessors as "Industrial Properties (Barton Hill) Ltd." The solicitors for AEI returned the draft, with the note: "Please confirm that the lessor is the freeholder and that there is no mortgage or charge restricting its power to grant this lease." The solicitors for the plaintiff company answered: "This is confirmed. The lease needs the consent of the lessor's mortgagees for which application has been made." That confirmation was a mistake. It was a misrepresentation. The lessors were not the freeholders. The
lessors were the plaintiff company, who were only the equitable owners. M
owners. It was, however, an innocent misrepresentation. The solicitor for the plaintiff company had got so used to treating them as owners of the estate that he had forgotten that they were not the freeholders but only the equitable owners. He also thought that the lessees' solicitors only asked the question because they wanted to be sure that there was no headlease in existence whose covenants would affect them. On October 24 1966 the lease was executed. It was expressed to be between the plaintiff company and AEI—called the "lessee." It was a demise of the factory and buildings for 21 years from December 25 1966 at £15,250 a year, determinable by

either party at the end of the first seven or 14 years. It contained a covenant by the lessee with the lessor "to keep . . . the demised premises . . . in good and tenantable repair and condition and so to yield up the same . . . at the end or sooner determination of the said term." It contained a covenant by the lessor for quiet enjoyment "without any interruption by the lessor or any person rightfully claiming under or in trust for it or by title paramount."

From that time forward AEI occupied the premises and paid the rent. In 1972 the plaintiff company wished to borrow money from the County Bank and proposed that the premises should be lodged as security. In the course of the negotiations the bank's solicitors discovered that the plaintiff company had not the legal title but only an equitable title. This discovery led to the case being put to another counsel to advise the bank and the industrial company. He did not agree with the previous counsel. He was particularly asked to advise "whether leases in the form of the standard lease have been validly granted: if not, what steps should now be taken to validate them." The new counsel recommended that further steps be taken. He said:

> In view of the decision in *Harrison v Wells* [1967] 1 QB 263 it seems that the position of the reversioner on the lease may not be quite as good as if the lease had been created at a time when the company had the legal estate: and a confirmatory deed executed by the trustees on the lessee would remove any doubts there might be on this point. So far as future leases are concerned, it would in my opinion be desirable for the trustees to join in the lease so as to ensure that it vests in the lessee a legal term of years good against all the world, and that no doubt can arise as to enforceability of the lessee's covenants after the lessee has ceased to occupy under the lease.

In view of this opinion the Parker family asked their solicitors to issue supplementary documents to put in order all the existing leases. But this was not done in time for the present lease. On June 13 1973 the solicitor for the plaintiff company gave notice to AEI to determine the lease on December 25 1973—at the end of the first seven years. AEI did not claim a new tenancy. They gave up possession on December 25 1973. The premises were then badly out of repair.

On January 23 1974 the plaintiff company issued a writ against AEI claiming damages for breach of the covenants to repair in the lease. AEI thought they had no defence. They were not aware of a possible defence on the basis that the plaintiff company had not got the legal title and so they submitted to judgment for damages to be assessed. In the course of discovery, however, they found out that that the plaintiff company were not the freeholders. AEI then raised the point that they were not liable. The judgment was set aside and AEI were given leave to defend. The proceedings were amended on both sides. The Parker trustees, the freeholders, were added as plaintiffs, and the defendants pleased they were not liable on the covenants. They relied on *Harrison v Wells.*

The case of *Harrison v Wells* [1976] 1 QB 263 was this: in 1949 a widow let business premises to a tenant on the usual repairing covenants. It was for a term of seven years, but it was continued under the Landlord and Tenant Act 1954. On June 24 1961 the tenant surrendered the premises to the widow. She died six months later. Her executor sued the tenant for damages for breach of the covenant to yield up in repair. The tenant had no defence save this: he proved that the widow was not the freeholder. The legal title was vested in trustees—of which she was one. Her husband had owned the property and had left it to the trustees, including his widow, on trust to pay the income to his widow for her life. She had managed the property and let it on lease to the tenant and had received the rent. When the tenant surrendered the tenancy the premises were much out of repair. So the tenant was clearly liable in damages unless he could find some technical defence. He found it. The Court of Appeal held that, as the widow was not the freeholder, she had no title herself to grant the lease; that the lease was void; and as the covenant "cannot stand naked by itself without any lease to support it" the covenant was no good. The court held that the doctrine that a "tenant cannot dispute his landlord's title" only applied while the tenant was in possession, and that it did not apply when the tenant had gone out of possession. The court recognised that "the defendant has not a shred of merit," but nevertheless decided that he was not liable. They added, too, that he succeeded "only on a technicality." But still he succeeded.

He (his Lordship) was aghast that in these days the Court of Appeal could reach such an unjust result. He knew that in days past the courts used to give effect to technicalities of the worst description, but he had hoped that the courts had outgrown all that. As it was, he was quite satisfied that the court in *Harrison v Wells* was wrong—and not only wrong but so wrong that it was now in a position itself to overrule *Harrison v Wells.* The truth was that the court had misunderstood a case over 100 years old called *Cuthbertson v Irving* decided in 1859. The court had had only the report in the Law Journal (1859) 28 LJ Ex 306. They ought to have looked at the report in 4 Hurlstone & Norman 742, which was much better, affirmed in the Exchequer Chamber in (1860) 6 H & N 135. The court must consider it in some detail.

John Biglands was at one time the legal owner of a water cornmill. In 1844 he borrowed money from his brother Benjamin, and mortgaged the mill to him as security. Under the mortgage deed the legal title was transferred to the mortgagee, Benjamin Biglands, but John Biglands remained in possession as mortgagor in possession, having only an equity of redemption. That was, of course, very common under the old conveyancing system. In 1848 John Biglands, being mortgagor in possession, let the mill on a lease for seven years from August 1 1848 to John Irving. The lease contained the usual covenants to repair during the term and to deliver up in repair at the end of the term. John Irving paid the rent to John Biglands for the first six years. But, in 1854, John Biglands sold his equity of redemption to Joseph Cuthbertson, subject, of course, to the lease. From 1854, for the last year of the term, John Irving paid the rent to Cuthbertson. On August 1 1855 the term of seven years expired. Two months later, on October 1 1855, Irving delivered up possession of the mill to Cuthbertson. The mill was then found to be in a very bad state of repair. The mortgagee, Biglands, never interfered in any way at any time with the mill, or any of these dealings. Some time later Cuthbertson brought an action against Irving for damages for breach of the covenant to yield up in repair. It was plain from the pleadings that this action was commenced after Irving had gone out of possession. In 1859 the case came for trial. Irving contended that Biglands had no legal estate in the mill at the time of the lease and, therefore, no estate in reversion to assign to Cuthbertson. This contention was rejected and judgment was given in favour of Cuthbertson against the tenant. In modern language, the issue was this: was Cuthbertson, the buyer, entitled to sue Irving on the covenants? It was admitted that the original lessor, Biglands, could have sued, even though he had not the legal title. But could Cuthbertson sue? The Exchequer Chamber gave the answer in a few sentences (6 H & N at p 139): the lessor (Biglands), being a mortgagor in possession at the time of the granting of the lease, had no legal title to the premises, but only an equity of redemption. His title, therefore, as between himself and his lessee was only by estoppel; and if the lessor assigned, as he could assign that which he had, his assignee would either have a title by estoppel as against the lessee, or no title at all. There was no sound reason why the assignee of the reversion should not establish his title by way of estoppel. The actual decision was, therefore, that the tenant was estopped from denying the title, not only of his lessor, Biglands, but also the title of the assignee, Cuthbertson. But the significant thing was that, on the facts, the estoppel operated against the tenant, *after* he had gone out of possession, on his being sued for dilapidations which were found to exist when he quitted and delivered up possession. The decision itself showed that the estoppel operated, not only while the tenant was in possession, but also when he had gone out of possession and was being sued on the covenants. It was, therefore, a direct contradiction of *Harrison v Wells.*

In the course of the Court of Exchequer judgment, Martin B at p 757 stated two propositions: "If the lessor have no title and the lessee be evicted by him who has title paramount, the lessee can plead this and establish a defence to any action brought against him, but . . . so long as the lessee continues in possession under the lease, the law will not permit him to set up any defence founded upon the fact that the lessor *nil habuit in tenementis.*" In *Harrison v Wells,* the Court of Appeal drew an inference from the last proposition. They treated it as carrying the corollary that once the lessee had gone out of possession he could dispute the landlord's title. He (his Lordship) considered that corollary was unfounded. It was contradicted by the actual decision in *Cuthbertson v Irving.* Those two propositions of Martin B only stated two sides of the same coin. The one proposition was simply the converse of the other. The first was that, if the lessee was evicted by title paramount, he could dispute the lessor's title. The second, that if he continued in possession—if he was not evicted by title paramount—he could not dispute it.

So far as concerned the position of the tenant who went out of possession of his own free will—without being evicted by title paramount or its equivalent—then the justice of the case required that he should be estopped from denying the landlord's title. This position was covered by the later words of Martin B. He said at p 758: ". . . . This state of the law in reality tends to maintain right and justice, and the enforcement of the contracts which men enter into with each other (one of the great objects of all law); for so long as a lessee enjoys everything which his lease purports to grant, how does it concern him what the title of the lessor, or the heir or assignee of his lessor, really is. All that is required of him is, that having received the full consideration for the contract he has entered into, he should on his part perform it." So here it seemed to him (his Lordship) that AEI, having received the full consideration for the contract in which they entered, should pérform their part of it. It did not matter to them what title the plaintiff company had—whether it was the legal title or an equitable title—it made no difference to them so long as they were not subject to any adverse claim by anyone else.

The court had been referred to many authorities old and new. He (his Lordship) had considered them all—and others too—but the result could be stated thus: if a landlord let a tenant into possession under a lease, then, so long as the tenant remained in possession undisturbed by any adverse claim, either during the term or holding over, then the tenant could not dispute the landlord's title. Suppose the tenant (not having been disturbed) went out of possession and the landlord sued the tenant on the covenant for rent or for breach of covenant to repair or to yield up in repair, the tenant could not say to the landlord "you are not the true owner of the property." Likewise, if the landlord, on the tenant's holding over, sued him for possession or for use and occupation or mesne profits, the tenant could not defend himself by saying "the property does not belong to you, but to another." But if the tenant was disturbed in his possession by being evicted by title paramount or the equivalent of it, then he could dispute the landlord's title. Suppose the tenant was actually turned out by the third person or if the tenant, without going out, acknowledged the title of the third person by attorning to him, or the tenant contested the landlord's claim on an indemnity from the third person, or there was anything else done which was equivalent to an eviction by title paramount, then the landlord was no longer estopped from denying the landlord's title: see *Wilson v Anderton* (1830) 1 B & Ad 450 at p 457 *per* Littledale J. The tenant, being thus disturbed in his possession, could say to the landlord "you were not truly the owner at the time when you demanded and received the rent from me. I am liable to pay mesne profits to this other man. So you must repay me the rent which I overpaid you. Nor am I liable to you on the covenants during the time when you were not the owner": see *Newsome v Graham* (1829) 10 B & C 234; *Mountnoy v Collier* (1853) 1 E & B 630; *Watson v Lane* (1856) 11 Exch 769. The tenant could also claim damages for the eviction if there was, as here, an express covenant for quiet enjoyment covering interruption by title paramount. Short of eviction by title paramount or its equivalent, however, the tenant was estopped from denying the title of the landlord. It was no good his saying "the property does not belong to you but to a third person" unless that third person actually came forward and successfully made an adverse claim—by process in the courts or by the tenant's attornment or acknowledgment of it as by the tenant defending on an indemnity. If the third person, for some reason or other, made no adverse claim or was debarred from making it, the tenant remained estopped from denying the landlord's title. This was manifestly correct, for without an adverse claim it would mean that the tenant would be enabled to keep the property without paying any rent to anybody or performing any covenants. That could not be right. That was the reasoning adopted by the Court of Queen's Bench in *Biddle v Bond* (1865) 6 B & S 225, a case of bailor and bailee, but the court treated it the same as landlord and tenant. If this principle had been applied in *Harrison v Wells* the decision would have been different. In that case the trustees (including the widow) were the legal owners. The widow was entitled to the rent for her life. The trustees had made no adverse claim on the tenant. He (his Lordship) could not see that they could ever have made one—at any rate not with any hope of success—since the widow herself was one of the trustees. So, there being no adverse claim, the tenant was estopped from denying the lessor's title.

In the present case AEI, the tenants, were not subject to any adverse claim whatever. The lessor to AEI was the plaintiff company, which was the equitable owner. The legal owners were the Parker trustees. They were also the directors and shareholders of the plaintiff company. They acquiesced in the lease being made by the plaintiff company to AEI. They could not by any possibility make any adverse claim against AEI on their own acount. Not only that. They had actually come in as plaintiffs in the proceedings jointly with the plaintiff company, so as to make sure that the benefit of these proceedings went to the plaintiff company only. Seeing that AEI were absolved from any adverse claim by the legal owners, it was a proper case for the doctrine of tenancy by estoppel. AEI had had the full benefit of the lease for the stipulated term of years. They should perform the covenants, or pay damages in lieu, to the only persons entitled to sue them—the plaintiff company. Even though AEI had gone out of possession, they could not avoid their responsibilities by reliance on a technical rule of law which on investigation was found to be groundless.

AEI sought to make something out of the innocent misrepresentation made by the solicitor. He represented that the plaintiff company was the freeholder when it was not. But that misrepresentation did no one any harm. If the truth had been known, all that would have happened would have been that the trustees would have joined in the lease and all this later trouble avoided. So the misrepresentation afforded no defence to AEI.

He (his Lordship) had considered the position at common law. But in equity there was a much shorter way to the

A decision. It was quite plain that, if the lease to AEI was defective in point of law, nevertheless it was good in equity, and for this simple reason. There were two agreements of which specific performance would be granted. One was the agreement by the Parker trustees to convey to the plaintiff company. The other was the agreement by the plaintiff company to grant a lease to AEI. In respect of each of these agreements equity looked upon that as done which ought to be done. It followed that, by combining the two agreements, the tenant, AEI, held upon the same terms as if a lease had actually been granted by the Parker trustees to AEI. This was, of course, an extension of the doctrine of *Walsh v*
B *Lonsdale* (1882) 21 Ch D 9, where there was only one agreement. But he (his Lordship) saw no reason why the doctrine should not be extended to a case like the present, where there were two agreements, each of which was such that specific performance would be granted.

Even at law all the discussions about estoppel might be unnecessary. Throughout the case he (his Lordship) had seen no reason why the plaintiff company should not sue AEI on the covenant to repair and to yield up in repair. It was plain that the plaintiff company had an interest in having the property kept in repair. They were the equitable owners
C of it. They could stipulate with anyone to repair it. It might be a builder, or a licensee, or anyone else. So why should they not stipulate with AEI that the latter should keep it in repair. It made no difference to AEI what the title was. It mattered not in the least whether the plaintiff company were the legal or equitable owners. So why should not AEI be liable on the covenant? It would be different, of course, if AEI had been evicted by title paramount or anything equivalent to it, but nothing of that kind happened. Seeing that AEI had had the whole of the consideration, they were liable in covenant: see *Hodson v Sharp* (1808) 10 East 350 at 364 per Bayley J; or, in the old phrase, on privity of contract:
D see *Baker v Gostling* (1834) 1 Bing NC 19. And the damages would not be nominal; they would be substantial.

For some years he (his Lordship) had tried to persuade others that the Court of Appeal should not be absolutely bound by a previous decision which was later found to be wrong. In this effort he had failed. The law had been stated, and the reasons given, by Lord Simon of Glaisdale in *Farrell v Alexander* [1976] 3 WLR 145. But it was still open to the court to depart from a previous decision if it was given *per incuriam*. The court could so find if it could fasten upon something in the previous decision and say of it "here was a manifest slip or error": see *Morelle Ltd v Wakeling* [1955]
E 2 QB 379. That could, he thought, be said of *Harrison v Wells* [1967] 1 QB 263. The court in that case had thought that there was "no case in the books where the estoppel has ever been held to operate against a tenant who has gone out of possession": see at p 278 per Salmon LJ. Whereas, in point of fact, the decision in *Cuthbertson v Irving* was just such a case, as the report in 4 H & N showed. They thought too that the "covenant cannot stand naked by itself," whereas there were cases which showed that it could; but these were not cited.

In his (his Lordship's) opinion, therefore, *Harrison v Wells* F
was wrongly decided *per incuriam*, and the court could, and should, overrule it. The doctrine of tenancy by estoppel had proved of good service and should not be whittled down. It should apply in all cases as between landlord and tenant, no matter whether the tenant was still in possession or gone out of possession, so long as he was not confronted with an adverse claim by a third party to the property. If the tenant was so confronted, his proper course nowadays was not to side with either claimant, but to let one or other of them sue him, and to bring in the other as a party under the wide powers now given in Order 15, rule 6(2)(b), which re-established the view taken by the Court of Appeal in *Re* G
Vandervell's Trusts [1969] 3 WLR 458, thus making interpleader available in respect of land. In the present case there was no adverse claim and no possibility of it. The tenant was estopped from disputing the landlord's title. He thought that AEI were liable on the covenants and would dismiss the appeal.

Agreeing, ROSKILL LJ said that in *Harrison v Wells*, although the lease had run its full term, the lessee was in theory at least at risk from the owner of the legal estate. But neither that case, nor *Levingston v Somers* [1941] Ir R 183 was binding authority for the proposition that, after the H
determination of a lease by effluxion of time, the determination of the estoppel left a lessee free to challenge his lessor's title so as to avoid all liability under a dilapidations covenant when at no time was he in peril of action by way of ejectment or otherwise by the owner of the legal estate; when the owner of that legal estate had acquiesced in the grant of the lease to him; and when he had at all times had the full benefit of the consideration for which he had paid rent to his lessor. The court was free to hold, and he (his Lordship) did hold, that a lessee or tenant who had had possession for the whole of the term was thereafter estopped from denying his lessor's or landlord's title in respect of the period for which he had J
had possession, unless after termination of that possession a claim was made against him by title paramount in respect of some part of that period so that he became in peril of an adverse claim. Even if his lordship had been of the opposite opinion on the estoppel question, he would unhesitatingly have held the "lease" to be effective as an agreement for a lease. The courts should not be astute to find reasons whereby lessees and tenants could escape easily from obligations into which they freely entered.

Also agreeing, LAWTON LJ said that AEI had occupied the premises until the agreement was terminated in accordance with its terms. They had used the premises for seven years. K
When they went out of possession the premises were in a worse condition than when they went in. They contended that the law relieved them of liability. If it did, it was a strange and seemingly unjust law. While *Harrison v Wells* on its facts might have been rightly decided, the wider statements in it were made *per incuriam* and need not be followed. AEI were estopped from denying the plaintiffs' title.

The appeal was dismissed.

LANDLORD AND TENANT

BUSINESS TENANCIES

CHANCERY DIVISION

November 1 1976

(Before Mr Justice GRAHAM)

COVELL MATTHEWS & PARTNERS v FRENCH WOOLS LTD

Estates Gazette March 26 1977

(1976) 241 EG 915

Application for new tenancy—Lease to withdraw sought—Action pending for specific performance of alleged binding agreement to grant lease—Leave to withdraw nevertheless granted on terms—Rent payable until " date of withdrawal " —Date of withdrawal normally date of judgment

This was an application by Covell Matthews & Partners, tenants, for leave to withdraw an application to the court to order a new business tenancy to be granted by French Wools Ltd, landlords of premises at Lexington Street, Piccadilly Circus, London.

Walter Blum (instructed by May, May & Merrimans) appeared for the plaintiff tenants; J C Harper (instructed by Theodore Goddard & Co) represented the landlord defendants.

Giving judgment, GRAHAM J said that the case related to business tenancies of premises at Lexington Street, Piccadilly Circus, and that some of the points involved were novel. The tenants were underlessees of the premises under the terms of an underlease of March 8 1960. The landlords gave notice to the tenants under section 25 of the Landlord and Tenant Act 1954 to terminate the tenancy as at June 24 1973. On February 5 1973 the tenants gave notice that they were not willing to give up possession of the premises, and on March 13 1973 the tenants issued an originating summons seeking an order for a grant by the landlords of a new tenancy under section 26 of the Act.

Negotiations between the parties as to the terms and conditions of a proposed new lease took place and the hearing of the originating summons was postponed from time to time. There was a dispute as to whether terms for a new lease were finally agreed or not. Some time later the tenants left the premises and on January 23 1976 served on the landlords an application for leave to withdraw their originating summons. On January 25 1974 the landlords had written a letter to the tenants stating that there was a binding contract to grant a new lease on the terms therein set out, including the very considerable rise in rent of £5,600 to the new rent of £30,000 a year. The landlords had recently begun proceedings for specific performance of the alleged contract. In addition to those proceedings the landlords had arranged for the issue of a summons on February 12 1976 to determine the interim rent in accordance with section 24A of the Act (introduced by the Law of Property Act 1969). The rent payable on that determination would be so payable in accordance with section 24A (2) from the date of the application, February 12 1976, since that was later than either the date specified in the landlords' notice under section 25 or in the tenants' request for a new tenancy under section 26. Rent would continue to be payable in accordance with the provisions of section 64 (2), if the application to withdraw the application for a new tenancy was granted, until " the date of the withdrawal." If leave were given the withdrawal date was important because the quantum of any interim rent payable would be much more if the date of withdrawal was the date of this judgment rather than January 23 1976, when the application to withdraw was made, or the date of the service of such application if that was the correct date.

A master on March 20 1976 gave leave to the tenants to withdraw the summons of March 13 1973 and gave general directions on the landlords' summons for the fixing of interim rent. This latter summons had been issued on February 12 1976, which was soon after the tenants' application of January 23 1976 for leave to withdraw their earlier application for a new tenancy.

The tenants argued that they were entitled to withdraw their application for a new tenancy if they did not wish to go on with it and, although there was a discretion under Order 21, Rule 3, which governed the case because leave was required, a plaintiff ought not to be compelled to litigate if he did not want to. But it was accepted that protection might have to be given to the defendant. It was further argued that the application could be dealt with separately from, and need not have any effect upon, the decision which the court would have to give on the question whether or not the parties had themselves entered into a binding agreement for a new lease. On the first point and on the difference in effect of discontinuance and dismissal, Mr Blum cited *Stahlschmidt v Walford* (1879) 4 QB 217, *Fox v The Star Newspaper Co Ltd* [1889] 1 QB 636 and [1900] AC 19, and *J T Stratford & Son Ltd v Lindley (No 2)* [1969] 1 WLR 1547. It was also pointed out that the rent in this case was not much more than £5,000, which would have enabled the matter to be dealt with in the county court, and that under Order 18 of the County Court Rules no leave to withdraw would have been required. The principles to be culled from the cases were that the court would normally allow a plaintiff to discontinue if he wanted to and if no injustice was caused to the defendant. It was not desirable that a plaintiff should be compelled to litigate against his will. The court should grant leave if it could without injustice to the defendant, but in so doing should be careful to see that the defendant was not deprived of some advantage which he had already gained in the litigation, and should be ready to grant him adequate protection to ensure that any advantage he had gained was preserved.

Mr Harper, for the landlords, argued that the tenants' application ought to be regarded as having been overtaken by the agreement for a new lease which the landlords contended had been entered into between the parties, and that the most convenient course would be to adjourn the proceedings until the trial of the action for specific performance and to leave all outstanding matters to be dealt with in that action. To do that, however, would deprive the tenants of their right to have their application for leave to discontinue decided if it was found that they could do so

A without doing injustice to the defendants. The real truth was that the tenants had some fear that a refusal of leave to withdraw might have some adverse influence upon the decision in the specific performance action, whereas the landlords feared the opposite. The specific performance action was a separate question, however, and no judge trying it would be influenced by the decision in the present application.

The master was correct in granting the tenants leave to withdraw their application for a new tenancy. At the same time, for the landlords' protection, the order being for leave to discontinue and not for dismiss, the tenants should be put on terms that: (1) they should not make any fresh appli-
B cation for a new tenancy; (2) the grant of leave should be expressed to be given without prejudice to the landlords' contention that a binding agreement had been reached between the parties and to the effect of such agreement if it was eventually decided that it was binding; (3) an order for costs against the tenants should be made.

The question remained what was the correct date to assign as the date of withdrawal within the meaning of section 64. The tenants said it should be the date of the application to discontinue the application for a new tenancy, ie January 23 1976, and the landlords said it should be the date of judgment. The landlords said that there could not be any
C withdrawal before leave to withdraw had been given and the date must be the date when leave was given. The tenants said that, as in any normal case where notice was required, it was the date of service of the notice which crystallised the D
position, never mind how much time elapsed before the decision actually granting such application was given. It was argued that the date of service was a certain date and the date of the decision uncertain and one which depended on innumerable factors outside the control of the parties, such as availability of counsel and the state of the list. If the correct date was that of the decision it was true that uncertainty was created and, further, that at a time of rising rents it would be to the landlords' advantage to find excuses for delaying the hearing of the application. However, he (his Lordship) thought that the argument calling for certainty of date could not prevail and that the date of withdrawal E
should be the date of judgment. The notice was not a notice to fix a time for anything but merely fixed the landlord with notice of the tenant's intention to ask for leave to withdraw. There was force in the argument of Mr Harper that in such a case, if the tenant thought that the landlord was trying to delay matters, he could come to the court, which had a discretion to fix the date of withdrawal at whatever date was just. Here there was nothing to show avoidable delay or unusual circumstance and the normal rule should apply. The date of withdrawal should be the date of the judgment.

The court was informed by counsel that it had been agreed F
that the date of the judgment should be deemed to be July 5 1976 because there had been an adjournment at the landlords' request.

LANDLORD AND TENANT

RENT ACTS

A

COURT OF APPEAL

July 7 1976

(Before Lord DENNING MR, Lord Justice ORR and Lord Justice WALLER)

MARCHANT v CHARTERS

Estates Gazette January 1 1977

(1976) 241 EG 23

Rent Act 1974—Occupant of bed-sitting room held a licensee and not a tenant—Observations on the distinction between a tenancy and a licence—Further point on the nature of the "attendance" which excludes a tenancy from control

B This was an appeal by George Albert Charters, occupier of a small top-floor room of a house in Clapton Common, against an order of Judge Willis at Shoreditch County Court giving the respondent, Mrs Anna Marchant, of William Mews, Westminster, London SW1, judgment for possession, arrears of rent and mesne profits.

D Gerrey (instructed by Clifford Watts, Compton & Co) represented the appellant. The respondent appeared in person. H K Woolf (instructed by the Treasury Solicitor) appeared as *amicus curiae.*

Giving judgment, LORD DENNING said: No 131 Clapton Common is an ordinary sort of house. It is owned by Mrs C Marchant. It was at one time a private dwelling-house, but for the last 23 years she has let it off in single rooms to men only. She has regularly put the same advertisement in the local newspaper: "Attractive bachelor service apartments. Superior house. Resident housekeeper, every convenience." There are seven bed-sitting rooms. Each room is simply but nicely furnished with a bed, table, a couple of chairs and so forth. Each room has, of course, its gas ring and grill, so that the residents can do a little cooking on their own, as meals are not included. The bathroom and lavatory are shared. The housekeeper cleans the rooms every day and provides clean linen every week. It is a D very usual sort of arrangement. Many young people live in bed-sitting rooms like this. The question is whether the occupant of such a bed-sitting room has security of tenure. Under the old Rent Acts before 1974 he had none. There was in those days a big difference between furnished and unfurnished lettings. The tenant of an unfurnished dwelling was protected from eviction. He had complete security of tenure. But the tenant of a furnished dwelling had not. He was provided with means of challenging the amount of rent he was charged, but these were not very effective, because he had no security of tenure, or at any rate, only for a very limited time if he challenged the rent. Now since the Rent E Act 1974 the tenant of a furnished dwelling is given security of tenure. He is protected from eviction just as much as the tenant of an unfurnished dwelling. But does this new protection extend to the occupier of one of those bed-sitting rooms? That is the question raised by this case. Obviously it is one of much importance.

First I must describe the facts of this particular case. In June 1972 Mr Charters took a room on the top of the house (the third floor). It was only a small room, 9 ft 5 in by 11 ft 6 in. He paid £6 a week at first, but this was afterwards increased to £7 a week. He thought this was F too high, and applied to the rent officer to fix a fair rent. The rent officer reduced the sum payable to £4.25 a week, and his decision was confirmed by the rent assessment committee. Mrs Marchant did not agree with this. She gave Mr Charters notice to quit, and took proceedings in the county court to turn him out. Mrs Marchant appeared in person; counsel appeared for Mr Charters. The judge made an order for possession. Mr Charters appealed to this court. Mrs Marchant also appeared in person before us. But the case is of such general importance that we are glad to have the assistance of Mr Woolf as *amicus curiae.* The relevant statutory provisions are these. Under the G Rent Act 1968, section 1 (1) says that:

> A tenancy under which a dwelling-house (which may be a house or part of a house) is let as a separate dwelling is a protected tenancy. . . .

Section 2 (1) (b) says that:

> A tenancy is not a protected tenancy if under the tenancy the dwelling-house is *bona fide* let at a rent which includes payments in respect of board, attendance [or any furniture].

Section 2 (3) says, however, that:

> a dwelling-house shall not be taken to be *bona fide* let at a H rent which includes payments in respect of attendance [or the use of furniture] unless the amount of rent which is fairly attributable to attendance [or use of furniture], having regard to the value of the attendance [or the use] to the tenant, forms a substantial part of the whole rent.

By the Rent Act 1974, schedule 4, the words which I have put in brackets, "or any furniture," "or the use of furniture," "or use of furniture" and "or the use" were repealed. So section 2 (3) must now be read without them. The result is that the use of furniture does not deprive the tenant of protection, but the giving of "attendance" may deprive him of security, depending on the amount of rent J attributable to the attendance.

Applying those provisions to the present case, there are two questions to be considered. First, was Mr Charters a tenant? Was this one room let to him as a separate dwelling? If not, he has no protected tenancy. Second, if he was a tenant, was the attendance given sufficiently substantial to deprive him of protection? The county court judge decided the first question in favour of Mr Charters. He held that he was a tenant. But he decided the second question against Mr Charters. He held that the "rent included services which were by no means minimal." So he made an order for possession. In order to describe the K facts, I will now read Mrs Marchant's evidence:

> I go to [the house] practically every day. I occasionally sleep there. [It] is a fine old house: a self-catering residential hotel for single men. Been running like this for 25 years. Each resident has a room of his own. They are completely furnished —wall-to-wall carpeting, curtains, bed, blankets, sheets, pillow, pillow case, towels—even tea towels—one arm chair and two wooden chairs, wardrobe, occasional tables, cooker, sink, with gas water heater, gas fire for heating, points for power. Each room has own meter for gas and electricity. Every resident gets

A daily cleaning of room, clean linen weekly, bathrooms are common. They can get an evening meal by arrangement with housekeeper; they pay her. She is engaged on basis that she will provide food if required. All rooms are provided with cutlery, crockery, and cooking utensils—all that is necessary. They can also get breakfast and lunch. Through the years, I find more and more people are providing their own meals. The housekeeper keeps the money for the food she provides.

On that evidence, the first question is, was Mr Charters a tenant or only a licensee? The law on this subject has been developed greatly in the last 25 years: I might almost say revolutionised. In 1952 I collected several of the cases in
B *Errington v Errington and Woods* [1952] 1 KB 290, and since that time there have been many more. The difference is very important under the Rent Acts (including the 1974 Act), because a "tenant" is given complete security of tenure but a licensee is not. There are four or five cases, somewhat like the present, to which we have been referred. The first is *Appah v Parncliffe Investments Ltd* [1964] 1 WLR 1064. A house had been split up into 17 different rooms, each with its own Yale lock and key. The rooms were furnished. No meals were provided, but the rooms were cleaned daily, and beds made. Fresh linen was provided. A lady took one of the rooms, but owing to the
C negligence of the owners someone was able to break into her room and steal her belongings. It was held that she was not a tenant but a licensee for reward, and so able to sue the owner for negligence. The next case is *Abbeyfield (Harpenden) Society Ltd v Woods* [1968] 1 WLR 374. A charitable society had converted a big house into an old people's home with 12 rooms. They let one of the rooms unfurnished to an old man of 85. He furnished it himself. The society provided food in a common dining-room, and so forth. There was a resident housekeeper. It was held that the old man was only a licensee and not a tenant.
D So he was not protected by the Rent Acts and had to go.

The nearest case to the present case is *Luganda v Service Hotels Ltd* [1969] 2 Ch 209. A student was reading for the Bar. He took a furnished room in a building called the Queensborough Court Hotel. There were 88 rooms. They were said to be "let" out to tenants. Every student had a Yale key for his room. It was a bed-sitting room with a double gas ring. He got his own meals and provided his own towels and soap. The chambermaids came in every day and made the bed and cleaned the room. Every week they changed the linen. It was held that he was a contractual licensee and not a tenant (see [1969] 2 Ch at 219C). But
E as he had applied to the rent tribunal, he was protected for a limited period from being evicted (see 219H). The next case arose under the planning Acts, *Mayflower Cambridge Ltd v Secretary of State for the Environment and Another* (1975) 30 P & CR 28. A building in Cambridge had been erected for students to stay, but it was not needed for them, so it was used for visitors who wanted to come to Cambridge. The question was whether the change of use was to an "hotel." Lord Widgery CJ said: "The essence of an hotel is that it takes transient passengers." That may well be a test under the planning Acts, but it is not a test for seeing whether there is a tenancy or a licence. The
F last case was reported on March 26 of this year in *The Times* newspaper. It is *R v South Middlesex Rent Tribunal ex parte Beswick* . A young lady lived in a single room at a YWCA hostel. It was a furnished room. It was her sole home. She was permanent, not temporary. In common with other residents, she had the use of a kitchen, dining-room, living-room, laundry-room, bathroom and toilet. It was held that she was not a tenant but a licensee. So the fair rent was to be fixed, not by the rent officer, but by the rent tribunal.

Gathering the cases together, what does it come to? What is the test to see whether the occupier of one room in a house is a tenant or a licensee? It does not depend on G
whether he or she has exclusive possession or not. It does not depend on whether the room is furnished or not. It does not depend on whether the occupation is permanent or temporary. It does not depend on the label which the parties put upon it. All these are factors which may influence the decision but none of them is conclusive. All the circumstances have to be worked out. Eventually the answer depends on the nature and quality of the occupancy. Was it intended that the occupier should have a stake in the room, or did he have only permission for himself personally to occupy the room, whether under a contract or not? In the latter case he is a licensee. Looking at the H
position in this case, in my opinion Mr Charters was not a tenant of this one room. He was only a licensee. A contractual licensee, no doubt, but still only a licensee. So he does not have security of tenure under the Rent Acts. He is not protected against eviction. On this point I differ from the judge. It is sufficient for the deciding of this case. Mr Charters has no right to stay. This decision has its bearing on another aspect of the case. Mr Charters, as I have said, applied to the rent officer to register a fair rent, and from the rent officer it went up to the rent assessment committee. That would have been the appropriate procedure if Mr Charters had really been a tenant, for then J
the case would have come under Part IV of the Rent Act 1968. But it was not the appropriate procedure if he was only a contractual licensee of a furnished room, for in that case he should have applied to a rent tribunal under Part VI of the Rent Act 1968. Mrs Marchant took this very point before the rent officer and the rent assessment committee, but they overruled her, holding that Mr Charters was a tenant. Mrs Marchant has, I understand, applied to the Divisional Court to get the matter put right. If he is only a licensee, it does appear that it was for the rent tribunal and not for the rent officer. But I would point out that, if he did go to the rent tribunal, that tribunal has K
power under Part VI to give him security of tenure for a time not exceeding six months.

There remains the other point about "attendance" which has been much discussed before us. The rent here clearly included payments in respect of attendance, but the question is whether the amount fairly attributable to it formed a substantial part of the whole rent, so as to take the tenancy out of protection under section 2 (3) of the 1968 Act. The word "attendance" was much considered by the House of Lords in *Palser v Grinling* [1948] AC 291. Lord Simon LC at pp 310 and 311 said that attendance means "services L
personal to the tenant provided by the landlord for the benefit or convenience of the individual tenant in his use or enjoyment" of the room. It does not include services in regard to the common parts, such as cleaning the common staircase, or the porter at the bottom. Applying that test, it is quite plain that the attendance here included these services: each day the room was cleaned, each day the rubbish was removed, and each week the dirty linen was removed and clean linen was supplied in its place. The fact that Mr Charters may have refused it on some occasions does not affect the matter. Then did the amount of M
rent attributable to these services form a substantial part of the whole rent, having regard to their value to the tenant? This is a difficult point, but it must be rememebered that in these days such attendance may cost quite a lot of money. The women who perform these domestic services command high wages. The judge found that in this case the rent attributable to it was "by no means minimal." He did not express it in those very words, but I think he must have intended to find that it did form a substantial part of the whole rent. On that account he held that it was not a protected tenancy and made an order for possession. It is a matter which depends very much on the facts, and I am

A not prepared to say that the judge was wrong. Therefore on these two grounds I think the appeal should be dismissed.

ORR LJ: I agree.

WALLER LJ: I agree.

The appeal was dismissed, with an order for payment of the respondent's costs out of the legal aid fund.

COURT OF APPEAL

May 13 1976

B (Before Lord Justice MEGAW, Lord Justice STEPHENSON and Sir John PENNYCUICK)

JESSAMINE INVESTMENT CO LTD v SCHWARTZ

Estates Gazette February 5 1977

(1976) 241 EG 383

A subtenant who acquires possessory title as against the mesne landlord does not thereby lose the protection of the Rent Acts as against the freeholder

This was an appeal by Jessamine Investment Co Ltd, freeholders of a house at 29 Evering Road, Stoke Newington, London N16, from a judgment of Judge Roger Willis at Shoreditch County Court on June 23 1975 dismissing their
C claim against their tenant, Mrs Miriam Schwartz, for possession of the premises.

L Bromley QC and R Campbell (instructed by Gale & Phelps) appeared for the appellants, and R Bernstein QC and R Tager (instructed by Rose & Birn) represented the respondent.

Giving the first judgment, SIR JOHN PENNYCUICK said that there were two issues before the court: (1) whether Mrs Schwartz as statutory subtenant had acquired a possessory title as against the mesne landlord, and (2) whether, if so, she had thereby lost the protection of the Rent Acts
D as against the freeholder. On November 19 1875 the freeholder's predecessor in title demised the house in question and another one in Evering Road to James Slater for a term of 99 years commencing on December 25 1874. The leasehold interest was severed in 1924, and the leasehold of the house with which the court was concerned was assigned to a Mrs David. The freehold was also severed at some time. In about 1937 Mrs David granted a weekly tenancy to Mrs Schwartz, then Mrs Levy, and her husband. The rent was 35s a week, which was increased to £2 shortly before or after 1939. At that point the contractual tenancy ceased and a statutory tenancy came into effect. During the war,
E Mrs David went to live in Cheltenham and made no further demand for the rent. Nothing turned on the precise date when rent ceased to be paid; it was by 1945 at the latest. Mr and Mrs Levy had not withheld the rent; they simply did not know where Mrs David was. In 1948 Mr Levy died and Mrs Levy became the sole statutory tenant. In 1973 Mrs Schwartz, as she had become by that time, applied to be registered as proprietor of the premises with a possessory title, as the period of limitation had expired in her favour in 1957. She had no knowledge of the existence of any leasehold interest. The registration did not take effect, and the headlease expired by effluxion of time on December 25
F 1973. After protracted correspondence between solicitors, the freeholder began an action for possession. The judge found for Mrs Schwartz, basing his decision on section 75(1) of the Land Registration Act 1925. An alternative basis which depended on Part I of the Landlord and Tenant Act 1954 had been abandoned by counsel.

Mr Bromley, for the freeholder, contended that a statutory tenant could acquire possessory title by non-payment of rent, as had been established in *Moses v Ludgrove* [1952] 2 QB 533, and it was plain that section 9(2) of the Limitation Act 1939, dealing with periodic tenancies without a lease in writing, squarely covered the facts of the case. By the operation of the proviso to the subsection, Mrs G
Schwartz acquired a possessory title against Mrs David 12 years after the date of the last payment of rent. Having regard to the decision in *Fairweather v St Marylebone Property Co Ltd* [1963] AC 510, Mrs David's title was extinguished as against Mrs Schwartz in 1957 but continued to exist as against the freeholder until 1973. The crucial question that thus arose was whether Mrs Schwartz's acquisition of the leasehold interest by way of possessory title destroyed her interest as a statutory tenant as against the freeholder. He (his Lordship) thought not. He agreed with Mr Bromley for the freeholders that section 18 (2) of the Rent Act 1968 was inapplicable to the case of a person who had H
once been a subtenant but no longer was so at the date when the headlease determined. Nevertheless the general provision in section 3 (1) concerning conversion of protected tenancies into statutory tenancies operated in Mrs Schwartz's favour. When she acquired Mrs David's lease, the headlease continued to exist; she was then no longer a subtenant, but her new direct relationship with the freeholder was insufficient to determine her statutory tenancy. The character of her occupation had not changed, and she had in no way lost the protection of the Rent Acts. The appeal should be dismissed.

STEPHENSON LJ, concurring, said that it would be unjust J
that the Limitation Act, which was intended to protect the tenant, should operate to do her harm.

MEGAW LJ, also concurring, said that he was not altogether satisfied that section 9 (2) of the Limitation Act applied where a landlord, by his own conduct, prevented a tenant from paying rent. For the purposes of this appeal, however, he would accept the decision in *Moses v Ludgrove*.

The appeal was dismissed with costs, and leave to appeal to the House of Lords refused.

K

COURT OF APPEAL

March 17 1976

(Before Lord Justice CAIRNS, Lord Justice SCARMAN and Sir Gordon WILLMER)

ROWE v TRUELOVE

Estates Gazette February 19 1977

(1976) 241 EG 533

Possession of rent-controlled flat really required by landlord to enable him to sell property and meet his debts—Not "reasonably required" for own occupation—Question of landlord's prospects and their relevance to his claim L

This was an appeal by Lewis Denton Rowe and his wife, Barbara Olwen Rowe, from a judgment of Judge McDonnell at West London County Court on June 19 1975 dismissing their claim against the respondent, Herbert Hallworth Truelove, for possession of the second-floor flat at 55 Eaton Place, London SW1.

R Fernyhough (instructed by Clintons) appeared for the appellants, and R Pryor (instructed by Macfarlanes) represented the respondent. M

Counsel for the respondent was not called on by the court.

Giving judgment, CAIRNS LJ said: This is an appeal from a judgment of His Honour Judge McDonnell given at the West London County Court on June 19 1975 dismissing a landlord's claim for possession. The learned judge expressed at the end of his judgment his sympathy for the landlords and his reluctance to dismiss this claim for possession. I share the feelings of the learned judge. If it were possible so to construe the law and to apply the evidence as to say that the judge ought to have been satisfied that the landlords were entitled to possession, I should be very happy.

A But notwithstanding the sustained argument of Mr Fernyhough on behalf of the plaintiffs, I have regretfully come to the conclusion that the judge's decision must be upheld.

The premises of which possession was sought were a second-floor flat at 55 Eaton Place, London SW1. The leasehold of that flat had been purchased in 1970 by a company, Trenbans Investments Ltd, of which the two plaintiffs are the shareholders. The plaintiffs are husband and wife. Whether it was necessary to join the wife as a plaintiff is open to question, and if in the course of this judgment I refer to "the plaintiff," I mean by that Mr Rowe. The leasehold having been bought by the company, there was a
B sublease to the plaintiffs for eight years of the second, third and fourth floors of the house. The second floor has been conveniently called "the flat," the third and fourth floors being referred to as "the maisonette," and that is the language which I shall use. The flat consists of a sitting-room, bedroom, kitchen and bathroom. The maisonette is much more spacious, having a drawing-room, a dining-room, three bedrooms, a kitchen and two bathrooms. To start with the plaintiffs went to live in the flat while the maisonette was made habitable for them. Then towards the end of 1971 they moved with their two sons into the maisonette, and by a lease of the flat dated October 5 1971 a term of
C three years from September 29 at a rent of £850 a year was granted to the defendant. He is a Sheffield businessman who, for convenience of his business, wanted to have a *pied-à-terre* in London and took this flat on lease from the plaintiffs. At that time, in 1971, in Greater London a house or flat was protected under the Rent Act 1965 if the rateable value on March 23 1965 was less than £400. The rateable value of this flat at that date was £472, and therefore the defendant was not at that stage a protected tenant. However, his lease did not expire till September 28 1974. In the meanwhile, the Counter-Inflation Act 1973 had come into
D force, and by section 14 of that Act the critical rateable value was raised to £1,500. The rateable value of the flats had by then risen to £742, but that being less than £1,500 meant that this had become a protected tenancy.

Mr Rowe, the plaintiff, is a man who has unfortunately in recent years got into financial difficulties. He is a director of a company in the Imperial Group at a salary of over £7,000. In the early 1970s he had a mortgage on the maisonette of £11,000. He also borrowed £6,500 from his employers. Early in 1973 he was intending to make a change. He first had the idea of purchasing some land in Portugal and developing it. For that purpose he approached his bank to
E get a loan, and had the prospect of getting a loan, on the security no doubt of the deeds of the land in Portugal, of £32,000. However, that plan went off, and he entered into negotiation with a firm called London & Midland Industries Ltd with a view to being employed by that company; and it was apparently a condition of his being employed that he should take a substantial block of shares in the company. Accordingly he bought 36,000 shares at a price of 90p to 95p each, which meant that the amount he had to pay for them was about £32,000. He originally borrowed that money from his mother, but then got his advance of £32,000 from the bank, depositing the share certificates as security, and
F was able to repay his mother. So there he was landed with a debt, in addition to his previous debts, of £32,000 with the bank at a time when the rate of interest on overdrafts was high, and he became saddled with a liability to pay interest on that overdraft in a sum of over £5,000 a year. Then his prospects of a job with London & Midland Industries dissolved. Whether they decided that they did not want him, or he decided that he did not want to go there, does not appear, but he did not go on with that. He stayed with the Imperial Group. He then wanted to sell the shares. The shares dropped in value from a little under £1 each to 19p, and he sold them at that price, which meant that he got only
G some £6,000 for them. So by the end of 1974 he was in a really desperate position financially. He not only could not repay any substantial amount of the overdraft from the bank—he did repay part with the proceeds of the sale of the shares—but his other resources were such that he could not even pay the interest on the overdraft.

In those circumstances, Mr Rowe decided that he would have to try to dispose of his interest in the house in Eaton Place. What he wanted to do was to sell the whole of that interest, and obviously he was going to get a much better price if he could get vacant possession of the flat. He would have been willing to sell the flat separately to the defendant
H if a satisfactory price could be arranged, and there were some negotiations between the plaintiff and the defendant towards that end. Alternatively, the plaintiff could have sold the whole of his interest at a much better price if he could get Mr Truelove, the defendant, out of the property. Letters were written in which the plaintiff intimated that if the defendant did not want to buy the flat, he wished the defendant to leave on the expiration of his lease in September 1974. The defendant then claimed that he was a protected tenant under the Rent Act. The plaintiff disputed that and brought proceedings in the county court for possession on the basis that it was not a statutory tenancy at that stage, the lease having
J come to an end. However, before the action came on for hearing the plaintiffs amended their particulars of claim to plead that in the alternative, if the defendant was a statutory tenant, the plaintiffs claimed possession under case 8 of schedule 3 to the Rent Act 1968, which provides that when a dwelling-house is reasonably required by the landlord for occupation by, among others, himself, that is a ground upon which possession may be ordered, subject to additional requirements with regard to greater hardship and reasonableness.

The hearing of the action occupied three days before the
K learned county court judge. He reserved his judgment, and he prepared an exceedingly careful reserved judgment on the basis of which he dismissed the plaintiffs' claim. He held that the defendant was a statutory tenant, and the plaintiffs had not made out their case under case 8 of schedule 3 to the Act. By the notice of appeal the finding that the defendant was a statutory tenant was challenged of the basis that his occupation of the flat was not such as to constitute the flat his residence. The evidence on the matter was, however, clearly such as entitled the learned judge to reach the conclusion that he was occupying the flat as his residence, and very sensibly and realistically Mr Fernyhough, on behalf of
L the plaintiffs, has not pursued that part of the appeal. He has however contended that, accepting the defendant as being a statutory tenant, the case under case 8 of schedule 3 has been made out and the judge ought to have made an order for possession accordingly. The issues are set out by the learned judge in his judgment at the foot of p 7, where he said: "There are therefore four issues for my determination, namely (1) whether the defendant is entitled to a statutory tenancy; (2) whether, if so, the plaintiffs reasonably require the flat for occupation as a residence by themselves; if so, (3) whether greater hardship would be caused by making an order for possession than by refusing it; and (4) whether in
M all the circumstances of the case it would be reasonable to make an order for possession." The judge answered the first question in favour of the defendant, but would have answered the third and fourth questions in favour of the plaintiffs. So it all depended upon whether he was satisfied that the plaintiffs reasonably required the flat for occupation as a residence for themselves. Putting it shortly, the conclusion that the learned judge reached was that they did not reasonably require the flat as a residence for themselves within the meaning of the statute, but that it was required with a view to the sale of the flat along with the maisonette in order to enable the plaintiff to pay off his debts.

A I think it is convenient to read the passage beginning near the top of p 15 of the judgment of the learned judge, where he deals with the evidence as to the intention of the plaintiff about selling any part of his interest in the premises. The learned judge says:

> In October 1974 Mr Rowe sent to the defendant an undated letter saying that he had in his possession "a signed draft contract for the sale of my third- and fourth-floor maisonette and all that is required is for me to sign it; if, however, I sign the contract I shall need the second-floor flat for myself, as I shall have nowhere else to live, and in those circumstances it will not be available for sale to you." On November 12 1974 B solicitors in Worcester acting for Mr Rowe wrote to the defendant stating that as he would require the flat "for his own family requirements" they would be obliged to know how soon the defendant would be able to give up possession. On November 25 1974 Mr Rowe wrote to the bank stating that 55 Eaton Place was in the hands of agents to sell at the following prices: "3rd/4th-floor maisonette £60,000, 2nd-floor flat £40,000, rear extension £10,000, head lease £5,000, total £115,000."

Then the judge explains the reference to the rear extension, and continues:

> The head lease had belonged to Tenbans Investments Ltd, and not to Mr Rowe. Mr Rowe has stated in evidence that the flat C was not in fact on the market, and that this letter had been written "to paint as favourable a picture as possible." He agreed that a board had been displayed at the premises stating that the second-floor flat was for sale, but said that a year ago this had been taken down and replaced by a board stating that the third- and fourth-floor flat alone was for sale. According to his evidence the flat was originally offered for sale, but to quote his own words, it was "withdrawn because of today."

The learned judge interpreted that as meaning "because of these proceedings." He (Judge McDonnell) then sets out the sundry liabilities of Mr Rowe, and continues in this way:

> Mr Rowe's financial position is that he has debts amounting D in all to about £47,000 and his only hope of payment is to realise the plaintiffs' interest in 55 Eaton Place. There was no evidence as to the market value of the maisonette save that Mr Rowe had refused an offer of £45,000 at some time in 1974. The plaintiff's evidence as to his need to recover possession of the flat run thus: "I only have 55 Eaton Place to pay off these loans. I can only pay them off by realising all three floors. The third and fourth floors are on the market. If I am able to sell the third and fourth floors for £45,000 we would have to try to find rented accommodation. If I had the second floor I would live in it. If I could get sufficient from the third and fourth floors to pay my debts I would live in the second floor. Otherwise I would have to sell the second floor as well." In cross-E examination the plaintiff said that if he obtained an order for possession on the ground the the defendant was not entitled to a statutory tenancy he would sell whatever he could sell most quickly: "If someone offered to buy the third and fourth floors I would sell and move to the second floor. If someone offered to buy the second floor I would sell it."

Mr Fernyhough has read to us some further passages from the evidence of the plaintiff in the context of the two passages that the judge quoted. It does not seem to me that these carry the matter substantially further. On pp 17, 18 and 19 of the judgment there are a number of passages which Mr Fernyhough criticises, and it will be convenient to F go through those. The first is a sentence which runs as follows:

> I hold that to succeed a landlord must establish more than intention to occupy the premises merely temporarily, and must establish not merely that he intends to occupy the premises as a residence, but a reasonable prospect that he will be able and willing to retain the premises as his residence indefinitely."

As I understand, Mr Fernyhough does not question that it is necessary, in addition to having an intention to occupy as a residence, that there should be a reasonable prospect of so occupying. Mr Fernyhough asked us to interpret the expression "reasonably requires" as connoting an element of need. I do not find it necessary to go into that question. If G it does include an element of need, I am quite willing to accept that for some purpose the plaintiffs had a need for the flat; it was not, as it has been put in some of the cases, "a mere whim or fancy" that led them to want possession. What is mainly criticised in this sentence is that the learned judge was saying that the landlord had to be able and willing to retain the premises as his residence indefinitely. I should agree that if the word "indefinitely" is to be construed as meaning that if the landlord had an intention of occupying the premises for a period of, say, 10 years, one could not say that because it was a definite period therefore he had not fulfilled the requirement of case 8. But I do not H think the learned judge meant anything of the kind. What he did mean is quite clearly indicated by the following sentence, where he said:

> In my judgment if all he established is an intention to realise his interest in the premises as soon as he can and to live in the premises temporarily until he is able to do so, he is not entitled to succeed under case 8.

That is the test which the learned judge applied in the latter part of his judgment, and it is one which to my mind is a perfectly apt test. The intention of case 8 is to enable a landlord to get possession of a dwelling-house when his J purpose in doing so is not with a view to a sale of the house, but is with a view to his living there for some reasonable period, definite or indefinite. The learned judge then went on to say on p 18 of his judgment:

> Mr Fernyhough presented an attractive argument that Mr Rowe's answers in the witness box were those of a harassed man, which I entirely accept, and that if the maisonette fetched £45,000 or thereabouts (less the costs of and incidental to the sales) he would be able to pay off sufficient to reduce his total indebtedness to £3,000 or thereabouts, and that this would be a manageable figure.

K Then a little later we come to another passage which is criticised:

> The difficulty about this argument is that it depends upon £45,000 or thereabouts being obtained for the maisonette in the very near future. Mr Rowe says that he refused such an offer in 1974 because he then had nowhere else to live. Bearing in mind that his employment requires him to be absent from London for long periods it might well have been prudent for him to have sought cheaper accommodation for himself and his family elsewhere. There was no evidence as to how serious the offer was or whether the offeror was in a position to pay the whole amount in cash. The restrictions imposed by the Finance L Act 1974, section 19, schedule 1, part II, on relief from income tax in respect of loans exceeding £25,000 for the purchase or improvement of land must considerably affect the market value of properties for which sums substantially in excess of that figure are sought.

One of the grounds of appeal in the notice of appeal was that the learned judge was not entitled to found upon that view of the effect of the Finance Act. That contention has not been developed today, and for my part I think that the learned judge, first of all, is bound to take into account the provisions of any Act of Parliament which he regards as being relevant, and secondly, is entitled as a matter of M general knowledge to know what the effect of such an Act is, or is likely to be, upon prices. But what has been criticised is that the learned judge did not treat the offer of £45,000 as being any real guide to the price that the plaintiff could have obtained for the maisonette in 1975. It is said that the offer that was referred to by the plaintiff in a letter was one as to which there was no suggestion that it was not a serious offer, and that the plaintiff was not cross-examined to indicate that it was not. But the evidence as to this offer was meagre in the extreme. It was merely that the plaintiff said that he had had the offer. No evidence was put before the court in

A the shape of a draft contract. (I should have said that the plaintiff said there had been a draft contract), and certainly no evidence was put before the court that the person who made the offer was in a position to pay the whole amount in cash. As to that the learned judge goes on at the foot of p 18:

> There was no evidence whatever before me as to the sort of price which might be expected to be realised on a quick sale of the maisonette in 1975.

That, indeed, was so: there was no evidence as to the price that would be raised at that time, and in the only way which would be useful to the plaintiff, for a quick sale. B Then the judge says:

> Unless there is a real prospect of selling the maisonette for more than £35,000 the argument breaks down. I am not satisfied that there is a real prospect.

I think probably that £35,000 is a slip or an error in transcription, and it ought to be £45,000. But whichever it may be, the judge, I think, on the evidence before him, was entitled to make the findings which he did in his next sentence: "I am not satisfied that there is a real prospect." Then he continued:

> Furthermore, I am of the opinion that a man in the financial C circumstances of Mr Rowe cannot really afford to live in the flat.

What I think Mr Fernyhough is justified in saying about that sentence is this: that if indeed the sale of the maisonette for something in the neighbourhood of £45,000 had been possible, then it could not be said that it was established that Mr Rowe could not reasonably afford to live in the flat. That may be so. But I do not regard this sentence about his not being able to afford to live in the flat as being really the basis of the final decision of the learned judge. That basis has already been provided in the previous sentence when he found that there was not a real prospect of making D a sale: and indeed, Mr Fernyhough does reasonably concede that unless we are in his favour in regard to the finding that there was not a real prospect of selling at that price, then he cannot succeed on the basis of this sentence dealing with the ability to live in the flat. Then the learned judge said:

> I have therefore come to the conclusion that although the plaintiffs reasonably and indeed desperately require the flat, they require it to relieve the serious financial difficulties of Mr Rowe, and I cannot find that it is reasonably required by them for occupation as a residence for themselves.

That is a finding of fact which, to my mind, it is quite impossible for this court to say that the learned judge was E not entitled to arrive at. It is said by Mr Fernyhough that the judge wrongly took into account the element of time in relation to the residence in the flat. That, I think, in the circumstances of this case, was an element which it was proper to take into account. If the learned judge was right in saying that the aim was to get possession of the flat with a view to residing there only up to the time when a sale of it could be brought about, then I do not think that that amounts to requiring the flat for residence within the meaning of the case in the schedule. I think it right to read the next passage in the learned judge's judgment. He says:

> I have reached this conclusion with great reluctance, since the F result of this case must be that the defendant, who can well afford to find suitable alternative accommodation and pay for it from his own free resources, is enabled to retain possession of a flat let to him on a tenancy which was unprotected at its inception, and that the plaintiffs, who let that flat on an unprotected tenancy, are deprived of the opportunity of relieving their desperate financial need by the intervention of Parliament during the currency of the lease. However, on the evidence before me no other conclusion is possible, and therefore this action must fail.

As I have already indicated, with equal feelings of reluctance I have reached the conclusion that this appeal must fail.

SCARMAN LJ: I agree with the judgment of my Lord and G do not wish to add anything.

SIR GORDON WILLMER: I also agree.

The appeal was dismissed with costs.

COURT OF APPEAL

November 1 1976

(Before Lord DENNING MR, Lord Justice ROSKILL and Lord Justice LAWTON)

HANSON v CHURCH COMMISSIONERS FOR ENGLAND AND LONDON RENT ASSESSMENT H PANEL

Estates Gazette March 5 1977

(1976) 241 EG 683

Rent Act 1968—Objection to registration of fair rent may be withdrawn if other party agrees and rent assessment committee sanctions withdrawal—Divisional Court's views on withdrawals not accepted—Natural justice—Position misunderstood by tenant—Certiorari issued

This was an appeal by John Hanson, the tenant of 21 Bramerton Street, Chelsea, London SW3, a house owned by the Church Commissioners, from a decision of the J Queen's Bench Divisonal Court on November 11 1975 dismissing his appeal under the Tribunals and Inquiries Act 1971 from the decision of a rent assessment committee and refusing his application for an order of certiorari to quash the committee's decision on the ground of a breach of the rules of natural justice.

Mr Hanson appeared in person, R Moshi (instructed by Radcliffes & Co) appeared for the first respondents, the Church Commissioners, and H K Woolf (instructed by the Treasury Solicitor) represented the second respondents, the London Rent Assessment Panel. K

Giving judgment, LORD DENNING MR said that Mr Hanson had rented his house from the landlords for the past 10 years. The rent in 1973 was £600 a year. The landlords applied to the rent officer for the registration of a fair rent of £900. Mr Hanson thought that was too high and in March 1974 the rent officer determined the fair rent to be £800. Mr Hanson objected and the rent officer, in accordance with his duty under paragraph 6 of Schedule 6 to the Rent Act 1968, referred the matter to the rent assessment committee. The committee notified both parties that the objection would be heard on August 15 1974. Meanwhile surveyors acting for Mr Hanson advised him that he L was unlikely to succeed in his objection, and although he still thought £800 was too high, he accepted their advice and said that he did not want them to take his objection any further. On August 8, the surveyors telephoned the clerk to the rent assessment committee. There was some conflict of evidence as to what took place, but as a result a letter was written the same day by Mr Hanson's surveyors referring to the telephone conversation and saying: "we confirm that we shall not be placing our objection before the London Rent Assessment Panel on Thursday August 15." The landlords were told nothing about the letter and the case was left in the list for hearing. Accordingly the com- M mittee and the landlords' representative attended on August 15, but Mr Hanson did not. The landlords' representative said that at the consultation with the rent officer he had accepted the £800 proposed, but that as he was before the committee he would like them to increase the fair rent to £900. In the afternoon the committee went to the house. Mr Hanson was at work but his wife was at home. The committee chairman telephoned Mr Hanson from the house and asked whether the committee could carry out an inspection. Mr Hanson replied that he understood that his objection had been withdrawn and that the hearing would not proceed. He could not allow an inspection in his absence. He asked if the committee could wait. The chairman said that the committee could not wait any longer, that the hearing had already taken place and that the purpose of the visit was a purely visual inspection. The committee subsequently notified Mr Hanson of their decision that the fair rent was £900 a year.

Mr Hanson appealed to the Divisional Court under the Tribunals and Inquiries Act 1971 and later applied for an order of certiorari. The Divisional Court dismissed the statutory appeal and also refused the application for certiorari on the ground that the application was made far too late. In relation to this ground he (his Lordship) wished to say that fresh evidence had shown that Mr Hanson had not in fact been guilty of any delay and the court would consider the application on its merits.

After dismissing as ill-founded a suggestion in the notice of motion that the Lord Chief Justice, being a Church Commissioner, should have disclosed his position before he heard the case, Lord Denning said that both the appeal and the application seemed to raise the same point, so the court did not need to go into any question of jurisdiction. The first point was whether Mr Hanson had any right to withdraw his objection. The court's attention had been drawn to textbooks in which it was suggested that neither party had any right to withdraw. That view had been accepted by the Divisional Court, but it went much too far. In fact in practice a limited right of withdrawal had always been given before the Divisional Court's decision. If an objecting tenant or landlord notified the rent assessment committee that he was withdrawing his objection the committee would get in touch with the other party and, if that party had no objection, the committee would allow the withdrawal. That practice was not only convenient but also in accordance with the true interpretation of the statute. In the ordinary way when there was a dispute before a tribunal on a civil matter either party had a right to withdraw his application or objection at any time before the decision was given: *Boal Quay Wharfingers Ltd v King's Lynn Conservancy Board* [1971] 1 WLR 1558. But when a dispute involved a public interest it might not be permissible for one party to withdraw on his own. Even if the other side agreed it might not be possible to withdraw unless the tribunal consented: *R v Hampstead & St Pancras Rent Tribunal, ex parte Goodman* [1951] 1 KB 541 and *R v West London Rent Tribunal, ex parte Napper* [1967] 1 QB 169. Applying that principle to this case, the fixing of a fair rent was a matter involving a public interest. It affected the premises *in rem* and could also affect neighbouring properties. No one party had the right on his own to withdraw an objection and stop the proceedings. If the other party objected he could not withdraw. But if the other party did not object the rent assessment committee could permit the proceedings to be withdrawn. It did not need to go on with the hearing unless it thought it proper to do so.

As to the question of natural justice, there were errors which vitiated the committee's decision. On receiving the letter from Mr Hanson's surveyors the clerk ought to have regarded it as a withdrawal, or, at any rate, he ought to have cleared up any ambiguity by telephoning Mr Hanson's surveyors. Then, when the chairman telephoned Mr Hanson and was told that Mr Hanson had understood that his objection had been withdrawn, the chairman should have realised that there had been a misunderstanding and given Mr Hanson the opportunity of being heard. It was one of the cardinal principles of natural justice that a matter should not be decided against a person unless he had had a fair warning of the case against him and a fair opportunity of dealing with it. That principle had been broken. There had been some discussion as to the proper remedy. It seemed to him (his Lordship) that the better remedy would be to issue certiorari to quash the decision.

Agreeing, ROSKILL LJ said that there could be no unfettered unilateral right to withdraw an objection referred to a rent assessment committee, but there was nothing which prevented the parties from agreeing to withdraw, provided that the withdrawal was sanctioned by the committee. It followed that in this case there had been no effective withdrawal. The Divisional Court had not dealt with the question whether the committee had acted contrary to natural justice, ie whether what they had done had operated unfairly to Mr Hanson. The letter from Mr Hanson's surveyors was far from clear, but the chairman must have realised from his telephone conversation with Mr Hanson that he thought he had effectively withdrawn the objection and did not know that the landlords were seeking an increase and that there was a possibility that the committee might grant it. He ought to have been told what the position was, asked if he wanted to be heard and been offered an adjournment. Certiorari should go to quash the determination.

Also concurring, LAWTON LJ said that, in the absence of a specific statutory provision, proceedings before a tribunal could be brought to an end if to do so would not prejudice any right, either public or private. It was for the tribunal to decide whether prejudice might arise from a withdrawal, and in general the simplest and best way would be by asking the parties.

The statutory appeal was dismissed, but certiorari was ordered to issue to quash the decision of the committee, the matter to be remitted to another committee to hear and determine the objection. No order for costs, provided that the appellant obtained his costs from the Legal Aid Fund.

COURT OF APPEAL

November 26 1976

(Before Lord Justice MEGAW, Lord Justice STEPHENSON and Lord Justice BROWNE)

KENNEALY AND ANOTHER v DUNNE AND ANOTHER

Estates Gazette May 21 1977

(1976) 242 EG 623

Rent Act 1968, Case 10 of Schedule 3—Question of landlords' (previous owner-occupiers') claim for possession of house let on regulated tenancy—Meaning of "required" in Case 10—Genuine desire and immediate intention implied, but no need to establish reasonableness of requirement—Landlords' appeal allowed

This was an appeal by George Kennealy and his wife, Catherine Bridget Kennealy, of St Ann's Court, Hayling Island, from the dismissal by Judge Honig at Willesden County Court on January 21 1976 of their claim for possession of a furnished flat in Thorpebank Road, Hammersmith, London W12, on the ground that he was not satisfied that they "required" the flat as their residence within the meaning of Case 10 of Schedule 3 to the Rent Act 1968. The respondent tenants were John Dunne and his wife, Mary.

Andrew Hillier (instructed by Barnett & Co) represented the appellants, and Miss Suzanne Shenton (instructed by Peter Ellis & Co) represented the respondents.

Giving the first judgment, STEPHENSON LJ said that the case concerned Case 10 in Part II of Schedule 3 to the Rent Act 1968, dealing with cases in which a court must order possession where a dwelling-house was subject to a regulated tenancy. Case 10 applied:

Where a person who occupied the dwelling-house as his residence (in this case referred to as the "owner-occupier") let it on a regulated tenancy and (a) not later than the relevant date the landlord gave notice in writing to the tenant that possession might be recovered under this Case, and (b) the dwelling-house has not, since December 8 1965, been let by the owner-occupier on a protected tenancy with respect to which the condition mentioned in paragraph (a) above was not satisfied, and (c) the court is satisfied that the dwelling-house is required as a residence for the owner-occupier or any member of his family who resided with the owner-occupier when he last occupied the dwelling-house as a residence.

There was no dispute that Mr and Mrs Kennealy had occupied the flat as their residence, that they had let it on a regulated tenancy to Mr and Mrs Dunne, and that paragraphs (a) and (b) of Case 10 were satisfied. The only dispute related to paragraph (c). The county court judge was not satisfied that the flat was "required" as a residence for Mr and Mrs Kennealy themselves or their family. The judge had said that he was entitled to interpret "required" in its ordinary dictionary meaning. It meant more than "wants" and bona fide "intention." There was something objective as well. Mr and Mrs Kennealy had a residence in Hayling Island. They were also entitled to use a flat which adjoined the disputed dwelling-house and which was occupied only by their daughter, and in his view they did not require the dwelling-house as a residence. Miss Shenton, for the tenants, had conceded that if the judge's construction of "required" was wrong the appeal must be allowed. Mr Hillier submitted that the judge had wrongly held that for the owner of a dwelling-house to establish that he required it for his own use as a residence within the meaning of Case 10, it was insufficient for him to establish merely that he wanted, and bona fide intended, to occupy it.

Case 10 was the first of the cases in Part II of Schedule 3 where the court *must* order possession where a house was required as a residence. That distinguished it from Case 8, in which the court *might* order possession where the dwelling-house was reasonably required by the landlord for occupation as a residence. It was an irresistible inference on the face of Cases 8 and 10 that under Case 8 the dwelling-house must be "reasonably required" by the landlord, but that he could bring himself within Case 10 without the dwelling-house being reasonably required, and the words "reasonably required" had been interpreted in a way not wholly unfavourable to the landlord. Mr Hillier had submitted that "required," taken out of context, could mean anything from "desired" to "indispensable," and he had relied on *G C & E Nuthall (1917) Ltd v Entertainments & General Investment Corporation Ltd* [1947] 2 All ER 384, a case under the Landlord and Tenant Act 1927. "Reasonably required" had been interpreted as connoting more than desire but something much less than absolute necessity in the case of *Aitken v Shaw* 1933 SLT 21, a Scottish sheriff court decision. Care had to be taken in applying the interpretation of "require" in other Acts to the present Act, but he (his Lordship) did get some help from two other cases cited by Mr Hillier. In *Armstrong v Vallance* (1951) 52 WALR 90, an Australian decision, the subject-matter was fairly near the subject-matter in the present case; and in *Ireland v Taylor* [1949] 1 KB 300 the court was considering the same provisions of the Landlord and Tenant Act 1927 as in *Nuthall's* case. The judge had rightly regarded himself as not bound by the interpretation put upon the words of the 1927 Act because it had a different purpose from the 1968 Act. But in *Ireland's* case, Somervell LJ, in a passage which the judge had quoted, said that "requires . . . in its present context . . . is satisfied if a landlord establishes . . . that he wants and intends to occupy the premises."

He (his Lordship) had come to the conclusion that "required" in Case 10 did not mean "reasonably required." It meant no more than bona fide wanting and genuinely intending to occupy as a residence within a reasonable time, but so wanting or intending whether reasonably or unreasonably, even from the landlord's point of view. There was force in Mr Hillier's submission that the purpose of Case 10 was to enable a landlord who was living in his own house to be free to take up a post in another part of the country or abroad and let his home secure in the knowledge that when he returned he could reoccupy his home on giving proper notice without the difficulties which a landlord under Case 8 had to overcome. If that were so, it explained the apparently deliberate omission of the qualifying word "reasonably." Under Case 10 there must be nothing more than a genuine requirement and a present intention. If that were shown, the landlord must have possession. The appeal should be allowed.

BROWNE LJ agreed.

MEGAW LJ, also agreeing, said that the legislature had clearly deliberately omitted "reasonably" in Case 10 and had deliberately inserted it in Case 8. There might be good policy reasons for drawing the distinction. Whether that be so or not the result was that in order to comply with Case 10(c) all that the person who had previously occupied the dwelling-house and now desired to occupy it again had to prove was that he desired to occupy it and intended so to do. The desire and the intention must be genuine but no more than that.

The appeal was allowed and possession ordered within eight weeks.

QUEEN'S BENCH DIVISIONAL COURT

July 28 1976

(Before Lord WIDGERY CJ, Mr Justice MELFORD STEVENSON and Mr Justice CAULFIELD)

R v CROYDON AND SOUTH WEST LONDON RENT TRIBUNAL, EX PARTE RYZEWSKA

Estates Gazette June 11 1977

(1976) 242 EG 879

Rent Act 1968—Tenant's application to fix rent under Part VI—Landlords claim holiday letting—Tribunal decided they had jurisdiction—Discretion of tribunals to decide preliminary question of jurisdiction—Principles to be observed by tribunals—Certiorari refused

In these proceedings Helena Maria Ryzewska moved for an order of certiorari to quash a decision of the Croydon and South West London Rent Tribunal dated January 22 1976 by which they purported to fix a fair rent, on the tenant's application, for a second-floor flat at premises owned by her at 30 Mulbrook Road, Putney, London.

Roger Bartlett (instructed by Davies, Topping & Watkins) appeared for the applicant; Guy Mansfield (instructed by the Treasury Solicitor) represented the respondent tribunal.

Giving the first judgment, CAULFIELD J said that the applicant owned a house in Putney which was divided into a number of flats. A Danish businessman, who was working in this country, entered into an agreement with the applicant to occupy one of the flats for 188 days at a rental of £3.50 per day. Endorsed on top of the agreement were the words "Short (working holiday) let." Subsequently the tenant applied to the tribunal to determine a reasonable rent under Part VI of the Rent Act 1968.

Once the tenant had applied to the tribunal the applicant raised the question of the tribunal's jurisdiction and claimed that the letting, as a holiday letting, was outside Part VI of the Act and therefore the tribunal had no jurisdiction. Before the tribunal proceeded to determine the rent it was at pains to offer both to the applicant and the tenant the opportunity of proceeding to the county court under section 105 of the Act to decide the jurisdiction point. The applicant's solicitor indicated that it was for the tenant to go to the county court. The tenant merely queried the cost of going there and this query was treated by the tribunal as a refusal to accept the invitation to apply to the court. The tribunal had therefore given both parties the opportunity of having the question of the tribunal's jurisdiction decided in the county court. The invitation having been declined by both parties, the tribunal had then gone on to decide the rent.

In his (his Lordship's) judgment every inferior tribunal, before determining the main issue for its consideration, had always to ask itself the preliminary question as to whether it had jurisdiction. The principles were fully explained in *R v London etc Rent Tribunal, ex parte Honig* [1951] 1 KB 641 and *R v Fulham, Hammersmith and Kensington Rent Tribunal, ex parte Zerek* [1951] 2 KB 1. The difficulty in the present case was that section 105 of the Rent Act 1968 provided in the form of the county court an alternative forum for deciding jurisdiction in certain instances. Once the question of jurisdiction was raised it was the duty of the tribunal to use its discretion in a certain way. A tribunal should look at all the circumstances. It should not deny a party the chance to test the jurisdiction point under section 105. In the present case the tribunal had in fact given both parties an opportunity of proceeding under section 105, an opportunity which was not taken. It had, in his (his Lordship's) judgment, proceeded strictly according to the principles which should be observed by a tribunal in a situation such as this.

What a tribunal should do where there is an alternative forum, as where section 105 applies, was to listen to the contentions of the parties, and to make up its own mind whether it is better to proceed with the determination as to rent, having first of all decided the preliminary question of jurisdiction. Even if it decided that issue wrongly the person aggrieved was not precluded from testing the decision in the county court under section 105. Nothing in the present judgment was in conflict with anything said by Lord Denning MR in *R v Westminster (City) London Borough Rent Officer, ex parte Rendall* [1973] QB 959 or with the judgment delivered by Park J in *R v Brent London Rent Officer, ex parte Ganatra* [1976] QB 576. He was of opinion that the application should be dismissed.

Melford Stevenson J agreed.

Lord Widgery CJ said that the new point brought out in this case was that jurisdiction under section 105 was exercisable either before or after consideration of the matter by the tribunal. That had the important consequence that a tribunal was entitled to decide the jurisdiction point if it thought that this was the right course in the particular circumstances and leave it to the aggrieved party to apply under section 105 to the county court if need be. The application should be refused.

The application was dismissed, no order being made as to costs.

QUEEN'S BENCH DIVISIONAL COURT

November 4 1976

(Before Lord WIDGERY CJ, Mr Justice EVELEIGH and Mr Justice SLYNN)

R v RENT OFFICER FOR KENSINGTON AND CHELSEA, EX PARTE NOEL

Estates Gazette June 11 1977

(1976) 242 EG 879

Fair rent—Motion for order of certiorari to quash rent officer's registration—Rent officer's jurisdiction challenged before hearing—Issue as to whether there was a tenancy or a licence—Rent officer's discretion to decide jurisdiction—Decision, however, based on submissions by parties' solicitors, not on evidence—Application for certiorari granted accordingly to quash rent officer's determination

In these proceedings Mrs Victoria Mary Noel moved for an order of certiorari to quash a decision of the rent officer of the London Borough of Kensington and Chelsea on January 16 1975 whereby he determined a fair rent for a three-room ground-floor flat at 34a Addison Gardens, London W14, at £70 per month exclusive of rates. She claimed that the rent officer had no jurisdiction because at the material time the person upon whose application the rent officer purported to act was not a tenant but a licensee.

Patrick Medd QC and Norman Banks (instructed by Arnold Fooks Chadwick & Co) appeared for the applicant; Harry Woolf (instructed by the Treasury Solicitor) represented the rent officer.

Giving the first judgment, Slynn J said that in November 1974 Miss Hilary Lowday applied to the rent officer for the registration of a fair rent of the premises in Addison Gardens occupied by her under the terms of an agreement made with the then owner, Arthur Younghusband. The rent officer notified Mr Younghusband of the application and on December 5 1974 solicitors acting for him wrote taking two points, only one of which was material, namely that from the terms of the document "misdescribed" as a tenancy agreement it appeared Miss Lowday was merely a licensee of the premises. That being the case, claimed Mr Younghusband's solicitors, there was no protected tenancy and the rent officer had no jurisdiction. They also stated that a High Court writ was to be taken out in the Chancery Division claiming a declaration that Miss Lowday did not have a tenancy or any other right or interest in the premises. On January 6 1975 the rent officer prepared a memorandum in which he set out what appeared to him to be the questions to be resolved. In that memorandum he described the agreement as an "artificial cloak" and went on to state that he was of the opinion that a tenancy had been created in favour of Miss Lowday. He then communicated with the parties and informed them he would discuss the matter with them on January 16. On that day solicitors for the owner and Miss Lowday attended and made submissions. At the proceedings solicitors acting for the owner served Miss Lowday with a High Court writ. The rent officer, after hearing the submissions, decided he had jurisdiction and he then went on to decide on the evidence before him that there was a tenancy and then purported to fix a fair rent.

Some time after the meeting before the rent officer Miss Lowday moved from London and the Chancery proceedings were abandoned. The property eventually changed hands and the present applicant now sought to quash the determination of the fair rent. Proceedings before a rent assessment committee had been adjourned pending the outcome of the present application to the court. It was against that background that Mr Medd now contended that the rent officer knew from an early stage that his jurisdiction was disputed. Accordingly, claimed Mr Medd, he should not have proceeded to determine a fair rent but should have adjourned so that a decision on the question whether a tenancy existed could be reached in the Chancery Division proceedings. Alternatively, if the rent officer had jurisdiction he had not acted in a judicial manner in that he had prejudged the issue; did not appreciate the difference between tenancy and licence; and should have dealt with the matter by hearing evidence as a judge would have done at a trial.

Mr Medd was not contending that the respondent did not have the power to decide on jurisdiction. He was right not to go so far. A rent officer was entitled, and might be bound, to decide in the first instance whether he had jurisdiction if the question arose. If the matter was such that he considered it more appropriate for a court to determine, he was entitled to invite the parties to take the issue to the county court. If he determined the question of jurisdiction himself the High Court could review his decision.

Section 105 of the Rent Act 1968 did not automatically require that any question of whether a tenancy was protected or not had to be referred to the county court. In a number

A of cases reference had been made to section 105 and this court had been told that it was apparently considered that, because of one case in particular, if an issue on jurisdiction arose the matter had to go to the county court. The court had been referred to the case of *R v Brent London Borough Rent Officer, ex parte Ganatra* [1976] 1 QB 576, but that case was not to be construed as saying that when there was a question falling to be decided under section 105 of the Act the rent officer *must* leave the issue to be decided by the county court.

In the case of *R v Croydon and South West London Rent Tribunal, ex parte Ryzewska* [1977] 1 All ER 312 at p 314 B Caulfield J said "the rent tribunal proceeded strictly according to the principles which should be observed by a tribunal in a situation such as this . . . to listen to the contentions of the parties and make up their own minds there and then whether it is better for them to proceed with the determination having first of all decided the preliminary point of jurisdiction." Rent officers could also gain assistance from *R v Fulham, Hammersmith and Kensington Rent Tribunal ex parte Zerek* [1951] 2 KB 1.

The basis of all these authorities was that a rent officer had a right to consider jurisdictional facts and that if he goes on to do so his decision will stand subject to an application C to this court. In the present case the rent officer was entitled to conclude on the material before him that it was proper for him to decide that he had jurisdiction to determine whether there was a tenancy at all. The court could not interfere with his decision simply because he had decided to go ahead and hear the application, but the court had authority to supervise the way in which a rent officer arrived at a decision and the way in which he conducted the proceedings. He (his Lordship) would not assume from the memorandum prepared by the rent officer in January 1965 that he had prejudged the issue before the meeting on January 16. Nor did what he wrote in the memorandum show he had failed to D understand the distinction between a tenancy and a licence.

On the other hand, when one considered what happened at the meeting on January 16 the position at the end of the day was not satisfactory. It was impossible to resolve an issue of the kind raised in the case merely on the basis of submissions and arguments put before a rent officer by the solicitors for the parties. The allegation was that the document relied upon by Miss Lowday as a tenancy agreement was a "sham" or "cloak" and a device to avoid the effects of the Rent Acts. That was a matter which could only be properly dealt with by the hearing of oral evidence and, if necessary, E by cross-examination. The rent officer was not, because of the procedure adopted, in a position to come to a determination on the issue of whether there was a tenancy in the way that he did. Accordingly the court should interfere and quash the decision.

Eveleigh J agreed.

Lord Widgery CJ, also agreeing, said that the county court was the appropriate court to decide these matters when the rent officer found them too difficult or was ordered to take them to court. Nowadays with universal legal aid there was a tendency to bring matters to the High Court which were really more appropriate to the county court under section F 105 of the Rent Act 1968. The court might have to take a stricter view about granting prerogative orders if there were too many applications which were really appropriate to the county court.

The application for certiorari was allowed with costs.

COURT OF APPEAL G

May 26 1976

(Before Lord Justice CAIRNS, Lord Justice ROSKILL and Sir JOHN PENNYCUICK)

SMITH v McGOLDRICK

Estates Gazette June 25 1977

(1976) 242 EG 1047

Action for possession—Regulated tenancy—Possession required by landlord for occupation as a residence for himself—"Admission" by tenant that she had no defence —Tenant under mistaken impression that she would be rehoused by council—County court judge refused leave to withdraw admission—Judge in error—Rent Act case—Reasonableness to be decided at date of hearing—Retrial ordered H

This was an appeal by the defendant in a possession action against the county court judge's order for possession. The defendant, Eileen Maud McGoldrick, was the tenant of a house at 7 Tollgate Road, Dorking, let to her by the plaintiff, Peter Philip Smith. The facts are set out in the judgment of Cairns LJ.

N J Forwood (instructed by Tuck & Mann, of Dorking) appeared on behalf of the appellant; R Rundell (instructed by Hart, Scales & Hodges, of Dorking) represented the respondent. J

Giving judgment, Cairns LJ said: This is an appeal from a decision of His Honour Judge Potter given in the Reigate County Court making an order for possession of a house. The defendant appeals and asks for a new trial.

The circumstances are very unusual and unfortunate. The defendant was the tenant of the plaintiff in a house, 7 Tollgate Road, Dorking, on a furnished tenancy from July 17 1973 to January 16 1974. It was a furnished tenancy at a monthly rent of £60. After the expiration of that tenancy on January 16 1974, the defendant continued in possession in K pursuance of a letter to her from the plaintiff dated December 30 1973 saying that he was prepared for her to continue for an indefinite period on a monthly basis. She did so continue. In due course notice to quit was served upon her. There is no question that it was a proper notice to quit served on May 30 1975 requiring possession on July 16 1976.

Admittedly this was a regulated tenancy under the Rent Act 1968 as amended by the Rent Act 1974. In due course the plaintiff started the action for possession, and for some reason it was set out in paragraph 2 of the particulars of claim that the plaintiff was an owner-occupier within the meaning of section 79 of the Rent Act 1968. That allegation L would only be relevant if possession were sought under Case 10 of Part II of the Third Schedule to the Act of 1968. The possession was not sought under that case, because paragraph 7 of the particulars of claim says expressly: "The plaintiff requires occupation of the said property under Case 8 of the Third Schedule to the Rent Act 1968." Case 8 is a case "Where the dwelling-house is reasonably required by the landlord for occupation as a residence for—(a) himself"—and I need not read paragraphs (b), (c) and (d) or the tailpiece of the section referring to the date when the landlord became landlord by purchase. I should refer to Part III of the schedule, which in reference to Case 8 provides M that the court shall not make an order for possession under that case if the court is satisfied that, having regard to the circumstances of the case, including the question whether other accommodation is available for the landlord or the tenant, greater hardship would be caused by granting the order than by refusing to grant it. I should refer also to section 10 of the Act of 1968, which is the section of the Act dealing with general grounds for possession, and providing under subsection (1) that an order for possession shall not be made unless the court considers it reasonable to make such an order.

Those particulars of claim having been served on September 12 1975 the defendant's solicitors wrote to the registrar of the county court a letter in these terms: "We have now taken our client's instructions. Our client does not have a defence to this action and will you kindly accept this letter as an admission of the claim. It is to be hoped that the court will give her not less than 28 days from October 7. You will appreciate that there are children involved here. We are sending a copy of this letter to the plaintiff's solicitors." And so they did. We are told that at the time when the defendant gave the instructions to her solicitors which led to that letter being written she was expecting to get alternative accommodation in a council house. Afterwards she found that she was not going to get that accommodation and, therefore, wished to defend the proceedings. Accordingly her solicitors on October 3, just four days before the return date, told the plaintiff's solicitors that the defendant had now changed her mind, and that message was confirmed by a letter written on October 6—that was a Monday—requesting an adjournment and enclosing a draft defence. The draft defence consisted mainly of admissions. In the first three paragraphs of it the only denial was a denial of the plaintiff being an owner-occupier, which was irrelevant to the claim under Case 8. But in paragraph 4 of the draft defence it was sought to be alleged: "The defendant denies the plaintiff is entitled to possession of the said dwelling-house as alleged and claims the protection of the Rent Act 1974." At the hearing counsel then appearing for the defendant asked for leave to file the defence out of time and to withdraw the admission contained in the letter of September 12. The learned judge, having heard arguments on both sides on that matter, rejected the application and proceeded to consider the case on the basis of the admission that had been made, and on the basis of that admission held that the plaintiff was entitled to his order for possession.

In considering whether the judge was right in taking that course it is of the utmost importance to remember that this was a case relating to a tenancy protected under the Rent Acts. Those Acts provide that where there is such a tenancy an order for possession can be made only in certain cases, and subject, as I have already indicated, to the overriding considerations in relation to the relevant case here of greater hardship and of reasonableness. It is quite clear that those questions have to be considered as at the date of the hearing, not at any earlier date. Putting it at its highest, it does not appear to me that the letter of September 12 could constitute such an admission as enabled the landlord without more to discharge the onus on him under the Rent Act in relation to the matter of reasonableness. It was impossible on September 12 to admit any more than that in the circumstances existing at that time it would be reasonable that an order should be made and that it would not then cause greater hardship to the defendant than any refusal of the order would cause to the plaintiff.

I, therefore, with great respect to this experienced and careful judge, am not able to accept the proposition which appears towards the end of his judgment in these terms: "As the letter of admission requested 28 days and as the plaintiff has offered twice that time, then that must be reasonable. It is like two advocates agreeing an order." In my view, it was necessary for the learned judge to consider at the time of the hearing before him, whatever may have been the situation three weeks or so earlier, whether at that time it was reasonable to make the order. It was very natural for the learned judge to take the view that, this letter having been written on behalf of the defendant, it was something which should be regarded as binding on her; and I can see that it might be said that the withdrawal of the admission would be prejudicial to the plaintiff inasmuch as the time for obtaining possession of the house, if he was entitled to get possession, would be deferred. But, for the reasons that I have given, I am of opinion that it was the duty of the judge to consider this matter afresh as at the date of the hearing, taking into account the letter that had been written on September 12 and taking into account also any fresh evidence that might be given by the defendant as to the circumstances at that time.

I am, therefore, of opinion that this appeal should be allowed and that the matter should go back to the county court for a rehearing, and I think it would be appropriate in all the circumstances of the case that the defendant should now have leave, if so advised, to file a fresh defence out of time. The matter should, in my view, be considered at the earliest possible date, and, without any sort of reflection on the learned judge who heard it the first time, it should go before a different judge.

ROSKILL LJ: I entirely agree. With great respect to a most experienced county court judge, I think on this occasion he went wrong. I think he went wrong because perhaps momentarily he overlooked what my Lord has stressed in his judgment, that it was his duty to consider the merits of the Case 8 claim in the light of the facts as they were proved at the trial. He would have been entitled to take into account the fact that the admission letter had been written earlier, but that fact of itself did not without more, in my judgment, in a Rent Act case, entitle the plaintiff to proceed to judgment for possession. It seems to me, therefore, that the judge was wrong in refusing leave to withdraw this admission. He should have given leave, and he should then have gone into the matter on the evidence, with an adjournment (if necessary)—and he obviously had that in mind—if he thought that that was required to do justice to both parties.

I see no escape from the fact, as my Lord has suggested, that there will have to be a new trial, and I respectfully agree that it should go back to the county court for a rehearing, and, without any disrespect to Judge Potter, before a different judge, the defendant having leave, if she so wishes, to file a defence. I think that should be done quickly, and I think the retrial should come on as a matter of urgency, because this case has hung about far too long. I would only add in conclusion my gratitude to Mr Forwood, who found himself in an unenviable position this morning with a brief in his hand in a case of which he had no knowledge, and if he will allow me to say so, I do not measure his skill by the fact that he has succeeded, but by the way he dealt with a difficult situation.

SIR JOHN PENNYCUICK: I agree with both judgments which have been delivered and do not wish to add anything.

The court expressed the view that the defence should be delivered within seven days. The costs of the appeal were ordered to be costs in the new hearing. No order had been made as to the costs of the original hearing and this position was not disturbed by the Court of Appeal.

LANDLORD AND TENANT

RENT REVIEW

A **CHANCERY DIVISION**

June 23 1976

(Before Mr Justice GRAHAM)

DEAN AND CHAPTER OF CHICHESTER v LENNARDS LTD

Estates Gazette January 29 1977

(1976) 241 EG 311

Rent review clause—Provision requiring statement of suggested new rent in landlord's notice held of the essence

This was a claim by the Dean and Chapter of the Cathedral Church of the Holy Trinity in Chichester, as trustees of
B the Custos and Poor of the Hospital of the Blessed Mary, a charity commonly called St Mary's Hospital, against Lennards Ltd, of Staple Hill, Bristol, a subsidiary of Great Universal Stores, for a declaration that a notice given by the plaintiffs to the defendants in accordance with the terms of a rent review clause in a lease of shop premises at the junction of East Street and North Street, Chichester, was not invalidated by the omission therefrom of any suggested figure as the new rent payable in respect of the property.

W Blum (instructed by Thomas Eggar & Sons) appeared for the plaintiffs, and J Harper (instructed by Paisner & Co) represented the defendants.

C Giving judgment, GRAHAM J said: This is another case relating to a rent review clause. Similar cases recently reported have all been mainly concerned with the question whether the times specified for giving a notice of intended increase and for operating the machinery for ascertaining the amount of such increase are of the essence of the contract. The present case is concerned not with time but with the form and substance of the notice itself. In particular it is concerned with the question whether in the circumstances a requirement in the terms of clause 4 (1) of the lease that such notice should state "the suggested new rent to be reserved" calls for the amount of such rent to be specified
D and is of the essence of the contract or not.

The lease in question is dated October 12 1965 and is between the landlords, the Custos and Poor of the Hospital of the Blessed Mary, a charity (commonly called St Mary's Hospital) in Chichester and the Dean and Chapter of the Cathedral Church of the Holy Trinity in Chichester, who are the trustees of the charity of the one part, and Lennards Ltd, which is a subsidiary of Great Universal Stores Ltd, of the other part. The lease relates to what was at that date a newly-built corner shop at the junction of East Street and North Street in Chichester. The term granted was for 21 years at a rent of £1,500 per annum and a down-payment
E of a premium of £1,200. Clause 2 (viii) contains a covenant not to assign without previous written consent, such consent not unreasonably to be withheld, and clause 2 (x) contains a covenant not to carry on any business at the premises other than that of retailers and repairers of footwear and similar goods without the consent of the landlords, but again such consent is not to be unreasonably withheld. Clause 4 contains the review provisions and reads as follows:

4. Provided always and it is hereby agreed as follows: (i) That after expiration of the 14th year of the term hereby created the landlords shall have power to vary the rent hereby reserved F and the tenant shall have power to apply to the landlords for a reduction of such rent to such a figure as shall then be the fair market rent of the property hereby demised (regard being given to the provisions of section 34 of the Landlord and Tenant Act 1954) upon either party giving to the other at least three months' written notice expiring at the end of the 14th year of the said term stating the suggested new rent to be reserved and in default of agreement between the landlords and the tenant the amount of the said new rent shall be such an amount as shall be fixed by a single arbitrator to be appointed by the parties should they agree upon one and otherwise by the President for the time being of the Royal Institution of Chartered Surveyors under the provisions of the Arbitration Act 1950 or any statutory
enactment for the time being in force in substitution of the same. G

Now the deadline before which the notice in question had to be given was September 25 1975 and it was in fact sent by recorded post in a letted dated September 17 1975 from the clerk to the trustees of St Mary's Hospital to Mr Burmester, a director of Lennards Ltd. That letter is exhibit CLH 2 to the affidavit of Mr Hodgett sworn on March 16 1976 and reads as follows:

Ref: 12/SMHE September 17 1975

Dear Mr Burmester,

St Mary's Hospital Estates H
1 East Street and 88 North Street
Chichester

I see from the lease of the above premises that the rent is due to be reviewed after the expiration of the 14th year of the term created on December 25 1961. The lease provides that notice of review must be given three months prior to the expiration of the 14th year and for arbitration if a new rent cannot be agreed.

I shall be grateful if you will accept this letter as formal notice of the landlord's desire to review the rent and confirm to me that it is so accepted in accordance with the terms of the lease.

I intend to instruct Messrs Wyatt & Son, of East Street, J Chichester, as valuers in this matter and anticipate that they would like facilities to inspect the premises at some time convenient to you. I shall be grateful, therefore, if you would let me know that you would have no objection to Messrs Wyatt & Son contacting you to this end, and also if you would be kind enough to let me know what firm of valuers you intend to instruct to act on your behalf, in the hope that the respective valuers can agree a figure for the new rent.

Yours faithfully,

Clerk to the Trustees of St Mary's Hospital

Reading clause 4 (i) it will be seen that it is an expressed K requirement that the notice should state the suggested new rent to be reserved, and that the letter in question does not in fact suggest any particular figure for the new rent but states that the lease provides for arbitration if a new rent cannot be agreed and that the landlords' valuers will be Messrs Wyatt & Son of Chichester and that they would like for the purposes of their valuation to inspect the premises at a convenient time. The question I have to decide is whether that is a good notice. If it is, then the matter must proceed to arbitration if not agreed, but if it is not, then the

A landlords fail and the tenants will continue to be liable to pay the rent of £1,500 for the next seven years. This is not therefore a "gap" case, in the sense in which that term has been used in some of these cases, in that no rent will be payable if a new rent is not fixed. On the other hand, if market rents continue at their present values, the tenants will, over the period of the coming seven years, as I am told, pay something of the order of £50,000 in total (equivalent to a discounted figure of some £35,000) less than the true market value at present rates. If the notice therefore is bad, there will be considerable financial hardship to the plaintiff charity. The defendant company on the other hand
B is responsible to its shareholders and naturally does not intend to pay a rent many times the present rent unless legally bound to do so. In any event, unless there is some reason for equity to intervene, the matter must be decided, in my judgment, on the basis of the legal position without prejudice to any hardship or benefits which may be caused to either party by the decision.

The arguments, which cannot be much elaborated, are in essence as follows. Mr Blum, for the plaintiff charity, contended that the rights to review here are mutual and not unilateral and that the same considerations should be applied
C to this notice as are applied to notices to quit (see *Woodfall's Law of Landlord and Tenant* 27th ed vol 1 at pp 944-5), namely that the notice must be clear and certain but need not normally be in any particular form. Here it is said the lessees were by the letter of September 17 left in no doubt that the landlords intended to ask for a revision. The correct principle in rent review cases dealing with questions of time is, as Roskill LJ recently reaffirmed in *United Scientific Holdings Ltd v Burnley Borough Council* [1976] 2 WLR 686 at 699G, as follows:

> The right question to ask in each case is: upon the true construction of the particular clause, did the parties intend that the
> D particular stipulations as to time must be strictly adhered to or not; or if, as happens in so many cases, the parties have not expressly dealt with this question, must there be imputed to the parties an intention that the particular stipulations as to time must be strictly adhered to or not? It is, I think, impossible to state a rule which will prescribe that every such clause must be interpreted so as to produce the same result in every case, however desirable that might ideally be, in order to avoid a multiplicity of disputes. But so to hold would be to defy the principle that every document must be properly construed according to its own terms and not according to the terms of some other document. Questions of construction of rent revision clauses are not resolved by asking, "Is this case like that case?"
> E They are to be resolved by asking the question I have just stated and answering it by the application to the particular clause of the proper principles of construction.

The argument proceeds that the requirement of "stating the suggested new rent to be reserved" is on all fours with a requirement as to time and should be treated in the same way. Here the intention of the parties was that a notice making clear the landlord's intention to ask for a rent review should be given, and the letter of September 17 1975 satisfies that intention. Mr Blum also somewhat mildly urged that the actual phrase "stating the suggested new rent
F to be reserved" could, without violence to the language, be read as meaning "how the new rent is to be ascertained." In this connection there is, as I suggested, a slight contrast between the phrase "stating the suggested new rent" and the words "the amount of such new rent" in the following passage in the clause. He did not, however, press this argument, and in my view there is nothing in it. I shall not therefore refer to it further.

The argument of Mr Harper for the defendant company is short and straightforward. He accepts the proposition that it is the intention of the parties which must govern the G
matter; but, he says, that intention as a matter of construction must be derived from and determined by the words in the clause which the parties have themselves chosen to use to express that intention. That intention from the wording of the clause is that a suggested figure for the new rent proposed by the landlord should be stated in his notice. There are, he said, good reasons for this requirement, in that an increase specifically stated in figures enables the lessee in the short space of time prior to the expiry of the first period to see what his maximum liability for the future will be and enables him either to decide to accept such increase or, if so advised, to take immediate action to try H
to find a likely assignee. Although it was argued that any figure in the landlord's notice, whether or not the result of deliberation and/or valuation, would satisfy the notice, nevertheless it must be assumed that any figure put in by the landlord would be so put in in good faith, and the tenant could act upon it accordingly. Mr Harper also argued that as a matter of construction the words should be taken against rather than for the landlord, who, he assumed here, had in the normal way put the lease forward and chosen the actual words of the clause in the first instance.

For my part I consider that here as a matter of construction the parties have, by the words used in clause 4 (i), made J
it of the essence of the contract that the notice should specify a figure for the proposed new rent. That, it seems to me, was their intention and there were good reasons for it. It would have been perfectly simple for the parties to have chosen general words for the notice, not requiring the amount to be specified, if such had been the intention. For example, the clause might have read "notice shall be given that the landlords intend to ask for an increase in rent, the amount thereof to be agreed or determined by arbitration in default of such agreement." As it is, they are bound by the words they have used. Though no doubt equity will in a proper K
case interfere where an action is sought to be brought at law on inequitable grounds or where an inequitable defence is sought to be put up, that does not mean that equity will rewrite a contract where it appears that the contract which the parties have entered into is clear on its face, although in the circumstances which have occurred it may work hardship on one of the parties. Equity acts on the conscience and there is nothing unconscionable in a party insisting on the strict terms of his contract where those terms are clear and where on their construction the proper view is that they were regarded by both parties as of the essence of such contract. L

The parties here by the words thay have used have defined some of the essential conditions and time-limits of the notice in question. They have defined the time-limit within which the notice must be given, the date when it must expire, that its form must be in writing and that one of the items it must contain is the amount of suggested new rent. There is here no question of any "unavoidable accident," "fraud," "surprise, or ignorance not wilful" (see the quotations from authority by Russell LJ in *Samuel Properties Ltd v Hayek* [1972] 1 WLR 1296 at 1303). There is no other possible reason, such as waiver or equitable estoppel dealt with by Lord Cairns LC in *Hughes v Metropolitan Railway Co* M
[1877] 2 App Cas 439 and developed in later cases such as *W J Alan & Co Ltd v El Nasr Export & Import Co* [1972] 2 QB 189, for giving equity ground to relieve against the consequence of mistake or inadvertence on the part of the plaintiffs. I cannot in the circumstances here see any reason for preventing the terms of the contract taking effect as, in my judgment, the parties intended that they should. I am unable therefore to make the declaration for which the plaintiffs ask.

MISREPRESENTATION

COURT OF APPEAL

February 6 1976

(Before Lord DENNING MR, Lord Justice ORMROD and Lord Justice SHAW)

ESSO PETROLEUM CO LTD v MARDON

Estates Gazette January 8 1977

(1976) 241 EG 82

Oil company's misrepresentation of potential throughput of petrol in negotiations for tenancy of filling station—Statement a warranty not of annual sales of petrol but of the fact that an expert estimate of such sales had been made—Company in any case liable under the Hedley Byrne doctrine, which extends to statements made during negotiations that end in contracts—Further points on assessment of damages in a case of the kind

This was an appeal by Philip Lionel Mardon, formerly tenant of Eastbank Service Station, Eastbank Street, Southport, from a judgment of Lawson J in the Queen's Bench Division on July 31 1974 ([1975] 1 QB 819), and a further judgment of the same judge on January 13 1975 deciding issues as to damages, by the combined effect of which the respondents, Esso Petroleum Co Ltd, were held liable to pay the appellant only some £9,000 in respect of loss suffered by him as the result of a misstatement made to him during the negotiations leading up to his tenancy agreement dated September 1 1964.

J A Hall QC and A Rawley (instructed by Batchelor Street Longstaffe, agents for Bellis, Kennan, Gribble & Co, of Southport) appeared for the appellant, and C Ross-Munro QC and J Peppitt (instructed by Durrant Piesse) represented the respondents.

Giving judgment, LORD DENNING said that this was a long story of "wasted endeavour and financial disaster," to quote the judge's words, and litigation had so far lasted for eight years. In 1961 Esso wanted an outlet for petrol in Southport, and they found a most suitable one on Eastbank Street, one of the busiest streets in the town. They made a careful forecast of what they called the "EAC," the estimated annual consumption of petrol, and concluded that the throughput of petrol would reach 200,000 gallons a year by the second year after development. Esso accordingly bought the site and proceeded with the construction of a filling station. But then something occurred which threw all their calculations out. The site had outline planning permission, and Esso had assumed that they could have their forecourt and pumps fronting on to the busy main street. The local planning authority made it a condition, however, of detailed permission that the station be built back to front, so that the showroom alone fronted on to the main street, and the forecourt and pumps were at the back of the site, accessible only from side streets and invisible from the main street. Esso had no choice but to build in compliance with these requirements, and they finished the work early in 1963. Now it was that they made what the judge had called their "fatal error": they failed to revise their original EAC of 1961 despite the fact that the throughput would be greatly affected by reason of passing traffic not being able to see the station. Lawson J found that this error was due to want of care on Esso's part, and there could be no doubt about it.

Still under the influence of this error, Esso began looking for a tenant, and they found an excellent man, the defendant. He was seen by their local manager, Mr Leitch, who had 40 years' experience in the petrol trade and on whose calculations and recommendations Esso had acquired and developed the site. He told Mr Mardon that Esso had produced an EAC of 200,000 gallons after two years. Mr Mardon said that he felt 100,000 to 150,000 gallons would be a more realistic estimate; however, he was convinced by Mr Leitch's far greater expertise, and on the basis of what he was told, he entered into a tenancy for three years at a rent of £2,500 for the first two years and £3,000 in the last year. He did everything desired of him, was an extremely good tenant, and tried every way he could to increase sales and profitability. He put all his available capital in, more than £6,000, and raised an overdraft which also went into the business. Yet when the accounts were taken for the first 15 months, the throughput was only 78,000 gallons and a net loss of £5,800 had been suffered. On July 17 1964 Mr Mardon wrote to the area manager giving notice to quit, explaining that he sought to salvage as much as he could in the face of the inevitable alternative, bankruptcy. The manager made representations to his superiors, and as a result a new lease was given to Mr Mardon on September 1 1964 for a year certain and thereafter determinable on three months' notice, the rent being reduced to £1,000 a year plus a surcharge of 1d or 2d per gallon, depending on the amount sold. Still a success could not be made of the site, which was simply not capable of selling more than 60,000 to 70,000 gallons, and Mr Mardon lost more and more of his money. He appealed to Esso, he consulted solicitors, all to no avail; Esso did nothing to help, but instead took steps, such as draining his tanks and cutting off supplies, which put him out of business as a petrol station. Ultimately they brought the present action for possession and for £1,100 odd for petrol supplied, and that defeated him. He gave up the site, having lost all his capital and incurred a large overdraft. By his defence and counterclaim, he sought damages for breach of warranty, the statement as to EAC, alternatively for negligent misrepresentation. No claim could be brought under the Misrepresentation Act 1967, because the representation now in issue was made before that Act took effect.

The first question, then, was whether the statement as to the EAC was a warranty. He (his Lordship) quite agreed with Mr Ross-Munro that it could not be regarded as a guarantee that the throughput would be 200,000 gallons. Nevertheless it was a forecast made by a party, Esso, with special knowledge and skill, a party in a much better position to make a forecast than was the defendant. In these circumstances, there was a warranty, as it seemed to him (Lord Denning), that the forecast as to EAC was sound, in the sense that Esso had made it with reasonable care and skill. That warranty had been broken, and for

A this Esso were liable in damages. Even were this not so, he (his Lordship) was of opinion that the principle of *Hedley Byrne & Co Ltd v Heller & Partners Ltd* [1964] AC 465 was not limited, as Mr Ross-Munro had argued, to cases in which no contract came into existence between the parties concerned, but extended to statements made in the course of negotiations which yielded a contract. So for example a professional man might give advice under a contract for reward; or without a contract, in pursuance of a voluntary assumption of responsibility, gratuitously and without reward. In either case, he was under one and the same duty to use reasonable care. For breach of that B duty he would be liable in damages, which should be, and were, the same whether he was sued in contract or in tort. Here Esso had clearly made a misrepresentation which was negligent and induced the defendant to enter into an agreement that turned out disastrous. For that misrepresentation Esso were liable in accordance with the *Hedley Byrne* doctrine.

The damages must be measured by the loss Mr Mardon suffered, not by reference to the prospect of a throughput of 200,000 gallons. They were not, however, to be confined, as the judge had confined them, to the period between April 1963 and the entry into the new agreement of Sep- C tember 1 1964. The effect of the original misstatement endured in September 1964, and laid a heavy hand on all that followed. He (the Master of the Rolls) agreed with the judge's conclusion that Mr Mardon could claim his original capital of £6,000 odd although technically that sum was put up by a private company, one in which he and his wife held all the shares. To that loss must be added the overdraft incurred, plus loss of earnings, plus interest. Damages for having to sell the house to pay off the overdraft were too remote. Actual figures should be agreed if possible, failing which counsel could come back with D figures for the court to consider. He (his Lordship) emphasised that the defendant was entitled to substantial damages.

Ormrod and Shaw LJJ delivered concurring judgments. The appeal was allowed and the assessment of damages, interest and costs adjourned. On February 27 the court approved the terms of a proposed order, indicating that the figure agreed as damages was in line with their Lordships' own approach to the matter. The defendant received all his costs in both courts.

E

CHANCERY DIVISION

January 14 1977

(Before Mr Justice FOX)

CREMDEAN PROPERTIES LTD AND ANOTHER v NASH AND OTHERS

Estates Gazette March 19 1977

(1977) 241 EG 837

Contracts for sale of two properties—Joint tender documents—Alleged innocent misrepresentation as to lettable office F space—Section 3 of Misrepresentation Act 1967—Preliminary issue before trial of action—Effect on liability of disclaimer of responsibility for accuracy of particulars and of provision that no error, omission or misdescription shall annul sale—Liability not necessarily excluded by disclaimer—Action must proceed—Decision of importance to estate agents—Leave to appeal given

This was a preliminary issue in which the applicant was George Percy Nash, the first defendant in an action for rescission of two contracts brought by Cremdean Properties Ltd and Compass Securities Ltd against the first defendant and against six trustees of the Mission House of St Paul in respect of properties at 16 and 17 Portland Square, Bristol. G The six trustees, the second defendants, took no part in the present interlocutory proceedings, which were concerned with the legal effect, having regard to section 3 of the Misrepresentation Act 1967, of a provision in the joint tender documents purporting to exclude liability for misdescription. In these proceedings the first defendant asked for an order placing a construction on the provision which would have disposed of the action in favour of the defendants.

G H Newsom QC and Spencer Maurice (instructed by Church, Adams, Tatham & Co, agents for Barnett & Leonard, of Bristol) appeared for the applicant (the first defendant); P V Baker QC and Gordon Hodgson (instructed by Portner H & Jaskel) represented the respondents (plaintiffs).

Giving judgment, Fox J said: This is a preliminary issue in an action in which the plaintiffs seek rescission on grounds of misrepresentation of certain agreements for the purchase by the plaintiffs from the defendants of some property in Bristol. The material allegations in the statement of claim are, briefly, as follows:

(i) Prior to November 27 1973 the first defendant owned 16 Portland Square, Bristol, and the second defendants, as trustees of a charity, owned 17 Portland Square.

(ii) Before October 1973 the defendants had obtained outline planning permission for the comprehensive development of both J properties and consequently decided to sell their respective interests by joint tender (but so that each defendant would enter into independent contracts with any purchaser) and to impose upon the purchaser a positive obligation to carry out the proposed development and to include therein a new church hall for the second defendants.

(iii) On October 17 1973 the defendants by their agents issued joint tender documents embodying the decision which I have mentioned and inviting tenders for the premises. The tender documents included a folder which contained, *inter alia*, special conditions of sale, a description of the premises, details of the planning consent and a scale drawing specifying in outline how the premises were to be developed. K

(iv) By the tender documents the defendants represented that approximately 17,900 sq ft of lettable office space, in addition to the new church hall, could be accommodated on the site of the premises within the elevations of the buildings required by the scale drawings and that such lettable office space be accommodated within a floor space index of 1.8 stipulated by the planning authority.

(v) In reliance upon the representations the plaintiffs on November 14 1973 submitted an offer, as required by the tender, for the premises for the sum of about £552,000.

(vi) The offer was accepted by both defendants on November L 27 1973 and the sum of £298,750 was allocated by the defendants as the consideration for the purchase of the first defendant's property.

(vii) The agreement with the first defendant was completed by a conveyance of February 5 1973. The agreement with the second defendants has not been completed.

(viii) The representations which I have referred to in paragraph (iv) were false in that on the true dimensions of the premises within a floor space index of 1.8 only 14,700 sq ft of lettable office space could be accommodated on the site. The scale drawing was false in that:

(a) It showed the listed buildings at the front to be smaller M than they in fact were thereby falsely representing that more new office space could be accommodated on the premises than was in fact so;

(b) It indicated in a schedule that 17,588 sq ft of lettable office space (allowing for the church hall) could be accommodated on the premises whereas in fact no more than 14,700 sq ft could be so accommodated;

(c) It indicated by measurement that within the elevations of the buildings shown thereon 18,203 sq ft of lettable office space could be accommodated within the site whereas within such elevations when corrected only 14,700 sq ft of lettable office space could be accommodated.

On these allegations the plaintiffs claim rescission of both agreements and ancillary relief. I should add here that while the folder included in the tender documents referred to the attached drawing LPSO 2A the drawings actually attached were numbered LPSO 2C. By a letter of November 7 1973 the defendants' agents asked that the plaintiffs, in submitting their tender, should on the folder alter the drawing number in ink. The folder contained on its back a form of tender whereby the person making the tender offered to purchase "subject to the foregoing conditions and I/we agreed that in the event of this offer being accepted in accordance with the above conditions on or before the date named in the conditions for the completion of the purchase I/we will pay the said purchase money and carry out and complete the purchase in accordance with the conditions." As I have already mentioned the conditions of sale are set out in the folder.

The first defendant by his defence in effect denies that any such representations as alleged were made or if made that they were false or, if false, that they materially affected the description of the property. The first defendant further relies upon the following statement (which I will call "the notice") which appears at the bottom of the final page of the special conditions of sale set out in the folder:

> Messrs Lalonde Brothers & Parham for themselves for the vendors or landlords whose agents they are give notice that (a) these particulars are prepared for the convenience of an intending purchaser or tenant and although they are believed to be correct their accuracy is not guaranteed and any error omission or misdescription shall not annul the sale or be grounds on which compensation may be claimed and neither do they constitute any part of an offer of a contract (b) any intending purchaser or tenant must satisfy himself by inspection or otherwise as to the correctness of each of the statements contained in these particulars.

In view of that provision the first defendant says that on no view is he under any liability. There was, he says, no misrepresentation; the plaintiffs were told that they must satisfy themselves as to the correctness of any statement.

The plaintiffs contend that the notice is void by reason of section 3 of the Misrepresentation Act 1967. That statute enables relief to be given to a purchaser in respect of a misrepresentation even though, as in the case of the sale by the first defendant, the contract has been completed. Section 3 of the Act provides as follows: "If any agreement (whether made before or after the commencement of this Act) contains a provision which would exclude or restrict: (a) any liability to which a party to a contract may be subject by reason of any misrepresentation made by him before the contract was made; or (b) any remedy available to another party to the contract by reason of such a misrepresentation; that provision shall be of no effect except to the extent (if any) that, in any proceedings arising out of the contract, the court or arbitrator may allow reliance on it as being fair and reasonable in the circumstances of the case."

By an order of the master made upon the application of the first defendant (and opposed by the plaintiffs) it was directed that the following question be decided as a preliminary issue:

> Whether on the true construction of the joint tender documents referred to in paragraph 3 of the statement of claim the provision contained in such joint tender documents and more particularly set out in paragraph 4 of the defence of the first defendant:
>
> (a) is (as alleged in paragraph 2 of the reply to the defence of the first defendant) void and of no effect by virtue of section 3 of the Misrepresentation Act 1967; or
>
> (b) takes effect so as to exclude any liability to which by virtue of the said Act the first defendant might otherwise be subject on account of any inaccuracy in the information contained in such tender document.

The first defendant asks for an order in the sense of paragraph (b) which would dispose of the action.

I come then to section 3 of the Misrepresentation Act 1967. By its terms it only applies to a provision in an agreement. In my view the notice is such a provision. It seems to me that the contract was entered into on the basis that the notice should be applicable to the contract. In my view, therefore, section 3 is capable of applying to the notice. The next question is whether the content of the notice is such as to make the section applicable to the whole or some part of it. The section applies to a provision which would exclude or restrict:

(a) any liability to which a party to a contract may be subject by reason of any misrepresentation made by him before the contract; or

(b) any remedy available to another party by reason of such misrepresentation.

In practice, therefore, the essential question is whether there was a misrepresentation. If there was, then a provision which seeks to exclude or restrict liabilities or remedies arising from that misrepresentation is within the mischief of the section and cannot be relied upon by the defendant. And I do not think it can make any difference what is the form of that provision. Section 3 is in the widest terms and strikes down any provision which excludes or restricts such liability or remedy as is specified in the section. I was referred to the decision of Brightman J in *Overbrooke Estates Ltd v Glencombe Properties Ltd* [1974] 1 WLR 1335. I do not think that case assists me here. It was concerned only with the ability of a principal publicly to limit the ostensible authority of his agent. The result, in my view, is that, if the true analysis in the present case is that there was a misrepresentation and the first defendant has to rely upon the notice to relieve him from the consequences of that misrepresentation, section 3 will prohibit such reliance.

I come then to the contents of the notice. They fall, I think, into the following parts:

(i) The particulars are prepared for the convenience of an intending purchaser. That I think is more or less neutral in its effect.

(ii) Although the particulars are intended to be correct their accuracy cannot be guaranteed and any error omission or misdescription shall not annul the sale or be grounds on which compensation may be claimed. In my view, if there was a misrepresentation, that provision comes within section 3 in that it excludes a remedy which would otherwise be open to the plaintiff by reason of the misrepresentation.

(iii) The particulars do not "constitute any part of an offer of a contract." This is not very happily drawn but I take it to mean simply that the sending out of the tender documents was not an offer by the defendants. That, it seems to me, is unobjectionable.

(iv) Paragraph (b) of the statement provides that "any intending purchaser or tenant must satisfy himself by inspection or otherwise as to the correctness of each of the statements contained in these particulars." The first defendant says that in view of this provision (at any rate when read with the rest of the notice) he can be under no liability. The action, it is said, is based upon misrepresentation. There never was any misrepresentation since the plaintiffs were, by the notice, put on inquiry and told to go and check the facts for themselves. Equally, it is said, there is no question of the application of the Misrepresentation Act 1967; the Act is concerned with misrepresentations and there were none.

As a preliminary to considering that contention I should observe that the issue now before me is, in effect, a proceeding by way of demurrer. I must, therefore, assume the facts alleged in the statement of claim. These facts so far as material are (a) that representations were made by the defendants (b) that the plaintiffs made their offers in reliance upon those representations (c) that the representations were false. In these circumstances it seems an odd result that the court should now, in effect, put an end to this action on the ground that there never was a misrepresentation and without ascertaining the facts at all. But putting that aside it seems to me to be going altogether too far, at this stage of the proceedings, to assume that the notice is conclusive evidence that the plaintiffs were not induced to enter into this contract by reason of false statements contained in the particulars. Theoretically it seems to me it would be open to the plaintiffs to prove at the trial that, notwithstanding the notice, they were induced to enter into the contract by such statements, though the notice will or may be a material matter to be taken into account in deciding that. One cannot, it seems to me, determine whether there has been a misrepresentation inducing a contract without ascertaining all the facts. The decision of the Court of Appeal in *Gosling* v *Anderson* (1971) 223 EG 1743, to which Mr Baker referred me, exemplifies, though on widely different facts, this principle. The plaintiff, Miss Gosling, wanted to buy a flat from the defendant. She was particularly anxious to have a garage. Doubt arose about planning permission. The vendor's agent wrote a letter indicating that there was planning permission. Miss Gosling's solicitor was not satisfied and wrote to the local authority but there was delay in replying. The defendant's agent then wrote again saying that there should not be any difficulty about planning permission but making it plain that so far as the defendant was concerned all that was being sold was a flat with a parking area. Despite the doubts Miss Gosling decided to go ahead and complete. Planning permission was refused. It was held that, notwithstanding that she deliberately took the risk of going ahead, she was entitled, on the facts, to establish that she was induced to enter into the contract by the vendor's misrepresentation. The facts, as I have said, are widely different from the present but the case shows, I think, that the question whether there was a misrepresentation is something to be decided in the context of all the facts. For example, in the present case, take the position as to the scale drawing. The plaintiffs allege, in effect, that the drawing, when measured, indicated lettable office space in excess of 17,900 sq ft but that the drawing was wrong as a drawing. There might, it seems to me, be questions, depending on expert evidence, as to what a purchaser could reasonably assume in relation to such a drawing.

In the circumstances I think that it would be wrong for me to assume that the plaintiffs could not at the trial establish that they were in fact induced to purchase by misrepresentation by the defendants. If there was such a misrepresentation then it seems to me that the existence of the notice is not an answer to the first defendant's liability. Accordingly, I think it would be quite unsafe, at this stage, to assume that the notice disposes of any question of liability. In my view therefore the first defendant is not entitled to the declaration that he seeks and the action must proceed to trial. I should add that the alternative declaration set forth in (a) of the issue as formulated (namely that the notice is invalid) does not seem to me to be appropriate on the view of the matter which I have taken as set out above. Parts of the notice are, on any view, unobjectionable. Further, as a purely factual matter at any rate, the notice may be material in determining whether there was a misrepresentation or whether the plaintiffs were induced to enter into the contract in reliance upon it. I quite appreciate Mr Baker's argument that the effect of the notice is to exclude or restrict liability or remedy for misrepresentation and is, therefore, simply invalid. But, as I have indicated, it seems to me that the notice may, at the least, have validity as an element in determining whether there was a misrepresentation upon which the plaintiffs relied. If parts are invalid, questions of severance may arise. Since the case must go to trial in any event it seems to me that the proper time to determine any general questions as to the validity of the notice is at the trial itself.

The form of the judge's direction was: "The court being of the opinion that the provision does not necessarily take effect so as to exclude any such liability as is referred to in paragraph (b) of the issue, and that accordingly the action must proceed to trial, the court does not think fit to make any order on the preliminary point." The respondents to the present application (the plaintiffs in the action) were given their costs of the application in any event. Leave to appeal was given in view of the importance of the issue as to the construction of the Misrepresentation Act 1967 and the validity of exclusion clauses.

NEGLIGENCE

A **COURT OF APPEAL**

February 10 1976

(Before Lord DENNING MR, Lord Justice ROSKILL and Lord Justice GEOFFREY LANE)

SPARHAM-SOUTER AND OTHERS v TOWN & COUNTRY DEVELOPMENTS (ESSEX) LTD AND ANOTHER

Estates Gazette January 29 1977

(1976) 241 EG 309

Aftermath of the Dutton Case—Limitation period begins to run against property-owner suing council in negligence B only when he knows of the defects relied on or could with reasonable diligence have discovered them—Lord Denning withdraws remark in Dutton's case about time running from date of bad construction

This was an appeal by Benfleet Urban District Council from a decision of Judge Norman Richards QC, sitting as an official referee, holding that the respondents, John Henry Sparham-Souter and his wife, Glenda June Sparham-Souter, of 3 Wincoat Close, Benfleet, and Eugene Ryan, of 4 Wincoat Close, were in time to sue the developers of the properties, Town & Country Developments (Essex) Ltd, and the appel- C lants for damages for negligence.

J A Tackaberry (instructed by Barlow, Lyde & Gilbert) appeared for the appellants, and J K Wood QC and T E Walker represented the respondents. The developers took no part in the appeal.

Giving judgment, LORD DENNING said that Town & Country Developments (Essex) Ltd were developer-builders. On May 6 1965 they began building 3 and 4 Wincoat Close, at Benfleet, according to plans approved by the local authority. The houses, part of a new estate of 11 houses and a new street, were completed on September 30 1965. On December 15 1965 the council issued a certificate that D inspections had been carried out and they found no reason to question the legality of the work carried out under the building by-laws. In October and November the developers had agreed to sell no 3 to the Sparham-Souters and no 4 to Mr Ryan. The conveyances were completed in November 1965 and January 1966. Two or three years later, several cracks appeared in the brickwork of the houses and they became uninhabitable. The alleged reason was that the foundations were inadequate to support the load, due to the negligence of the developer-builders and also to the council surveyor's negligence in passing the work as satisfactory when he ought not to have done so. In October 1971 the E plaintiffs issued the writ in the action. There was a decision of Cusack J which went to appeal, ***Dutton v Bognor Regis Urban District Council*** [1972] 1 QB 373, which the plaintiffs relied on in so far as it established that a claim lay against a council in such circumstances, but which contained a remark of his (Lord Denning's) to the effect that "the damage was done when the foundations were badly constructed." Judge Richards held, on a preliminary issue, that the plaintiffs' action was in no wise barred either against the developers or against the council. The developers had accepted that decision, but the council now appealed, sub- F mitting that two of the acts of negligence alleged against them occurred more than six years before action was brought, and that in respect of those acts the claim was statute-barred.

On the authorities, it was clear to him (his Lordship) that a cause of action for negligence accrued not at the date of the negligent act or omission, but at the date when the damage was sustained by the plaintiff: *Cartledge v E Jopling & Sons* [1963] AC 758 at 777. Time did not begin to run in a case of the kind until there was a party capable of suing and a party liable to be sued. Where a builder had negligently made foundations for a house which were quite G insufficient and in breach of the by-laws and the council's inspector had negligently passed them as sufficient, the cause of action against the builder and the council accrued not at the time of the negligent making or passing of the foundations, nor at the time when the latest owner bought the house, but at the time when the house began to sink and cracks appeared. Only then could the man who at that stage owned the house reasonably be expected to know that he might have a cause of action, and time therefore should not begin to run against him until he knew of, or could with reasonable diligence have discovered, the defective foundations. H That principle underlay the Limitation Acts of 1963 and 1975, and it was the principle the court should adopt in regard to the new cause of action introduced by *Dutton*. In his judgment in that case, he (his Lordship), following a dictum of Diplock LJ in *Bagot v Stevens Scanlan & Co* [1966] 1 QB 197 at 203, had indeed said that the damage was done when the foundations were badly constructed. Having thought it over time and again—and been converted by his brethren—he had now come to the conclusion that, when building work was badly done, and covered up, the cause of action did not accrue, and time did not begin to run, until such time as the plaintiff discovered that it had done damage or ought, with reasonable diligence to have J discovered it. Further, the only owner who had a cause of action was the owner in whose time the damage appeared. He alone could sue for it unless, of course, he sold the house with its defects and assigned the cause of action to his purchaser. It might seem hard on the builder or the council surveyor that he might find himself sued many years after he left the work, but it would be harder on the householder that he should be without remedy, seeing that the surveyor had passed the bad work and the builder had covered it up and thus had prevented it being discovered earlier. K

On arriving at that conclusion, he (Lord Denning) owed an apology to two judges at first instance who, in reliance on what he had said in ***Dutton***, had with regret held that actions before them on similar facts were statute-barred (Mars-Jones J in *Higgins v Arfon Borough Council* [1975] 1 WLR 524 and Judge Fay in *Anns v Walcroft Property Co Ltd* (unreported), October 24 1975). He now recanted and thought that both those cases were wrongly decided. In the present case, assuming that the builder and council surveyor had been negligent, the "damage" occurred when the cracks first appeared in the brickwork. In the case of each house, that was less than six years before the writ was issued. The claims were therefore not statute-barred and the appeal should be dismissed.

Roskill LJ delivered a judgment concurring in the result, and Geoffrey Lane LJ also delivered a concurring judgment. The court adjourned an application for leave to appeal until after the trial, Lord Denning indicating that the court thought it would be a proper case for leave.

COURT OF APPEAL

March 1 1976

(Before Lord DENNING MR, Lord Justice SCARMAN and Lord Justice GOFF)

ANNS AND OTHERS v WALCROFT PROPERTY CO LTD AND ANOTHER

Estates Gazette January 29 1977
(1976) 241 EG 311

Limitation point in Dutton's case and Sparham-Souter's case to go to the House of Lords

This was an application by M R Anns and seven other plaintiffs, lessees of maisonettes in Beaconsfield Court, Devonshire Road, Colliers Wood, London SW19, for leave to appeal out of time against a decision of Judge Fay, sitting as an official referee on October 24 1975, holding that an action brought by the applicants against the owners and builders of the maisonettes, Walcroft Property Co Ltd, and against the London Borough of Merton, for damages for negligence, was barred by lapse of time.

M H Johnson (instructed by W H Matthews & Co, of Sutton) appeared for the plaintiffs, and J A Tackaberry (instructed by Barlow, Lyde & Gilbert) represented the second defendants, the London Borough of Merton.

The plaintiffs, who were lessees for 999 years, issued a writ on February 21 1972 claiming (1) against Walcroft Property Co Ltd an order that they remedy defects in the building which had first appeared in or about February 1970 by reason of their alleged negligence when they built the block in 1962, and (2) against Merton London Borough Council for damages for alleged negligence in that by their servants or agents they had approved the foundations on which the building was erected even though they had not been taken down to a sufficient depth and/or in failing to inspect the foundations. Judge Fay based his decision on the view that he was bound by the dictum of Lord Denning MR in *Dutton v Bognor Regis Urban District Council* [1972] 1 QB 373 at 396 that "the damage was done when the foundations were badly constructed" and that "the period of limitation then began to run." In *Sparham-Souter v Town & Country Developments (Essex) Ltd* [see previous report] the Court of Appeal had reinterpreted what was said in *Dutton's* case and expressed the view that the present case had been wrongly decided.

Giving judgment, Lord Denning said that the sensible thing to do was to extend the time for appealing and, by agreement, in view of the decision in the *Sparham-Souter* case, the appeal would be allowed and the preliminary point answered in favour of the plaintiffs by saying that their cause of action ran from the time when they discovered or ought to have discovered the damage. Leave would be given to the council, the second defendants, to appeal to the House of Lords from that decision.

Scarman and Goff LJJ agreed.

QUEEN'S BENCH DIVISION

July 20 1976

(Before Sir DOUGLAS FRANK, sitting as a deputy judge of the division)

RUMSEY v OWEN WHITE & CATLIN

Estates Gazette February 26 1977
(1976) 241 EG 611

Measure of damages in action for negligence against solicitors—Advice to vendor selling property subject to tenancies—Inappropriate notices under Landlord and Tenant Act 1954—Failure to ensure that vendor could deliver vacant possession in time—Claim by purchasers—True measure of damages in action against solicitors difference between value immediately before contract and value at date of breach

This was an action for damages against a firm of solicitors, Owen White & Catlin, for negligence or breach of duty arising out of a contract between the plaintiff, Derek Leonard Rumsey, and Portgate Securities Ltd, for the sale by the plaintiff to the company of properties at 60, 60a and 61 High Street, Egham, Surrey, consisting of two shops and five units of offices of which the plaintiff was the freehold owner.

R L Johnson (instructed by Rooks & Co) appeared for the plaintiff, and A L Price QC and H A P Picarda (instructed by Hewitt Woollacott & Chown) represented the defendants.

Giving judgment Sir Douglas Frank said: On April 26 1973 the plaintiff contracted to sell the property to Portgate Securities Ltd for £120,000 payable in instalments of £30,000 each on the signing of the contract, on or before July 3 1973, on or before October 26 1973, and on completion. The date for completion was to be May 10 1974 "or such earlier date as the vendor shall require upon giving 28 days' notice in writing." It was a condition of the contract that vacant possession should be given on completion. In April 1974 one of the shops was occupied by the plaintiff as a pet shop and the other by S J C Stanley under a tenancy agreement for five years from March 25 1971, but subject to the lessor being at liberty to determine by giving six months previous notice in writing should the lessor desire to demolish, reconstruct or alter, or add to the demised premises or any substantial part of them. One of the offices was let to Wren & Co, a firm of insurance brokers, under an agreement for five years from September 29 1971 subject to the condition that, should the premises be required for redevelopment, "the landlord shall be able to determine this agreement subject to a minimum of 12 months' notice." Two other office units were let, but, as they gave rise to no difficulty, they call for no further comment. The plaintiff was also lessee of adjoining land occupied by Clarks (Ammonia Specialists) Ltd in which company he originally held about half the shares and later all, or nearly all, of them. However, it was his intention to move the activities of that company to a factory to be built for it at Basingstoke by the Greater London Council.

The years 1972 and 1973 were, it is commonly agreed, a time of boom in the property market. The climax was reached in 1973, but in the following year there was a rapid fall off in demand and in prices. The property was a prime site for redevelopment and the plaintiff received offers for it even before he had decided to sell it. However, by the end of 1972 he had decided to close his retail business at Egham and to concentrate his efforts and capital in moving Clarks to Basingstoke. A number of offers for the property were considered but the plaintiff decided to accept that which became the subject of the contract of April 26 1973. He instructed the defendants in this matter. They were already acting for him in other matters. B P White, a legal executive in the defendants' firm and whose duties were those of a conveyancing clerk (and that alone until September 1973) had the handling of this matter for the defendants. He was, of course, aware that it was essential to obtain vacant possession and therefore that time must elapse before completion. But it was also to the advantage of the landlords to delay payment of the purchase money, because they would require to obtain planning permission for the redevelopment of the property. Mr White was aware of the Landlord and Tenant Act 1954, so clearly had formed the view that the Act did not stand in the way of obtaining possession of the parts let. Accordingly at a meeting held on April 26 (at which there were present the plaintiff, Mr White, M N Henley of the estate agents who had effected the sale, A P Baker, a director of Portgate Securities Ltd, and their solicitor) the contract to which I have already referred, and which had been drafted by Mr White, was executed.

On May 11 1973 Mr White served notices under the Landlord and Tenant Act 1954 on each of the tenants. That to Wren & Co gave notice terminating their tenancy on May 16 1974, and that to Mr Stanley terminating his tenancy on November 16 1973. Each of the notices stated that the plaintiff would oppose an application to the court for the grant of a new tenancy on the ground that: "On the termination of the current tenancy I intend to demolish or reconstruct the whole, or a substantial part, of the premises, or to carry out substantial work of reconstruction on the whole or part of them" and that "I could not reasonably do so without obtaining possession of the premises." Counter-notices were served and on September 7 1973 Mr Stanley applied to the Slough County Court for the grant of a new tenancy pursuant to Part II of the Act of 1954. It then occurred to Mr White or one of the partners in the defendants' firm that the notices served in May were invalid in that there was no question of the plaintiff demolishing and reconstructing the property, but that, on the contrary, that was to be done by the purchasers. It cannot be said that it was the service of those notices which gave rise to the problem which has now arisen, but the procedure of overcoming the protection afforded by the 1954 Act failed *in limine*, the *limen* being the form of the April contract. Accordingly A M Burns, a partner in the defendants' firm, considered what steps could be taken to secure vacant possession. The upshot was a supplemental agreement made on January 28 1974 between the plaintiff and Portgate Securities Ltd whereunder the company agreed to complete the purchase for a total of £90,000 and to take over the action in the county court, but subject to a provision to the effect that if vacant possession was not obtained by June 7 1974 then the company would be entitled to require the plaintiff to repurchase the property by way of reimbursement at the price of £90,000, and to repay certain other moneys by way of reimbursement. Leave was given to substitute the company for the plaintiff in the county court proceedings and on March 26 1974 Mr Stanley's application for a new tenancy was dismissed on the undertaking by the company that as soon as possession was obtained of the property it would be demolished as soon as practicable, with liberty to apply on any question arising on that undertaking. Nevertheless, having regard to section 64 of the Act of 1954, Mr Stanley could not be required to vacate until the expiration of three months, that is to say until after the date for completion, namely, June 7 1974 specified in the supplemental agreement. Moreover, Wren & Co had not complied with the notice to quit by that date. Accordingly on June 20 1974 Portgate Securities Ltd served on the plaintiff a notice requiring him to repurchase the property in accordance with the terms of the supplemental agreement dated January 28 1974. Mr Rumsey did not, and could not, comply with that notice, so the company have instituted proceedings in the Chancery Division, in which the company claims £35,000, being the difference between the contract price under the supplemental agreement and the value of the property on July 31 1974, and about £23,000 for reimbursements and interest.

There are two issues between the parties, one being liability and the other quantum of damage, and I shall deal with each of them in turn. The defendants concede that there is no doubt that the plaintiff did not receive all appropriate advice, in particular that the plaintiff could not ensure vacant possession of the property either on May 10 1974 or at all. They concede further that the plaintiff was not advised about the time-table in section 64. Indeed, Mr White was ignorant of the effect of those two sections, and Mr Price frankly and properly conceded that the defendants' conduct must amount to a breach of the duty, unless "there is some other feature which indicates that that breach has no relevant value on the question whether the contract was entered into or not." Put shortly, his case was that the plaintiff represented that the detailed procedure for possession did not matter, as the plaintiff knew the tenants and was confident that he could get them out. That is to say, he was content to rely on the prospect of negotiations if the legal processes failed. Mr White said that he had advised the plaintiff of the risks involved, but that the plaintiff made the decision and told him he knew his tenants and felt sure he could come to terms if necessary. The plaintiff denies that he made any such representations and says that he was advised by Mr White that possession could be obtained and that it was safe to go on with the sale, and that if Mr White's advice had been otherwise he would not have proceeded. Further, he had no cause to proceed with a sale which was risky, because he was being inundated with other offers. He alleged that Mr White said in terms that he had checked the leases and that there would be no difficulty in getting rid of the tenants provided that he served the notices. I have no doubt that the plaintiff's version is correct. I do not think Mr White can be heard to say that he advised of the risk involved, when, at the same time, he confesses to having been ignorant of the provisions of the sections of the Act which contain those risks. Moreover, Mr White said it was his practice to make notes of interviews and conversations but he made no notes of the assurances and representations which he says had been given and made by the plaintiff. I find it incredible that the plaintiff would have taken upon himself such a very great burden, and that no record was made, having regard to the fact that vacant possession was a vital element in the transaction. Further, the plaintiff was not (and could not have been) told that the Landlord and Tenant Act, which Mr White proposed to invoke and sought to invoke, was an empty vessel. In the circumstances I have no difficulty in holding that the defendants were negligent.

I now turn to the more difficult question of damages. The plaintiff claims £65,000, being the difference between £115,000, the market value of the property with sitting tenants in April 1973, and £50,000, its value with sitting tenants in August 1974. Alternatively he claims £40,000, being the amount of the liability for repurchase less the market value in 1974, plus £25,000 for being precluded from selling the property elsewhere in April 1973. He also claims approximately £23,000, being the amount claimed against him in the Chancery action for interest and disbursements. Mr Price contended that damages for professional negligence are to be measured at the date of the breach of duty and that in this case the breach of duty was completed when the plaintiff entered into the contract, for thereafter it would have been too late to have tendered the right advice. He said—relying on the evidence given by R T R Jenkins, a chartered surveyor—that on April 26 1973 the plaintiff could have sold the property, subject to the benefit and burdens of the contract of sale, for £112,000. He then invoked the discounted value of the purchase-price of £120,000 as at April 26 1973, namely, £110,000 or £109,000 according to whether Mr Jenkins' evidence or Mr Henley's evidence was accepted. Thus, he concluded, there was no damage. Alternatively he argued that the damage should be the difference between the discounted sale price and the market value of the property subject to the subsisting tenancies, assessed by Mr Jenkins at £109,000 and by Mr Henley at £115,000.

Mr Price referred to *Sykes and Others v Midland Bank Executor & Trustee Co Ltd* [1969] 2 QB 518, where solicitors had failed to point out that in a proposed underlease the superior lessor could arbitrarily withhold his consent to a change of use in the premises. Mr Price had argued that damages should be based upon the chances of what would have happened if such warning had been given, and on that basis no damage accrued while the plaintiffs were actually in possession. Further, that there should have been a set-off for the rise in the value of the property during the tenancy. That argument was not accepted. Mr Justice Paull said: "In my judgment there is a fundamental fallacy in these arguments. Damages are assessed at the date of the breach. Damage in this case is the capitalised sum of the difference between the rent payable and that which would have been payable. What happened thereafter is immaterial." Mr Price then referred to *Philips v Ward* [1956] 1 WLR 471, in which the plaintiff bought a property for £25,000 relying on the defendant's report in which he had negligently overlooked the fact that much of the timber in the house was badly affected by death-watch beetle and worm. The work of repair would have been £7,000, but the value of the house having regard to the bad condition of the property at the time of sale was £21,000. It was held that the proper measure of damage was the difference between the fair value of the property if it had been in the condition described in the defendant's report, namely £25,000, and its value in its actual condition, £21,000. Accordingly the amount recoverable was £4,000. Mr Price then referred to *Ford and Another v White & Co* [1964] 1 WLR 885, where the plaintiffs bought land at a price reflecting the existence of a restriction against building on part of it, their solicitors having negligently stated that the land was not so restricted. In an action against their solicitors the plaintiffs claimed the difference between the value of the land subject to the restriction and the value free from restriction, but it was held that the proper measure of damage was the difference between the market value and the price actually paid.

I do not find that any of those cases is authority for the proposition that in all cases the measure of damage is the difference in values at the time of the breach, but, rather, that they are applications of the basic principle that, as far as possible, he who has proved a breach of a bargain to supply what he contracted to get is to be placed, as far as money can do it, in as good a situation as if the contract had been performed. The fundamental basis is thus compensation for pecuniary loss naturally flowing from the breach: per Lord Haldane in *British Westinghouse Electric & Manufacturing Co Ltd v Underground Electric Railways Co of London Ltd* [1912] AC 673, at p 689. The three cases cited by Mr Price have one feature in common, namely, that an award of damages on any other basis would have put the plaintiff in a better position than he would have been but for the professional negligence. Moreover, Mr Price admitted that he was unable to find a case in which the advice had been tendered to a vendor. He submitted that there is no difference in principle and that the formula he proposed had the advantage of convenience and consistency. Even if I were able to accept the proposition as put by Mr Price (which I do not) I would nevertheless reject the evidence that a purchaser would have been willing to pay £112,000 for the property subject to the contract of sale. Mr Jenkins admitted that it would not be an easy transaction to undertake and that he had never dealt with that kind of operation and could not even suggest who might be a potential purchaser. As I see it, any potential purchaser would take proper advice, which advice could only lead to the conclusion that he would be buying an expensive law-suit.

In my judgment the correct approach to the assessment of damages in this case is to apply the rule in *Hadley v Baxendale* as explained by Lord Reid in *Koufos v C Czarnikow Ltd* [1969] 1 AC 350, namely, to ascertain what loss the defendants ought to have realised was not unlikely to result from a breach of their duty, but the words "not unlikely" connoting a degree of probability considerably less than an even chance, but nevertheless fairly usual and easily foreseeable. In applying that test I reach the conclusion that an inability to obtain vacant possession and therefore a rescission of the contract was a probability of such a degree as to bring the loss resulting therefrom within the rule as stated by Lord Reid. I think it follows from this that the true measure of damage is the difference in value immediately before the contract for sale and the value at the date of the breach. There is no difficulty over the latter amount. The parties agree it as £58,000 because by mid-1974 there had developed a depression in the property market. However, the parties are at variance as to the value on April 26 1973. Mr Jenkins' valuation as between a willing vendor and a willing purchaser on the open market, subject to the subsisting tenancies, is £102,000, whereas Mr Henley's is £115,000. Mr Henley's figure was derived from an offer of £100,000 which had been made but which he thought he could in negotiation have raised to £115,000 notwithstanding that that amount was £5,000 or £6,000 in excess of the discounted price in the contract of sale upon the basis of vacant possession. I do not find that evidence very helpful. On the other hand Mr Jenkins did not think that the fact of the tenancies was a disadvantage, because a purchaser would know that he would be able to obtain possession under the Act of 1954. He further said that if £120,000 was the right price when payment was spread over 12 months (as in the contract for sale) then £112,000 would have been right for occupation in April 1973. I find that evidence acceptable notwithstanding that Mr Jenkins still felt that £102,000 was right. Accordingly I adopt the figure of £112,000. It follows that I assess the principal sum of damages to be the difference between £112,000 and £58,000, that is £54,000.

It is not in the circumstances necessary for me to deal with the plaintiff's alternative claim, but I should mention that it arises in this way. As I have said, by September 1973 Mr Burns realised that things had gone wrong, although he was seeking to repudiate liability on the ground that the plaintiff knew the risk he was taking. Accordingly in order to escape from the situation in which the parties found themselves it was he who suggested and arranged for completion to take place in January 1974 at the reduced price. There can be no doubt that this transaction was entered into for the defendants and at their instigation. In *Banco de Portugal v Waterlow & Sons Ltd* [1932] AC 452 Lord Macmillan, at p 506, said: "Where the sufferer from a breach of contract finds himself in consequence of that breach placed in a position of embarrassment, the measures which he may be driven to adopt in order to extricate himself ought not to be weighed in nice scales at the instance of the party whose breach of contract has occasioned the difficulty,". *a fortiori* where those measures are taken at the instigation of the party in breach in order to escape the consequence of his breach. Accordingly I would have been prepared if necessary to have adopted the plaintiff's alternative method of valuation. Accordingly the damages are £54,000.

The plaintiff was given judgment for £54,000 and costs, with directions as to the disposal of a sum paid into court, the payment of interest on the balance and a stay of execution on terms.

A **QUEEN'S BENCH DIVISION**

March 22 1977

(Before Mr Justice MAY)

LEES AND ANOTHER v ENGLISH & PARTNERS

Estates Gazette April 23 1977

(1977) 242 EG 293

Survey and report for purchase on 1880 house with 1961 additions—Bad "tie" between old and new brickwork—No cracking visible but no warning to purchaser of possible effects of bad tie—Breach of duty—Measure of damages—
B **Difference between purchase price and value on the basis of a correct surveyors' report**

This was an action by Brian Hubert Lees and his wife, Irene Elsa, of Flexlands Farmhouse, Station Road, Chobham, Surrey, for damages against English & Partners of Southlands, Portsmouth Road, Esher, Surrey, practising as surveyors at Wood Street, Kingston upon Thames, alleging failure to exercise due care in surveying Flexlands Farmhouse and making a report of the kind which was to be expected from a reasonably competent and careful surveyor.

Charles Welchman (instructed by Caporn, Campbell, Clare & Clare, of Surbiton) appeared for the plaintiffs, and P H
C Ripman (instructed by Berrymans) represented the defendants.

Giving judgment, MAY J said that in early 1973 Mr and Mrs Lees saw Flexlands Farmhouse in the desirable neighbourhood of Chobham which they liked and ultimately offered £38,000 for it subject to contract and survey. They instructed the defendants to make a survey and report and a partner, Mr Atkinson, and his assistant, Mr Bond, inspected and made a report dated March 13. Overall it was a satisfactory report with one or two matters that Mr Lees queried—in particular a bulge mentioned on the front wall and a tie rod. He discussed the matters with Mr Atkinson, who put his
D mind at rest and the sale went through. Completion was in May 1973. By Christmas cracks had appeared on the exterior of one wall and on the plaster opposite inside. Matters were investigated and in consequence this action was brought, Mr Lees alleging that the defendants failed to exercise that due care and skill in surveying property and giving him their survey report which was to be expected from a reasonably competent and careful surveyor.

The house was probably originally a couple of farm cottages built about 1880. At some stage they were amalgamated and about 1961 a side addition was added and the right-hand side of the front wall was rebuilt in order that windows
E and doors might be inserted, and the right side re-rendered.

It was not disputed by the defendants that the new brickwork was inadequately tied into the old and that there was a jagged line from top to bottom where the two did not properly meet. The new brickwork was in header bond of 9 inch bricks whereas the old brickwork was in Flemish bond. The plaintiffs' experts said that was an unusual and improper method. The bonding was not as good as either English or Flemish bonding and a houseowner would be best advised to have the whole wall rebuilt. The defendants' experts said it was unusual but there was nothing wrong with it from a stability point of view. It was said that although cracks
F developed after Mr Lees bought in 1973 they had not grown any worse to any appreciable extent, which spoke well for the stability of the wall, separated though it might be from G
the old brickwork. He (his Lordship) preferred the evidence of the defendants' experts that although it did not look very nice, particularly on the inside, there was nothing inherently wrong from the point of view of stability with a 9-inch brick wall built in header bond, such as this was. The lack of a proper tie was shown running all the way down to ground-floor level. A poor tie between the new brickwork in 1961 and the old brickwork had failed to prevent relative movement between the two slabs, resulting in the rendering on the outside and the plaster on the inside being cracked.

He (his Lordship) did not think it mattered whether cracks H
had appeared between 1961 and 1973 or not. On all the evidence he was satisfied that there were no cracks visible to Mr Atkinson and Mr Bond at their examination in 1973 and that there was no evidence which they ought to have picked on to show them that there had been previous cracks. In their survey report they wrote of the metal tie rod and the slight bulge in the front wall, which was not considered serious as there had not been movement for some years, and none of the small cracks by the windows were considered serious. What they did not say, and it was accepted they ought to have said, was "there is this junction between new and old brickwork which can be seen in the loft space and which in all probability goes all the way to ground level. J
It is a poor tie, indeed a very bad tie, and although it may not have shown any signs yet—and we can give no guarantee about that—it seems possible that you will get movement between these slabs of brick in the future and, if you do, this will cause cracks inside and outside of the rendering and plaster work." That was very different from "the condition of the wall is good for a property of this age."

It was not disputed by the defendants that failure to draw Mr Lees' attention to the bad tie constituted a breach of their contractual obligations to exercise due care and skill in surveying the property and advising him about it. The K
measure of Mr Lees' damage was the difference between the £38,000 paid and the value of the property on a surveyors' report as it should have been reported, with the mention of a very bad tie between old and new brickwork. Though Mr Lees had said he would "not have touched the house with a barge pole," the vendor might have been able to get the money from another purchaser when the remedial work had been done. The evidence about this had been unsatisfactory, but the finding must be that the defect could have been made good by "stitching in" bricks where the tie was bad, costing between £500 and £1,000, depending on what was found when the job was done. L

He (his Lordship) could not accept that in 1973 a defect costing up to £1,000 to put right (and which had not then shown itself) would have enabled a willing purchaser to get the house for £2,000 to £2,500 less than the price actually paid. A willing and reasonable vendor and purchaser would realise that the defect would detract to the extent of a relatively small figure from the £38,000, a figure which his Lordship fixed as £500.

There would be judgment for the plaintiffs for £500.

When the court was told that the defendants had paid £500 into court in April 1975 the judge gave the plaintiffs judgment for £500 with costs up to date of payment in, from which M
time the defendants were awarded costs.

A

HOUSE OF LORDS

January 28 1976

(Before Lord WILBERFORCE, Viscount DILHORNE, Lord SIMON OF GLAISDALE, Lord EDMUND-DAVIES and Lord FRASER OF TULLYBELTON)

VANDYK v OLIVER

Estates Gazette January 22 1977

(1976) 241 EG 231

Rating—No relief for disabled solicitor's Kensington flat—"Predominant purpose" of person providing accommodation is relevant to the interpretation of section 45, General Rate Act 1967—Here the predominant purpose was the B **provision of a home**

This was an appeal by a valuation officer against a majority decision of the Court of Appeal holding that a flat occupied by the respondent, Neville David Vandyk, a disabled person, at 7 Addisland Court, Holland Villas Road, London W14, qualified for derating under section 45 of the General Rate Act 1967, as being a structure provided for the care or after-care of a person suffering from illness.

G Rippon QC and A Fletcher (instructed by the Solicitor of Inland Revenue) appeared for the appellant, and D Widdicombe QC and M Horton (instructed by N D Vandyk) represented the respondent. C

In his speech, LORD WILBERFORCE said that the respondent was the lessee of a large flat on the second floor in a building in Kensington, London. It had five rooms, two bathrooms, a dining hall, kitchen and balconies. The present appeal related to the rateable value of this hereditament. The respondent in 1968 and again in 1971 made proposals for alteration in the valuation list so as to show the value as nil, as against the previous gross value of £500, rateable value £388. Objection was taken to both proposals, and on December 30 1971 the local valuation court ordered that the assessment should be amended to £425 gross, £326 net. Both sides D then appealed to the Lands Tribunal, which on June 28 1973 fixed the assessment at site value, making £80 gross, £50 net. Then a case was stated for the Court of Appeal which by a majority dismissed the present appellant's appeal [1975] 2 WLR 797. The appellant sought to have the original assessment restored. There was an alternative contention on the part of the respondent that in any event the assessment should be reduced as regards two rooms in the flat.

The respondent was a severely disabled person, as the result of poliomyelitis suffered in 1947, which had left him paralysed in both legs and his left arm. He was confined to a wheel-chair, and it was not disputed that he was in need of E after-care, both in an ordinary sense and within the statutory meaning of those words. In plain terms he needed constant attention; this was provided by his wife, who lived in the flat, and, when she was away, by professional attendants. The respondent acquired the hereditament in 1968, having previously lived in another flat. Among the reasons why he chose it were the facts that the doorways were wide enough to allow easy passage for his chair and that a lift, porterage and central heating were all available. After moving in, he had certain adaptations made to his bedroom and one bathroom (the two rooms that had been mentioned above), F adaptations that were necessary or suitable for a man suffering from his disability. In general, it was fair to say that the flat was a normal residential flat which suited the circumstances and requirements of Mr Vandyk.

Should he, in these circumstances, be relieved from rates in respect of his hereditament except as to the site value? The answer must be found in statute, but before he (his Lordship) entered the labyrinth, he thought it advisable to state some general considerations. Everyone had to live somewhere, but there were alternatives. A man might find his own accommodation, or he might go to a housing authority and ask it to provide a house; or if he was seriously ill—mentally or physi- G cally—he might go to a health authority and ask to be put in accommodation where he could be looked after. Health authorities had special powers as to persons requiring care or after-care and these would have to be considered. If a man provided his own home, but had to spend more on it because he was ill or because he needed after-care, one might expect to find in the legislation provision for relieving him from having to pay more rates on account of the extra expense. If, for example, he had to provide a garage for an invalid chair, it would be very unfair to charge him rates on it, and the same principle could be applied to adaptations. One might go so far as to expect that if, exercising his choice, a man, H on account of a disability, selected a dwelling which was more expensive than one which he would acquire if he were not disabled, he should get some relief on account of this excess.

Apart from this type of case, however, one might question whether a man in need of care or after-care should be relieved from rates on his residence altogether. This seemed to introduce a different principle, and if it were to be admitted one would expect to find clear words. The consequences of giving such relief would be very considerable. It would mean relieving from rates all hereditaments (ie buildings) occupied by persons in need of care or after-care, and this whether or not J the particular hereditament needed any adaptation. The relief could not be confined to persons like the respondent, but would equally have to be available for persons suffering from heart disease, or mental illness, if they could claim to be in need of care, and regardless of whether their condition made adaptation necessary. This was pointed out by MacKenna J in the Court of Appeal, and he (his Lordship) could see no answer to it. The majority in the Court of Appeal saw the difficulty and tried to meet it, but he did not think that they succeeded. Cairns LJ said:

> I do not consider that the physical characteristics of the structure can be wholly disregarded. K

This was a negative formula, and it was not clear what positive test it intended to apply. If it was saying that rating relief depended on some physical characteristics, it created a distinction, for which no justification appeared in the legislation, between those who needed a specially-adapted residence and those who did not. Moreover, it could not be enough, in order to entitle the occupier to relief, merely that the hereditament had *some* physical characteristics, but if these had to be specially relevant to the illness or disability, it was difficult

A to see how the present residence possessed them. Then Lawton LJ seemed to apply a different test:

> In order to attract rating relief a structure must be of a size and type reasonably suitable for the ratepayer's needs for after-care.

These were very general words, and it was hard to see how they excluded any residence the ratepayer might reasonably choose. To say that he might not choose a castle was hardly a significant restriction, and he (Lord Wilberforce) did not think that the present respondent could succeed without the decision carrying with it the consequence that anyone in need of after-care who acquired a residence could claim total tax B exemption for the building so long as he could say that it was suitable for the purpose of after-care being provided.

Did Parliament intend so far-reaching and indiscriminate a bounty at the expense of other ratepayers? Though Parliament could, if it wished, have gone so far as this, one must look carefully at the provisions relied on. The relevant provision was section 45 of the General Rate Act 1967, first enacted as section 9 of the Rating and Valuation (Miscellaneous Provisions) Act 1955. There were several things that struck one about the section. First, the introductory words did not suggest the total derating of a hereditament. They suggested that there was a hereditament, but that in valuing C it for rates you were not to take account of special features related to illness or disability, in other words that the purpose was to grant the limited relief of the kind he (his Lordship) had mentioned, relief for special features needed on account of the disability. Secondly, the section throughout referred to "structure"; that was no doubt a general enough word, but the fact that it was used after a reference to the hereditament suggested that the structure was not itself the hereditament but was something added to the hereditament. It was this addition which was to be "disregarded," but one was then left with the hereditament to value. Thus in paragraph D (a) "structure" clearly referred to a garage for an invalid chair.

Thirdly, the section referred to section 28 of the National Health Service Act 1946 and section 29 of the National Assistance Act 1948, and neither of these sections referred to accommodation. So far as the first Act was concerned, it was pointed out that this had now been replaced by the Health Services and Public Health Act 1968, section 12 of which (replacing section 28 of the 1946 Act) did now refer to the provision, equipment and maintenance of residential accommodation for after-care, and it was said that section 28, through more general language, had the same effect. But E what was relevant was the interpretation of the words used in the General Rate Act 1967, the question being whether, when section 45 (b) referred to structures, it included residential accommodation. Even if the earlier Act covered residential accommodation as well as "structures," it seemed a *non sequitur* to say that the later Act (that of 1967) when it referred to structures referred to residential accommodation. As for the Act of 1948, it seemed clear that section 29 did not authorise the provision of residential accommodation. Section 21 of the same Act did so authorise, and the reference in paragraph (c) to section 29, not to section 21, seemed significantly to suggest that "structures," at least in that F paragraph, did not include residential accommodation.

One would thus so far be inclined to accept a restrictive interpretation of the word "structures," and that was what, in the dawn of innocence, all three members of the Court of Appeal did in *Jewish Blind Society Trustees v Henning* [1961] 1 WLR 24. In *Almond v Birmingham Royal Institution for the Blind* [1968] AC 37 the approval of the House was given to an unrestricted meaning of "structure" and that must be accepted as the law. But the very fact that "structure" now bore a wide meaning might make their Lordships more inclined—it certainly permitted them—to find some limitation on the width of the section in other provisions. The fact that Parliament for its own purposes chose to legislate in an G obscure manner did not force their Lordships to be the blind led by the blind. Moreover, it was at least arguable that even through the literal approach, a result less far-reaching than that contended for by the respondent might follow. Assume, as was required by paragraph (d), an identical or similar hereditament supplied by the respondent instead of by the local health authority. In accordance with section 45 paragraph (b) and the incorporated legislation, accommodation could only be supplied if the dominant purpose of the supply was to provide after-care or to enable after-care to be given. But the dominant purpose for which this hereditament was supplied was to provide a residence for the respondent. H

He (his Lordship) agreed with this argument, but whether right or not it was clearly a maintainable argument, and he rested on the principle that if one possible interpretation of a statute led to a result contrary to the evident general purpose, one should prefer to it an interpretation also reasonably possible which did not lead to such a result. He would therefore hold that the respondent did not bring himself within paragraph (d) and that his claim failed. He was also of opinion that the alternative claim for relief in respect of the two rooms failed for the reasons given in the opinion of Viscount Dilhorne. He would add that if the scope of J section 45 was to be extended far beyond local authorities and charities to many thousands of citizens, many of them ill or handicapped, or if the contrary was the case, Parliament really ought to make its dispositions intelligible. He hoped, too, that it might make them fair and rational. He would allow the appeal.

LORD DILHORNE said that the object sought to be secured by the enactment of section 9 (1) of the Rating and Valuation (Miscellaneous Provisions) Act 1955, now replaced by section 45 of the General Rate Act 1967, appeared to him to have been to secure that the occupier of a hereditament should not have its gross value increased in consequence of it con- K taining a structure put there for the use of someone suffering from a disability or illness. Paragraph (a) of the section told a valuation officer to ignore a structure belonging to the Minister of Health and supplied by him or (before August 31 1953) by the Minister of Pensions for the accommodation of an invalid chair or other vehicle constructed or adapted for use by invalids or disabled persons. So, to decide whether a structure came within this paragraph, a valuation officer must not only ascertain to whom the structure belonged and who supplied it, but also its function or purpose. It would have been very unfair if an occupier of a hereditament who supplied such a structure for such a purpose at his own expense were L to have the gross value of his hereditament increased in consequence, when that would not have happened if the structure had belonged to and been supplied by a ministry; and paragraph (d) when read with paragraph (a) operated to prevent such unfairness. Similarly it exempted by paragraph (b) any structure belonging to a local authority, or to certain voluntary organisations, supplied for the use of any person in pursuance of arrangements made under section 28 (1) of the National Health Service Act 1946, those arrangements being for the prevention of illness, for the care of persons suffering from illness or for the after-care of such persons. In relation to this paragraph, a valuation officer had again to determine M not only to whom the structure belonged and by whom it was supplied, but also that it was supplied for use by a person in pursuance of the arrangements mentioned. If so supplied, it was to be inferred that the structure was supplied for use for the prevention of illness or for care or after-care; and the valuation of the hereditament was not to be increased in consequence of the presence of a structure which would not be there but for a person's disability or infirmity.

Examination of the section led him (his Lordship) to the conclusion that the first question to be considered in relation to its application was whether there was any structure in or

A upon the hereditament which would not be there but for some person's illness, disability or infirmity or if it was not required for the prevention of illness. The section did not define "structure," but it followed from *Almond v Birmingham Royal Institution for the Blind* [1968] AC 37 that a structure however large and however constructed might be regarded as coming within the section. It might be something in or on the hereditament; it might indeed constitute the hereditament. The respondent claimed that if his main argument as to the flat as a whole failed, the bathroom and his bedroom as altered should be regarded as a structure to which the section applied. He (his Lordship) regretted that he could not come to the B conclusion that by reason of these alterations, the bedroom and bathroom were to be regarded as a "structure" which would not be there but for his incapacity. Even if they were to be so regarded, he (Lord Dilhorne) doubted very much if the alterations made would increase the gross value of the flat, and it was the increase in the gross value due to the presence of a structure to which the section applied which had to be ignored. In his opinion, the respondent's claim in relation to these two rooms failed.

As to the respondent's main claim that in ascertaining the gross value of the hereditament the value of the flat must C be ignored as being a structure to which paragraph (b) or (c) or (d) of the section applied, he (his Lordship) could not regard the flat as a structure for the purpose of care or after-care or for the prevention of illness or for the welfare of a handicapped person. It appeared to be a perfectly ordinary flat of five rooms, two bathrooms and dining-room, similar to the other 37 flats in the building. The flat was the respondent's home. In it he received after-care, but it was not a building or structure designed for his after-care or welfare. It was not a hereditament of which the gross value would, if the section did not apply, be increased due to the presence there of a structure required for his after-care or D welfare as a handicapped person. If it had been the intention of Parliament that persons in the position of the respondent, and other handicapped and ill persons, should be relieved of liability for rates in respect of their homes, he (his Lordship) would have expected that to have been clearly stated. The appeal should be allowed.

Lord Simon of Glaisdale agreed with the speech of Lord Dilhorne.

In a dissenting speech, **Lord Edmund-Davies** said that in *Almond v Birmingham Royal Institution for the Blind* [1968] AC 37 Lord Reid said of section 9 (1) (d) of the Rating and Valuation Act 1955 that "it raised . . . difficult problems." E That observation was equally true of its successor, section 45 of the General Rate Act 1967, as the present appeal amply demonstrated. He (his Lordship) could find no warrant for the appellant's submission that to get rate relief, accommodation had to be of an *institutional* character. Furthermore, he thought that the proper question was not whether a local health authority or voluntary organisation would be likely to provide such private accommodation for private after-care as the flat with which the House was presently concerned, and still less whether any local authority or voluntary organisation had in fact made such provision. The relevant question was whether the authority or organisation could lawfully provide F it. The Lands Tribunal and all three members of the Court of Appeal held that they could, and in his (Lord Edmund-Davies's) judgment that conclusion was correct. He did not think it correct to say that a local health authority could do no more than provide care (defined by Denning LJ in *Minister of Health v Royal Midland Counties Home for Incurables* [1954] Ch 530 at 547 as "the homely art of making people comfortable and providing for their well-being"), or that if this care was provided by members of the family of the person being cared for "accommodation" could never be provided.

Then did the foregoing conclude the inquiry? In other words, if the local health authority could have supplied this flat to Mr Vandyk for the purpose of his after-care by members G of his family, did that automatically entitle him to rating relief? Or were (a) his own purpose in taking the flat, and (b) the purpose for which it was being used, of relevance? As to (a), Cairns LJ in the Court of Appeal rejected the submission for the valuation officer that the relevant question was whether Mr Vandyk's "sole or dominant purpose" was for the purpose of after-care, and added:

> Now there is not in the legislation any adjective qualifying the word "purpose." This may be contrasted with the Rating and Valuation (Apportionment) Act 1928, the proviso to section 3 (1), where an industrial hereditament is defined so as not to include a factory if it is primarily used for non-industrial purposes. H

In similarly rejecting the dominant purpose test, Lawton LJ said:

> The question which should be asked is this: why did the ratepayer provide himself with this flat, being a structure for the purposes of section 45 of the General Rate Act 1967?

He (Lord Edmund-Davies) thought that such inquiry was relevant, and in his opinion both purposes for which the hereditament was at the date of the proposal being used bore upon the entitlement to be derated. MacKenna J considered that Mr Vandyk's dominant purpose "was to provide a home for himself and his family," and accordingly held against his J claim to be derated. As such a purpose was inevitably of considerable importance whenever accommodation was being sought, and might indeed have chief prominence and importance, the adoption of the dominant purpose test would (as counsel for the valuation officer candidly conceded) for all practical purposes mean that no person seeking private accommodation for himself and his family would ever establish entitlement to be derated under section 45, whatever the type of accommodation sought and however specialised his needs. He (his Lordship) could not accept this, and thought that the proper way of looking at this case was that adopted by Lawton LJ in dealing with the question he had posed: K

> On the evidence the answer is clear. He wanted somewhere to live which was not an institution; and if he was to live outside an institution his place of abode would have to have certain physical characteristics or be capable of being adapted to have those characteristics which would enable him to receive after-care, for example, the absence of steps, doors wide enough to take a wheelchair, wall or ceiling devices to enable him to move in bed, washbasins at a convenient height, specially adapted wc seats, and reasonable accommodation for his personal attendant, without whose help he could do very little.

The appeal should be dismissed. L

Lord Fraser of Tullybelton said that the purposes which qualified under paragraph (b) of section 45 were those specified in section 12 of the Health Services and Public Health Act 1968. They were three in number: the prevention of illness, the care of persons who were ill, and the after-care of persons who had been ill. The provision of residential accommodation must be for one of those purposes if it was to be *intra vires* a local health authority or voluntary organisation acting under section 12, and it was implicit in the section that one of them must be the predominant purpose in the mind of the authority making the provision. Though the word "predominant," or any equivalent, did not appear in the M section, it was necessarily to be implied in order to give effect to the scheme of the section. Section 12 was dealing, as the side-note showed, with "prophylaxis, care and after-care," and subsection (1) began by conferring on a local health authority a general power to make arrangements for those three purposes. Then it proceeded, without prejudice to the generality of the opening provision, to confer some more specific powers, one of which was the power to make arrangements for the provision of residential accommodation. All the specific powers were intended to be used for one of the three purposes, and there could be no doubt of that so far

A as the power of providing residential accommodation was concerned, because it was expressly so stated (section 12 (1) (a)). If those were not intended to be the predominant purposes, the powers of a local health authority would be extended far beyond the bounds of public health.

He (his Lordship) agreed with the statement of MacKenna J in the Court of Appeal that nothing less than a dominant purpose of the supplier to prevent illness or to provide for the care or after-care of persons suffering from illness would suffice to bring a structure within paragraph (b) of section 45 of the Act of 1967. If that was the meaning of paragraph (b), then when paragraph (d) referred to a structure which was of
B a kind "similar to" that mentioned in paragraph (b), it must refer to one which was supplied for the predominant purpose of care or after-care. The predominant purpose must in each case be that of the supplier, who in this case was the respondent himself. The suggestion that the respondent's flat had been "supplied" for his use by the owner who let it to him seemed quite untenable. It was supplied for the use of the respondent by himself when he took a lease of it. It was therefore the predominant purpose in his mind that was material, and it seemed clear that the predominant purpose of the respondent in taking this flat was to have somewhere, other than an institution, for himself and his family to live.
C He chose this particular flat, after a long search, because it suited his requirements by having, among other things, wide doors through which he could pass in a wheeled chair, and he had made some alterations in it to make it more convenient. But it was not possible to treat these matters as showing that his main reason for taking it was to obtain after-care. His main reason was to have a home. The claim for derating under paragraph (d) read with paragraph (b) must accordingly fail.

The appeal was allowed. The valuation officer had previously undertaken to bear the costs in any event.

D

HOUSE OF LORDS

November 17 1976

(Before Lord WILBERFORCE, Lord DIPLOCK, Lord SIMON OF GLAISDALE, Lord KILBRANDON and Lord EDMUND-DAVIES)

UNION COLD STORAGE CO LTD v PHILLIPPS (VO)

Estates Gazette February 19 1977

(1976) 241 EG 531

Rating of cold store divided internally by insulated walls—
E **Lands Tribunal properly directed as to law—Tribunal's finding that each internal chamber was "a chamber for . . . refrigerating" not open to challenge on appeal—Conceded by ratepayers that on this basis each chamber was "in the nature of a . . . structure"—Whole insulation assembly deemed part of hereditament**

This was an appeal by Union Cold Storage Co Ltd from a decision of the Court of Appeal on July 17 1975 holding that the insulation assembly in a cold store at Fairbrother Street, Salford, was deemed under section 21 (1) (a) of the General Rate Act 1967 to be part of the hereditament for the purpose of ascertainment of rateable value. The Lands
F Tribunal's decision was reported at (1973) 226 EG 1965 and the Court of Appeal's decision at (1975) 236 EG 125.

M Chavasse QC and F A Amies (instructed by R A Roberts) appeared for the appellants, and P Medd QC and A Fletcher (instructed by the Solicitor of Inland Revenue) represented the respondent valuation officer.

In a brief speech, LORD WILBERFORCE said that he had had the benefit of reading Lord Diplock's opinion in advance. He entirely agreed with it, and would dismiss the appeal.

LORD DIPLOCK: The question in this appeal is whether the insulation assembly in a cold store occupied by the ratepayers is to be deemed to be part of the hereditament under section 21 (1) (a) of the General Rate Act 1967 for the G
purpose of ascertaining the rateable value of the hereditament This depends upon whether the insulation is plant or machinery falling within class 4 of the Plant and Machinery (Rating) Order 1960. Class 4 contains a lengthy list of various items of plant or machinery introduced by the words:

> The following items, except:
> (a) any such item which is not, and is not in the nature of, a building or structure; and
> (b) any part of any such item which does not form an integral part of such item as a building or structure or as being in the nature of a building or structure. H

The list which follows includes "chambers, vessels and containers for . . . refrigerating." That part of the hereditament in which the insulation assembly is housed consists of an outer shell of brick and concrete walls, enclosing an insulated floor and covered by a corrugated asbestos roof. Within this is an inner shell of insulated walls attached to rails fixed to the outer walls so as to leave a space between the two walls, and an insulated ceiling suspended from steel girders attached to the roof joists of the outer shell. The insulated space thus formed is divided by internal insulated walls into 16 separate chambers for refrigerating, two of which were used for freezing and the remainder as cold stores, and J
includes a corridor providing a means of access to some of the chambers. Full details of the method of construction are to be found in the case stated by the Lands Tribunal and the various drawings attached thereto. A more condensed description is to be found in the judgment of Buckley LJ in the Court of Appeal. Before the Lands Tribunal, the argument for the ratepayers was put in two ways. One was on the basis that each of the 16 rooms into which the cold store was divided constituted a separate item under class 4 of the 1960 order. The alternative argument was on the basis that the outer shell constituted a single chamber for refrigerating. On the 16-items basis, the ratepayers contended K
that none of the individual chambers formed by the insulating walls, ceiling and floor which enclosed it, was in the nature of a building or structure. On the single-item basis, they argued that the insulating walls lining the outer shell, the ceiling and the partition walls did not form an integral part of the chamber formed by the external walls and the roof and floor. Before your Lordships' House, however, the contention that if there were 16 chambers the insulating walls and ceilings which constituted each chamber were not in the nature of a building or structure was abandoned by the ratepayers. Moreover, it was never contended on the basis of L
there being not one but 16 chambers that the panels and other components from which they were built were not integral parts of the walls and ceilings. In other words, if it be right that there are 16 chambers for refrigerating, it is conceded that they do not fall within either of the exceptions to class 4 of the 1960 order.

The initial question for the Lands Tribunal to determine was whether the disputed part of the hereditament comprised 16 chambers for refrigerating or only one. This is in my view a question of fact for the Lands Tribunal, and provided that the tribunal did not misdirect itself in law as to the tests to be applied in identifying what constitutes an "item," M
its finding on this matter is conclusive for the purposes of an appeal to the Court of Appeal. The finding of the Lands Tribunal was expressed as follows:

> For my part, having seen the premises, I think there are on the appeal hereditament 16 chambers for refrigerating which are "items" included in class 4 of the 1960 order. Each of the rooms seems to me to be properly described as a chamber and each chamber is used for refrigerating. It is first necessary, therefore, to consider the disputed "bits" in order to decide whether they by themselves can properly be said to constitute the chambers for refrigerating which are on the hereditament. In my opinion they clearly do, and I can think of no other words more appropriate

to describe the space enclosed by these insulating panels put together in the manner described than a "chamber."

It is evident from the reference to "bits" that the Lands Tribunal was following the direction given by Lord Reid in *Shell-Mex & BP Ltd v Holyoak (VO)* [1959] 1 WLR 188 at 198:

> . . . looking to the descriptions of a number of the items in it [sc class 4] such as burners, crane gantries and stagings, it seems to me to be clear that they need not be, and in some cases cannot be, capable of working as self-sufficient units. So the order requires one to pick out of the whole installation of plant and machinery those bits which correspond to any of the items mentioned in the list irrespective of whether or not they require to be associated with other things in order to produce any useful result. Then the next step is to consider whether each bit of plant so picked out is or is not, taken by itself, a building or structure or in the nature of a building or structure.

Lord Reid was dealing with an earlier order, the Plant and Machinery (Valuation for Rating) Order 1927, in which the wording of the introductory words in class 4 was not identical with the introductory words in the 1960 order which I have set out above, but in my view the passage which I have cited from Lord Reid's speech is applicable also to the 1960 order.

The Lands Tribunal accordingly did not misdirect itself in law. With its findings that the disputed part of the hereditament comprised 16 chambers for refrigerating, neither the Court of Appeal nor your Lordships' House has any jurisdiction to interfere. Since it is now conceded that the Lands Tribunal was right in finding that each of the chambers was in the nature of a structure and so do not fall within exception (a) of class 4, it follows that this appeal must fail. The Court of Appeal did not consider whether the Lands Tribunal had misdirected itself in law in finding that the disputed part of the hereditament did not constitute a single chamber for refrigerating, but comprised 16 chambers. The court dismissed the appeal upon the grounds that even assuming that the whole space within the outer shell constituted a single chamber for refrigerating, the internal insulating walls and ceiling were structurally an integral part of it and so would not fall within exception (b) of class 4. In the absence of any finding by the Lands Tribunal as to the structural integration of the internal insulating walls and ceiling with the outer shell, Mr Medd, for the valuation officer, has not sought in your Lordships' House to rely upon this ground. I would accordingly dismiss the appeal upon the ground that the Lands Tribunal's finding that there were 16 chambers for refrigerating has not been shown to have been wrong in law. It is accordingly conclusive, and in the light of the concession that the chambers are a structure, or in the nature of a structure, is sufficient to dispose of this appeal.

LORD SIMON OF GLAISDALE: I have had the advantage of reading in draft the speech delivered by my noble and learned friend, Lord Diplock. I agree with it, and would therefore dismiss the appeal. The crucial issue is to identify the "item" for the purpose of the words introducing class 4 in the schedule to the Plant and Machinery (Rating) Order 1960 (SI 1960/122). There was an unimpugnable finding of fact by the Lands Tribunal that each of the 16 insulated chambers within the hereditament was such an "item," each being a "chamber . . . for . . . refrigerating," one of the "items" in the class 4 list. Do any of these 16 "items," then, fall within either of the exceptions in the introductory words? It is conceded that each was a structure or in the nature of a structure: thus they cannot fall within (*a*). Exception (*b*) is concerned with excepting part of any such "item" (here each of 16 insulated chambers) which does not form part of such "item" as a structure or something in the nature of a structure. So far as concerns any matter in this appeal, there was no part of any of these 16 insulated chambers which did not form part of it as a structure or something in the nature of a structure. It follows that each of the 16 insulated chambers was, for the purposes of rateability, "deemed to be part of the hereditament" by virtue of the schedule to the statutory instrument.

I would only add that in my respectful opinion, contrary to the contention on behalf of the appellant, *Manchester Marine Ltd v Duckworth (VO)* [1973] 1 WLR 1431 correctly distinguished *Inland Revenue Commissioners v Barclay, Curle & Co Ltd* [1969] 1 WLR 675, though both were concerned with consideration of a dry dock as "plant." But the former was a rating, the latter a revenue, case. The revenue case was concerned with an "initial allowance" when the taxpayer incurred expenditure on the installation of plant. The purpose of the revenue provisions—to encourage investment—demanded that the whole installation in question should be regarded as a "functional entity." The prime purpose of the rating legislation is, however, to promote fairness in the valuation list between one ratepayer and another. Both this general approach and the statutory language itself show, in the words of Lord Reid, that it is "the purpose of this legislation to split up for rating purposes installations which are for practical purposes indivisible" (*Shell-Mex & BP Ltd v Holyoak (VO)* [1959] 1 WLR 188, 199). Thus the Lands Tribunal was fully entitled to consider that each of the 16 insulated chambers within the hereditament, each being a "chamber for . . . refrigerating," was a separate "item" within class 4.

LORD KILBRANDON: For the reasons given by my noble and learned friend, Lord Diplock, in his speech, which I have had the advantage of seeing in draft, I would hold that the decision of the Lands Tribunal was right, and that this appeal must accordingly be dismissed.

LORD EDMUND-DAVIES: In my judgment, the decision of the Lands Tribunal was clearly right, and this for the reasons given in the speech of my noble and learned friend, Lord Diplock. I therefore concur in holding that this appeal should be dismissed.

QUEEN'S BENCH DIVISIONAL COURT

December 15 1976

(Before Lord WIDGERY CJ, Mr Justice TALBOT and Mr Justice SLYNN)

R v LIVERPOOL CITY JUSTICES *EX PARTE* LANCKRIET

Estates Gazette March 5 1977
(1976) 241 EG 685

General Rate Act 1967—Motion to set aside justices' committal orders—Lack of inquiry under section 103 as to reason for failure to pay rates—Order of certiorari to quash orders

In these proceedings Mrs Sheila Margaret Lanckriet moved for an order of certiorari to bring up and quash two orders made by Liverpool City Justices on November 2 1976, on an application by the rating officer to Liverpool City Council, for her committal to prison for 90 days for non-payment of general rate arrears and costs totalling £122.

D M Evans (instructed by Woolwich Lander & Co, of Liverpool) appeared for the applicant, and R B Martin (instructed by the Liverpool City Solicitor) represented the respondents.

Giving judgment, LORD WIDGERY said that the modern method of collecting rates was governed by the General Rate Act 1967. By section 97 of the Act the rating authority could apply to justices for a warrant to levy distress on the ratepayer's property to the value of the unpaid rates. No question of committal to prison could arise until an application was made under section 102 on the ground that there

A were no goods sufficient to satisfy the distress warrant. Section 103 contained an important safeguard in that justices, on an application for committal, had to inquire, in the ratepayer's presence, whether failure to pay was due to either wilful refusal or culpable neglect. What was therefore contemplated was that, before the final sanction was put into effect, the justices had to be satisfied that there had been wilful refusal or culpable neglect.

What happened in the present case was that the matter before the justices was dealt with quite briskly. The applicant was asked by the chairman of the bench if she could pay and, when she said she could not, the chairman said B she would have to go to prison. It was clear beyond argument there had been no proper inquiry by the justices as to the reason for failure to pay and certiorari should go to quash the two orders.

Talbot and Slynn JJ agreed, and the applicant was granted an order of certiorari to quash the committal orders.

QUEEN'S BENCH DIVISIONAL COURT

C November 26 1976

(Before Lord WIDGERY CJ, Mr Justice EVELEIGH and Mr Justice SLYNN)

DIXON v HARDING AND OTHERS

Estates Gazette March 12 1977

(1976) 241 EG 765

Rating surcharge on property not used for purpose for which it was constructed—Unoccupied premises—No rates payable—Question of liability to surcharge—Liability not contingent on occupation—Case sent back to justices

This was an appeal by way of case stated against the dis- D missal by Suffolk justices of a complaint by Eric Kendall Dixon on behalf of Ipswich Borough Council that Lenore June Harding and three others had failed to pay a rating surcharge under section 17A of the General Rate Act 1967 on commercial property at 21 St Margaret's Green, Ipswich.

A Fletcher (instructed by Sharpe, Pritchard & Co, agents for the town clerk of Ipswich) appeared for the appellant, and N Nardecchia (instructed by Gotelee & Goldsmith, of Ipswich) represented the respondents.

Giving the judgment of the court, Slynn J said that this was an appeal by way of case stated in respect of the adjudication of justices for the county of Suffolk sitting at E Ipswich on February 10 1976. On December 17 1975 a complaint had been made by the appellant, acting on behalf of Ipswich Borough Council, that the respondents had failed to pay a rating surcharge totalling £1,048.17 alleged to be owed in respect of certain unused commercial property at 21 St Margaret's Green, Ipswich, for a period from February 8 1974 until September 30 1975. Liability to a progressive surcharge was imposed by section 17A of the General Rate Act 1967, which section was added to the General Rate Act by the Local Government Act 1974, section 16. For the purposes of the hearing it had been admitted that the property was a commercial building that had not been used for the purpose F for which it was constructed or adapted for a continuous period of six months. It was further admitted that the respondents were the persons liable to pay any rates due in respect of the property, that neither the respondents nor any other persons had been in occupation of the property during the relevant period and that no rates were payable in respect of the property during the relevant period.

On these facts the appellant had contended that the respondents were liable to the surcharge in accordance with the section and that the surcharge fell to be calculated by reference to the notional figure of the rates which would have been payable had the building been occupied. The respondents' case was that, since no rates were payable G during the relevant period, there was no normal rate which could have been taken in order to achieve the figure by way of surcharge. The essence of that case had been that no surcharge could be made under section 17A in respect of unoccupied property unless a resolution had been passed by a rating authority, under section 17, bringing into operation the provisions of Schedule 1. No resolution had been passed during the relevant period in respect of this area, although subsequently it was resolved that Schedule 1 to the Act should apply from April 1 1976.

In their Lordships' judgment the effect of section 17A was that, so far as a surcharge was concerned, the amount was H to be the same whether or not the owner was in occupation and was to be based upon the rates ordinarily payable when the owner was in occupation. The respondents' argument was that, since no rates had been payable, the normal rates were in effect nil, so that there could have been no surcharge. This ignored the provisions of subsection 4 (a) that when an owner is not in occupation during the period of non-use the amount of the surcharge shall be the same as if the owner had been in occupation. Their Lordships did not consider that it was a precondition to liability under section 17A in respect of unoccupied property that a resolution should have been passed under section 17. The two J sections were, in their judgment, dealing with different matters. Accordingly, in their judgment, whether section 17A was looked at in isolation or in conjunction with section 17, the premises in the present case were liable to the surcharge despite the fact that no rates were payable because the property was unoccupied.

The appeal should be sent back to the justices with a direction for the hearing to continue.

K

QUEEN'S BENCH DIVISIONAL COURT

February 9 1977

(Before Lord WIDGERY CJ, Mr Justice MICHAEL DAVIES and Mr Justice GOFF)

LONDON ELECTRICITY BOARD v LONDON BOROUGH OF TOWER HAMLETS

Estates Gazette April 16 1977

(1977) 242 EG 205

General Rate Act 1967—Whether Electricity Board's unoccupied showroom liable to be rated—Effect of rating of L unoccupied property—Exception for showrooms not occupied or used unaffected by empty property rate—Distress warrant quashed

This was an appeal by the London Electricity Board from a decision of a stipendiary magistrate sitting at Thames Magistrates' Court on October 13 1975, granting the London Borough of Tower Hamlets a distress warrant for the recovery of £386 rates due on premises owned by the board in Roman Road, Bow.

D Widdicombe QC and A Alesbury (instructed by Waltons & Morse) appeared for the appellants; C Joseph and C Magill (instructed by the Solicitor to the London Borough M of Tower Hamlets) represented the respondents.

Giving judgment, Michael Davies J said that the appellants' complaint was that the magistrate erred in law because the period for which the rates were claimed was a time when the premises were both unoccupied and unused.

Section 34(1) of the General Rate Act 1967 provided that, subject to the provisions of section 34(2), no premises which were, or formed part of, premises occupied by an electricity board were liable to rates or to be included in any valuation list. Section 34(2)(b) provided that section 34(1) should not apply to a showroom or any other place occupied and used

A by an electricity board either wholly or mainly for the display of apparatus.

There was no dispute that the premises in the present case were used by the appellants until November 1974 as showrooms and accordingly section 34(1) did not apply and they were rated in the ordinary way. It was now said on behalf of the appellants that when the premises were empty and not occupied or used by the appellants they were not taken out of section 34(1) and there was therefore no question of their being liable for rates. That submission had been accepted by the learned stipendiary and he (his lordship) agreed with his finding in that respect. On the face of it that would
B seem to end the case, but matters had gone further, because what had happened was that the respondents, by virtue of powers in section 17 of the 1967 Act, decided that the provisions of Schedule 1 to the Act should apply and unoccupied premises should be liable for rates. For the respondents it was contended before the magistrate, who accepted it, that para 1 of Schedule 1 should be construed as meaning that unoccupied premises should be deemed to be occupied and liable for rates. He (his lordship) found that argument unacceptable. It seemed to him that if it had been intended to provide for the situation now urged by the respondents the schedule could easily have "deemed" that this situation
C existed for the purposes of section 34. There was, however, nothing in the schedule which caused him to think that the conclusion which one reached from section 34(1) and (2)(b) was affected in any way. The situation was not changed in any way by Schedule 1. It did not provide that the premises should be deemed to be occupied and used for the purpose of a showroom.

The stipendiary magistrate reached a conclusion on the facts which was correct, namely, that the occupation and use by the board had to be established in order to bring the premises into section 34(2) but was in error in his conclusion
D that Schedule 1 enabled the respondents to rate and assess the premises during the period of non-occupation. For those reasons the appeal succeeded and the distress warrant should be quashed.

LORD WIDGERY CJ and GOFF J agreed. The appeal was allowed and the distress warrant quashed.

QUEEN'S BENCH DIVISIONAL COURT

February 19 1977

E (Before Lord WIDGERY CJ, Mr Justice MICHAEL DAVIES and Mr Justice GOFF)

PEARCE v CROYDON JUSTICES

Estates Gazette April 16 1977
(1977) 242 EG 207

General Rate Act 1967—Application to quash order for warrant of commitment—Duty of justices to inquire into means before issue of warrant—Application dismissed—Full inquiry made by justices

In these proceedings counsel moved on behalf of David Pearce, of Heathfield Road, Croydon, for an order of
F certiorari to quash a decision of Croydon justices on October 21 1975 whereby they ordered the applicant to pay £665.20 rate arrears or go to prison for six weeks.

Stephen Walsh (instructed by Cohen & Naicker) appeared for the applicant. The respondent justices were not represented.

ROBERT GOFF J said that the background to the case was that the applicant bought the house 17 Heathfield Road, Croydon, in April 1972. Apparently at that time part of the premises was used as an office for the business of selling coin laundry equipment. Shortly after purchasing the house on mortgage the applicant ran into financial difficulties and the office ceased to exist, but it was not until 1975 that the local authority finally agreed to change the entry on the valuation list relating to the property. G

By section 96 of the General Rate Act 1967 provision was made for enforcement of rate demands by distress and by section 103(1) justices were required to make an inquiry as to whether failure to pay was due to wilful refusal or culpable neglect. By that section justices should make an inquiry as to means before issue of a warrant of commitment. The applicant now sought to quash the order made against him on the ground that there had been an error of law. The main contention was that there had been a denial of natural justice in that the justices had failed to inquire into the applicant's H
means before making the order. When one turned to the facts of the case it was clear from the affidavits sworn on behalf of the justices that a full inquiry did take place. The more one looked at the case the clearer it became that the justices did not err in law. What the applicant was seeking to do was to persuade the court to review the justices' factual findings. That the court would not do. The application should be dismissed.

LORD WIDGERY CJ and MICHAEL DAVIES J agreed and the application was dismissed.

J

QUEEN'S BENCH DIVISIONAL COURT

March 18 1977

(Before Lord WIDGERY CJ, Mr Justice EVELEIGH and Mr Justice WIEN)

ROYAL BOROUGH OF KENSINGTON AND CHELSEA v VICTORIA WINE CO LTD

Estates Gazette April 16 1977
(1977) 242 EG 207

General Rate Act 1967—Appeal from dismissal by justices of claim for unpaid rate surcharge on unused commercial building—Definition of "commercial building"—Justices' "bewilderment"—Building not a commercial building, but a dwellinghouse—Appeal dismissed K

This was an appeal by way of case stated by justices sitting at Kensington on February 12 1976 when they dismissed a complaint preferred by the appellants, the Royal Borough of Kensington and Chelsea, against the respondents, the Victoria Wine Co Ltd, for non-payment of £852.42 rating surcharge made on premises comprising the first, second and third floors of premises at 18 Notting Hill Gate, Notting Hill, L
pursuant to section 17A of the General Rate Act 1967 (inserted by section 16 of the Local Government Act 1974).

Guy Roots (instructed by A G M Ellery, Borough Solicitor) appeared for the appellant local authority; J Grove (instructed by Lovell, White & King) represented the respondents.

Giving judgment, EVELEIGH J said that by section 17A(1) of the General Rate Act 1967 "If for a continuous period exceeding six months a commercial building is not used for the purpose for which it was constructed or has been adapted, its owner shall pay in respect of that period (the 'period of non-use') a surcharge additional to the rates. . . ." Section M
17B defined "commercial building" as referring to a hereditament not being a dwellinghouse. Section 115, the definition section, stated that "'dwellinghouse' means a hereditament which, in accordance with Schedule 13 to this Act, is used wholly for the purposes of a private dwelling or private dwellings." Schedule 13, paragraph 2(1), made reference to a building which was used for the letting of rooms: "If in the case of a hereditament which is used for the letting of rooms singly for residential purposes there is used for such lettings the whole, or substantially the whole, of the available accommodation . . . that hereditament shall be

A deemed not to be used for the purposes of a private dwelling or private dwellings. . . ."

The relevance to the letting of single rooms emerged from the facts found by the justices. They found that the building in the present case had not been used for a period of six months; that it was built as a single dwellinghouse in about 1900; that the interior state of repair and decoration was poor and that, in the condition as they found it, there was no reason why it could not be used either as a dwellinghouse or for the letting of rooms. The property had been purchased by the respondents in 1974 when they became lessees. The ground floor was separately occupied as a shop. The lease
B of the first, second and third floors provided that the respondents should not underlet except with the agreement of the landlords. In the course of its life the building had been used for the purpose of letting single rooms. The date when that ceased was not known. At the date the respondents took over the lease there were gas meters in each of the rooms including the bathroom. Since taking over the respondents had, at all times and in good faith, intended to carry out works necessary for the building to be used as a dwelling-house or two flats and they had done their best to procure the carrying out of such works but, at the time of the hearing before the justices, the designs and costs had not been
C finalised. On May 28 1975 the respondents applied for, and were granted, planning permission for conversion of the premises so as to form one one-bedroomed flat on the first floor and two two-bedroomed flats on the other two floors.

In these circumstances the justices had to consider whether it was a commercial building and, on referring to the Act, they found themselves in difficulties. One could summarise the effect of the Act in the following way: section 17A imposed a surcharge upon commercial property which was not in use. Section 17B defined "commercial building" as one which was not "a dwellinghouse." The definition section
D defined "dwellinghouse" as in effect a house for dwelling in unless let in single rooms. One could understand the bewilderment of the justices when they found they had to determine whether to impose a surcharge on a building not in use and had to ask themselves the question "what was it used for?"

It was not surprising that the justices decided to have nothing to do with the statutory definition but held, as a matter of commonsense, that the building was not commercial. But the definition in the Act had to be complied with and the words had to be given their proper meaning. The definition section was, in truth, describing the kind of building
E concerned generally. Thus, the phrase "is used" meant "existing for the use." It did not mean that the use was actually being carried on. The magistrates had come to the correct decision. They had considered all the circumstances and there was ample evidence to support their conclusion. The appeal should be dismissed.

Lord Widgery CJ and Wien J agreed and the appeal was dismissed with costs.

F **QUEEN'S BENCH DIVISION**

February 4 1977

(Before Mr Justice THOMPSON)

CAMDEN LONDON BOROUGH v THE POST OFFICE

Estates Gazette April 23 1977

(1977) 242 EG 289

Unpaid rates on empty property—General Rate Act 1967, section 17 and Schedule 1—Borough's claim against Post Office—Completion notice served did not relate to the actual hereditament let to Post Office—Condition precedent to liability to empty property rate not satisfied—Borough's claim dismissed

G In this action Camden Borough Council claimed unpaid rates of £64,550.74 levied as empty property rates upon a part of Euston Tower in Camden let to the Post Office. The Post Office denied liability to pay the rates claimed which covered the period December 11 1970 to September 11 1971.

Gerald Moriarty QC and G R G Roots (instructed by B H Wilson, Solicitor, Borough of Camden) appeared for the plaintiffs, and Guy Seward (instructed by Saul Rothstein, Solicitor, the Post Office) represented the defendant.

Giving judgment, Thompson J said: This claim by the plaintiffs is brought in their capacity of rating authority for the Borough of Camden. It relates to a claim for rates for
H the period December 11 1970 to September 11 1971. In the demand for payment it is described as "empty property rate." It is made upon the defendant as occupier of what are described as "Offices and Premises 20th/33rd Floors Part 34th Floor and 43 Car Spaces Basement" at Euston Centre. I have had no oral evidence. The evidence consists of a statement of agreed facts and various documents exhibited to that statement. The part of Euston Centre with which I am concerned is Euston Tower. This 36-storey tower block rises to a height of some 400 ft and is reputed to be the tallest office building in London.

Certain dates are agreed to be important:

J (1) By resolution dated July 19 1967 the plaintiffs adopted Schedule 1 to the General Rate Act 1967. Section 17 of the Act, which has the side note "Liability to be rated in respect of certain unoccupied property," provides in section 17, sub-section (1): "A rating authority may resolve that the provisions of Schedule 1 to this Act with respect to the rating of unoccupied property—(a) shall apply . . . to their area, and in that case those provisions shall come into operation . . . in that area on such a day as may be specified in the resolution." I have not heard whether the resolution specified a date other than the date of its passing, but it does not matter, for it is
K accepted that the provisions of Schedule 1 were in operation consequent upon a properly passed resolution by the time of the next date which initiated the matter before me.

(2) The next date is November 21 1969. On that date the plaintiffs served a completion notice on the owners of the Euston Centre development to the address of their appointed agents, Herring, Daw & Manners. The completion notice was enclosed with a letter to the agents which said—correctly—that the completion notice was in respect of the 36-floor tower block, and this letter confirmed arrangements for a meeting at the premises when two gentlemen from the plaintiffs were
L to meet two from the agents. The notice is exhibit 5 and is in these terms:

> Completion of newly erected or altered buildings. Re: 286 Euston Road (Euston Centre—36-storey Tower Block). The council is of the opinion that the work remaining to be done on the above-mentioned building is such that the erection of the building can reasonably be expected to be completed within three months and that when so completed it will be a hereditament liable for unoccupied property rate in accordance with section 17 and Schedule 1 of the General Rate Act 1967 and the council's resolution made under the said section 17. Notice is hereby given to the effect that the erection of the building is to be treated for the purposes of Schedule 1 of the General Rate Act 1967 as
> M completed on February 20 1970.

That was sent to the Balgray Investments Ltd, who were the owners of this building. This notice was served under paragraph 8(1)(b) of Schedule 1, and that provides:

> Where a rating authority are of opinion . . . (b) that the work remaining to be done on a building within their area is such that the erection of the building can reasonably be expected to be completed within three months, and that the building is, or when completed will be, comprised in a relevant hereditament, the authority may serve on the owner of the building a notice (hereafter in this paragraph referred to as "a completion notice") stating that the erection of the building is to be treated for the

A purposes of this Schedule as completed on the date of service of the notice or on such later date as may be specified by the notice.

The completion notice appointed February 20 1970 as the date at which the erection of the building of the 36-storey tower block was to be treated for the purposes of Schedule 1 as completed. A person served with a completion notice can appeal to the county court on the ground that the erection of the building to which the notice relates cannot reasonably be expected to be completed by the date specified in the notice. This right of appeal was exercised in this case and within the time limited by paragraph 8(4) of the Schedule. B In fact, it was exercised on the third date which is important:

(3) December 3 1969. The appeal was due to be heard on February 5 1970 in Clerkenwell County Court. Before that date the parties to the appeal were having discussions and, as a result of these discussions and consequent upon a letter of January 19 1970, on:

(4) January 26 1970, by a letter from the plaintiffs to the appointed agents, an agreement was concluded. The letters of January 19 and 26 1970 are exhibit 6.

What was agreed provided for the substitution of a different and later completion date than the notice had done. Had C the appeal gone forward the county court could have determined the date on which the erection of the building should be treated as completed. As it was, the parties having agreed as to substituted dates, the plaintiffs' notice was withdrawn and the owners' appeal was abandoned. So far as material to the problem before me, the agreement provided: ". . . the whole of: Euston Tower, 286 Euston Road, NW1, with the exception of the ground-floor shops shall be deemed to be complete on September 11 1970, and the empty property rate will be payable from December 11 1970." Agreement was also come to at the same time as to the completion date for the shop units, and the date for the empty property rate to D become payable on the shop units. The shop units do not concern me.

(5) The next important date is December 7 1970. Between January 26 1970, when the agreement was concluded between the plaintiffs and the owners' agents, and December 7 1970 the owners had let to the Minister of Public Building and Works " the whole of the building known as Euston Tower . . . but excluding parts of the basement, ground and first floors." This letting was by a lease dated June 9 1970. On December 7 1970, the Department of the Environment (the new and then current title for the Ministry of Public Building and Works) sublet to the defendant the whole of the E 20th to 33rd and part of the 34th floors of Euston Tower, together with 43 car-parking spaces in the basement for a term commencing on March 25 1970. In pursuance of this subletting the defendant took possession on December 7 1970, that is to say, four days before the date from which empty property rate is claimed.

(6) The next important date is March 26 1971. On that date the valuation officer for the area made a proposal for the inclusion in the valuation list of an entry in respect of "Offices and Premises 20th/33rd floors, part 34th and 43 Car Spaces in Basement" at Euston Tower. A year later, F on March 29 1972, the valuation officer made a further proposal in respect of the same premises, but with slightly different figures, for gross and rateable values, from those proposed in March 1971. The defendant objected to both proposals and in the end an agreement was reached substituting in respect of both proposals an agreed gross and an agreed rateable value for the defendant's premises. These agreed values were thereafter entered on the valuation list, but not, I think I was told, until September 1973.

(7) Meanwhile the plaintiffs and the defendant had agreed that the defendant took normal occupation, for rating purposes, of the premises sublet to it, on September 12 1971. The effect was that from that date full rates became payable G by the defendant and they have been paid from that date.

It is an agreed fact that, and I quote from page 7 of the agreed statement of facts: " At no time has there appeared in the valuation list an assessment for the whole building described as Euston Centre—36-storey tower block—286 Euston Road as referred to in the London Borough of Camden's completion notice dated November 21 1969, nor at any time has a proposal been made by the valuation officer to bring such property into assessment."

To those accustomed to rating law it is a new and revolutionary idea that rates should be paid on an unoccupied hereditament. But, none the less, Parliament decided that under H certain circumstances an unoccupied or empty rate could be levied. The Act lays down in Schedule 1 certain conditions that have to be satisfied before such a rate is exigible. The completion notice procedure has to be followed whether the building to which the notice relates is completely erected or is nearing completion so as to permit the naming of a date by which the erection can be said to be complete. In the present case the completion notice related to the whole building, Euston Tower, and as to that building a date in the future for its completion was agreed between the rating authority and the building owner. It was also agreed between them that unoccupied rates should become payable on a date J three months thereafter. That agreement was based on the acknowledgement that Euston Tower as a whole when complete would constitute a hereditament. Indeed, that is what the notice (exhibit 5) says and I quote again: ". . . when so completed" (that is, the whole building) "will be a hereditament liable for unoccupied property rate. . . ."

In fact, before the agreed date, Euston Tower had been divided into separate hereditaments of which the defendant's hereditament is one. The argument for the defendant is that since the notice contemplated an undivided hereditament and no notice had been served which related to the defendant's actual hereditament, being part of but less than the whole K building, the defendant is not liable to a rate under the empty property provisions. The valuation list does not show the whole building Euston Tower as a single hereditament. It does show the part so let to the defendant as a hereditament. The defendant expressly does not advance any argument that the agreement between the rating authority and the building owner is *res inter alios acta* and for that reason does not affect them. Expressly the defendant accepts that if there had been a coincidence of hereditament between that to which the agreement related and that which it took, it would be bound by the agreement. The defendant's case L is simply that there has to be a coincidence between the hereditament actually created and the hereditament described in the notice. The defendant acknowledges that at the time when the notice procedure is begun, and properly begun, neither the rating authority nor even the owners may know what hereditaments within the whole building may later be created. None the less, the defendant submits that the notice procedure has to be followed, and that, on a true construction of the schedule, there has to be coincidence between what ultimately appears in the valuation list as the hereditament and that which is described in the completion notice. The defendant's case is that it is not liable because there has M never been a notice relating to its hereditament.

The defendant acknowledges that if this submission is correct the empty property procedure cannot be safely or surely initiated in the case of an office building like this until the actual hereditaments have been determined. That submission must involve the submission that even though a date has been determined for the completion of the erection of the buildings:

1. no rating consequences follow until the actual hereditaments are determined, and

2. if the hereditaments actually created differ from the hereditament mentioned in the completion notice, though they are hereditaments carved out of the whole hereditament to which the completion notice relates, the rating authority has to start again and serve a fresh completion notice or notices in order to be able to demand empty property rates.

The contention of the defendant is very clearly and succinctly stated in paragraphs 3 and 4 of the defence.

Before I move to a consideration of the issue thus raised there are two matters that are relevant but not contentious:

1. By the Post Office Act 1969 the defendant is not to be regarded as exempt from any rate (see section 6, subsection (5)) and
2. There is no dispute that if the defendant is liable, then the amount claimed is the amount for which it would be liable.

I turn now to the issue which has been most lucidly and helpfully argued, and I can state at once that I have not found in any of the provisions to which I have been referred any escape from the conclusion that the defendant's submission is correct.

I turn to the schedule and I read paragraph 1, subparagraph (1), which is in these terms:

> Where, in the case of any rating area in which, by virtue of a resolution under section 17 of this Act, this Schedule is in operation, any relevant hereditament in that area is unoccupied for a continuous period exceeding three months, the owner shall, subject to the provisions of this Schedule, be rated in respect of that hereditament for any relevant period of vacancy; and the provisions of this Act shall apply accordingly as if the hereditament were occupied during that relevant period of vacancy by the owner.

The paragraphs upon which the plaintiffs principally relied and to which my attention was directed—emphatically directed—are the group of paragraphs 7-10, which has a cross heading: "Completion of newly erected or altered buildings," and paragraph 7 reads:

> For the purposes of paragraph 1 of this Schedule, a newly-erected building which is not occupied on the date determined under the subsequent provisions of this Schedule as the date on which the erection of the building is completed shall be deemed to become unoccupied on that date.

And then paragraph 8(1)(b):

> Where a rating authority are of opinion . . . (b) that the work remaining to be done on a building within their area is such that the erection of the building can reasonably be expected to be completed within three months, and that the building is, or when completed will be, comprised in a relevant hereditament, the authority may serve on the owner of the building a notice (hereafter in this paragraph referred to as "a completion notice") stating that the erection of the building is to be treated for the purposes of this Schedule as completed on the date of service of the notice or on such later date as may be specified by the notice.

I have in mind, in the context of these paragraphs, the terms of the notice itself, which I have now referred to more than once: ". . . when so completed *it* will be a hereditament liable for unoccupied property rate. . . ."

Under paragraph 8(1) the rating authority have, as a precondition of serving a completion notice, to be of opinion that the erection of a building has been completed or can reasonably be expected to be completed within three months. That, however, is not the only precondition. The rating authority have also to be of opinion that the building is or when completed will be "comprised in a relevant hereditament." The words: "comprised in a relevant hereditament" are acknowledged by counsel on each side to be difficult to construe and awkward to apply. I have accordingly wondered whether these words might have been deliberately chosen to give a degree of elasticity which would impose rating liability on the owner of an actual hereditament carved out of a whole building hereditament at a date subsequent to the determination by the completion notice procedure of a date for the commencement of empty property rate liability for the whole building, but in the end I have not been able to satisfy myself that that would be a correct conclusion.

Accordingly I have to say that I do not find in the terms of the Schedule or of the Act any clear or persuasive answer to the defendant's submission or any warrant for the alternative conclusion proposed by the plaintiffs. That suggested alternative conclusion is that since the 36-storey tower block, Euston Centre, could have been a hereditament (and the completion notice expressed the opinion that when its erection was completed it would be) and since the empty property rate on that hereditament would have become payable on December 11 1970 (as the owners' agents agreed) the completion notice procedure should be considered to have been complied with in regard to any hereditament created within the whole building and the appropriate rate for that lesser hereditament should become payable on that date. I have not been able to find any warrant in the terms of the schedule for applying the completion notice and its consequences in that way to a hereditament different from the one identified in the completion notice. The hereditament contemplated by the completion notice never did appear in the valuation list or even become the subject of a proposal by the district valuer. On the other hand, the part of the 36-storey tower block which was let to the defendant did form the subject of a proposal and did appear in the valuation list as a hereditament. That hereditament, however, was never the subject of a completion notice.

In my judgment the empty property rate did not become payable by the defendant, since the condition precedent to liability for such rate was never complied with in that the notice served did not relate to the hereditament in respect of which the rate is claimed.

For these reasons the claim fails and must be dismissed.

Judgment was given for the defendant with costs.

COURT OF APPEAL

February 18 1977

(Before Lord Justice CAIRNS, Lord Justice ROSKILL and Lord Justice BRIDGE)

HARWICH DOCK CO LTD v INLAND REVENUE COMMISSIONERS

Estates Gazette May 21 1977

(1977) 242 EG 623

Rateable value of dock wharf — How to be determined — General Rate Act 1967, section 35 — Docks and Harbour (Valuation) Order 1971 — Special basis of valuation prescribed — Applicable only to named undertakers — Present case governed by general law — Profits basis — Appeal dismissed

This was an appeal by Harwich Dock Co Ltd from the refusal of the Divisional Court to grant an order of mandamus requiring the Commissioners of Inland Revenue to calculate the rateable value of a hereditament known as Navyard Wharf, occupied by the company in Harwich harbour, in accordance with the Docks and Harbour (Valuation) Order 1971, and for the valuation officer for Tendring, Essex, to make any consequent alterations in the valuation list.

George Dobry QC and F A Amies (instructed by Middleton, Lewis & Co, of Horley, Surrey) appeared for the dock company; Brian Davenport (instructed by the Solicitor of Inland Revenue) represented the commissioners and the valuation officer.

Giving judgment, Bridge LJ said that the wharf in question comprised a complex of jetties with berths for five ships. At all material times the hereditament as a whole had been an approved wharf for the purposes of section 14, and had included two transit sheds approved within section 17, of the Customs and Excise Act 1952. The sole question was how the rateable value was to be determined. Under section 35 of the General Rate Act 1967 the Secretary of State for the Environment was empowered to make provision by order for determining the rateable value of " any hereditament occupied by the persons carrying on, under authority conferred by or under any enactment, a dock or harbour undertaking." The company claimed that the hereditament should be valued under the Docks and Harbour (Valuation) Order 1971, and sought to establish the claim by mandamus. The commissioners and the valuation officer contended that the rateable value should be assessed after 1971 as it had been before 1971 — under the general law of rating on the so-called profits basis. The company contended that the dock undertaking was carried on under authority conferred by the Customs and Excise Act 1952 by virtue of the approvals given under sections 14 and 17. The commissioners contended that the statutory authority envisaged by section 35 of the General Rate Act 1967 was authority conferred on a particular named undertaking to carry on the totality of the undertaking.

The rival contentions were finely balanced. He (his Lordship) found nothing in the context in which the words " under authority conferred by or under any enactment " appeared in section 35 to afford any clue to the underlying legislative purpose. Nor was there any decisive pointer in the arguments for either side. It was said that the commissioners' construction gave no effect to the words " under any enactment," since no example could be given of a statutory dock or harbour undertaking whose powers were not derived directly from statute. On the other hand, it was said that the company's construction would render purposeless the whole phrase, since in practice every dock undertaking must operate at a place where goods could be handled in connection with their export, import, or coastwise transportation.

In the end his Lordship's impression was that it was marginally more probable that Parliament intended the result which followed from the narrow rather than the wide construction, and he would dismiss the appeal.

Roskill LJ and Cairns LJ delivered concurring judgments and the appeal was dismissed.

COURT OF APPEAL

February 24 1977

(Before Lord DENNING MR, Lord Justice ORR and Lord Justice SCARMAN)

SHEFFIELD CITY COUNCIL v GRAINGERS WINES LTD

Estates Gazette May 28 1977

(1977) 242 EG 687

General Rate Act 1967—Council's resolution to rate unoccupied properties—Resolution did not expressly "specify" date when provisions were to come into operation—Date nevertheless clear from surrounding circumstances—Resolution not invalidated—Authority's appeal allowed

This was an appeal by Sheffield City Council from the dismissal by Melford Stevenson J of its claim for £1,188 from Graingers Wines Ltd, of Allen Street, Sheffield, in respect of rates for unoccupied premises in their ownership.

Francis Radcliffe (instructed by Sharpe, Pritchard & Co, agents for Samuel Jones, Sheffield Council Legal Department) represented the appellants, and Brian McCarthy (instructed by Lucas Styring & Appleby, of Sheffield) appeared for the respondents.

Giving judgment, Lord Denning said that the City of Sheffield had got into some trouble. As from April 1 1974 they determined to rate empty commercial properties. There were some 1,108 unoccupied properties representing a rate income to the council of £373,289. About £269,997 had been collected for 677 buildings. There was a balance of £103,292 to be paid. Someone connected with Sheffield Chamber of Commerce said that it was unlawful, that the council had not got the procedure right and might have to repay what they had collected. The judge had said that the chamber of commerce were right. The question was whether the council had complied with all the legal requirements when they set out to rate unoccupied properties.

Section 17 of the General Rate Act 1967 as amended provided:

> (1) A rating authority may resolve that the provisions of Schedule 1 to this Act with respect to the rating of unoccupied property—(a) shall apply . . . to their area, and in that case those provisions shall come into operation, or, as the case may be, cease to be in operation, in that area on such a day as may be specified in the resolution. (2) The day to be specified in a resolution under subsection (1) of this section shall be—(a) in the case of a resolution providing that the said provisions shall apply to the area in question, the first day of a rate period for that area beginning after the day on which the resolution is passed. . . . (3) As soon as may be after a resolution is passed by a rating authority under this section, the authority shall cause a copy of the resolution to be published in the *London Gazette* and in one or more newspapers circulating in the area of the authority. (4) A document purporting to be a copy of the minutes of a resolution passed by a rating authority under section 20 of the Local Government Act 1966 and to be certified under the hand of the clerk of the authority as a true copy of the minutes of the resolution shall be evidence that the resolution was passed by the authority.

So the section said that unoccupied premises could be rated and that the provisions came into operation " on such day as may be specified in the resolution." It was said that the council had not specified the day in the resolution. If one examined the resolution very formalistically that might be said to be right. The judge had so held. But should the resolution and all that followed be interpreted so formalistically? There had been much discussion as to whether the provision about the resolution was mandatory or directory. That was a matter of impression. His Lordship's impression was that it was required for the purposes of certainty.

Did the resolution make sufficiently clear the day from which it was to operate? On March 6 1974 the council's policy committee had met and it was pointed out that the resolution to be passed was to come into operation on April 1 1974. After a debate and a division along party lines the resolution was passed. The resolution did not include the key words " as from April 1 1974." That was an error by the draftsman. They ought to have been put in. The press had been present and realised that the resolution would operate from April 1 1974. The local papers came out with the pronouncement that the rating of unoccupied premises was to take place straight away. By the time it came for the publication of the statutory notices under section 17(3) of the Act it was made quite clear that the resolution would operate from April 1 1974.

The resolution should not be construed so technically as was suggested. Documents and transactions were to be construed so as to give them validity and not so that they should perish or fall—so far as that could reasonably be done. Because of the mistake it might be that all the ratepayers who had not paid would not have to pay. Those who had paid might get back what they had paid. The burden would fall on the ratepayers of Sheffield. The resolution had been passed by members of the committee and it had been

A deliberately determined that it should come into operation from April 1 1974. The formal resolution should be read in the context of the minutes from which it was clear it was to operate from April 1. That had been made clear in every publication to the general public and the press. There had been sufficient compliance with the statute. The appeal should be allowed.

Agreeing, ORR LJ said that it was the clear intention of the policy committee that the rating of unoccupied premises should begin on April 1 1974.

Also agreeing, SCARMAN LJ said that it was perfectly clear that if a power was given to impose a financial burden on
B a member of the public, the donee of the power must act strictly in accordance with the conditions laid down by Parliament. That principle lay behind the Bill of Rights 1688. It showed the sort of approach which the courts should make where there was to be a levy of money on the public. "Specified" in section 17(1) and (2) meant "made clear"—see *Reid v Dawson* [1955] 1 QB 214, 220. The date on which the resolution was to come into operation must be made clear. The obligations on the local authority under subsection (3) had been carried out. The statutory advertisements had described the resolution as operating from April 1 1974. The resolution was in the present tense. He (his Lordship)
C had come to the clear conclusion that the resolution, properly construed in its surrounding circumstances, specified the date from which it was to operate and the day from which the provisions were to be imposed.

The appeal was allowed. Leave to appeal to the House of Lords was refused.

QUEEN'S BENCH DIVISIONAL COURT

March 3 1977

D (Before Lord WIDGERY CJ, Lord Justice GEOFFREY LANE and Mr Justice CAULFIELD)

BRENT LONDON BOROUGH v ALFA ROMEO (GREAT BRITAIN) LTD

Estates Gazette June 25 1977
(1977) 242 EG 1053

**Rating surcharge on unused commercial property—Section 17A of General Rate Act 1967—Non-user exceeding six months alleged—Effect of assignment of lease during period of alleged non-user—Whether assignor "tried his best to
E let the building"—Test is the efforts to get a tenant into possession—Non-user while assignor "tried his best" to be disregarded—Assignee's non-user less than six months—Rating authority's appeal dismissed, but leave to appeal further given**

This was an appeal by case stated from a decision of magistrates sitting at Harrow who had held, on a complaint by the London Borough of Brent, that the respondents, Alfa Romeo (Great Britain) Ltd, had a good defence to a demand under section 17A of the General Rate Act 1967 (inserted by section 16 of the Local Government Act 1974) for the payment of a rating surcharge in respect of the non-use of
F a commercial building for a period exceeding six months.

Charles Fay (instructed by K B Betts, Town Clerk, Brent) appeared on behalf of the appellants; Matthew Horton (instructed by Titmuss, Sainer & Webb) represented the respondents.

Giving judgment, LORD WIDGERY CJ said: This is an appeal by case stated by justices of the peace for the Middlesex Area of Greater London sitting as a magistrates' court at Harrow on two days in May and June 1975. On that occasion the magistrates had before them a complaint that the respondents, Alfa Romeo (Great Britain) Ltd, being a person duly rated and assessed by a general rate made on March 14 1974, had failed to pay in respect of premises at 256 G
Water Road, Wembley, the surcharge rate of £1,649.92 due for the period from November 11 1974 to February 22 1975. That complaint relates to a comparatively recent innovation in the law on rating—the surcharge on unused commercial property. One finds the provisions in section 16 of the Local Government Act 1974, which introduced a new section in the General Rate Act 1967.

What this section says, so far as presently material, is this:

> 17A If for a continuous period exceeding six months a commercial building is not used for the purpose for which it was constructed or has been adapted, its owner shall pay in respect of that period H
> (the "period of non-use") a surcharge additional to the rates (if any) payable apart from this section.

One therefore has this conception of a surcharge payable if for a period exceeding six months the building, being a commercial building, is not used for the purpose for which it was constructed or adapted. This building, 256 Water Road, Wembley, was a commercial building, a purpose-built warehouse with ancillary office accommodation, and it stood on an industrial estate.

The respondents, Alfa Romeo, are the English trading company for the import and servicing of Alfa Romeo cars
in Great Britain. Prior to the Alfa Romeo company becoming J
owners of the building it was owned by another Italian organisation called Italvini Ltd, an English company dealing with Italian wines. Italvini Ltd was the owner of the building for the purposes of the rating statutes from March 1973 until November 1974. The period of non-use which was relied upon by the rating authority as the appellants in this case is a period from May 1974 to February 1975. It is to be observed that Italvini were owners for part of that period, namely, from May 1974 until November 11 1974. On November 11 1974 the lease vested in Italvini was assigned to
the present respondents, and they became owners from that K
date. Again it becomes obvious that for a substantial part of the period of non-use relied upon the owners were the present respondents, Alfa Romeo. It is convenient to regard this period of non-use as being divided into two unequal parts—one when Italvini were the owners from May until November 1974, and the second part from November 1974 until February 1975 when the owners were Alfa Romeo (the present respondents). During the first of those periods the material findings of the justices are that the premises were empty. Italvini had gone out and left the premises empty. Italvini was anxious, I think I can properly say, to let the
premises or to dispose of the premises in some way so as not L
to have an unproductive asset. At quite an early stage Italvini discovered that Alfa Romeo might be interested in acquiring a tenancy of these premises because they wanted a service station in London. Through most of 1974 one has rather desultory contacts between Italvini and Alfa Romeo on the question of whether Alfa Romeo would take over Italvini's lease, and in the end a contract to that effect was made.

It is not disputed that during the period when Italvini were owners they came within the terms of section 17A, namely, there was a period in which a commercial building was not used for the purpose for which it was constructed,
but they contend that they were entitled to a special defence M
or exemption provided in the same section in subsection (2), where it is provided: "Subsection (1) of this section shall not apply where—(a) the owner has tried his best to let the building."

One can usefully pause there for a moment just to remind oneself of the mischief at which this provision was aimed. It was concerned with such notorious cases as Centre Point where large commercial buildings were left empty, and under the law as it then stood not contributing to general rate fund at all, while their owners waited for a more suitable occasion upon which to put them into use. The purpose of the legislation with which we are concerned in this case is to control or prevent that kind of activity, and so one finds, not surprisingly, that an owner of a building which has been empty for a period may claim as a special defence to the surcharge that he has tried his best to let the building.

As far as the second half of the period is concerned—that is the period when the respondents Alfa Romeo were owners—the situation was quite different. They made no attempt to let the building because they wanted it for themselves. They had acquired by assignment the lease from Italvini. They wanted the premises for themselves and they made no attempt to dispose of them, and they set about getting the necessary consents to use the premises as a service station. They were not altogether successful. They did not get permission to do vehicle repairs on the premises, but they were advised that they could set up a spares organisation for Alfa Romeo cars without planning permission, and this they set about doing. At what seems to have been a reasonable tempo between November 11, when Italvini moved out, and February 22 1975, when on any view the premises were used and occupied again, they brought racking from their other premises to set up what I call the spares organisation and generally got the property ready for its intended future as a spares organisation for Alfa Romeo. Here of course the premises were not used in the active sense during the period of Alfa Romeo's occupation, but Alfa Romeo nevertheless sought to allege that they were using them for the purpose of section 17A, and that was because Alfa Romeo clearly could not claim the special defence under section 17A(2)(a), namely, that they had tried their best to let the building. What they had to contend on their own account was that in the period November 1974 to February 1975 they were using the premises for the purposes for which they were constructed. They were arguing that while the premises were being adapted for use to their new role they would be used for the purposes for which they were constructed.

On the argument of Alfa Romeo thus briefly summarised there was at least one considerable difficulty in the shape of prior authority in the House of Lords. I refer to the case of *Arbuckle Smith & Co Ltd v Greenock Corporation* [1960] AC 813. I do not propose to go into the authority in detail because in my judgment it is unnecessary finally to pronounce on any of the important points which this authority leaves for decision at the present day. Suffice it to say that in a closely related problem, where the question was raised as to whether premises were occupied for a particular purpose, the House of Lords decided that premises were not used for a particular purpose merely because they were being prepared for use for that purpose. As I say, that seems to me to present a very formidable difficulty to Alfa Romeo in their primary argument, which was that they were using the premises for the particular purposes between November 1974 and February 1975.

However, that is not an end to the matter because Alfa Romeo had an alternative way of putting their case. They seek in the alternative to say that the period when Italvini were owners is to be excluded from the computation altogether. If that is a proper argument, then at a stroke Alfa Romeo have destroyed the period of at least six months' non-use upon which the whole of this case turned. The way in which Alfa Romeo set about it is this. They say: "The period of six months relied upon in this case was partly when Italvini were owners and partly when we were owners. Italvini's period of ownership is to be totally disregarded for present purposes by virtue of section 17A(2)," which I have already read. The argument is that the effect of section 17A is to exclude altogether the period in which Italvini were the owners of the premises and thus to cut the relevant period down to the period from November 11 1974 to February 1975, which is demonstrably not six months. If that can be done, if that can be achieved on the case and the statutes, in my judgment the respondents have the answer which they want. Of course one must not overlook the fact that this is a magistrates' case. We are not here to decide what the answer ought to be. We are here to advise on the legal problems which were raised. The magistrates, one should notice, with some enthusiasm decided that the defence now raised was open to the respondents and Italvini, and therefore it is only for us to say whether as a matter of law they were wrong in that view.

There are two points that have to be argued to decide whether this defence is open to Alfa Romeo. The first one is this. Section 17A, it is said, would not apply at all to the facts of this case when the owners had sought not to let the property but to assign an existing lease, and when that has to be considered against the background of the statutory language that the owner has tried his best to let the building. The argument obviously is that assigning a lease is not what one would normally describe as "letting the building." Normally the phrase "to let the building" implies "to create a new leasehold interest in the building." It is argued competently by counsel in this case that here there is no question of the efforts of Italvini to assign their existing lease being relevant for consideration under section 17A(2)(a) at all. This has given us all some cause for concern. Certainly I have pondered it considerably since the case began. But I think that this is in particular a statute phrased in not altogether precise language where one is entitled to take a certain amount of liberty with the specific language if it is necessary to do so to prevent the undoubted mischief which lies behind this legislation. If one looks at it from that point of view the first point which comes to mind at once is that merely letting the premises would have got nowhere in regard to the mischief in this matter. If Italvini had sought to grant a sublease of the premises with a view to that sublease being put in a drawer and forgotten and no key of the building being turned, then the mischief behind this legislation would not have been met at all. Merely letting the property is not enough. Furthermore, looking at the other side of the coin, if the object to be achieved is to get a tenant into possession, and that I believe to be the object to be achieved here, then it matters not in the slightest whether that tenant is a tenant by virtue of assignment or a tenant by virtue of the grant of a sublease. It literally matters not at all. In each case a tenant is produced and a tenant goes into the building, and the mischief is met.

For those reasons I think that the proper approach to section 17A(2)(a) is to say that what the owner has to plead is that he has tried his best to get a tenant into possession. If he proves he has tried his best to get a tenant into possession, then in my judgment he can satisfy the requirement of section 17A(2)(a). There is no doubt whatever that if the justices had been told to approach the matter from that point of view they would have found enthusiastically that he had done his best to get a tenant into possession.

But that of course is not the end of it, because although that approach justifies the contention that the magistrates were right in holding the special defence of section 17A(2)(a) to be available to the respondents in this case, it is still necessary to show what the effect of that would be in our present situation. But here I find great help from the realistic and helpful concession made by Mr Fay that, if once we have found against him on the point that I have just been dealing with, he would accept that the effect of setting up that successful defence would be that the period during which Italvini were trying to let the property would be totally disregarded for the purposes of this legislation. In other words, one would have then to consider the period of non-use starting on November 11 1974 and running on to February 22 1975. One would totally disregard for all purposes what had been happening between May 1974 and November 1974. On that footing there would be less than the six months' period of

A non-use for consideration and the essential and vital contention of the appellants would fall to the ground. I think that is the right way to do it, and I am encouraged by the fact that we are told there have been certain changes in the law relating to these matters which produce a somewhat more realistic basis than that upon which this case has had to be decided.

In my judgment this appeal should be dismissed.

Agreeing, GEOFFREY LANE LJ said: It would have been great hardship in the circumstances of this case had the surcharge been exigible from Alfa Romeo because Alfa Romeo B have throughout, whatever else may be the case, acted with great speed and promptness so far as they were able. What surprises me is that the appellants have declined to exercise their undoubted powers to remit to this case.

CAULFIELD J also agreed.

The appeal was dismissed with costs. Leave to appeal to the Court of Appeal was granted.

REAL PROPERTY AND CONVEYANCING

CHANCERY DIVISION

June 24 1976

(Before Judge THOMAS, sitting as a deputy judge of the division)

ROBINSON v ROBINSON

Estates Gazette January 15 1977

(1976) 241 EG 153

Man and mistress buy house in contemplation of marriage—Parties subsequently separate without marrying—House in substance purchased by man—Conveyance to parties "as beneficial joint tenants" not conclusive—Transfer not executed by man, not an effective declaration of trust by him—Mistress entitled on facts to a share of one-fifth

This was a summons by Phyllis Gertrude Robinson seeking a determination of the extent of her beneficial interest in 14 Alexandra Road, Sheerness, occupied by the defendant, Albert John Robinson.

M Thorpe (instructed by Vanderpump & Sykes, of Enfield) appeared for the plaintiff, and Clarissa Dickson Wright (instructed by Weld & Beavan, of Enfield) represented the defendant.

Giving judgment, JUDGE THOMAS said that in February 1970 the plaintiff left her husband and set up home in a flat at Enfield with the defendant. They lived there as man and wife with the intention of marrying when her marriage was dissolved, and she changed her name to Robinson by deed poll. The defendant was a bachelor now aged 62; the plaintiff was somewhat younger. In August 1970 they decided to move from Enfield, and on August 14 1970 they jointly contracted to buy 14 Alexandra Road, Sheerness, for £2,500. It was transferred to them by a land registry transfer dated August 28 1970, and subsequently registered in their joint names. The major part of the purchase-price came from the sale of the Enfield flat, which belonged to the defendant. The net proceeds of that sale were £3,284. In 1972 the relationship between the parties deteriorated. At the end of May 1972 the plaintiff left the defendant, and there were proceedings before the magistrates in which the defendant was bound over to keep the peace. In August 1972 there was a reconciliation, and the parties resumed cohabitation at 14 Alexandra Road, but at the beginning of 1973 there was further strife, caused partly by divorce proceedings involving the plaintiff, her husband and the defendant. There was a hearing of those proceedings in February 1973, and relations between the plaintiff and the defendant then deteriorated to the point where they decided to separate. She finally left 14 Alexandra Road on May 2 1973.

The transfer of the property to the parties was expressed to be to them "as beneficial joint tenants." It was executed by the transferor, but not by the plaintiff or by the defendant. It appeared that before signature of the contract to purchase the house, the plaintiff expressed fears to the defendant as to what might happen to her if he died before they were able to marry. He too was concerned that she should be provided for if anything happened to him. He therefore instructed the solicitors who acted for them in that purchase to have the house transferred to them jointly. No thought was given as to what should happen if they did not marry. Of the purchase-price of £2,500 and costs amounting to £42 the plaintiff provided £140. The couple borrowed £110 from Barclays Bank on a joint loan account, but as between the parties the defendant was considered to be liable to repay this loan and in due course he did so. The balance of £2,292 required for the purchase was provided by the defendant; thus in all he provided £2,402. It was not necessary to raise any part of the purchase-price by means of a mortgage. When the parties moved to Sheerness neither was employed. The defendant drew social security benefits and was in receipt of a disability pension from the GPO. In September 1970 a joint general account was opened at the bank. The opening balance was £837, which was provided by the defendant. In August 1971 the defendant got a job which lasted for six weeks or so, and then from November 1971 until November 1972 he was fully employed. The plaintiff was not employed for the period of some 2½ years during which they lived together at 14 Alexandra Road. Thus the outgoings and living expenses were paid directly by the defendant or out of the joint current account, although no doubt the plaintiff did the housekeeping.

After moving into the house the parties carried out substantial works of improvement and modernisation of the property. The plaintiff stripped off wall-paper and paint and painted a large part of the interior, and she helped the defendant to hang new wall-paper. She also helped him to remove fireplaces from the three bedrooms. He bricked up the fireplaces and she carried bricks and mortar up a ladder to one of the bedroom windows. She also painted the frames of new windows which were installed. Together they made one large room out of two small ones on the ground floor. He put false ceilings in the lobby and passageway, did carpentry work to the stairs and landing and treated the house for woodworm. He pointed the outside brickwork and built a summer-house. In short, the parties set about improving their intended matrimonial home by their joint efforts. The raw materials appeared to have been paid for partly out of the joint current account. The parties also together carried out works of painting and redecoration at a neighbour's house for which they were paid about £120, and this money went towards the purchase of raw materials for no 14. Since the plaintiff left the defendant had carried out further improvements to the house. The parties agreed that the house would now sell for something between £8,000 and £9,000. There was no expert evidence either as to this or as to what the value was when they separated. He (his Lordship) could only assume that the value was then something near that figure. How much of the increase was attributable to the improvements, and how much to inflation, he did not know.

The primary question he (Judge Thomas) had to resolve was what were the respective beneficial interests of the parties in the house. Mr Thorpe, on behalf of the plaintiff, argued that the case was determined by the form of the registered transfer to the parties, which was declaratory of their beneficial interests and behind which the defendant was not entitled to go. This, he said, meant that the parties were in the end entitled in equal shares. In support of his argument Mr Thorpe cited what Lord Upjohn had to say in *Pettit v Pettit* [1970] AC 777 at 813. He (his Lordship) did not accept counsel's argument. Lord Upjohn was concerned to state what the position would be where a document contained an effective declaration of trust of beneficial interests by someone with a title to declare those interests. That was not the case here. Neither the plaintiff nor the defendant executed the transfer, and therefore neither effectively declared or settled their respective beneficial interests in the property. In other words, neither the plaintiff nor (in particular, for the purposes of Mr Thorpe's argument) the defendant was estopped by the deed of transfer from going behind it. In his (Judge Thomas's) view, this let in evidence as to circumstances which dictated how the beneficial interests should be determined, and the proper approach to this problem was that adopted by the Court of Appeal in *Cooke v Head* [1972] 1 WLR 518, where the court followed the guidelines laid down by Lord Diplock in *Gissing v Gissing* [1971] AC 886 at 909.

In *Cooke v Head* the court, in the light of the circumstances there present, concluded that the mistress's equity merited evaluation at one-third. In the present case Mr Thorpe had submitted that even if the approach in *Cooke v Head* was the proper approach the plaintiff should have an equal one-half share. That was too much. Miss Dickson Wright, on behalf of the defendant, said that the plaintiff should have no more than one-sixth. That he (his Lordship) felt was too little. Making the best assessment that he could in the circumstances of the present case, he valued the plaintiff's share in the equity at one-fifth. It had been agreed that he should take the matter no further than that, so that the parties should have an opportunity to try to reach agreement as to how those interests should be satisfied. If they were unable to reach an agreement, then the case could come back and he would deal with the position as he then found it.

COURT OF APPEAL

October 24 1975

(Before Lord Justice STAMP, Lord Justice ROSKILL and Lord Justice ORMROD)

MENTEL v MENTEL

Estates Gazette January 29 1977
(1975) 241 EG 307

Matrimonial Causes Act 1973, section 24—Adjustment of property between spouses in a case in which wife had remarried and had a home elsewhere—Court decides against straight 50 per cent approach despite substantial contributions of wife—A matrimonial home is not in a true sense a mere capital asset

This was an appeal by Mrs Marianna Lee (formerly Mentel) from a decision of Dunn J on February 19 1975 transfering to her former husband, Henryk Mentel, her interest in the matrimonial home, 33 Western Road, Plaistow, London E13, on payment by him of £1,750.

G Stephens (instructed by Gersten & Co) appeared for the appellant, and J Hamilton (instructed by Winstanley Burgess, agents for Dawson & Co) represented the respondent.

Giving the first judgment, ROSKILL LJ said that the parties were married in April 1961, when the husband was 28 and the wife 19. At the beginning of the marriage the couple lived with the wife's parents and both were earning. They managed over a comparatively short period of time to save a substantial sum, of the order of £2,000 to £3,000. Their purpose was to acquire a home. In 1963 they acquired the house which was the subject of the dispute, the £3,600 purchase price being provided as to £1,600 from their joint savings and as to £2,000 on mortgage, repayment amounts being £14.53 per month. Unhappily the marriage did not last long. Dunn J had referred to increasing tension between the parties from 1966 to 1968. Ultimately the appellant met a Mr Lee. She left Mr Mentel in 1968, marrying Mr Lee after divorce proceedings. Mr Lee had become a member of the metropolitan police, and as a result had secure employment and the benefit of a police house, which one assumed would remain his for the foreseeable future. Mr Mentel was still living in the former matrimonial home, to which the appellant had contributed. She had apparently put up £800, that was to say half the down-payment of £1,600, and had also made payments towards the mortgage interest out of her own earnings while she was still living with the respondent in the early years. Between 1966 and 1968 she did not contribute towards the mortgage interest, though Dunn J accepted that she contributed an amount by way of paying for food and other household necessities.

The court was told that at the time of the hearing before Dunn J the value of the house was something over £9,300. At that time there remained owing about £1,650 on the original mortgage. On these figures the equity was of the order of £7,700. Mr Mentel, who from the middle 60s onwards had paid the whole of the mortgage interest and principal himself, had paid off up to the time of the trial £1,423, and counsel had told the court that between that time and the hearing of the appeal he had paid a further £130. The appellant originally took proceedings under section 17 of the Married Women's Property Act. It was not disputed that for the purpose of proceedings under that section her share of the interest in the former matrimonial home was 50 per cent, subject to crediting the respondent with interest payments made since 1966. The court was not, however, concerned with section 17 proceedings, because the respondent started proceedings under section 24 of the Matrimonial Causes Act 1973. Dunn J made an order transferring the appellant's interest in the house to Mr Mentel upon payment by him of £1,750. The appellant contended that that was not sufficient. There was no cross-appeal by Mr Mentel, and it was therefore accepted on his behalf that that payment at least would have to be made. The short question was how the court should make a property adjustment order under section 24.

The right way to approach this question was to start by asking what the assets of this couple were at the time the marriage broke up. What property had they got? The only property each had was a one-half interest in this house, subject to the mortgage. But, as had been said in the Court of Appeal recently, it was all very well in these cases to talk about the matrimonial home as if it were a capital asset. In actual practice, anything less realisable than a matrimonial home was difficult to imagine. True, it consisted of bricks and mortar. But once the home was realised on the break-up of a marriage, one person anyhow had to be re-housed, so that one was in no true sense realising a capital asset. How was the house to be dealt with? Was it sufficient that the respondent should be expected to accept the additional burden enabling him to buy out the appellant's interest for £1,750, or should she have a larger cash payment in return for her share, causing him to accept a much bigger burden?

He (his Lordship) doubted whether the court derived much help from other cases. Mr Hamilton, for the respondent, had helpfully taken the court through section 25 of the 1973 Act. The respondent was undoubtedly worse off in the matrimonial home than he had been: he was having to pay the whole of the mortgage interest, whereas while the

A marriage subsisted the wife was contributing. On the other hand, he had kept the house for his home. From her point of view, she had been freed from the burden of contributing towards the house; on the other hand, she had lost the benefit of her original capital contribution of £800 assuming that the £1,600 came from them jointly, and she had lost the benefit of her original contributions. On the other side of the scale she had re-married, and re-married a man in a secure occupation with a police house for which no rent was being paid. She had been re-housed without having to make any contribution towards that re-housing.

B As always in these cases, there were factors which had to be taken into account on both sides. Had the wife not re-married, the respondent would have been hard put to it to resist an order for a particular provision, and indeed the house would probably have had to be sold to realise the requisite cash payment to the wife. But that was not the position. If the appellant had a full half share made over to her, or its equivalent in terms of cash, she would be getting half the value of the house at a time when she had the benefit of a completely new home rent-free as the result of her re-marriage. The respondent was less favourably placed. In the end the court had to do what was just to put the parties in the finan-
C cial position in which they would have been if the marriage had not broken down and each had properly discharged financial obligations towards the other. In his (Roskill LJ's) judgment it would not be right in this case to say that the appellant should have anything approaching a full one-half interest. Whether or not the court would, had it heard her case at first instance, have awarded her rather more than £1,750, there was great force in Mr Hamilton's argument that if the figure were substantially raised by requiring the respondent to borrow a much larger capital sum, he would be asked in terms of mortgage interest and repayment of capital to accept an unfair burden. He (his Lordship) had reached the
D conclusion that the judge's figure was one with which the court should not interfere, and he would dismiss the appeal.

Agreeing, Ormrod LJ said that the court under section 24 was making what was called a property adjustment order. That, in his judgment, was a very apt description of the nature of the order to be made. The court was involved in adjusting the property rights of the spouses to take account of the new situation following on the dissolution of the marriage. The court was dealing with homes and not with capital assets in any realistic sense of that word. It was impossible to talk about shares being worth £3,000 or more
E without being wholly unrealistic. Moreover, income consequences had to be borne very much in mind. There was no doubt that if the respondent had to raise a substantial amount to pay off the appellant the income effect on him would be very considerable. It had been estimated that instead of paying £14 a month he would have to pay £40 to £50 a month. The right approach was to look at the situation of both parties, apply all the items set out in the subsections of section 25, and at the end of the day arrive at a figure which was just in the broadest sense.

Also agreeing, Stamp LJ said that, regarding the interests of the parties in the house as capital assets, it would seem right
F that the appellant should receive one-half the value of the equity. But he was not satisfied that this was the correct approach. The truth of the matter was that it was impossible to regard this house as a realisable asset in the hands of the respondent. It was now costing him more than would have been the case had the wife continued to pay a share of the mortgage instalments. To require him to increase the mortgage in order to buy out the full value of the wife's equitable interest, or the greater part of it, would place a heavy burden on Mr Mentel. He would be paying a second time for the home merely because the marriage had broken down. On the other hand the wife had a home and would have £1,750, which incidentally was more than twice the amount of her G
contribution to the original purchase.

The appeal was dismissed.

QUEEN'S BENCH DIVISION

January 20 1976

(Before Judge Brian GIBBENS, sitting as a deputy judge of the division) H

CENTRAL COMMERCIAL PROPERTIES v ROSE

Estates Gazette February 5 1977
(1976) 241 EG 391

New Malden House developers sue engineer for delivery-up of designs etc—Engineer's fear of being "pilloried" as responsible for unsafe building without a judicial finding against him—Undertaking given by developers in response to judge's observations—No publication of designs etc in any newspaper or magazine

This was a claim by Central Commercial Properties Ltd, J
of Frederick's Place, Old Jewry, London EC2, against Gordon Rose, a structural engineer, of Heathgate, Golders Green, London, for delivery-up of designs, schedules and calculations relating to New Malden House, New Malden, Surrey, made by him while in the plaintiff company's employment, and for damages.

R Bernstein QC and R Jacob (instructed by Forsyte, Kerman & Phillips) represented the plaintiffs. The defendant appeared in person.

Opening the case, Mr Bernstein said that New Malden House was an 11-storey office block evacuated in 1975 K
and well known as an unsafe building. The plaintiffs had been associated with Ravenseft Properties Ltd and City Centre Properties Ltd in the development, and had employed the defendant as an engineer engaged on (*inter alia*) the construction of the building. They sought delivery-up of the documents produced by him as their servant acting in that behalf.

Mr Rose said that he was very concerned about the possibility of adverse publicity in the Press. *The Sunday Times* had already published an article about New Malden House, and was understood to be keen to publish his plans. L

His Lordship: Mr Bernstein, Mr Rose does not want to be vilified and held up as the engineer whose fault it was without any judicial finding on the matter. Could your clients not give him an undertaking of no publicity? They can make this building safe with the documents in their hands without putting Mr Rose in the pillory.

Mr Bernstein said that wood wool had been used in the construction of New Malden House, and the companies concerned had received many inquiries from people who had also experienced difficulties with that material. They (the companies) felt they had a public duty to pass on information. M

His Lordship said that the proposed undertaking need not prevent the companies from passing on information at second hand.

After a short adjournment, a settlement was announced. The defendant agreed that the three companies were entitled to the designs, schedules and calculations concerned, or copies thereof, on terms that they would not knowingly cause or permit them to be published in any newspaper or magazine. The plaintiffs withdrew their claim for damages for delay in handing over the documents, and made no application for costs.

CHANCERY DIVISION

November 5 1976

(Before Mr Justice FOSTER)

Re YATELEY COMMON: ARNOLD v DODD AND OTHERS

Estates Gazette February 19 1977

(1976) 241 EG 537

Commons registration—Common rights not abandoned by mere non-user—Intention never to use again essential—Rights not lost by period of requisition, by some house building or by absence of mention in conveyances—Effect of copyhold enfranchisement—"Common land" either land subject to rights of common or waste land of a manor

This was an appeal by Douglas Arnold, owner of Blackbushe Airport, Hampshire, from the decision of a Commons Commissioner, A A Baden Fuller, that the whole of Yateley Common (including the airport) was properly registered as common land under the Commons Registration Act 1965.

W J Mowbray QC and J Byng (instructed by Stallard & Gurney, agents for Jackson, Kent & Hodges, of Littlehampton) appeared for the appellant, J W Mills QC and J Bradburn (instructed by Gouldens) represented the individual claimants, and M Essayan QC (instructed by the Hampshire County Council solicitor) represented the authority.

Giving judgment, FOSTER J said that Hampshire County Council had registered the whole of Yateley Common as common land, and persons who claimed common rights over the whole of the land had submitted a large number of applications. The appellant, Douglas Arnold, the owner of Blackbushe airport, was now opposed by 23 persons. Yateley Common was said to be of about 1,218 acres, a little more than 3 miles long and was crossed by the A30 road. It was divided into four parts, the Blackbushe part, of 300 acres owned by Mr Arnold, which was north of the A30 and west of the Old Vigo Lane; the Calthorpe part, now owned by Sir Richard Anstruther-Gough-Calthorpe, south of the A30; the defence part owned by the Ministry of Defence, south of the A30 and east of the Calthorpe part; and the county part, owned by Hampshire county council, to the north of the defence part and east of the Blackbushe part. The appeal concerned only the Blackbushe part. Since time immemorial the commoners had exercised rights of common over Yateley Common until 1942. In the 16th century there was an intense dispute between the lord of the manor and the copyholders which resulted in the "Crondal Customary" in 1567, which could only be described as a peace treaty. The copyholders claimed then, and still claimed, rights of common not only by prescription or by user but as a legal right granted from time immemorial. Mr Arnold claimed that those rights had been abandoned, but, if not, that some claimants had lost their rights by conveyancing defects. It was difficult to prove that a person having such a right had abandoned it, especially if he had no reason to use it. It required something more than a lengthy period of non-user to amount to abandonment. It had to be proved not only that a person had not used it but that he intended never to use it again. In considering whether or not the legal rights had been abandoned it was immaterial whether the present user as an airfield was of importance as being in the public interest and it was equally immaterial for what motive the commoners sought to maintain their rights.

From 1942 to 1960 the Blackbushe part was requisitioned as a wartime aerodrome, and subsequently had been used as a civil airport, under licence from the Civil Aviation Authority, which brought into play restrictions of rights of access to the land to avoid the obstruction of flying. The county council had wished to buy the Blackbushe part, and Mr Arnold had brought an action against them to prevent that. In those circumstances it could not be said that the future of the land as an airfield had been secure. The fact that there was no user during the requisition period did not show any intention by the commoners to abandon their rights—they were merely prevented from exercising them. Although the land had been split and small houses had been built on parts of it, that was not such a fundamental change of character as the building, say, of a factory or garage which would have prevented a claim to rights of common.

It was also submitted that there had been many conveyances and sets of auction particulars which never mentioned the common rights, but the rights passed by operation of law. Such rights may have been thought to be of no value, but that did not mean that they had been abandoned. As there was no abandonment, the claimants whose copyhold lands had been enfranchised by statute were not affected and their claims must be confirmed. If the grantee of a deed of enfranchisement had a right of common at the date of the deed, there was no need to re-create the right or regrant it. The deed merely had to show that the right should not cease when the copyhold interest disappeared. In this case the grantees had a right from time immemorial and the words in the deed of enfranchisement were sufficient to confirm them. It was further submitted by Mr Arnold that the rights of common were exaggerated and should be reduced, but that point was not taken before the Commons Commissioner and so it was not open to the appellant to raise it in the appeal. Accordingly the decision of the Commons Commissioner was correct and the appeal must be dismissed.

In case of further appeal and if it should be found that there were no rights of common there was a point raised as to whether the appellant's land was nevertheless "common land" as being "waste land of a manor." The Commons Commissioner had decided that point by finding that the Blackbushe part was in fact waste land of a manor. The appellant submitted that that was not now so, as it was an aerodrome and also that it ceased to be waste when the Ecclesiastical Commissioners conveyed the land to Lord Calthorpe in 1891. Regarding the first point, the land was requisitioned for an aerodrome and nothing done during or after its derequisition could affect the position. The land could easily revert to common land when it stopped being used as an aerodrome. On the second point, the manor ceased to exist as a geographical entity after 1925 and had become an incorporeal hereditament. There was no reason why land should cease to be registrable because it was no longer owned by the lord of the manor. The Commons Commissioner had been correct on that point also.

The appeal was dismissed, but leave to appeal to the Court of Appeal was granted.

COURT OF APPEAL

July 7 1976

TRELOAR v NUTE

(Before Lord Justice STAMP, Lord Justice ORMROD and Sir John PENNYCUICK)

Estates Gazette March 12 1977

(1976) 241 EG 759

Claim for trespass on land—Possessory title set up—Limitation Act 1939—Whether adverse possession established—No requirement that owner must be inconvenienced by possession—Appeal by possessor allowed

This was an appeal by a defendant, Ernest Rodney Nute, from an order of Judge Chope at Penzance County Court upholding a claim by the plaintiff, Mrs Joanna Treloar, who sought, in addition to damages and a declaration as to title, an injunction restraining trespass by Mr Nute upon a small plot of land at Halwin, Wendron, Cornwall. Mr Nute appealed on the grounds (i) that the judge misdirected himself in holding that none of the acts of possession had been adverse, and (ii) that the judgment was against the weight of evidence.

The appellant appeared in person and P R Dunkels (instructed by White & Leonard & Corbin Greener, agents for Thrall, Llewellyn & Pearce, of Truro) represented the respondent.

Giving the judgment of the court SIR JOHN PENNYCUICK said that in 1961 the plaintiff was the freehold owner of a number of parcels of land at Halwin in the parish of Wendron. The land in question in the present case, consisting of about one-seventh of an acre, was part of one parcel and had previously been used by her in connection with her farm but in 1961 was derelict. In 1961 Mr Nute's father purchased some land adjacent to that of the plaintiff and believed that the disputed one-seventh of one acre was included in his purchase. Between the date of the purchase in 1961 and the date of the conveyance in September 1962 the defendant's father and the defendant did certain acts on the disputed land, such as grazing animals upon it, depositing soil in a gully running across it, storing materials and (in the case of the defendant and a brother) riding motor cycles on it. In 1963 the defendant's father erected a fence on one of the boundaries of the disputed land. The plaintiff protested and had the fence removed, but it was re-erected by the defendant's father. In 1965 the defendant's father conveyed the disputed land to the defendant by way of gift. In 1966 the plaintiff complained of the dumping of earth on the disputed land. In 1974 the defendant, having in the meantime filled up the gully, began to erect a bungalow on the disputed land. This led to the commencement of the plaintiff's action for an injunction and other relief. The defendant contended that by virtue of the Limitation Act 1939 the plaintiff's title had been extinguished since he (the defendant) and his father together had had possession of the land for 12 years. Judge Chope took the view that the possession was not adverse to the plaintiff's title because it did not inconvenience her as she had no need for the land then.

The question whether a person has taken possession of land was a question of fact. Apart from the minor acts of grazing and so on, which were of little weight, the deposit of soil in the gully and the preparation of the surface for a bungalow were sufficient evidence of possession. Adverse possession, however, meant possession exclusive of the owner. The normal rule was that if someone takes possession of land belonging to another and remains in possession of it for 12 years to the exclusion of the owner adverse possession was established, so that at the end of the 12 years the owner's title became extinguished. This rule had been adapted by the court to meet the special case where the owner had no immediate use for the land, and left it unoccupied, but intended to use it for a particular purpose in the future. In such a case the owner did not lose his title to the land despite temporary possession by another person. This had been established by a series of cases from *Leigh v Jack* (1879) 5 Ex D 264 to *Wallis's Cayton Bay Holiday Camp Ltd v Shell-Mex & BP Ltd* [1975] QB 94. These cases, however, depended on the intention of the owner to use the land for a particular purpose in the future. That was not this case. There was no evidence of special purpose here.

So far as the present case was concerned, section 10 of the Limitation Act 1939 defined adverse possession as possession by some person in whose favour the period of limitation can run. There was no requirement that, in order to constitute adverse possession, the owner must be inconvenienced. Indeed, if inconvenience to the owner was a requirement it would be difficult to acquire a possessory title, since the owner, if put to inconvenience, would be likely to take proceedings. In the present case time began to run from the taking of possession and the defendant had acquired a possessory title.

The defendant's appeal was allowed with costs.

COURT OF APPEAL

October 25 1976

(Before Lord Justice CAIRNS, Lord Justice ORMROD and Lord Justice WALLER)

WILLDAY v TAYLOR

Estates Gazette March 19 1977

(1976) 241 EG 835

Complaint by purchasers against builder alleging defective construction of bungalow—Dampness—Two contracts—Warranty in second contract giving purchaser rights under National House-Builders Registration Council certificate scheme—Purchasers unsuccessful in arbitration on second contract—Purchasers estopped from claiming under first contract—Substantially same claim—Purchasers' appeal dismissed

This was an appeal by Mr and Mrs Willday from the dismissal by Judge Heron at the Evesham County Court, on a preliminary issue, of their action for breach of contract based on alleged defects of construction in a bungalow built by the respondent, Harry Taylor.

D G Readings (instructed by Cross, Son & Hodgetts, of Evesham) appeared for the appellants, and N W Budgen (instructed by New & Saunders, of Evesham) represented the respondent.

The court did not call on counsel for the respondent.

Giving judgment, CAIRNS LJ said: The parties had entered into two contracts with regard to this bungalow. The first was dated October 29 1970, and the second was dated January 4 1971. The purpose of the second agreement was to supplement the first one and to give the purchaser certain additional rights in connection with an organisation called the National House-Builders Registration Council. Each of the agreements contained an arbitration clause. The plaintiffs, who were the purchasers, proceeded to arbitration under the second agreement. An arbitrator was duly appointed and heard the case, and gave an award in which he dismissed most of the plaintiffs' claims, though he did award in their favour a small sum of £28. Included in the complaints which the arbitrator dismissed was a complaint of defects in construction, giving rise to dampness. The plaintiffs afterwards started this action, which is an action under the first of the contracts, claiming (and claiming only) damages in respect of defects in construction causing dampness. The preliminary issue which the learned judge heard was as to whether, by reason of the terms of the two contracts and of the matter having gone to arbitration and the arbitrator's award having been made, the plaintiffs were estopped from bringing this action. The learned judge held that they were so estopped. The contention on this appeal on behalf of the plaintiffs has been that the arbitration was an arbitration under the second contract and not under the first one, and that despite the award against the plaintiffs in respect of this dampness, it is still open to them to claim under the first contract.

The first contract provided in its third clause: "The vendor"—that is the builder, the defendant—"will in proper and workmanlike manner complete the erection of the said bungalow in accordance with the approved plans deposited with the local authority and whereof the plan and specification are annexed to this contract." The arbitration clause was in these terms: "In case any question shall arise between the parties hereto as to the workmanship and/or materials in connection with the erection of the said bungalow the same shall be referred to a chartered architect to be appointed by the president for the time being of the Royal Institute of British Architects and the decision of such referee shall conclusively settle the matter." The later agreement provided in its second recital: "This agreement is supplemental to the said contract"—that was the one that I have just referred to—"and is intended to ensure that the dwelling

A will be or has been erected and completed at least in accordance with certain minimum standards and to afford a guarantee and remedies to the purchaser additional to any guarantee or remedies contained in the said contract." Clause 3 contains warranties by the builder in these terms: "The builder hereby warrants that the dwelling has been built or agrees that it will be built: (1) in an efficient and workmanlike manner and of proper materials and so as to be fit for habitation and (2) so as to comply in all respects with the Council's requirements and (3) so as to qualify for the certificate." That was a certificate of the NHBRC. Then it goes on: "Provided that (i) if the Council's requirements B conflict with any drawings or specification contained or referred to in the said contract then to the extent that the Council's requirements provide for a higher standard of construction or for additional work to be done the Council's requirements shall prevail and (ii) the builder shall during office hours show to the purchaser upon his request a specimen of the certificate and the Council's conditions of registration and rules." I should refer to clause 5 of this contract, because some reliance was placed on it by the plaintiffs' counsel on this appeal:

"The builder shall make good within a reasonable time after receipt of the purchaser's report and at his own expense any C defects in the dwelling which are consequent upon any breach by the builder of the Council's requirements and which are reported in writing to the builder within the initial guarantee period. Nothing in this agreement shall oblige the builder to make good defects in the dwelling caused by wear and tear or by normal shrinkage. . . ."

The latter part of it is not relevant here. The "initial guarantee period" is defined in the definition clause as being a period of two years from the date of the certificate of the National House-Builders Registration Council. Clauses 10 and 11 provide:

"10 The rights conferred on the purchaser by the terms of D this agreement are in addition to any rights conferred on the purchaser by any other contract between the purchaser and the builder.

"11. Nothing contained in any other contract made between the purchaser and the builder relating to the dwelling shall restrict or over-ride in any way whatsoever that which is contained in this agreement and subject to proviso (i) to clause 3 hereof in so far as any term in any such other contract is inconsistent with the terms of this agreement the terms of this agreement shall prevail provided that a higher standard or additional requirements called for by the said contract shall not be deemed inconsistent."

E Then the arbitration clause, clause 12, is in somewhat different terms:

"If any dispute shall arise between the purchaser and the builder concerning any matter or thing arising hereunder or in connection herewith such dispute shall be and is hereby referred to the arbitration and final decision of an arbitrator to be appointed at the request of either party by the presidents of the Royal Institute of British Architects and of the Royal Institution of Chartered Surveyors from a panel of arbitrators nominated by them and such reference shall be deemed to be a reference to arbitration within the meaning of the Arbitration Act 1950 or any subsequent re-enactment or modification thereof."

Then there is a sentence about the awarding of damages; F then a rather curious proviso:

"Provided that if there shall be any other dispute or difference between the purchaser and the builder in any way connected with the contract to which this agreement is supplemental and which the builder does not agree to being referred to the arbitration of the same arbitrator in the same reference as any dispute arising hereunder is referred the purchaser shall not be obliged to refer any dispute arising hereunder to arbitration but may take such other proceedings in connection therewith as he may be advised."

Now, I have said that an arbitrator was appointed under that second contract. He was appointed in pursuance of two applications duly made by the plaintiffs for the appointment of such an arbitrator, and in each of those applications G there appeared the sentence: "As a condition of the appointment I hereby agree as follows: (1) That the arbitrator's award will be in complete satisfaction of my claim."

It will be observed from the parts that I have read from the two contracts that the provisions of the later contract are decidedly more favourable to the purchaser in relation to the undertakings as to the quality of the workmanship and so forth by the builder than the provisions of the earlier one. It will further be observed that it is expressly provided in the later contract that the terms thereof shall prevail H where there is any inconsistency between the two, provided that if there is a higher standard under the earlier contract it shall not be deemed to be inconsistent. It seems to me that it is impossible for the plaintiffs, having gone to arbitration under that second contract, to say that they can pursue the same claim under the first contract as has already been unsuccessfully arbitrated by them under the second contract. I am referring, of course, to the claim in respect of the dampness. First of all they set out in particulars of claim the express terms of the first contract, that the defendant would in proper and workmanlike manner complete the erection of the bungalow, and then allege the breach of that J contract in negligently carrying it out because he so constructed and fitted the lintels etc as to allow damp to enter, and in the premises that the defendant was in breach of the contract. Those are the very matters which were alleged in relation to the dampness in the arbitration. In respect of constructing the bungalow in a proper and workmanlike manner, in effect the provisions of the two contracts were the same. If they did differ, then the higher standard was provided by the second contract and that had been arbitrated on, and on this part of the case the plaintiffs had been unsuccessful. It seems to me clear beyond a peradventure that the plaintiffs having gone to arbitration K under the second contract, having expressly agreed that the award in that arbitration should be in complete satisfaction of their claim, this action is misconceived.

Reliance was placed on a recent decision of Forbes J in *Purser & Co (Hillingdon) Ltd v Jackson* [1976] 3 WLR 700 where that learned judge held that where there were two different defects or classes of defect under a building contract, there can be two separate arbitrations. That, of course, is quite a different matter from the present case, where it is exactly the same defects which are being complained of. Reference was also made to *Conquer v Boot* [1928] 2 KB 336, L a decision of the Queen's Bench Divisional Court on appeal from a county court. That, if it can have any bearing on the present case, would be against the plaintiffs, because it was held in the particular circumstances of that case that when there had been one action for breach of contract in completing a building in a good and workmanlike manner, there could not be a second action of the same nature, even though some different defects were alleged. This case is fundamentally different from that, but, in so far as it is different, it is not to the advantage of the plaintiffs but to their disadvantage.

The one point in this appeal which did seem as if it might have some possible chance of success was based upon this M fact, that either accompanying his award or a little later the arbitrator addressed a letter to the plaintiffs or their advisers, saying that the reason why he had rejected the claim in respect of this dampness was that it was out of time under the two-year provision in the second contract. Mr Readings did not feel that he could base any argument upon that communication, having regard to the simple terms of the award dismissing that part of the plaintiffs' claim. I think he was right in taking that view. If any argument had been founded upon it, if it were open to the court to consider the letter in addition to the award itself, I think that the further answer

A to the point would be that the two-year provision being contained in the second contract, that, if it is to be regarded as inconsistent with the first contract, would prevail; the two-year period would apply to both of them. There is no relevant provision of the first contract which is in any way saved by the latter part of clause 11 of the second contract. For these reasons, I would dismiss the appeal.

ORMROD LJ: I agree, and would only add one word. It seems to me probable that the two plaintiffs in this case feel, or have felt, that their complaint about defective workmanship, which they say was the cause of this dampness, was never in fact adjudicated upon by the arbitrator, because of
B the two-year limitation period. No doubt, as a matter of simple fact, that is right. But if they had brought this action in the county court they would have been faced with the same difficulty, having not complained within the period limited by the contract, and so they would have failed in that litigation, apart altogether from the fact that if they wanted to go back and try to litigate on the first contract, they would have found themselves once more prevented by the arbitration clause in that contract. So the proceedings, in my judgment, in the county court were totally misconceived for a number of quite different reasons. I agree that this appeal should be dismissed.

C WALLER LJ: I also agree.

The appeal was dismissed with costs.

QUEEN'S BENCH DIVISIONAL COURT

December 13 1976

(Before Lord WIDGERY CJ, Mr Justice PARK and Mr Justice TALBOT)

BASSETLAW DISTRICT COUNCIL v F B DYSON (HAYTON) LTD

D Estates Gazette March 19 1977
(1976) 241 EG 837

Public Health Act 1961—Alleged breaches of building regulations—Informations laid more than six months after alleged offences were committed—Informations out of time—Council's appeal dismissed

This was an appeal by way of case stated by East Retford Justices who, on December 16 1975, held they had no jurisdiction to hear and determine 17 informations preferred by the appellants, the Bassetlaw District Council, against the respondents, F B Dyson (Hayton) Ltd, under section 4 (6) of
E the Public Health Act 1961, alleging breaches of the Building Regulations in houses built by the respondents in Retford.

R Sears QC and R Hunt (instructed by Sharpe, Pritchard & Co, agents for R W N Broadhead, of Worksop) appeared for the appellants; R D Naish (instructed by Ward Bowie, agents for Saffman & Co, of Leeds) represented the respondents.

Giving judgment, LORD WIDGERY said that the basic facts were that the respondents completed the houses in question in December 1974 and the justices found as a fact that the appellants must have known that month that the houses were completed. The houses were inspected by the appellants in
F May 1975 and the informations were issued on December 3 1975, just within the six months from the inspection. The respondents were not seeking to deny the existence of the alleged defects. What was being said was that the informations were out of time because the cause of action had arisen before May 1975, ie more than six months before the informations were laid. For the appellants it was said that the breaches arose as soon as it appeared that the regulations had not been complied with and there was no intention to comply. For the respondents it was said that the breach arose when the defective material was put in or the defective work done.

One could not solve such problems by some convenient G
rule. One had to look at each case. In the present case all that was known was that the houses had been constructed by December 1974, and that the appellants must have known that fact. He would dismiss the appeal on the facts found. There was no justification for the suggestion that the houses were not completed by December 1974.

PARK and TALBOT JJ agreed, and the appeal was dismissed with costs.

H

COURT OF APPEAL

October 29 1976

(Before Lord Justice STAMP, Lord Justice SCARMAN and Lord Justice GOFF)

BERKLEY v POULETT AND OTHERS

Estates Gazette March 26 1977 and April 2 1977
(1976) 241 EG 911; (1976) 242 EG 39

**Important decision on fixtures—Claim by subpurchaser of Hinton House to pictures and other objects—Pictures fixed in recesses of panelling—Test as to whether objects are fixtures—Degree and purpose of annexation—Is the purpose better enjoyment of object itself?—Objects not J
fixtures in opinion of majority—Goff LJ's dissenting views—Remedies available to a subpurchaser—Subpurchaser's appeal dismissed**

This was an appeal by Rowland John Berkley, the subpurchaser of Hinton House, part of Earl Poulett's estate at Hinton St George, Somerset, from the decision of Sir Anthony Plowman, then Vice-Chancellor, dismissing his claim for the delivery up of pictures and other objects, which he alleged to be fixtures passing to him on the sale of the house, and for compensation for loss and damage, and other relief. The respondents were the 8th Earl Poulett (and after his death his K
personal representatives), Sothebys and T R G Lawrence & Son, Crewkerne, Somerset.

P J Millett QC and B Marder (instructed by Ward Bowie, agents for Clarke, Willmott & Clarke, of Yeovil) appeared for the appellant, and N C H Browne-Wilkinson QC and L A Tucker (instructed by Eland, Hore, Patersons) represented the respondents.

Giving the first judgment, SCARMAN LJ said: In opening the appeal for the plaintiff Mr Millett QC said that the basic issue in the case was whether some pictures, some prints, a statue and a sundial were fixtures or chattels. If fixtures, they passed, he submitted, to the subpurchaser when Hinton House L
was sold: if chattels, he conceded they did not. The case has many complexities, which counsel have done their distinguished best to unravel, but ultimately we have to return to the issue identified by Mr Millett in the first few words of an address which, though it took five and a half days, lasted not a moment too long.

On November 11 1971 Mr Berkley (the plaintiff and appellant) issued a writ naming Earl Poulett, Sothebys, and a firm of surveyors, T R G Lawrence & Son, who practise in Crewkerne, Somerset, as defendants. Put briefly, the claim was for delivery up of certain fixtures which it was claimed passed with Hinton House when Lord Poulett sold the house, M
compensation for the loss of, or damage to, the fixtures, together with appropriate consequential relief. The claim was, and remains, strongly contested by all the defendants. Lord Poulett died in 1972, since when his personal representatives have carried on as defendants in his place. The plaintiff was a subpurchaser of Hinton House. He bought from Lord Poulett's purchaser. The history of the sale was as follows. Lord Poulett, the 8th Earl, decided to put up for sale by public auction his ancestral estate at Hinton St George in Somerset. Hinton House, with its gardens and grounds, comprising some 23 acres, was part of the estate and included

in the sale. The Poulett family had lived in the house for 500 years. The south wing, said to have been designed by Inigo Jones, was built for the reception of a queen who never arrived—Queen Anne. The wing includes two rooms which feature in the case, the Queen's Dining Room and the Queen's Ante-Room. The Queen's Dining Room is described in the auction particulars as follows: "The Queen's Dining Room, facing south and measuring about 39 ft 3 in by 29 ft, with oak panelling around picture recesses; open stone fireplace; magnificent moulded and decorated ceiling with egg and dart cornice and large casement windows overlooking the sunken garden." The Queen's Ante-Room is described in the same particulars as follows: "The Queen's Ante-Room, facing south and measuring about 28 ft 9 in by 27 ft 6 in, with oak panelling around picture recesses; large open stone fireplace; three-quarter glazed doors to sunken garden; doors to staircase hall and two communicating doors."

In the Queen's Dining Room there were six pictures, which the plaintiff claims passed with the house on its sale to him. These pictures were firmly fixed into the recesses in the panelling of the room. They were: a framed oil painting of Marquis of Hamilton, attributed to Mytens; a framed oil painting of Elizabeth, Countess of Essex, attributed to Mytens; a framed oil painting, said (but not believed) to be of James II, attributed to Kneller; a framed oil painting of the 1st Duke of Leeds, attributed to Gibson; a framed oil painting of Montague, Lord Willoughby, attributed to Old Stone; a framed oil painting of "The return of the Poulett family from the war," artist unknown.

In the Queen's Ante-Room, there were a number of pictures, only two of which were affixed to a wall, namely two portraits by Wood, of the 7th Earl and his countess respectively, each in coronation robes. The plaintiff claims them because, like the pictures in the Queen's Dining Room, they were fixed into the recesses of a panelled wall. On the first floor of the house there is a suite of rooms known as the Chinese Suite. These rooms contained a number of Chinese prints done on rice paper and stuck to the wall. It is conceded that these were fixtures and passed with the house. Finally there were in the grounds two objects which the plaintiff claims: a white marble statue of a Greek athlete, weighing approximately half a ton and standing on a plinth which was (and remains) fixed into position on the West Lawn; and a sundial which rested on a stone baluster (or pedestal) outside the South Wing. Neither the statue nor the sundial nor any of the eight pictures came to the plaintiff. He did get the Chinese prints all tattered and torn—severely damaged by someone who had tried unsuccessfully to remove them intact from the wallpaper to which they were stuck. The plaintiff's case is that all these items were fixtures; that they passed to him, the purchaser, by subpurchase of the house, garden and grounds; that he is entitled to their delivery up in good condition, damages for their detention, and compensation for their loss or damage.

The Hinton St George Estate was put up for sale by auction on August 2 1968. The auctioneers were Knight Frank & Rutley, who prepared the particulars of sale. There were 24 lots. Hinton House with its gardens, grounds and entrance lodge was lot 1, approximately 23 acres. The plaintiff was interested in lot 1 and also lot 4 (the park, 40 acres). A company, Effold Ltd, was prepared to bid for the whole estate, but was particularly interested in the forestry. A tenant farmer was interested in buying his farm. Such was the coincidence of their interests that, not surprisingly, a syndicate was formed. Mr Simon Lawrence was instructed to bid on their behalf. Mr Lawrence is a partner in the third defendants but at the auction sale on August 2 was acting for the syndicate which included the plaintiff and Effold. On the morning of the sale (which itself took place in the afternoon) the plaintiff contracted with Effold to buy lots 1 and 4 from them if their bid for the whole estate was successful. Later, by mutual agreement, he withdrew from the purchase of lot 4. Effold's bid for the estate was successful at a figure of £225,000 whereupon the plaintiff's contract to buy the house and grounds (lot 1) from Effold became unconditional. Effold were due to complete on September 30, but completion was delayed until November 4.

Meanwhile there occurred other events of some importance. On August 5 the plaintiff, having told Lord Poulett that he was the purchaser of the house, paid a visit to the house. The plaintiff's reason for buying the house and grounds was that he hoped to be able to develop Hinton as a tourist attraction; Longleat was in his mind's eye. Naturally he wanted to secure as many of its historic ornaments as he could. On his visit he sought to discuss with Lord Poulett the possibility of buying the contents of the house, or some of them. But the Earl would not be drawn.

The Earl had in mind not only an auction of the contents but also the sale by private treaty or otherwise of such of the treasures as he might be advised to sell separately, that is to say, otherwise than at a contents auction. He arranged for an auction on October 23, with October 22 as the viewing date. The plaintiff did not visit the house between August 5 and October 22—almost certainly because Lord Poulett did not wish him to do so. On August 14 upon the Earl's instructions, five of the six paintings were taken from the Queen's Dining Room, namely: "Lord Willoughby" removed by Rogers instructed by Sothebys (subsequently, that is to say on February 19 1969, sold by Sothebys at auction for £100); "Duke of Leeds" removed by Rogers instructed by Sothebys (who subsequently, on June 11 1969, sold it at auction for £180); the alleged "King James II" removed by Rogers instructed by Sothebys (who subsequently, on June 11 1969, sold it by auction for £250); "Elizabeth Countess of Essex" removed by Rogers instructed by Sothebys (not sold); "Marquis of Hamilton" removed by Rogers instructed by Sothebys, who sold it by auction on March 5 1969 for £550. "The Return" was left in position.

On October 22 the plaintiff, armed with a sale catalogue, the contents of which had disturbed him, came to view. He found the two Poulett coronation portraits in place in the Queen's Ante-Room but the five pictures gone from the Queen's Dining Room: only "The Return" was in its place. The statue was on its plinth on the West Lawn, but the sundial had been removed from its baluster. In the Chinese Suite he noticed that several prints were missing. The catalogue upset him because it listed for sale "The Return," the two coronation portraits, the statue, the sundial, and a chandelier (about which there is no longer any dispute). He made complaint and sought advice the upshot of which was that Mr Simon Lawrence, now acting for Lord Poulett as auctioneer of the contents, withdrew from the sale the two coronation portraits and the plinth but not the statue which stood on it. The sundial and the statue were sold at the auction on October 23. So also was "The Return"; its frame was also sold on the same day to the same purchaser, but privately. A few days later the plaintiff went again to Hinton House only to find that all he claimed had gone save the Chinese prints, and these he found in bad shape. He refused to complete his subpurchase on the day agreed for completion, that is to say October 31. We now know that "The Return" and its frame were removed on October 24 by a local builder, Mr Ash, after being sold (total price £700) by Mr Simon Lawrence on the 23rd; that the coronation portraits were also removed by him on Mr Lawrence's instructions; that the statue was sold for £225 at the auction and removed shortly thereafter by the purchasers; and that the sundial was sold at the auction for £40 and taken away by the buyer.

Despite the plaintiff's claims, of which Lord Poulett and Effold had notice before their completion, Effold completed their purchase of the whole estate without reservations or

A conditions on November 4, and the plaintiff, likewise, his purchase of the house and grounds on February 19 following. Thus all the objects (other than the Chinese prints) which the plaintiff claims as his property were removed on Lord Poulett's instructions after sale and before completion by Effold. The conveyances which followed were silent as to the matters in dispute.

I have recited enough of the history to show that, if the pictures, statue or sundial, or any of them, be fixtures, very difficult questions arise as to the entitlement of the plaintiff to sue the head vendor, Lord Poulett. But if none of them was a fixture, the plaintiff is non-suited at the outset, save B for his claim in respect of the Chinese prints, which are admitted to have been fixtures and did come into his possession. The first question, therefore, is whether the objects (other than the Chinese prints) were fixtures. On this preliminary, but fundamental, issue I confess that my mind has wavered. But at the end of the day, after listening to prolonged and painstaking argument upon the facts and reflecting upon counsel's submissions in the course of the tranquillity obtained by reserving judgment, I find myself in agreement with the Vice-Chancellor. None of the objects (other than the Chinese prints) was, in my judgment, a C fixture. I shall now endeavour to explain how I have reached this conclusion. Lord Poulett sold as an absolute owner. He was in a position to decide what he would offer for sale and what he would exclude from his offer. One turns therefore first to the contract of sale for the terms of his bargain. They are to be found in the following documents: a memorandum of agreement signed by the purchasers, Effold Ltd, and dated August 2 1968; printed stipulations which were bound up with the auctioneers' particulars of sale; some printed special conditions of sale; and the National Conditions of Sale (17th ed) with one or two express modifications. Effolds bought the whole estate (lots 1 to D 24). Lot 1 was the house, garden and grounds. The particulars of sale contained a description of the lot, but, as is usual, the particulars were not incorporated into the contract. The printed stipulations, which were of contractual force, included the following clause:

> Fixtures and fittings. Fixtures and fittings usually denominated tenant's fixtures and fittings belonging to the vendor shall be taken by the purchaser, according to an inventory to be prepared by the auctioneers and paid for at a valuation to be made in the usual way and the decision of the valuer or umpire as to what are tenant's fixtures and fittings shall be conclusive in the event of any dispute. Certain items are however specifically excluded E from the Sale—see Special Conditions of Sale No 14.

I need not read the rest of the stipulation. Special condition no 11 reserved to the vendor the right to sell by auction at Hinton House "household goods, effects, chattels or . . . other possessions of the vendor" prior to completion. Special condition no 14 specifically excluded certain items from the sale—two sets of tapestries, two statues, some ornamental stonework, some bookcases and the cornice cupboards and curtain rails of the Grand Saloon. None of the items in dispute was specifically excluded under this condition. The contract was therefore for the sale of the realty —Hinton House, garden and grounds: certain specifically F mentioned fixtures were excluded, but an inventory was to be prepared of "tenant's fixtures," which were to be taken by the purchaser and paid for at a valuation. No inventory has in fact ever been taken and, though the point is mentioned in the respondents' notice, no submissions based on this provision have, as I understand the litigation, been advanced to this court by either side. In my judgment, since the contract was silent as to the items in dispute, it is necessary to determine whether under the general law they are to be considered as fixtures, which pass with the realty, or as chattels, in which event they remain the property of the vendor.

As so often, the difficulty is not the formulation but the G application of the law. I think there is now no need to enter into research into the case law prior to *Leigh v Taylor* [1902] AC 157. The answer today to the question whether objects which were originally chattels have become fixtures, that is to say part of the freehold, depends upon the application of two tests: (1) the method and degree of annexation; (2) the object and purpose of the annexation. The early law attached great importance to the first test. It proved harsh and unjust both to limited owners who had affixed valuable chattels of their own to settled land and to tenants for years. The second test was evolved to take care primarily of the limited owner, for example a tenant for H life. In *Leigh v Taylor* the House of Lords invoked it to protect the interest of the tenant for life who had affixed large and valuable tapestries to the walls of the house for the purpose of adornment and enjoyment of them as tapestries. As I read that decision, it was held that she had not made them fixtures. "They remained chattels from first to last," said Lord Lindley at p 164 of the report. In the law of landlord and tenant the law's protection went further: even if the chattel affixed by the tenant must be held to have become a fixture, that is to say part of the realty, a rule was evolved that it was to be treated as the property of the tenant and could be removed by him if it fell into a class J recognised by law as "tenant's fixtures," that is to say if it be a trade, agricultural, or an ornamental fixture. We are not concerned, on the view I take of the case, with "tenant's fixtures." The governing relationship with which this case is concerned is that of a beneficial owner of the legal estate selling the freehold to a purchaser. Such a seller can sell as much or as little of his property as he chooses. Lord Poulett excluded certain named objects from the sale, but the contract was silent as to the objects claimed by the plaintiff. I think it was conceded by the defendants—certainly I so read the contract of sale—that, if the pictures, statue, and K sundial were fixtures at the time of the contract, they were included in it as part of the freehold (subject of course to a valuation if they should prove to be tenant's fixtures). The preliminary, and basic, question is therefore whether these objects were at that time fixtures.

Since *Leigh v Taylor* the question is really one of fact. The two tests were explained in that case by the Lord Chancellor (see the report at pp 158 and 159), who commented that not the law but our mode of life has changed over the years; that what has changed is "the degree in which certain things have seemed susceptible of being put up as mere ornaments whereas at our earlier period the mere construc- L tion rendered it impossible sometimes to sever the thing which was put up from the realty." In other words, a degree of annexation which in earlier times the law would have treated as conclusive may now prove nothing. If the purpose of the annexation be for the better enjoyment of the object itself, it may remain a chattel, notwithstanding a high degree of physical annexation. Clearly, however, it remains significant to discover the extent of physical disturbance of the building or the land involved in the removal of the object. If an object cannot be removed without serious damage to, or destruction of, some part of the realty, the case for its having become a fixture is a strong one. The M relationship of the two tests to each other requires consideration. If there is no physical annexation there is no fixture. *Quicquid plantatur solo solo cedit.* Nevertheless an object, resting on the ground by its own weight alone, can be a fixture, if it be so heavy that there is no need to tie it into a foundation, and if it were put in place to improve the realty. *Prima facie,* however, an object resting on the ground by its own weight alone is not a fixture: see *Megarry and Wade*, p 716. Conversely, an object affixed to realty but capable of being removed without much difficulty may yet be a fixture, if, for example, the purpose of its affixing be that "of creating a beautiful room as a whole" (Neville J in *In Re Whaley* [1908] 1 Ch 615 at p 619). And in the famous instance of *Lord Chesterfield's Settled Estates* [1911] 1 Ch 237 Grinling Gibbons carvings, which had been affixed to a suite of rooms 200 years earlier, were held to be fixtures. Today so great are the technical skills of affixing and removing objects to land or buildings that the second test is more likely than the first to be decisive. Perhaps the enduring significance of the first test is a reminder that there must be some degree of physical annexation before a chattel can be treated as part of the realty.

When one seeks to apply the law to the facts of this case, it is necessary to discriminate between what is relevant but not decisive and what is decisive. Investigatory expertise of a high order has been devoted by both sides to discovering how the pictures were affixed to the panelling and the wall in the two rooms. Fascinating though the investigation was, its conclusion (whether it be, as Mr Millett contended, that the pictures were really put up as an integral part, with the panelling, of the wall-covering, or, as Mr Browne-Wilkinson contended, were put into recesses left for them) is not, in my judgment, decisive. It is enough to say that the pictures were firmly fixed and that their removal needed skill and experience if it were to be done without damage to the wall and panelling. Certainly they were firmly enough fixed to become fixtures if that was the object and purpose of their affixing. But, if ordinary skill was used, as it was, in their removal, they could be taken down, and in the event were taken down, without much trouble and without damage to the structure of the rooms. The decisive question is therefore as to the object and purpose of their affixing. Pictures had hung in the two rooms for centuries. "The Return" had been in the ante-room for a very long time—perhaps ever since it was painted. The 7th Earl decided in the early part of the 20th century to install in the two rooms the panelling and so designed it that there were recesses for pictures. It is this feature which lends plausibility to the suggestion that the pictures, fitted into the recesses left for them, were not to be enjoyed as objects in themselves but as part of the grand architectural design of the two rooms. The Vice-Chancellor rejected this view. So do I. When the panelling was installed in the two rooms the design was either panelled walls with recesses for pictures to be enjoyed as pictures, or rooms having walls which were a composite of panelling and pictures: in other words, the pictures were to be part of a composite mural. I think the former was the truth. The panelling was Victorian, the pictures a heterogeneous collection. According to Sothebys' expert they were of different dates in the 17th and 18th centuries, of different styles, by different hands, the sort of set anyone could put together at any time—very different, I would comment, from that unity of design, the "Elizabethan Room" in the case of *In re Whaley*. There was a particular Poulett family interest in "The Return" and in the two coronation portraits, but this interest focused attention not on the design of the room but on the pictures themselves. Notwithstanding the painstaking and attractive arguments of Mr Millett for the plaintiff, I find, applying the second test, that the pictures were not fixtures. They were put in place on the wall to be enjoyed as pictures. The panelling presented a technical problem in putting them up. The way the carpenter, or whoever it was, solved the problem is not decisive in determining their legal character. But the purpose in putting them there is.

The statue and the sundial give rise in my judgment to no difficulty. Neither was at the time of the sale physically attached to the realty. The sundial was a small object and, once the Earl had detached it (as he did many years earlier) from its pedestal, it ceased to be part of the realty. The statute was heavy. It weighed 10 cwt and stood 5 ft 7 in high on its plinth. There is an issue as to whether it was cemented into the plinth or rested on its own weight. The question is not decisive, for, even if it was attached by a cement bond, it was (as events proved) easily removable. However, upon the balance of probability, I agree with the Vice-Chancellor in thinking it was not attached. The best argument for the statue being a fixture was its careful siting in the West Lawn so as to form an integral part of the architectural design of the west elevation of the house. The design point is a good one so far as it goes: it explains the siting of the plinth, which undoubtedly was a fixture. But what was put upon the plinth was very much a matter for the taste of the occupier of the house for the time being. We know that at one time the object on the plinth had been a sundial. At the time of the sale it was this statue of a Greek athlete. The plinth's position was architecturally important: it ensured that whatever stood on it would be correctly positioned. But the object it carried could be whatever appealed to the occupier for the time being. Sundial or statue—it did not matter to the design, so long as it was in the right place—a result ensured by the plinth which was firmly fixed into the ground. Being, as I think, unattached, the statue was, *prima facie*, not a fixture, but, even if it were attached, the application of the second test would lead to the same conclusion.

As for the Chinese prints, I agree with the Vice-Chancellor. Even if the plaintiff could prove a title to sue in conversion, detinue or trespass, there is no evidence that it was by the hand or upon the instructions of Lord Poulett that the prints were damaged. And, if the plaintiff's case is put in equity as a breach of trust by a vendor when he was constructive trustee for the purchaser, the plaintiff (always assuming he can prove his entitlement to sue the head vendor, a question upon which I express no opinion) has failed to prove any act or omission by the defendant which could amount to a breach of trust.

Accordingly, I agree with the Vice-Chancellor. The action fails *in limine*. The plaintiff cannot show either that the pictures, statue and sundial were fixtures or that Lord Poulett, his executors, or anyone acting on their behalf were guilty of any acts or omissions in respect of the Chinese prints which, assuming the plaintiff could make title, could amount either to trespass, conversion or detinue or to breach of trust. Accordingly, I would dismiss the appeal.

After I had prepared this judgment, I had the advantage of reading the judgment about to be delivered by Goff LJ. I would make two comments. First, without in any way disagreeing with the possibility, in logic, of drawing the distinction I understand my brother to be making between chattels and removable fixtures, I think it would be illogical to overlook the possibility that an object fixed to realty may yet remain a chattel. In my judgment, the pictures, statue and sundial did so remain. Secondly, upon my understanding of the way the case has been fought, Mr Millett for the plaintiff submitted that these objects were fixtures and part of the realty, while the defendants submitted that they were chattels. Having declared myself in favour of the defendants on this issue, I have not thought it necessary to consider the possibility of the objects being fixtures but not part of the realty, that is to say removable fixtures.

Dissenting, GOFF LJ said: In this case I have had the advantage of seeing the judgment already delivered by Scarman LJ and that about to be delivered by Stamp LJ, and I regret to say, as I do with due respect and diffidence, that I find myself in a large measure in disagreement with them. I need not restate the facts, as they are very fully recited in those judgments, and I will proceed to my view of the law and the effect of those facts. I must, however, observe that as pleaded the case against the first defendant was based upon his duty as a vendor to preserve the trust property (see para 7 (7) B and para 18 (1) of the statement of claim) and detinue, trespass and conversion (para 18 (2)). There was no claim for specific performance of the contract for the sale to Effold, to which I will refer as the head contract, or of the subsale by Effold to the plaintiff, to which I will refer as the sub-

A contract. The case, however, as opened by Mr Millett proceeded on the basis of his right to specific performance of the head contract and that was fully argued before us. This is a question of law; all the relevant facts were pleaded and in my view we must consider and decide upon that claim. I can dispose of the Chinese prints at the outset, as the evidence wholly failed to establish the case pleaded, and we ruled on that during the hearing. I turn therefore to the other disputed items, and in using that expression in this judgment I must, of course, be understood as excluding the Chinese prints.

In my judgment anything which is attached to the land B by physical connection, or in some cases by simply resting on its own weight, is a fixture, although it may be such that a tenant for years or other limited owner is entitled to remove it. Until severed it is not a chattel; see per Rigby LJ in *De Falbe* [1901] 1 Ch 523 at p 530, where he described such things as "removable fixed things" and again at p 531, where he said: "In all such cases the object (whatever it may be) which is affixed for the purpose of ornamentation is affixed to the freehold, but the exception allows it to be removed." It seems to me that all the items in dispute except the sundial, as to which special considerations apply, must be fixtures, and the question is not, I think, with C respect, whether they are fixtures or chattels, but whether they were removable after conclusion of the head contract, which unless therein otherwise provided would preclude the first defendant, albeit an absolute owner, from removing anything not within a recognised exception to the rule *quicquid plantatur solo solo cedit,* which exceptions include trade fixtures and ornamental objects; see per Lord Justice Vaughan Williams in the same case at p 535: "It will not be denied, and indeed it has not been denied in argument, that among the exceptions to the rigid rule, *quicquid plantatur solo solo cedit,* there are these two—one in respect D of trade fixtures, the other in respect of ornamental objects, which have been annexed in some way to a freehold." Unless the vendor by his contract reserves some special right, he is, as I see it, in this respect in the same position *vis-à-vis* the purchaser as is a tenant for life *vis-à-vis* the remainderman, and as Rigby LJ said at p 531, substituting as I do the words "the disputed items" for "these tapestries" and "the vendor" for "the tenant for life" the question in the present case is reduced to this, whether the disputed items were so affixed as to become part of the house, or so affixed as to continue removable at the will and pleasure of the vendor who affixed them.

E It follows in my judgment that the disputed items except the sundial were either "so affixed as to become part of the house," in which case they were included in the land sold without addition to the price, or they were "fixtures and fittings usually denominated tenants' fixtures" within the meaning of the provision in that behalf in the general stipulations, so that the plaintiff became entitled to them, but was bound to pay for them at a valuation. That provision is part of the contract and cannot be ignored. In my view it postulates the hypothesis of landlord and tenant. I agree that the proper test for determining whether anything is a fixture passing with the inheritance or a removable fixed F thing is now well settled (see *Leigh v Taylor*) and the difficulty is not what is the test but what conclusion one should reach when applying it to the facts. The test is in two parts. It is partly the degree of physical annexation, though that is generally a subordinate consideration, and more important the "object and purpose of the annexation as it is to be inferred from the circumstances of the case." This in my view appears very clearly from the judgments of both Vaughan-Williams and Stirling LJJ in the *De Falbe* case, and as the former pointed out at p 536 a fixture may be removable although its nature is such that it has necessitated a large degree of physical annexation. That being so, I would agree that the two pictures in the Queen's Ante-Room were removable G fixtures. I would take the same view of those in the Queen's Dining Room were it not for the description in the particulars "with oak panelling around picture recesses." It is conceded, and I think rightly so, that in the face of that description the first defendant cannot claim that the panelling was a movable fixture, and if so, then I do not myself see how the pictures can be any different, as I think the panelling and pictures were made one integral whole.

The statue of the Greek athlete and the sundial were both placed at focal points in the grounds and the inference is, I think, that they were not placed there for their better enjoyment as chattels, but for the permanent enhancement of the H beauty of the grounds. *Prima facie,* therefore, I would regard these also as not removable. It is uncertain whether the statue was cemented to the plinth or whether it was merely cemented in part by way of a repair when it became broken, but in my view it matters not, since having regard to the size and situation of the statue, I would regard this as one of those cases where resting upon its own bulk was a sufficient annexation, and so I adhere to my *prima facie* view; see *D'Eyncourt v Gregory* (1866) LR 3 Eq 382 at p 396 where Lord Romilly MR said: "I think it does not depend on whether any cement is used for fixing these articles, or J whether they rest on their own weight, but upon this—whether they are strictly and properly part of the architectural design for the hall and staircase itself and put in there as such, as distinguished from mere ornaments to be afterwards added"; and *Bulkeley v Lyne Stephens* (1895) 11 TLR 564. I am of course not overlooking the criticisms of *D'Eyncourt v Gregory* in *De Falbe* [1901] 1 Ch 523 at pp 532 and 537, but those criticisms were as to the inference which Lord Romilly drew from the facts, not the principle that a thing may be a fixture because it is part of the architectural design. The sundial is more difficult because the first defendant had unfastened the actual dial before the date of the head con- K tract. True this was not in order that it might be removed and cease to be connected with its pedestal, but in order that it might be adjusted to take account of summer time. Nevertheless, as it was in no way physically annexed to the land and was not a heavy object resting on its own weight, I have finally come to the conclusion that this was not a fixture at all, whether removable or not, but a chattel, and in my judgment the plaintiff cannot on any showing have any right in respect of this item. Apart from the sundial, therefore, and the two portraits in the ante-room, I would have been for holding that all the disputed items were irremovable fixtures. However, L this is not a difference of view on the law or in principle, but a matter of impression, and as both my brethren think otherwise I am prepared to proceed on the basis that they were removable. The consequence, however, in my judgment must be that the fixtures clause applies and Effold, and through it the sub-purchaser, were entitled to all the disputed items save the sundial but bound to pay for them upon a valuation.

Then I turn to consider whether the plaintiff as sub-purchaser is entitled to relief against the first defendant as head vendor. It is clear that after Effold paid the contract price and took a conveyance without the disputed items, without M compensation, and without reserving any rights in respect thereof, it could have no claim against the first defendant. It is said that the plaintiff is a stranger to the head contract and so cannot in any event claim under it, and even if he could have done so at one time he cannot now, as he cannot be in any better position than the original purchaser, Effold. I think it is also suggested that in any case he cannot now pursue any remedy against the first defendant, since he has completed the sub-contract. I cannot accept these arguments, and I regret to say that I find myself in disagreement with my brother Stamp LJ as to the law on this question. I agree that the sub-contract did not make the first defendant

A a trustee of the land or fixtures for the plaintiff, but with all respect in my view Effold became a trustee, in the sense that a vendor is a trustee, of the benefit of the head contract, which entitled the plaintiff to specific performance of the head contract subject to notice of the sub-contract having been given to the first defendant, which it clearly was. In the present case he had express notice of the sub-contract and of the plaintiff's claim to the disputed items. This being so, although the subsale was not an assignment of the head contract, the plaintiff acquired an equitable interest in it, which the first defendant was bound to recognise and protect, and he and Effold could not, with notice of the plaintiff's B claim, agree to rescind the head contract and make new terms, or vary the terms, to the prejudice of the plaintiff, leaving him with no more than a remedy against his own vendor, Effold. In my judgment, therefore, notwithstanding the first defendant and Effold purported to complete the head contract, and Effold could not thereafter maintain any claim against the first defendant, nevertheless, as they excluded part of the property agreed to be sold, the plaintiff remained entitled to sue the first defendant for specific performance in the form of an order for delivery up of the one picture formerly in the Queen's Dining Room which remains unsold (see *Phillips v Lamdin* [1949] 2 KB 33 at p 41) C and compensation for the rest of the disputed items, except of course the sundial.

In my judgment this is well supported by authority, which begins with the very ancient case of *Dyer v Pultenay* (1740) Barn Ch 160 at p 169. With respect I cannot limit that case, as I think Stamp LJ does, to a special case of getting in an outstanding legal estate. A large part of the purchase money under the head contract in that case had not been paid. What the sub-purchaser sought, but could not get, was to stand in the purchaser's shoes to compel the vendor to perform not the original contract but his own contract, but D it was made quite clear that the sub-purchaser, had he been willing to complete the original contract, could have had a specific performance of it. But if the matter would remain doubtful on that authority it becomes clear in the later case of *Fenwick v Bulman* (1869) LR 9 Eq 165. True, in that case the vendors had sued not only their own purchaser but the sub-purchaser in an action for specific performance, and this was held against them when they resisted his suit, but the rights of the sub-purchaser were laid down in perfectly general terms. Sir John Stuart VC said at p 168: ". . . can it be successfully contended that the plaintiff is not entitled to a right arising out of the original contract? I think not, for E if that contract be not performed, he can have no performance of his contract with the original purchaser," and again: "However, if I entertained any doubt in this case, it would be dispelled by the decision of Lord Hardwicke in *Dyer v Pultenay*, in which the obvious equity of a subpurchaser is clearly recognised." Next in my respectful view, the speeches in the case of *Shaw v Foster* (1872) LR 5 HL 321, to which Stamp LJ has drawn attention, support my view of the legal position and not his. The decision itself is, of course, completely distinguishable on the following grounds: (1) the transaction with the third party was only an agreement to assign conditional on request which the House thought F had never been made, for good reason in that the property was leasehold and the third party did not wish to be saddled with its burdens; (2) even if a request had been made, no notice of that circumstance had been given to the vendor, so that he had no notice of any absolute interest in the third party; (3) the third party was impeding performance of the head contract by refusing to advance the balance of the purchase money, yet requiring a conveyance to himself (see pp 347/8); (4) the vendor performed the head contract according to its terms. In the present case, however, there was an actual binding contract of subsale which was an assignment in equity of the benefit of the head contract, or at least gave the subpurchaser an equity in it, and the first defendant was given G express notice of the equity and that the plaintiff claimed the disputed items. The first defendant did not perform the head contract, but broke it by retaining the fixtures. The speeches, however, in my view clearly recognised the right of a subpurchaser to sue the head vendor for specific performance where proper notice of the subcontract has been given. Thus at p 333 Lord Chelmsford said:

> According to the well known rule in equity, when the contract of sale was signed by the parties Sir William Foster became a trustee of the estate for Pooley, and Pooley a trustee of the purchase-money for Sir William Foster; and it was competent to Pooley to assign the benefit of his contract, or to charge his H equitable interest in the property in favour of another person, and upon notice given to Sir William Foster of such assignment or charge, he would have been bound to protect and give effect to it.

Again, at p 338 Lord Cairns said:

> Under these circumstances I apprehend there cannot be the slightest doubt of the relation subsisting in the eye of a court of equity between the vendor and the purchaser. The vendor was a trustee of the property for the purchaser; the purchaser was the real beneficial owner in the eye of a court of equity of the property subject only to this observation, that the vendor, whom I have called the trustee, was not a mere dormant trustee, he was a trustee having a personal and substantial interest in the property, J a right to protect that interest, and an active right to assert that interest if anything should be done in derogation of it. The relation, therefore, of trustee and cestui que trust subsisted, but subsisted subject to the paramount right of the vendor and trustee to protect his own interest as vendor of the property.
>
> My Lords, in that state of things Mr Pooley, the purchaser, being the real and beneficial owner, I apprehend that there cannot be any doubt of the rights of Mr Pooley with regard to the property of which he had thus become the beneficial owner. He had a right to devise it; he had a right to alienate it; he had a right to charge it. There are various ways which might be suggested in which, for valuable consideration, he might have created a charge more or less affecting the property. I apprehend that he might K have contracted, for valuable consideration, with any person to whom he was indebted, that he (Pooley) would complete the purchase, and that when the purchase was completed and the property assigned to him, he would then make it over to the person to whom he was thus indebted. That would have been one way of dealing with his interest. Another way might have been this— he might have contracted with any person to whom he was indebted that he would pay the purchase-money remaining unpaid, and that then, the purchase-money being thus paid, and the time for the assignment having arrived, he would authorise and require the vendor to assign, not to him, Pooley, but to the person to whom he was indebted. A third and a simpler way in which he might have effected his interest would have been to contract with L any one to whom he was indebted to assign to him the contract which he had entered into in whole, making the person to whom he was indebted assignee of the contract. Any one of those modes might, in my opinion, have been resorted to; and the only qualifications to which all or any of them would have been subject are these: first, that by none of these modes could anything have been done by Mr Pooley derogating from, or impeding, or delaying the rights of the vendor to require the fulfilment of his contract according to its terms; and, secondly, whatever course was taken by Mr Pooley and any person with whom he contracted to charge his interest, notice of the particulars of that charge, and the mode and form of the charge, would be required to be given to the vendor, in order that the vendor might shape his course according M to the notice he had thus received.

It is true that the instant case is a subsale and thus different from the examples given by Lord Cairns, but the authorities already cited show that the subpurchaser has an equity, which is in my judgment equivalent to the position in those examples.

Finally, there is the case of *Harmer v Armstrong* [1934] Ch 65. There the purchaser entered into the contract as a trustee, and then, claiming that he had been acting on his own behalf, purported to agree with the vendor to rescind the contract, and the beneficiaries successfully sued for specific performance. There the purchaser offered, if he were held to be a

A trustee, to perform the contract, but that was too late, unless the beneficiaries had an independent right which they could enforce against the vendor. This case in my judgment is a complete answer to the proposition that because the purchaser had lost its right to sue in respect of the disputed items the subpurchaser's right was also defeated. I agree that a subpurchaser can only have a specific performance on the terms that he performs the purchaser's part of the head contract, so far as it remains to be performed. Here, however, Effold paid the price, and so the only question in this respect arises with regard to the inventory and valuation, and it is true that it never offered to pay. In my judgment, however, that does B not prevent him from obtaining specific performance now on submitting to be charged with the value of the disputed items. If the first defendant had required payment and the plaintiff had refused, that no doubt would have prevented him obtaining specific performance, at all events once the first defendant had accepted it as a repudiation, but the first defendant refused to allow the plaintiff to have the disputed items on any terms, nor did the first defendant himself take any steps towards having the requisite valuation.

There remains the question whether it makes any difference that the plaintiff was a subpurchaser of part only of C the property, and in *Fenwick v Bulman* (1869) LR 9 Eq 165 at p 167 Sir John Stuart did suggest that this might be a valid objection. It may well be where there are a lot of subpurchasers, as where a large estate is sold off in lots, that the court will refuse relief by way of specific performance to a subpurchaser, either because it will be found that sufficient notice of the relevant subsale has not been given to the vendor, or because the circumstances make the case one in which the equitable remedy is unsuitable and the subpurchaser must be left to his remedy against his own vendor: see *Harmer v Armstrong* [1934] Ch 65, particularly at p 88 and pp 93 and 94. That case can be considered if D and when it arises, but for my part in a case such as we have here, where there are few subpurchasers, the plaintiff has subpurchased a substantial part of the estate, and clear express notice of the subcontract has been given to the vendors, I cannot think that the fact that he has not purchased the whole of the estate should be allowed to defeat his equity.

Where a subpurchaser seeks to enforce this equity he should make the original purchaser a defendant, or obtain leave from the court to sue in his name, and neither course was pursued in this case; but instead, a few months after the conveyance from Effold, the plaintiff took an assignment E containing the following provisions: recital (6):

> Some of the said fixtures have been removed from Hinton House aforesaid or as the case may be the said grounds without the authority or consent of the Assignor or the Assignee and some of the same have suffered damage and for the purpose of better enabling the Assignee to take all such proceedings in respect of such fixtures as aforesaid as he may be advised or desire to take the Assignor has agreed with the Assignee in manner hereinafter appearing Now This Deed Witnesseth that in consideration of the premises the Assignor as Beneficial Owner hereby assigns unto the Assignee All That the benefit of the Agreement and Conveyance hereinbefore first recited together F with all the Estate and interest of the Assignor in the fixtures which at any material time were annexed to Hinton House aforesaid and the said grounds and all rights of action choses in action and other rights benefits advantages or demands which the Assignor has or may have in the same or with respect thereto or with respect to any matter or thing pertaining thereto arising thereout or connected therewith To Hold all the same unto the Assignee absolutely.

I am not sure that that was the correct way to deal with the matter, but as I understood it no objection was raised on the score of parties if as a matter of equity the plaintiff could in any form enforce the head contract.

G The plaintiff is in my judgment, therefore, entitled to an order for delivery up of the unsold portrait formerly in the Dining Room, and to an order for payment of the value of the remaining disputed items, except the sundial, and as this is by way of specific performance that value should in my judgment be ascertained on the equitable principle as at the date of judgment, see *Re Massingberd* (1890) 63 LT 296 (where the date taken was that of the commencement of the proceedings) and *Grant v Dawkins* [1973] 1 WLR 1406 at p 1410 citing *Wroth v Tyler* [1973] 2 WLR 405. This, however, is subject to the plaintiff submitting to be charged with the value as at the date fixed for completion by the H head contract. I would therefore allow the appeal so far as concerns the first defendant; order delivery up of one picture and direct an inquiry as to the value of the disputed items other than that picture and the sundial at each of these dates and order that the plaintiff do recover from the first defendant the excess, if any, of the present value over the completion date value, and reserve the costs of the inquiry.

I now turn to the case against the other defendants, and it is submitted that they are liable at law in conversion or detinue, or alternatively in equity as constructive trustees. In my judgment none of these claims can be supported. The claim at law was based first upon the Effold conveyance, J but like my brother Stamp LJ I fail to see how that could possibly give the plaintiff any right to sue the agents at law, or for that matter, as was at one time suggested in argument, the first defendant himself, because the disputed items were severed before it was executed. Alternatively, Mr Millett relied upon the case of *Healey v Healey* [1915] 1 KB 938. That case, if rightly decided, was a special one depending upon the right of a married woman to maintain an action against her husband in respect of her separate estate, and the trustees so far as the right to immediate possession was concerned were mere bare trustees. In my view it has nothing to do with the present case, where the K plaintiff's only right to possession was through an order for specific performance of a contract to which he was not even a party. The case in equity ran thus: the first defendant was a constructive trustee for the plaintiff, and therefore the agents in dealing with the property on his instructions in derogation of the plaintiff's rights became liable to him as constructive trustees. I am not satisfied that the premise is correct for the present purposes. The first defendant was no doubt a constructive trustee for Effold under the head contract until the Effold conveyance, but he did not hold the fixtures on a constructive trust for the subpurchaser L albeit the latter had an equity which he was entitled to protect by an order for specific performance. Even, however, if the premise be right, still in my judgment the conclusion does not follow. The first defendant and Effold took it upon themselves to construe the head contract and to apply it to the facts and decided that there was nothing in the plaintiff's claim. They acted upon that and the agents acted on their instructions. They were wrong and therefore there was a breach of contract, but there was not, as I see it, any allegation, let alone proof, of such dishonesty or want of probity as must be shown to render the agents liable in accordance with the principles laid down by Sachs and M Edmund Davies LJJ in the *Carl Zeiss* case (*Carl-Zeiss-Stiftung v Herbert Smith & Co* [1969] 2 Ch 276 at pp 299 and 301).

Agreeing with Scarman LJ, STAMP LJ said: I agree that the items in dispute were not part and parcel of Hinton House and that this action accordingly fails. But lest it be thought that I had accepted the plaintiff's submissions on the question whether the plaintiff's action is well founded in law and because this case has been so fully argued on all points I will give a judgment of my own.

On August 2 1968 the Hinton St George Estate was put up for sale by auction by Lord Poulett. The auction

A particulars divided the estate into 24 lots of which lot 1 comprised Hinton House and its outbuildings, gardens and grounds. By agreement made at the auction, Effold Properties Ltd in fact agreed with Lord Poulett to purchase the whole estate—not in separate lots—for a sum of £225,000. I will call this contract the "estate contract." On the same day the plaintiff agreed to purchase and Effold agreed to sell the property described as lot 1 and lot 4 in the particulars of sale for the sum of £28,928. I will call this latter agreement "the subcontract." There were other subcontracts for the sale of other lots to other subpurchasers. Between the date of the subcontract and the date of completion of the
B sale of the estate, on October 31 1968, various items of property which the plaintiff claims were attached to and had become part and parcel of Hinton House and were accordingly comprised in both sales were removed. If these items (which I will call collectively "the disputed items") were ever part and parcel of Hinton House, they had by October 31 1968 been severed therefrom. On October 31 1968 the sale by Lord Poulett to Effold of the Hinton St George Estate was, as I have indicated, completed and on that day the estate was conveyed to Effold in consideration of the sum of £225,000 the receipt of which was acknowledged. Effold was at the date of completion aware that
C the disputed items had been removed from Hinton House. Effold nevertheless made and makes no claim for compensation in respect of the disputed articles or in respect of their removal. On February 5 1969 Effold and the plaintiff entered into an agreement supplemental to the subcontract whereby lot 4 and a part of lot 1 were excluded from the sale to the plaintiff and the purchase price was reduced to £17,928. The sale by Effold to the plaintiff was completed on February 19 1969, when Hinton House (less the part excluded by the agreement of February 5 1969) was conveyed to the plaintiff. The plaintiff was then, and had been prior to October 31 1960, aware of the removal of the dis-
D puted articles. By a deed of assignment dated October 2 1969 and made between Effold of the one part and the plaintiff of the other part, after reciting the contracts and conveyances to which I have referred and reciting that at all times there were annexed to Hinton House certain fixtures all of which the plaintiff claimed ought to have passed under the recited transactions first to Effold and then to the plaintiff, reciting that some of the fixtures had been removed from Hinton House without the authority of Effold or the plaintiff and some of them had suffered damage, and reciting that for the purpose of enabling the
E plaintiff to take proceedings in respect of such fixtures Effold had agreed with the plaintiff in manner thereinafter appearing, Effold as beneficial owner assigned to the plaintiff the benefit of the estate contract and the conveyance of October 31 1968 by Lord Poulett and all the estate and interest of Effold in the fixtures which at any material time were annexed to Hinton House and all rights of action and other rights, benefits, advantages or demands which Effold had.

In this appeal it was common ground that after the conveyance by Lord Poulett to Effold, which had been accepted by Effold with knowledge of the relevant facts and without
F any claim for compensation, Effold could make no claim against Lord Poulett either at law or in equity in respect of the disputed items. At law, as I understand it, the contractual rights of Effold to have a conveyance of the property agreed to be sold merged in the conveyance; and it was conceded that since Effold completed the contract knowing of the removal of the disputed articles and thereafter made no claim against Lord Poulett in respect thereof, it must be taken to have waived any right founded on the breach of the vendor's fiduciary duty to take reasonable care to preserve the property in a reasonable state of preservation: see and compare *Clarke v Ramuz* [1891] 2 QB 456, where the purchaser was at the time of completion unaware that G
surface soil had been removed after the date of the contract and where this court left unresolved the question whether a right of action would have survived if the purchaser had known of the removal when he completed the contract. Effold in any event is not a party to this action and makes no claim to the disputed items. Among the submissions made on behalf of the plaintiff is the submission that pending completion of the estate contract it was the duty of Lord Poulett, *as trustee for Effold*, to use reasonable care to preserve the property agreed to be sold, including the disputed items, in a reasonable state of preservation. The
case was so pleaded. As I have indicated, Lord Poulett indeed H
had a duty to Effold to use reasonable care to preserve the property agreed to be sold in a reasonable state of preservation; but because of the edifice which is sought to be built on the proposition that he had that duty "as trustee for Effold" I find it necessary to examine that proposition.

A vendor under a contract for the sale of land is bound on completion to convey what he has agreed to sell. And so (I take his duties from the statement in *Fry on Specific Performance* 6th ed at p 638) he is bound on completion to show a good title to the property contracted to be sold and is bound to take reasonable care of the property and to pay
the outgoings until the purchaser takes, or ought to take, J
possession and is bound upon payment of the money payable by the purchaser to execute and procure the execution by all other necessary parties of a proper conveyance vesting the legal estate in the purchaser and to put him in possession of the property agreed to be sold. The purchaser has corresponding rights. These duties and rights arise from the contract of sale and it is because of their existence that the vendor is said to be a constructive trustee, or a trustee *sub modo*, of the estate for the purchaser from the time when the contract is constituted. But to say that it is the
duty of the vendor *as trustee for the purchaser* to care for K
the property is to put the cart before the horse and may lead you into error. He is said to be a trustee because of the duties which he has, and the duties do not arise because he is a trustee but because he has agreed to sell the land to the purchaser and the purchaser on tendering the price is entitled to have the contract specifically performed according to its terms. Nor does the relationship in the meantime have all the incidents of the relationship of trustee and cestui que trust. That this is so is sufficiently illustrated by the fact that *prima facie* the vendor is until the date fixed for the completion entitled to receive and retain the rents
and profits and that as from that date the purchaser is L
bound to pay interest. And you may search the Trustee Act 1925 without obtaining much that is relevant to the relationship of vendor and purchaser. Thus, although the vendor because of his duties to the purchaser is called a trustee, it is wrong to argue that because he is so called he has all the duties of or holds the land on a trust which has all the incidents associated with the relationship of a trustee and his cestui que trust. With these considerations in mind I pass on to consider the submission which is foreshadowed in the statement of claim that because, as is the fact, Lord Poulett had on August 2 notice of the subcontract he there-
after had a duty *as trustee for the plaintiff* pending completion M
of the estate contract to use reasonable care to preserve lot 1, including Hinton House and the disputed items, in a reasonable state of preservation. In my judgment that submission is not well founded. The plaintiff was not, and I emphasise this, an assignee of Effold's rights and interest under the estate contract and was no party to that contract. Lord Poulett could not have required the plaintiff to pay a single penny for Hinton House and could not in the events which happened have compelled the plaintiff to accept a conveyance of Hinton House. It follows in my judgment that as between Lord Poulett and the plaintiff there did not come

into being those mutual rights and obligations of a vendor and purchaser of land the existence of which enables one to describe the vendor as a constructive trustee or a trustee *sub modo* for the purchaser. Putting it more concisely, because there was not between Lord Poulett and the plaintiff the relationship of vendor and purchaser the incidents of such a relationship were absent. Lord Poulett did not hold Hinton House as a trustee for the plaintiff and was under no fiduciary or contractual obligation to him to take reasonable care of Hinton House. The plaintiff's remedy in respect of any failure by Lord Poulett in his duty to Effold was against Effold in respect of the consequences of Effold's failure to enforce that duty. I am fortified in my conclusion by the speeches in the House of Lords in *Shaw v Foster* (1872) LR 5 HL 321.

I would add this in relation to the proposition that a vendor of land who has notice of a subsale becomes a trustee for the subpurchaser. The proposition, if it was a correct statement of the law, would have to apply as well in favour of all the subpurchasers as in favour of any of them. In the instant case the position is simplified by the fact that it so happens that the several subcontracts of the several lots were, as I understand it, entered into by reference not only to the parcels described in the estate contract but on the terms and conditions set out in the estate contract with a substitution only of a different price. But suppose an agreement for sale of a large estate followed by division into and subsale of say 100 parcels, the contract for sale of land containing varying conditions, can it be the law that the original vendor, on learning that fact, comes under a fiduciary duty to each of the 100 purchasers? I would require authority before I would so hold. If such were the law I would ask the rhetorical question: does the fiduciary duty to the subpurchasers continue after the head purchaser ought to have taken possession but failed to do so, that is to say after a point of time at which the head vendor ceases to owe the duty of care to the head purchaser? Or suppose that immediately prior to completion of the head contract the subcontract be rescinded, to whom prior to such rescission was the duty owed? I do not doubt that other difficult questions would arise; and the law relating to the rights and obligations which arise as incident to a contract for the sale of land is complicated enough without introducing into it the conception that a vendor becomes a trustee for the subpurchasers of each part of it.

It was submitted, relying on the authority of the judgment of Lord Hardwicke in *Dyer v Pultenay* (1740) Barn Ch 160 and *Fenwick v Bulman* (1869) LR 9 Eq 165, that a subpurchaser willing to perform the head contract is entitled as against the vendor to specific performance of that contract. And, so the argument runs, the subpurchaser being entitled to specific performance of the head contract is in the same position vis-à-vis the head vendor as is the purchaser, and is as regards the land which he has agreed to purchase entitled, pending completion of that contract, to the same rights as the purchaser himself. This submission is not in my judgment well founded. A purchaser of land is against the vendor entitled on completion to the execution by all necessary parties of a conveyance vesting the legal estate in him. Thus if there is an outstanding legal estate not vested in the vendor he is bound to get it in. So as between purchaser and subpurchaser the subpurchaser is entitled to require the purchaser, if he can, to get in the legal estate from the head vendor. If the purchaser will not or cannot do so, *Dyer v Pultenay* is authority for the proposition that the subpurchaser by joining his vendor and the head vendor as defendants to an action brought for the purpose may, standing in the shoes of his vendor, obtain an order for the conveyance of the land. But the head vendor cannot be required by the purchaser to convey any part of the land comprised in the head contract except upon performance by the purchaser of all the terms of the head contract, and the subpurchaser standing in the shoes of the purchaser can be in no better position. Thus the specific performance which the subpurchaser obtains is specific performance of the head contract and in my judgment, far from establishing the right of a subpurchaser to specific performance of the subcontract against the head vendor, establishes the absence of such a right. Nor is it because the head vendor is a trustee for him that he can obtain the land which he has contracted to purchase but because of his right to insist that his vendor gets in the legal estate and, if he will not do so, to stand in his shoes to seek performance of the head contract. In the instant case the plaintiff never offered to perform the estate contract. Nor can I equate the right of the several purchasers to insist that their vendor gets in the legal estate to the parcels which they have agreed to purchase to that of beneficiaries under the head contract. As I see it, it was in the instant case open to Effold consistently with its duty to the plaintiff to agree with Lord Poulett, before the removal of the disputed items, that the estate contract should be rescinded on the terms that Lord Poulett would sell Hinton House to Effold at a new negotiated price, completion to take place on completion of the subcontract. Provided Effold was in a position on completion of the subcontract to convey or procure the conveyance to the plaintiff of what he had purchased, Effold would have complied with the terms of the subcontract. If it were not in a position to convey the disputed items to the plaintiff his remedy in respect of such of them as had been severed from the freehold could only be against Effold for damages or specific performance with compensation. But the plaintiff was content to complete his contract without pursuing his claim against his vendor.

I turn to consider another aspect of the plaintiff's case. In the amended statement of claim it is alleged that the conveyance of October 31 operated to pass the disputed articles to Effold and that the conveyance of February 19 operated to pass them to the plaintiff. In view of the fact that the disputed articles had been severed from Hinton House and removed prior to the execution of either of the conveyances and what was expressed to be conveyed by each of those conveyances was property of which the vendor was seized "for an estate in fee simple," that contention is in my judgment clearly wrong. The conveyances could not by their terms convey a severed chattel.

Since in my judgment Lord Poulett never became a trustee for the plaintiff, and since he never entered into any contract with the plaintiff and did not assign the disputed articles to the plaintiff, the claims of the plaintiff against Lord Poulett, whether founded in breach of trust, conversion or detinue, must in my judgment fail. And since the plaintiff has never acquired a right to possession of the disputed articles the claims against Messrs Sotheby & Co and T R G Lawrence & Son, each of whom is sued for conversion or detinue, must likewise so fail.

If the views I have expressed are well founded, that is the end of this appeal. But because a great part of the argument in this court has been directed to the question whether any, and if so which, of the disputed items had in law become part of the freehold of Hinton House at the date of the estate contract, it is right that I should also express my conclusions on this question. I deal first with the disputed pictures. They are described very fully as well in the judgment of Plowman VC as in that of Scarman LJ. So far as is relevant one may, I think, summarise those descriptions in this way. They are pictures of a conventional character such as may be found adorning the walls of the rooms in countless great and medium-sized houses, town halls, museums and other public and private halls throughout this country, and indeed throughout the civilised world. With the exception of "The Return" which depicted the Poulett family, and which I should perhaps

A add was on a substantially larger scale than described by the Vice-Chancellor, they were portraits of individuals. They were painted in oils on canvas mounted on stretchers and framed in a conventional manner in heavy gilt frames. They are large pictures, "The Return" being larger than most family portraits. It cannot be doubted that when first painted and framed each of them was a chattel. Nor can it be doubted that when first set up on the wall which it adorned, whether hanging from a picture rail or from hooks or nails or screws set in the wall, or fastened by mirror plates such as you see in miniature in the corridors of this building, each picture, like the vast majority of framed pictures, remained a chattel and B did not become part of Hinton House. Framed pictures are hung on or fixed to the walls for their better enjoyment as pictures, however much they may beautify the rooms in which they are found. Pictures, like other furnishings and chattels in private houses, must almost invariably serve two purposes: to be enjoyed as pictures in the only way in which a picture can be enjoyed in a private house, that is to say by being fixed to or hung on a wall, and to embellish and beautify great rooms which without decorations would by most people be regarded as uninhabitable. Nor does, nor could, counsel on behalf of the plaintiff submit otherwise. What is said is that when the rooms in which the disputed pictures were at the C date of the estate contract were panelled prior to the 1914-18 war, the pictures ceased to be chattels and became part and parcel of the freehold of Hinton House. In view of the Vice-Chancellor's finding this submission is not in my judgment well founded. The submission is, as I understand it, based on two propositions. As part of that submission, it is said first that when the walls were panelled the pictures were fixed in such a way that they could not be readily removed and second that they were so fixed as part of a plan under which they were to become a part of the walls of the room and not for their better enjoyment as pictures. The second part of D the submission must I think be dependent upon the first: because if the pictures had been screwed by mirror plates to the face of the panelling or hung over the panelling, the submission that they thereupon became part of the structure of the house would have been no more maintainable than if the walls had never been panelled. Framed oil paintings in my judgment are, and remain, chattels whether they be hung upon, or over, or screwed by mirror plates to, the panelling of a room, or hung from, over, or screwed by mirror plates to, the walls of an unpanelled room. Had the frames of the pictures been removed so that the panelling itself formed the frames to which the pictures were attached I would find E the submissions on behalf of the plaintiff less difficult to accept; and had this been so it would no doubt have been necessary to consider whether pictures and panelling, forming as it were a single piece, had or had not become part of the freehold prior to the date of the estate contract. In this connection I would observe in parenthesis that the terms of the contract can throw no light on the question what was then part and parcel of Hinton House. And because the pictures remained in their frames I do not find it necessary to express a view whether, as Plowman VC thought to be the case, the panelling itself remained a chattel available to be removed and replaced elsewhere. The submissions on behalf of the F plaintiff must I think rest on the fact that the pictures were not hung or fixed over the panelling but at the relevant time were found set in recesses in the panelling in such a manner that their removal was very difficult and so that the panelling appeared to frame the pictures and their frames, except of course that the embellishment of the frames of the pictures protruded over the edges of the panelling. The rooms were conveniently, and I think accurately, described in the auction particulars as "with oak panelling around picture frames." It is clear that however the pictures were kept in place the panelling was constructed so as to fit round them. A great deal of evidence was given in the court below and I think most of it must have been read to this court G during the course of a long hearing, as to how the pictures were or could have been affixed. The judge described much of the evidence as pure speculation, as indeed it was. The Vice-Chancellor in resolving this part of the case said this:

> One firm fact is that the pictures had mirror plates attached to the back of their frames, the number and position of those plates depending on the size of the picture. There are then three theories as to how these mirror plates were attached to the walls. One is that pictures and panelling were assembled on the floor, the mirror plates being screwed first to the back of the picture frames and then to the back of the panelling, the pictures and panelling together then being affixed to studs in the lath-and-plaster H walls. A second is that the pictures, with their mirror plates already affixed to the frames, were screwed to the walls and the panelling erected around them. The third is that the mirror plates on the frames were secured to the front of the panelling. I reject the third theory, principally for the reason that, except in a single instance, no screw holes were to be found in front of the panelling, but as between the first two, there is little to choose and the evidence is inconclusive.
>
> If a finding were essential to the determination of this case, I should conclude that on the balance of probability the second theory is the correct one.
>
> The result, however, whichever theory may be right, was that the pictures, when hung, hung in recesses in the panelling, made J to fit the pictures, and in one of the cases the lath and plaster had been slightly cut away to accommodate the stretchers of the picture frames.

Clearly one end of each mirror plate must have been screwed to the back of the picture which it served before the picture found itself with its back to the wall; and having given the evidence the best consideration I can I am quite unable to conclude that the Vice-Chancellor erred when he found that on the balance of probability the other end of the mirror plate was affixed by screws in the wall. Had the other end been screwed to the back of the panelling the removal of the pictures could not as I see it have been accomplished in K the way it was but would have required the removal of the panelling, which was not done, so as to get at the heads of the screws, or a degree of force which I would have thought would have broken up the picture frames.

The Vice-Chancellor, however, found the following facts regarding the removal:

> Although the precise method of affixation of the pictures is obscure, the evidence about their removal is less so. Mr Browning, of the firm of Rogers, who removed the pictures which Sothebys were taking, said that he and his men (of whom there were about half a dozen) got them down in, he supposed, half to three-quarters of an hour—a time substantially confirmed by Lady Poulett—using only a 10- or 12-in screwdriver and pincers, L except in the case of one screw the head of which had to be sawn off. He denied that it was necessary to use force, and was obviously aggrieved by the suggestion that a skilled picture-remover with a very high class firm like Rogers would contemplate such a thing.
>
> Mr Ash who, with two men, removed "The Return" and the portraits of the 7th Earl and his wife, said that it took them between two and three hours to get "The Return" down because it was hard to find out how it was fixed. This is understandable because the panelling was over the screws; eventually he found that it was fixed by mirror plates to the wall, not to the panelling. The other two pictures, he thought, took about three-quarters of M an hour each to remove and resulted in some minor damage. Lady Poulett, who watched those two pictures being removed, said that they came down very easily and took about 15 minutes each. I prefer her evidence about this to that of Mr Ash.

How Mr Browning and Mr Ash got at the screws in the wall to unscrew them must remain something of a mystery. The evidence was, I must assume, inconsistent with the simple explanation that the heads of the latter screws were sufficiently small to enable the eyes of the mirror plates to be hooked over them after they had been screwed in. The heads of the screws which were handed up to the Bench were larger than the eyes of the plates.

A Accepting, as I do, the Vice-Chancellor's finding that the pictures were not fixed to the panelling but to the wall and fixed indeed in a conventional way, I find it impossible to conclude that on the setting up of the panelling around them they lost their character as chattels and became part of Hinton House. They were still enjoyed as theretofore as pictures albeit in a different setting and removed with little more difficulty than any large framed portrait fastened to a wall by mirror plates. As the Vice-Chancellor indicated towards the conclusion of his judgment, just as the pictures had previously been displayed elsewhere so they could be taken down with no damage to the painting and no more than minor damage B to the frames and displayed elsewhere again. And, as he indicated with great force, a chattel cannot become part of the freehold because it is displayed to the best advantage. I agree with the Vice-Chancellor that this latter consideration is not to be forgotten in relation to the statue of the Greek athlete. That statue was no doubt sited at a key point in the most advantageous position in the grounds; and just as a picture is placed in a room not merely for the purpose of its enjoyment as a picture but also to embellish and beautify the room in which it is placed and may be positioned for that purpose, so no doubt was the Greek athlete statue sited. This particular statue if fixed at all to the plinth on which it stood C was removed with minimal damage and in my judgment it did not lose its character as a chattel to be enjoyed as such by reason of being so fixed. Similar considerations in my judgment apply to the sundial, which was removed and was removed without any damage at all. I entirely agree with what Scarman LJ has said regarding the Chinese prints.

I wish to make it absolutely clear that at no stage of the litigation did either party claim relief on the basis of the stipulation in the head contract that "fixtures and fittings usually denominated tenant's fixtures or fittings should be taken by the purchaser . . . at a valuation." Neither party at any relevant time has asked for an inventory or valuation. D Almost all the disputed items were sold at an aggregate price exceeding £1,950 and for all I know the plaintiff would not have been willing to produce that not inconsiderable addition to his purchase price of £17,928. And for all I know Lord Poulett would have been willing, had he been asked to do so, to accept the view that the disputed items fell within the terms of the stipulation and so relieved himself of the trouble and expense of removing them and putting them up for sale and paying the auctioneer's commission.

I would also emphasise that although for the purpose of determining whether Lord Poulett had a duty to the plaintiff E in relation to the disputed items to take reasonable care of them, I have had to consider whether the plaintiff had a right to some form of specific performance, this is not an action for specific performance, but, as against Lord Poulett, for breach of fiduciary duty or detinue, trespass or conversion.

I would dismiss the appeal.

The appeal was dismissed with costs.

F

QUEEN'S BENCH DIVISION

February 9 1977

(Before Mr Justice GRIFFITHS)

BERNSTEIN v SKYVIEWS & GENERAL LTD

Estates Gazette March 26 1977

(1977) 241 EG 917

Landowner's claim against aerial photographers for trespass and invasion of privacy—Extent of owner's rights to airspace above his property—Maxim "Cujus est solum ejus est usque ad coelum et ad inferos"—Rights in airspace not unlimited—Restricted to airspace necessary for ordinary use and enjoyment of the land and structures on it—Civil Aviation Act 1949—Landowner's action fails G

This was a claim by Lord Bernstein for damages against Skyviews & General Ltd, who, he alleged, trespassed in the airspace above his property, Coppings Farm, Leigh, Kent, and invaded his right to privacy when they took an aerial photograph of the property in August 1974.

Charles Gray (instructed by Turner, Peacock) appeared for the plaintiff, and L D Lawton QC and Gerald Lumley (instructed by Levi & Co, Leeds) represented the defendants.

Giving judgment, GRIFFITHS J said Skyviews were a company who took aerial photographs of properties. For this purpose they sent photographers on survey flights. H During these flights a number of properties would be photographed and later the photographs would be offered for sale to the owners. Lord Bernstein complained when he was offered an aerial photograph of his country home, Coppings Farm, in Kent. He said it was taken without his consent and was a gross invasion of his privacy. He demanded that the prints and negative should be handed over to him or destroyed. Mr Arthur Ashley, managing director of Skyviews, said if he had known of this request he would have undertaken to destroy the photograph and negative and not to take another similar photograph in the future. J However, Lord Bernstein's letter was answered by an 18-year-old secretary, who offered to sell him the negative for £15. After further letters Lord Bernstein started proceedings claiming that Skyviews wrongfully entered the airspace above his premises in order to take the photograph of his house. He claimed they trespassed and committed an actionable invasion of his right to privacy. Skyviews, who admitted taking the photograph, denied entering the airspace to do so. They claimed it was taken while the aircraft was flying over adjoining land. Skyviews, who instructed their pilots to photograph buildings, the owners of which might prove likely customers, claimed that their photograph was K taken at a height of some 630 ft and 30 metres outside the boundary of Coppings Farm. Lord Bernstein, who had sent an aerial photographer up in a helicopter to try to recreate the Skyviews photograph, claimed that it would have been taken at around 1,000 ft from 30 ft inside the boundary. The pilot of the plane from which the Skyviews photograph was taken in 1974 was not available to give evidence. However, on any view of the evidence the plane would have flown close to the border of Coppings Farm at the time the photograph was taken and the probabilities were that at some time it had flown over the land. He (his Lordship) found that the land had been overflown for the purpose of L photography without Lord Bernstein's consent.

Relying on the old Latin maxim *Cujus est solum ejus est usque ad coelum et ad inferos* (whose is the soil his is also that which is above and below it), Lord Bernstein claimed the right to exclude any entry into airspace above his property. The maxim, first coined in the 13th century in Bologna, had been used by English judges in a number cases concerned with structures attached to adjoining land—overhanging buildings, signs, telegraph poles and the like.

It was well established by authority that an owner had certain rights in the airspace above his land. The Court of M Appeal did not doubt the right of a landowner to cut a wire placed over his property in *Wandsworth Board of Works v United Telephone Co* (1884) 13 QBD 904. In *Gifford v Dent* (1926) 71 SJ 83 it was held by Romer J that it was a trespass to erect a sign projecting 4 ft 8 in over the plaintiff's forecourt. In *Kelsen v Imperial Tobacco Co* [1957] 2 QB 334 McNair J granted a mandatory injunction ordering the defendants to remove a sign which projected only 8 ft over the plaintiff's property. Strong reliance was placed on the last case by Lord Bernstein. But his Lordship doubted if McNair J's intention was to hold that the plaintiff's rights in airspace continued to an unlimited height. He had been considering whether the sign, which was at a very low level, was a trespass or nuisance. This had been indicated by the judge's reference to *Winfield on Tort* (6th ed p 380) in which the author held that trespass was limited to the height at which it was contemplated that an owner might be expected to make use of the airspace as a natural incident of the user of his land.

Lord Ellenborough in *Pickering c Rudd* (1815) 4 Camp 216 said it would not be a trespass to pass over a man's land in a balloon. Lord Wilberforce in *Commissioner for Railways v Valuer-General* [1974] AC 328, at 351 said that it was unlikely that such a sweeping, unscientific and unpractical doctrine as that "land" meant the whole of the space from the centre of the earth to the heavens would appeal to the common law mind. If the latin maxim were applied literally it would lead to the absurdity of trespass being committed every time a satellite passed over a suburban garden. The problem in this case was to balance the rights of a landowner to enjoy the use of his land against the rights of the general public to take advantage of all that science now offered in the use of airspace. The best way to strike that balance in our present society was to restrict the rights of an owner in the airspace above his land to such height as was necessary for the ordinary use and enjoyment of his land and the structures upon it, and to declare that above that height he had no greater rights in the airspace than any other member of the public.

Applying this test to the facts of this case, his Lordship found that, even though Skyviews' aircraft had flown over Lord Bernstein's property, it did not infringe any of Lord Bernstein's rights to airspace, and thus no trespass was committed. Lord Bernstein had complained, not that the aircraft had interfered with his use of his land, but that a photograph was taken. There was no law against taking a photograph, however, and the taking of a photograph could not turn an act which was not trespass into trespass. Even if Lord Bernstein had succeeded in establishing that the infringement of the airspace above his land constituted a trespass he would only have achieved a sterile remedy. For there would still be nothing he could do to prevent Skyviews taking a virtually identical photograph of his property from adjoining land, provided they took care not to cross his boundary and were taking it for an innocent purpose.

A further defence had been raised under the Civil Aviation Act 1949. He (his Lordship) read section 40 (1) as giving protection from actions of trespass or nuisance to all flights which were at a reasonable height and complied with the statutory requirements.

At the same time, however, the present judgment should not be understood as deciding that in no circumstances could a successful action be brought against an aerial photographer to restrain his activities. The judgment was far from saying that if a plaintiff was subjected to the harassment of constant surveillance of his house from the air, accompanied by the photographing of his every activity, the courts would not regard such a monstrous invasion of his privacy as an actionable nuisance for which they would give relief.

The action failed and there would be judgment for the defendants.

CHANCERY DIVISION

March 17 1976

(Before Mr Justice GRAHAM)

PEFFER v RIGG AND ANOTHER

Estates Gazette April 9 1977

(1976) 242 EG 123

House purchased in pursuance of family agreement conveyed to A to be held in trust for A and B in equal shares as tenants in common—A transfers the house to his wife for the nominal consideration of £1 in the course of divorce proceedings—Wife fully aware of trust—Wife a constructive trustee—Not protected by sections 20 and 59 of Land Registration Act 1925—Wife takes subject to B's interest under trust—B granted appropriate relief accordingly

This was an action in which the plaintiff, Stanley George Peffer, of Palmer's Green, London, claimed that the first and second defendants, Reginald James Edward Rigg, of Wood Green, London, and his former wife, Irene Rigg, had been guilty of breaches of trust in respect of a house at 103 Leighton Road, London NW5, which the second defendant was occupying at the date of the action. The plaintiff sought a declaration that the second defendant held the house in trust as to a half share thereof for the plaintiff, and asked for consequential relief.

W Poulton (instructed by Fensom & Co) appeared for the plaintiff; R Reid (instructed by Piper, Smith & Basham) represented the first defendant, and N Banks (instructed by Pierron, Goodwin & Co, of Harrow) represented the second defendant.

Giving judgment, GRAHAM J said: This is an action for breach of trust in which the plaintiff, Mr Peffer, claims that both the first defendant, Mr Rigg, and the second defendant, Mrs Rigg, have committed breaches of trust in respect of the title to and letting of a house at 103 Leighton Road, London NW5, in the parish of St Pancras in the borough of Camden. The title No 425862 was first registered in the Harrow and District Land Registry on January 8 1932. The main actors in this unfortunate dispute are the plaintiff and his wife, and the first and second defendants who were for part of the material time married but who were divorced in 1971, the decree nisi being dated July 8 1971. The second defendant and the plaintiff's wife are sisters and daughters of a Mrs Bingley who, until her death on August 10 1970, lived on the ground-floor and basement flat in the house.

No 103 Leighton Road had been lived in by Mrs Bingley since about 1935 when her husband and she first became tenants. She had been a widow for some years prior to the events with which we are concerned. Their two daughters of course lived with them in their childhood days. In the year 1962 the position was that the house was divided into three parts, the owners, whose name was Hutchinson, occupying the top-floor flat and Mrs Bingley occupying part, but not I think the whole, of the area below. In 1962 the owners decided to sell and the family, as they have been called, mainly to provide security for Mrs Bingley but also to provide a useful investment, decided to purchase the freehold. In order to obtain the highest possible mortgage and as big an improvement grant as possible matters were arranged between them as follows. The house was purchased in the name of the first defendant for £2,500. He was at that time a warrant officer, first class, and a conductor of ordnance, and was more likely to get a mortgage on the basis of being a new owner-occupier than the plaintiff, who was already the owner of a house in the same borough. They were in fact given a mortgage of £2,000. The plan was to reallocate the accommodation in the house and to turn it into two flats, the upper floor, and the first or ground floor and basement respectively. This, incidentally, enabled them also to get two improvement grants of £400 each, one for each flat. The conversion work was put in hand and executed for £1,815 and a further advance on mortgage of £400 was obtained. The council, it seems, was left with the impression that the first defendant was to be in possession as owner-occupier of the top flat. This was not in fact the case at any time, and the necessity to keep the facts from the council may have led to a number of the difficulties which arose. The upshot of the whole transaction was that they bought and converted the house into two flats, subject to the repayment of the mortgage, at a cost to themselves of some £115, as shown on page 1 of the accounts bundle. From

A their point of view it was clearly a very satisfactory piece of business.

The agreement was that the freehold property should be held by the first defendant on trust for the plaintiff and himself as tenants in common in equal shares and that both of them would contribute to future outgoings and maintenance in equal shares, a balance between them being struck from time to time. At the end of 1962 Mrs Bingley was the only resident at no 103, but when the conversion had been completed tenants by the name of Stranahan were found for the top floor, their choice, partly at any rate, having been determined by their willingness to help Mrs Bingley in B practical matters if she needed it. From then on the plaintiff and first defendant each at various times paid money for outgoings and they had a reconciliation generally at the end of the year or at some other convenient time. All went smoothly until late in the year 1967 when matrimonial difficulties arose between Mr and Mrs Rigg. He had been working in a bomb disposal unit in London, but when the trouble arose he arranged for his wife and daughter to go to live with Mrs Bingley and he returned to an army post in Germany. He says he did not see his wife again until the hearing of the case, though he did have communications with her as C appears from the evidence and correspondence. The plaintiff, about this time, and no doubt partly because of the knowledge of the matrimonial troubles between the Riggs, became concerned about the safety of his investment in no 103 and approached Mrs Rigg and suggested that, in case the first defendant might do something silly with the property, she should obtain an acknowledgment from him that he held the property in trust for the plaintiff and himself. She telephoned her husband and he first sent her the acknowledgment on page 3 [of the bundle of correspondence], dated November 20 1967. Later, on May 30 1968, the first defendant and the plaintiff formally entered into the trust deed (pages 6 and 7). D This deed makes it clear that the first defendant is registered as proprietor of the freehold of no 103 but holds the property on behalf of himself and the plaintiff as tenants in common and equal shares. It also records the covenant by the plaintiff to be responsible for half the cost of the mortgage payments and other outgoings.

Now late in 1967 or early in 1968 the upper-storey flat again became vacant and in 1968 the Rent Act of that year creating protected tenancies was passed. The plaintiff stated that he thought about the matter and realised that if the top flat was again let the tenancy would be a protected tenancy under the Rent Act with the result that the house could not be E sold with vacant possession. This would mean that the value of the investment in the house would be lessened. In view of these uncertainties as to the future, having regard particularly to the domestic difficulties of the first and second defendants, he came to the conclusion that the flat ought not to be let again and he said he made these views clear to the first and second defendants. The Stranahans had moved out, according to the plaintiff's evidence, after the first defendant had signed the deed on May 30 1968, when he was on leave in this country. The flat had apparently fallen into some disrepair during the Stranahans' occupation and it was repaired and redecorated after they had moved out and these repairs were F completed by September 1969 or thereabouts. The financial position in general had been that if the top flat was let the whole operation was largely self-financing because the rent received for the flat would more or less balance the mortgage payment· and at the end of 1969 there was about £2,600 outstanding on the mortgage, the house being worth perhaps something of the order of £8,000 at that time. On October 12 1969 (see p 9) the plaintiff wrote to the first defendant, who had returned to this country after discharge from the army in April 1968 and was then, in October 1968, living at 22 Woodleigh Court, Stuart Crescent, Wood Green. After referring in the first paragraph to his "disagreement" with Mrs Rigg G ("Irene") about the future of the house, he went on to make three alternative suggestions for the future of the property—none of these involves reletting of the top floor. This letter, to my mind, clearly indicates that there had been a prior discussion between the plaintiff and Mrs Rigg which had probably been heated, but certainly had resulted in disagreement. The plaintiff says that all he had done was to make clear that he wanted to sell the house and for that purpose did not want a relet but would like Mrs Bingley and Mrs Rigg to find other accommodation which he thought and suggested they would be able to do by using money out of the proceeds of sale. Mrs Rigg in effect says that the plaintiff was very H unpleasant and threatened to turn her and her mother out of the flat. I think the proper inference from the evidence as a whole and the correspondence is that the meeting was heated and that there was stalemate at the end of it, the plaintiff saying that if they would not go out then, until the situation was resolved, they could not expect help from him with more money as in the past, and Mrs Rigg saying that she and her mother most certainly were not going to move out so that the whole property could be sold with vacant possession. What then happened was that the defendants, without further discussion with the plaintiff and without telling him what they were doing, did in fact find new tenants for the top flat, Mr J and Mrs Lewis. They took possession of the top flat on September 26 1969. Rumours of the letting may have come to the ears of the plaintiff and this is no doubt why he wrote the letter of October 12 (p 9) in the terms in which he did. One of the issues between the parties is whether or not this letting without the plaintiff's approval, and indeed without reference to him at all, with the result that a protected tenancy was created, is a breach of trust by the first defendant. The first defendant says, as is recorded in his letter of November 5 1969 (p 13) that he had a telephone çonversation with the plaintiff in which the plaintiff had stated that the second defendant had turned down all his suggestions (referring to K those made at his interview with her) and went on, "In that case I am not prepared to spend any more money on the property and wash my hands of the situation."

This is an important part of the story, because the second defendant's defence to the relief claimed against her was that she thought the plaintiff had given up any legal or other interest in the property by the time the beneficial interest in it was purported to be conveyed to her. Her belief was based, she said, on what the plaintiff had said to her and on what her husband had told her that the plaintiff had said in the telephone conversation with him. This was, as she said she L thought, confirmed by several letters (nos 9 to 13) which she agrees were sent to her by her husband. I shall return to this in due course.

Matters proceeded with little if any communication between the plaintiff and the defendants. In 1971 the first defendant filed divorce proceedings. On March 11, from paragraph 7 of the petition, p 18, it is seen there is set out a proposed agreement by the first term of which the first defendant was to "transfer to the respondent the freehold property, being the former matrimonial home at 103 Leighton Road . . . and she shall receive all rent from the part of the home which is let and she will undertake the repayment of the mortgage instal- M ments due and the existing mortgage advance." This was followed by a letter (p 23) from the second defendant's solicitors to the first defendant's solicitors to the same effect and an acceptance from Mrs Rigg's solicitors appears on p 25. The decree nisi embodying the agreement was granted on July 8 1971. The transaction is a peculiar one because, from all the evidence, I feel quite sure, and find to that effect, that both the first and second defendants were perfectly well aware throughout that the first defendant, though the house was in his name, only had a half share in it and that he held the other half in trust for the plaintiff. However, neither of the

A defendants apparently told their solicitors that this was so, and in view of the language used, though the words in the decree nisi are "his interest," which could of course mean a half interest just as much as the whole interest, I suspect the court may well have thought that the whole interest was being transferred. This finding as to the defendants' knowledge seems to me to be fundamental to the determination of the questions of transfer and ownership. Whatever therefore the transfer of March 21 1972 may say, and it is entitled "Transfer of Whole" and describes the first defendant as transferring as "beneficial owner," I am sure both defendants at the time of the divorce and of the later transfer, knew perfectly B well what the true position was and that there was no representation by the first defendant to the second defendant on which she can found her claim to contribution. I have some sympathy for her because, though one of course does not know all the facts, she was undoubtedly placed at the time in a very uncomfortable position by the divorce proceedings, and, as she said, all she was interested in was "to ensure that she and her daughter continued to have a roof over their heads." I think it may well be that she has now persuaded herself that in fact her husband did represent to her that he had acquired the whole beneficial ownership of no 103 and that she was going to get the whole property in the house for C herself. I have to look at the matter, however, as of the time in question when the events took place and I cannot accept that this was the position then.

Equally, I suspect the first defendant, in view of his disagreement with the plaintiff as to the future of the house and his decision to obtain a divorce because his marriage had completely broken down, was anxious to divest himself of the trust and any further responsibility for arranging for the maintenance of the house and for paying the outgoings, including the mortgage repayments. From this point of view the agreement of his former wife to step into his shoes as D trustee was an excellent arrangement, leaving him free of troubles in this connection, apart from having to pay maintenance at an appropriate rate for the future. From his point of view so long as he could get her to assume his responsibility *qua* the plaintiff, that was good enough, and even if he had thought his wife was not clear as to the ownership I doubt if he would have been scrupulous to emphasise to her what the true position was. I was not impressed by the way in which he gave his evidence and he had a tendency, when he thought a question was awkward, to try to avoid giving a straight answer and to start excusing himself for doing what he had done. Nevertheless, on balance, whatever he may E have said or left unsaid, I do not think Mrs Rigg, at the time, thought that he was free to convey the whole property to her. After the hearing had concluded it occurred to me that on findings which I might make and which I have in fact set out above, there were possible legal conclusions upon which I ought to hear argument by counsel since they had not so far been dealt with. After informing counsel of the position and of the points upon which I desired to hear such argument there was a further hearing.

As a result a number of matters now remain to be dealt with. Firstly, was the letting to Mr and Mrs Lewis by the defendants a breach of trust *qua* the plaintiff? I do not think so. F It seems to me quite clear that the plaintiff had told the defendants that he was not going to pay out further money in respect of this house until the argument as to its future was resolved. Faced with this refusal the defendants were in the position of having to provide for the mortgage repayments without there being any money coming in. The alternatives were to find somewhere else for Mrs Bingley to live and then to sell with vacant possession after turning her out, or to sell subject to her continuing to live there, a course unlikely to succeed or unlikely at any rate to result in a high price, or to relet the top-floor flat and carry on as before, with the rent for this flat balancing the mortgage repayments. In my judgment, the last course seems to have been at least as sensible G as, if not more sensible than, either of the others and I do not think that in the circumstances Mr Rigg can be blamed for following it. In so doing he seems to me to have acted honestly and reasonably and as prudently as he would be likely to have done in relation to his own affairs. I am not therefore prepared to hold that any breach of trust was committed by this transaction.

The purported transfer, however, of the whole of the beneficial interest in the property by the first to the second defendant on the occasion of the divorce agreement, in the light of their knowledge of the true facts as I have found them, seems to me to be in a very different position. It was H argued by Mr Banks for the second defendant that the property was transferred to her for valuable consideration as part of the divorce agreement and that, therefore, the combined effect of sections 20 and 59 of the Land Registration Act 1925 protected the second defendant against any claim or interest of the plaintiff because there is no entry on the register in his favour prior to the transfer to the second defendant. This argument would be convincing if it were not for my finding that the second defendant at the time knew perfectly well that the first defendant could not transfer to her more than a half share of the property. It is this knowledge J which seems to me to cause great difficulty to her and prevents her argument succeeding for a number of different reasons put forward by Mr Poulton at the second hearing. He argues firstly that the purported transfer (correspondence, p 56) from the first to the second defendant of the beneficial interest on the whole of the property of 103 Leighton Road was expressed to be for the consideration of £1. This is nominal consideration and not valuable consideration and it follows that the second defendant is not protected by section 20 of the Land Registration Act 1925. In accordance with the provisions of subsection (4) she can only take subject to any minor interests subject to which the first defendant held the same. He was K party to the trust deed of May 30 1968 (correspondence, p 6), and clearly had notice of the plaintiff's half interest in the property. The second defendant can therefore only take subject to the minor interest of the plaintiff in the property subject to which the first defendant held it.

The argument to the contrary is that the transfer was only part of the whole agreement entered into by the first and second defendants on the occasion of the divorce and it is not therefore right to limit the consideration for the transfer to the £1 expressed to be therefor. The consideration there was a great deal more and included all the obligations undertaken L by the second defendant. Such consideration was therefore not nominal but valuable within the section and the second defendant received the protection of the section. I do not see why, when the parties have chosen to express a transfer as being for a nominal consideration, the court should seek to hold that the consideration was in fact otherwise than as agreed and stated. If, however, the proper view is that there was valuable consideration for the transfer here, then it is argued as follows. There is a contrast between sections 20 and 59 of the Act. Section 20 (1) protects any "transferee" for valuable consideration. By section 18 (5) "transfer" and "transferee" in relation to freehold land have very wide M meanings but are not specifically defined in section 3. It is to be noted, however, that section 20, though it mentions valuable consideration, does not mention "good faith" as being necessary on the part of the transferee, nor does it mention notice. It can be argued therefore that the section seems to be saying that a transferee whether he has good faith or not and whether he has notice or not, takes free of all interests (other than overriding interests) provided he has given valuable consideration. This at first sight seems a remarkable proposition and, though undoubtedly the law of property legislation of 1925 was intended to simplify such matters of title as far as possible, I find it difficult to think that section

20 of this Act can have been intended to be as broad in scope as this. Similar doubt is expressed in Brickdale & Stewart-Wallace on the *Land Registration Act 1925* 4th ed, p 107, note (1). The provisions for rectification in section 82 as against a proprietor in possession who has been a party to a fraud, mistake or an omission in consequence of which rectification of the register is sought also seems to me to show that section 20 must be read with some limitations—see also Ruoff and Roper's *Law and Practice of Registered Conveyancing,* 3rd ed, p 417. Section 59 (6) on the other hand speaks of a "purchaser" not being affected by matters which are not protected by a caution or other entry on the register. By definition, however (see section 3 (XXI)), "purchaser" means a purchaser *in good faith* for valuable consideration. It seems clear therefore that as a matter of construction a purchaser who is not in fact one "in good faith" *will* be concerned with matters not protected by a caution or other entry on the register, at any rate, as I hold, if he has notice thereof. If these sections 20 and 59 are read together in the context of the Act they can be reconciled by holding that if the transferee spoken of in section 20 is in fact a "purchaser" he will only be protected if he has given valuable consideration and is in good faith. He cannot in my judgment be in good faith if he has in fact notice of something which affects his title as in the present case. Of course if he and, *a fortiori,* if a purchaser from him has given valuable consideration and in fact has no notice, he is under no obligation to go behind the register and will in such a case be fully protected. This view of the matter seems to me to enable the two sections to be construed consistently together without producing the unreasonable result of permitting a transferee purchaser to take advantage of the Act, and divest himself of knowledge of defects in his own title, and secure to himself a flawless title which he ought not in justice to be allowed to obtain. This view of the Act produces a result which is also produced by applying the principles applicable in the case of a constructive trust which I will now consider.

On the evidence in this case I have found that the second defendant knew quite well that the first defendant held the property on trust for himself and the plaintiff in equal shares. The second defendant knew this was so and that the property was trust property when the transfer was made to her, and therefore she took the property on a constructive trust in accordance with general equitable principles—see *Snell's Principles of Equity,* 27th ed, pp 98-99. This is a new trust imposed by equity and is distinct from the trust which bound the first defendant. Even if, therefore, I am wrong as to the proper construction of sections 20 and 59, when read together, and even if section 20 strikes off the shackles of the express trust which bound the first defendant, this cannot invalidate the new trust imposed on the second defendant.

On this assumption it seems to me that the ground is properly laid for granting rectification of the register under section 82. The second defendant, even though in possession, comes within the exceptions of subsection (3) and this would in my judgment be a case where rectification could properly be ordered against her. Mr Reid, for the first defendant, supported the propositions of Mr Poulton and adopted his argument. In addition he referred to *Jones v Lipman* [1962] 1 WLR 832, and which he submitted could only have been decided on the basis that the company in that case could not escape from or divest itself of its knowledge by reason of sections 20 and 59. It seems that this must be so and Russell J (as he then was) mentions and rejects the argument at p 837. There is one further matter I should mention. At a late stage in his submission on the last day of the adjourned hearing Mr Banks took the point that the plaintiff ought to be refused relief anyway because he had not come to the court with clean hands. This was based on the suggestion that the plaintiff and the first defendant at the start had deceived the Camden Council. They had, it was said, by false pretences deliberately led the council to believe that the first defendant was to be the owner-occupier of the property. This was untrue; the first defendant never had occupied the property and it was never intended that he should. Had this been known to the council, it is said, they would never have granted to the parties the mortgage which they did. The argument went as far as suggesting there was fraud and a false pretence under section 32 of the Larceny Act 1916. The truth of such an allegation could obviously not be determined without considerable investigation and detailed evidence which was not called before me. If the point was to be taken it should have been taken at the start and it would clearly have had to have been pleaded and particularised in detail. As things stood, it was taken too late and appeared to be an afterthought. I should, however, say I was not favourably impressed by the conduct of any of the parties in the action, but appreciate that I may not have had the full story about all the matters which seemed on the face of them to require further explanation. In the circumstances I am certainly not prepared to stigmatise the conduct of the plaintiff as being any more blameworthy than that of the defendants; still less am I prepared to accept that the plaintiff's conduct amounted to the commission of a criminal offence. I therefore reject the submission that the court ought not to assist the plaintiff for this reason.

It follows that in my judgment the second defendant holds the property in question in trust for herself and the plaintiff and that the latter is entitled to appropriate relief. I will hear the parties' submission as to the form this relief should take.

There was a declaration that the second defendant held half the property in trust for the plaintiff and that the land register should be rectified to show that fact. There was also an order for the sale of the property by the second defendant. No order was made as to costs.

CHANCERY DIVISION

October 4 1976

(Before Mr Justice OLIVER)

SAMPSON AND ANOTHER v CALDOW AND ANOTHER

Estates Gazette April 9 1977

(1976) 242 EG 129

Right of pre-emption in conveyance—Right normally assignable—Clause expressed to benefit "the purchaser or his personal representatives" but no reference to successors in title—Whether successors excluded on the ground "expressio unius est exclusio alterius"—Held that right of pre-emption passed to successor in title

This was a summons for a declaration in which the plaintiffs, Hugh Sampson and his wife, Hilary Rosalind Sampson, sought to establish that a right of pre-emption in a conveyance from them to Lady Ingram of part of Bruns Farm in Buckinghamshire was not assignable and had accordingly not passed to the defendants, David Thomas Cunningham Caldow and his wife, Marion Rose Caldow. The plaintiffs also asked for consequential relief in the form of the vacation of entries in the register of land charges.

P Talbot (instructed by Lee, Bolton & Lee, agents for Horwood & James, of Aylesbury) appeared for the plaintiffs, and A Taussig (instructed by Paisner & Co) represented the defendants.

Giving judgment, OLIVER J said: This is a case in which the plaintiffs seek against the defendants a declaration that a right of pre-emption in a certain conveyance to which I will refer in a moment is personal to the purchaser under that conveyance, Lady Ingram, and her personal representatives and is incapable of assignment, and for consequential relief in respect of the registration of an estate contract under the Land Charges Act 1925 in respect of that right of pre-emption. The right of pre-emption with which we are concerned is contained in a conveyance dated May 30 1969, made between the plaintiffs in the present proceedings of the one part (they were the vendors) and Lady Ingram (who is therein defined as the purchaser) of the other part. It was a conveyance under which the plaintiffs conveyed to Lady Ingram a property then forming part of what was known as Bruns Farm in Buckinghamshire in the parishes of Ashton, Clinton and Windover. It was a tract of some 308 acres of land and it adjoined some land retained by the plaintiffs on which was erected the farmhouse.

There was an exception and reservation in the conveyance to which I had better refer briefly. There was excepted and reserved to the vendors the free passage and running of water and soil to and from the vendors' adjoining property through certain sewers and pipes to a particular septic tank, and the words to which some importance has been attached are the following words: "with power for the vendors and their successors in title owner or owners of such adjoining property or any part of parts thereof to enter for the purpose of cleansing and renewing the said pipes sewers drains" and so on.

Clause 3 of the conveyance is the one on which the present question arises. It is, so far as material, in the following terms: "If at any time hereafter during the lives of the vendors and the survivor of them and two years after the death of such survivor the vendors or the survivor of them should desire to sell the property edged blue on the plan annexed hereto," and that, I would mention, is the property on which is situate the farmhouse to which I have referred, "the purchaser or his personal representatives being the estate owner in fee simple of the whole or some part of the land hereby conveyed shall have the option of purchasing the same in the manner and upon the terms following." Then there is a provision for the vendors or survivor of them to make an offer in writing to the purchaser to sell the property and that offer was to be open for acceptance for one month from the date of its receipt. The purchaser was given the right before the expiration of the month to accept the offer and on that being done a contract would come into being on the terms which were thereinafter contained. The vendors were not to be at liberty to sell the property until the expiration of the month or the offer having been unconditionally declined by the purchaser in writing, and if the offer was so declined or determined by lapse of time they were free to sell elsewhere on such terms as they thought fit. Then there were certain administrative provisions which perhaps I ought to read. "An offer under this clause should be sufficiently made if sent by registered post or recorded delivery service addressed to the purchaser at his last known address. Time is to be of the essence of the contract." There were provisions for fixing the price. There was to be an arbitration in the event of disagreement, and a deposit of £3,000 was to be paid to the vendors or the survivor of them upon the acceptance in writing of the offer. Then there was a provision that "The vendors or the survivor of them shall not be required to execute a conveyance in favour of any person other than the person exercising the option, and the person exercising the option shall not be entitled to make any inquiry or requisition or take any objection to title of the vendors or the survivor of them save in respect of matters arising after the date hereof."

In fact the land comprised in that conveyance, or at least the major part of it, came to be vested in the defendants in the present summons, Mr and Mrs Caldow. That came about in this way, that on October 16 1972 Lady Ingram sold to them the property which she had purchased under the conveyance, and in the conveyance to them she conveyed the property together with the benefit "so far as the vendor can convey the same" of the rights granted to the vendor by clause 3 of a conveyance made the 13th day of May 1969 between the said Hugh Sampson and Hilary Rosalind Sampson of the one part and the vendor of the other part. So there was on the face of it an express assignment of the rights conferred upon Lady Ingram by clause 3 of the conveyance. It should perhaps be mentioned that the conveyance to Lady Ingram, although it in fact recites an agreement for the purchase by Lady Ingram of the property, was in fact in pursuance of an agreement between the plaintiffs and Lady Ingram's son, Robin Ingram, and that contract provided that the conveyance to the purchaser should contain a covenant in the *ipsissima verba* of the covenant which in fact went in in clause 3 of the conveyance which was in fact executed. That no doubt accounts for the reference to the purchaser and "his" personal representatives, although in fact the ultimate purchaser was Lady Ingram. After the purchase by the defendants of the property from Lady Ingram they sought to register the right of pre-emption conferred by clause 3 under the Land Charges Act and in fact they did so. There was a certain amount of correspondence between the solicitors acting for them and the solicitors acting for the plaintiffs in which it became evident that the plaintiffs were alleging that the right conferred by clause 3 of the conveyance was a right that was personal to Lady Ingram and her personal representatives. It was asserted, therefore, that the defendants did not have the right and did not have any title to register an estate contract against the plaintiffs. Disagreement having arisen on that point these proceedings were ultimately commenced.

The point really is a relatively short one of construction. The plaintiffs, through Mr Talbot, accept that clause 3 of the conveyance created a contract the benefit of which was vested in Lady Ingram and they accept, as indeed they have to in the light of such cases as *Re Button's Lease* [1964] Ch 263, that the right was *prima facie* assignable. And they also accept, I think, that the burden of demonstrating that the right is not assignable rests upon them, because the benefit of a right of pre-emption such as this is always assignable unless there is something in the context which indicates the contrary, something to indicate that there is some restriction on assignability. The plaintiffs rely for their construction of the conveyance upon a number of points. They point out first of all that, in the conveyance itself, the plaintiffs on the one hand and Lady Ingram on the other are referred to as "the vendors" and "the purchaser" respectively without any reference at all to successors in title. But Mr Talbot accepts, as indeed he must, that the mere absence in the definition of a reference to successors in title is not sufficient to displace the *prima facie* rule, namely, that a right such as this is assignable. But here, he says, it is quite evident, when you look at the conveyance itself, that the draftsman had successors in title in mind at least in one part and that is the reservation of the right of drainage, because he refers there explicitly to the vendors and their successors in title. This is, says Mr Talbot, significant when you come to clause 3 because you find the person there referred to as having the benefit of the covenant is not "the purchaser or his personal representatives or successors in title" but simply "the purchaser or his personal representatives." There is nothing, he says, in this case to indicate that the omission there was accidental, and he relies upon the maxim *expressio unius est exclusio alterius*.

There is no authority which expressly helps very much here. I have been referred by Mr Talbot to a report of a case of *Williams v Andrews* [1972] Estates Gazette Digest 158 where, in the case of a right of pre-emption in a lease the purchaser (the original purchaser of the lease) was granted a right of pre-emption, and the grant was a grant under which the purchaser covenanted that if she or her personal representatives should desire to sell then the appropriate

A machinery should be gone through for offering the property to the landlord. Ungoed-Thomas J found, as a matter of construction, that the only event giving rise to the right of pre-emption there was an offer by or desire to sell by the purchaser and her personal representatives. But, as Mr Taussig points out, not very much can hinge upon that case, which was a case where what was sought to be established was the passing of the burden of the covenant, whereas what we are concerned with here is the question of the passing of the benefit. I do not find that that case helps me very much as a matter of construction.

It seems to me that really at the end of the day the
B indications in this conveyance that the right was intended to be one that was purely personal to Lady Ingram are of a fairly tenuous nature. Mr Talbot's case, as it seems to me, relies really in the ultimate analysis almost entirely upon the application of the maxim *expressio unius*. Mr Taussig has referred me to two cases in which a caution was expressed as regards that. In *Colquhoun v Brooks* (1887) 19 QBD 400 at 406 Wills J stated that the maxim was one the application of which had to be watched rather carefully. "Perhaps," he said, "few so-called rules of interpretation have been more frequently misapplied and stretched beyond their due limits. Failure to make the *expressio* complete very often
C arises from accident, very often from the fact that it never struck the draftsman that the thing supposed to be excluded needed specific mention of any kind." In the very much later case of *Lowe v Dorling & Son* [1906] 2 KB 722 the Court of Appeal approved this and Farwell LJ in his judgment at p 785 said: "It is not enough that the express and the tacit are merely incongruous; it must be clear that they cannot reasonably be intended to co-exist." And he then quoted the judgment of Wills J in *Colquhoun v Brooks* and approved it.

Now Mr Taussig, on behalf of the defendants, points out
D that really here one is dealing with a position where, if nothing is said, if you have simply got a right of pre-emption conferred upon Lady Ingram, that would *prima facie* be an assignable right. If that stood alone, it might be argued that it was a right which was confined to Lady Ingram personally. To obviate that difficulty the draftsman put in the words "or his personal representatives" in order to make it perfectly clear that the right is not one which is simply existing for the benefit of the lady herself. Mr Taussig points to some of the rather curious results that might ensue if Mr Talbot's construction is correct. I do not think that these are really anything more than, in a sense,
E straws in the wind, but it does, as it seems to me, become very difficult to construe this conveyance as one conferring a purely personal right when one bears in mind, first of all, section 78 of the Law of Property Act, under which the covenant is, *prima facie*, with the purchaser and his successors in title and assigns; secondly, that what is created is a *prima facie* assignable right; thirdly, that it would have been extraordinarily easy for the draftsman, if he wanted to make it a personal right, merely to say that in terms; and fourthly, that the persons seeking now to say the right is restricted to one that is personal to Lady Ingram are the grantors under the conveyance. I am reminded by Mr
F Taussig of the maxim that the conveyance must be construed most strongly against the grantor. In all the circumstances I cannot express myself as being satisfied that Mr Talbot's construction of this conveyance is correct, and I must, I think, therefore dismiss this claim and, I think, dismiss it with costs.

The plaintiffs' claim was dismissed with costs.

COURT OF APPEAL G

October 28 1976

(Before Lord DENNING MR, Lord Justice ROSKILL and Lord Justice LAWTON)

JONES v JONES

Estates Gazette April 30 1977

(1976) 242 EG 371

Son entitled by virtue of contributions to equitable interest in one-quarter of house purchased by father—House occupied by son and family—House vests in step-mother after father's death—Claim by her that son shall pay three-quarters of a
fair rent, failing which the house should be sold—Held that son and step-mother were tenants in common and one tenant in common cannot demand rent from another—Also held that step-mother was estopped from defeating by sale son's reasonable expectation of living in house as his home H

This was an appeal by Frederick William Jones from a decision of Deputy Judge Sebag-Montefiore at Lowestoft County Court on October 24 1975, in favour of an application by Mr Jones's stepmother, Alice Eva Jones, that he should pay rent in respect of a house at Blundeston, Suffolk, in which he had a one-quarter equitable interest or, in default, that the house be sold.

J G Ross-Martyn (instructed by Mears, Hobbs & Durrant, of Lowestoft) appeared for the appellant, and J H G Sunnucks (instructed by Lucas & Wyllys, of Lowestoft) represented the respondent. J

Giving judgment, LORD DENNING MR said that old Mr Jones was a scrap merchant who carried on business at Kingston upon Thames. His two sons were partners in the business. He did reasonably well. Like other merchants he conducted his business in pound notes and his accounts did not reflect the true state of the business. He owned three or four houses in Kingston. He lived in one himself and each of his sons lived in one, too. His first wife died and he
married his second, Alice, in April 1964. He made a will on October 25 1964 under which he gave his new wife his share in the partnership business. He gave his son George the house in which George was living, and his son Frederick the house in which Frederick was living. He gave the remainder of his property to his new wife. So he made fair provision for his new wife and for his sons. About 1967 there was a compulsory purchase order and the local authority took over all his properties. The scrap business was given up. Old Mr Jones then went up to Blundeston in Suffolk and bought a house, "St Alban's," in which he lived with his new wife. He K
was then concerned about his son Frederick who had been left behind at Kingston. He wanted Frederick, with his wife and children, to go up to Blundeston. He found a house there, "Philomena," which was suitable for them, and Frederick and his wife went up to see it. They liked it very much, so the father bought it for £4,000. Frederick gave up his job at Kingston and went with his wife to live at Blundeston. Frederick made two payments of £500 each to his father towards the cost. But the house was taken in the father's name and remained in his name. The son believed that the father had given it to him and the judge accepted his evidence about that. The judge had said, ". . . But there is a ring of L
truth about the account of the transaction he had with his father. His father did not intend a gift of 'Philomena' to him lock, stock and barrel." M

The son said that he gave his father the first £500. The trial judge said, "His father did not resist the payment of £500 to him." When Frederick said "What about the money for the property?" his father replied "It is your place." He also said "This is because I owe you a lot." The son then paid the father a second £500 and the father then said of the house "So far as I am concerned, it is yours." When Frederick or his wife asked from time to time about the property the father said "I don't know what you are worrying

A about. Alice knows." The son had always trusted his father and never asked for a receipt. There were two other witnesses who supported the view that the father had said to the son that it was to be the son's house. At all events on this assurance the son and his wife and family went into the house. They had not paid any rent to the father, but the father said they must pay the rates, and they did so. Things went on in that way from 1968 until February 1972, when old Mr Jones died. Thereupon the stepmother claimed that the house, "Philomena," was hers. She took out letters of administration with the will annexed and by an assent dated August 18 1972 got the house, "Philomena," vested in her. She also had B the house, "St Alban's," in which she had lived with old Mr Jones when he was alive, and other property. When she had "Philomena" vested in her she said that the son, Frederick, ought to pay her rent. When he did not do so she gave him notice to quit and took proceedings for possession in the county court.

After hearing the evidence the judge found that £1,000 had been given by the son to the father when the house was acquired. As the house cost £4,000 altogether, the judge thought that the son in effect contributed £1,000 towards the £4,000. On that account, so far as any equitable interest in the property was concerned, he held that it was to be one- C quarter for the son and three-quarters for the stepmother. The judge made a declaration accordingly, but he also went on to say that he was not going to order the son out of possession. He said, "I am not disposed to make an order for possession, because it would be inequitable to do so. I find that the plaintiff, the stepmother, has a three-quarters share in the value of 'Philomena,' and the defendant (Frederick) a one-quarter share." He added "A proper arrangement would be for the defendant to pay some rent." That was the end of the first action, which was in 1973. After that decision the stepmother brought another action. In it she wanted some payment in the nature of rent, which had never been fixed. D She said that a fair and reasonable rent for the whole house would be £10 a week, and she wanted three-quarters of it—£7.50 a week. If it was not paid, she asked that the house be sold. The judge, at that second hearing, came to the conclusion that he was bound by the previous case of the three-quarters to one-quarter division. He decided that the only thing he could do was to ask the rent officer to fix a fair rent, and he ordered the son to pay a rent of 75 per cent of that figure, with arrears right back to the time of the assent on August 18 1972. If Frederick did not pay this amount, then the house was to be sold and Frederick evicted.

E The appeal raised some interesting points. First of all, whatever doubts one might have about the first decision of the judge, whether it was really right to hold that the stepmother had a three-quarters interest, or whether she had any beneficial interest, the court should nevertheless not disturb the decision of the judge at the first hearing. It was not appealed against, and no appeal was now open. So the court must hold that the stepmother had a three-quarters proprietary interest in the house "Philomena," and Frederick a one-quarter interest. But the question was whether the son could be compelled to pay rent to her, or whether she could turn him out of the house if he did not.

F Regarding the claim for rent, it was quite plain that these two people were in equity tenants in common, having a three-quarters to one-quarter share respectively. One was in occupation of the house, the other was not. The common law had stated clearly that one tenant in common was not entitled to rent from another tenant in common, even though that other occupied the whole. That appeared from *McMahon v Burchell* (1846) 2 Ph 127. Of course, if one of the tenants let any premises at a rent to a stranger and received the rent, there would have to be an account. But the mere fact that one tenant was in possession and the other out of possession did not give the one who was out any claim for rent. It did not do so in the old days of legal tenants in common. Nor G did it in modern times of equitable tenants in common. In *Bull v Bull* [1955] 1 QB 234 he (his Lordship) had said at p 239, ". . . The son, although he is the legal owner of the house, has no right to turn his mother out. She has an equitable interest which entitles her to remain in the house as a tenant in common with him until the house is sold." As between tenants in common, they were both equally entitled to occupation and one could not claim rent from the other. Of course, if there was an ouster, that would be another matter. Or if there was a letting to a stranger for rent, that would be different. But there could be no claim for rent by one tenant in common against the other, whether at law or H in equity.

The second point was the order for sale. Here came into play the doctrine of proprietary estoppel. It had been considered by the Appeal Court in the cases of *Inwards v Baker* [1965] 2 QB 29 and *Crabb v Arun District Council* [1975] 3 WLR 847. It was quite plain that the principles of these cases applied here. Old Mr Jones's conduct was such as to leave the son Frederick reasonably to believe that he could stay there and regard "Philomena" as his home for the rest of his life. On the basis of that reasonable expectation the son gave up his work at Kingston upon Thames and moved to Blundeston. J He paid the £1,000, too, in the same expectation. He did work on the house as well. It was all because he had been led to believe that his father would never turn him out of the house. It would be his family's home for the rest of his life. He and the rest of the family thought that the father would alter his will or give over the house to the son. The father did not do so, but nevertheless he led the son to believe that he could stay there for the rest of his life. On the two cases it was clear that old Mr Jones would be estopped from turning the son out. After his death his widow, the stepmother, was equally estopped from turning him out. Similarly she was not entitled to order the property to be sold, nor any payment K to be made by the son to her pending sale. Even though there was an implied trust for sale, nevertheless the courts would not allow it to be used so as to defeat the purposes contemplated by the purchase. That appeared from *Bedson v Bedson* [1965] 2 QB 666: see his Lordship's judgment at p 679 and Russell LJ at p 697-8. No order for the sale of the property should be made because that would defeat the very purpose of the acquisition—that the son would be able to live there for his life and remain in it as his home.

The two doctrines went hand in hand to show that no order should be made so as to disturb the son in his possession of the house. Nor should he be made to pay anything for L staying there. The ultimate result of the case was that the son had a proprietary interest of a one-quarter share in the house. He was able to stay there for his life by virtue of his interest in it, and the stepmother was estopped from turning him out. Nor could it be sold without his consent. The appeal should be allowed and an order made accordingly.

Roskill and Lawton LJJ agreed.

M

CHANCERY DIVISION

March 17 1977

(Before Mr Justice BRIGHTMAN)

RE THE RYE, HIGH WYCOMBE, BUCKS

Estates Gazette June 4 1977

(1977) 242 EG 811

Commons—Registration—Meaning of "allotment" of land within section 23(1) of the Commons Registration Act 1965

This was an appeal by way of case stated under section 18(1) of the Commons Registration Act 1965 whereby the

A High Wycombe Rye Protection Society contended that a Commons Commissioner was wrong in deciding that High Wycombe Rye was not a "town or village green" within the meaning of section 1(1)(a) of the Act. The society contended that the commissioner was wrong in holding, on an objection by Wycombe District Council, that an "allotment" of land under an Act for the exercise or recreation of the inhabitants of an area, within the meaning of section 23(1) of the Act, cannot take place without a formal transfer of the land. The society asked that the Rye be placed on the Register of Village Greens. The district council, owners of the land, opposed the application.

B Dennis Pugh (instructed by G B & L Ellis) appeared for the society; R A Campbell (instructed by the solicitor to Wycombe District Council) represented the local authority.

Giving judgment, Brightman J said that section 22(1) of the Commons Registration Act 1965, the definition section, defined "town or village green" as "land which has been allotted by or under any Act for the exercise or recreation **C** of the inhabitants of any locality or on which the inhabitants of any locality have a customary right to indulge in lawful sports and pastimes or on which the inhabitants of any locality have indulged in such sports and pastimes as of right for not less than twenty years."

The society contended that the land was allotted for this purpose under the Chepping Wycombe Corporation Act 1927, whereby the land was deemed to be a public park and pleasure ground, and was required to be retained by the corporation for that purpose. But he (his Lordship) did not think it right to read the 1927 local Act as "allotting" lands to the corporation "for the exercise or recreation of the inhabitants of **D** any locality." The land was included in the local Act for quite a different reason: to define the purposes for which it had to be held by the corporation, and to define the powers exercisable by the corporation over it.

The appeal was dismissed with costs.

TAXATION

A **CHANCERY DIVISION**

February 18 1977

(Before Mr Justice FOX)

McGREGOR (INSPECTOR OF TAXES) v ADCOCK

Estates Gazette April 23 1977

(1977) 242 EG 289

Section 34 of Finance Act 1965—" Retirement relief " against capital gains tax—Sale of part of farmland—Not sale of part of farm business—Land only asset of business—Crown wins appeal

This was an appeal by the Crown from a decision of
B General Commissioners for Income Tax that a farmer, Edward Adcock, aged 71, of Brandon Road, Watton, Norfolk, was entitled to " retirement relief " against capital gains tax under section 34 of the Finance Act 1965 in respect of the sale of part of the land farmed by him.

Brian Davenport (instructed by the Solicitor of Inland Revenue) appeared for the Crown; David Milne (instructed by Daynes, Chittock & Back, of Norwich) represented the respondent taxpayer.

Giving judgment, allowing the appeal, Fox J said that in 1973 Mr Adcock sold nearly 5 acres of his farm for £72,000, carrying on farming on the rest of his land. He was assessed
C for capital gains tax in respect of the sale, but the commissioners decided that he was entitled under section 34 of the Finance Act 1965 to relief of £10,000 against the assessment of £64,481. Section 34 provided that where an individual had attained the age of 60 years and disposed by sale or gift of the whole or part of a business he had owned throughout the period of 10 years ending with the disposal, then relief was to be given in respect of his gain. The amount of the relief at the material time was £10,000 if he had attained the age of 65. On behalf of Mr Adcock it was contended that the sale of the land constituted a sale of part of the farming business, thus entitling him to relief. At the time of the sale Mr
D Adcock had attained the age of 65 and had owned the farm for more than 10 years.

The commissioners had said that without the occupation of land there could be no farming business and disposal of part of the land was " disposal of part of the business." That reasoning appeared to him (his Lordship) to be wrong. It did not follow that a sale of part of the land constituted a sale of part of the business and the commissioners had misdirected themselves. The business and the assets of the business were different things.

Section 34 required the taxpayer to establish a disposal of " the whole or part " of his business. In ordinary language
E " land " was not the same thing as a business. It was merely an asset of the business. There was a clear distinction between the business and the individual assets used in the business. The mere sale of farmland was not a disposal of part of the farm business. The sale was merely a factor which the court had to consider in deciding whether there had been such a disposal. There was nothing to justify saying that farming was in a special position. Nothing in the section justified that. It did not suggest that land was to be treated differently from other assets.

He (his Lordship) had to decide whether the sale was a part F
of Mr Adcock's farming business. The sale did not constitute a disposal of any part of the business—the nature of the business, a mixed farming business, was the same before and after the sale—and he saw no reason for inferring that the sale of part of the farm was necessarily disposal of part of the farming business. The only reasonable conclusion on the facts was that there had been a sale of an asset of the business; that the business continued unaltered; and that no part of it was disposed of.

Appeal allowed. The Crown agreed to pay the taxpayer's costs.

G

CHANCERY DIVISION

December 1 1976

(Before Mr Justice TEMPLEMAN)

DAVIS (INSPECTOR OF TAXES) v POWELL

Estates Gazette April 30 1977

(1976) 242 EG 380

Capital gains tax — Compensation for disturbance under section 34 of Agricultural Holdings Act 1948 — Compensation not derived from the disposal of an asset — Sum which H
Parliament says is to be paid for loss and expense unavoidably incurred on termination of tenancy by notice to quit — No liability to capital gains tax

This was an appeal by the inspector of taxes against a decision of General Commissioners, who had upheld an appeal by the present respondent, Robert James Powell, against an assessment to capital gains tax in respect of compensation for disturbance under section 34 of the Agricultural Holdings Act 1948. The compensation had been paid by the respondent's landlord, Milton Keynes Development Corporation, in respect of part of Southside Farm, Milton
Keynes, of which the respondent had been the tenant. J

Brian Davenport (instructed by the Solicitor of Inland Revenue) appeared for the appellant; the respondent appeared in person.

Giving judgment, Templeman J said: For the purposes of this appeal I must assume that Mr Powell, the respondent, was paid £591 by way of compensation for disturbance under section 34 of the Agricultural Holdings Act 1948. The question is whether that sum is liable to capital gains tax in his hands.

Section 34 applies where the tenancy of an agricultural holding terminates by reason of a notice to quit, and in con-
sequence the tenant quits the holding. If these conditions K
are satisfied, then subsection (1) continues, " compensation for the disturbance shall be payable by the landlord to the tenant in accordance with the provisions of this section." Subsection (2) defines the amount of the compensation as " the loss or expense directly attributable to the quitting of the holding which is unavoidably incurred by the tenant upon or in connection with the sale or removal of his household goods, implements of husbandry, fixtures, farm produce or farm stock on or used in connection with the holding,

A and shall include any expenses reasonably incurred by him in the preparation of his claim for compensation."

Then there is a proviso that, "(a) compensation shall be payable . . . equal to one year's rent of the holding at the rate at which rent was payable immediately before the termination of the tenancy without proof by the tenant of any such loss or expense as aforesaid; (b) the tenant shall not be entitled to claim any greater amount than one year's rent of the holding unless before the sale of any such goods, implements, fixtures, produce or stock as aforesaid he has given to the landlord a reasonable opportunity of making a valuation thereof; . . . (d) the tenant shall not in any case be entitled B to compensation in excess of two years' rent of the holding." The plain object of this section is to provide by statute that a landlord shall make good to a tenant the loss or expense which he unavoidably incurs in connection with the matters mentioned in the section; and to avoid any quibbling and quarrels about matters it is said that he is not to get less than one year's rent or more than two, and if he wants more than one year he must prove his loss. That is a very sensible provision for giving a tenant compensation for loss or expense unavoidably incurred. Looking at section 34 by the light of nature and asking "Can that amount be liable to capital gains tax?," the answer is, "Nonsense; you cannot C make a gain out of a sum of money which is given to compensate for loss or expense which is unavoidably incurred."

However, that is not the end of the matter, because one must go to the Finance Act 1965 and find what, in that Act, the legislature thought and decided was a capital gain for the purposes of the Act. Section 19(1) of the Finance Act 1965 provides: "Tax shall be charged in accordance with this Act in respect of capital gains, that is to say chargeable gains computed in accordance with this Act and accruing to a person on the disposal of assets." Section 22(1) provides: "All forms of property shall be assets for the purposes of this Part of this Act"; and subsection (3) states that for the D purposes of the Act there is "a disposal of assets by their owner where any capital sum is derived from assets." What is said in this case is that Mr Powell had a lease and that lease was an asset, it was property of some form. He disposed of that asset by accepting the notice to quit which was given and by getting out, and he derived a capital sum from the asset when he did so. The capital sum was the amount of the compensation under section 34, which, as I have said, was £591. It does not seem to me that the compensation paid under section 34 is derived from the asset, namely, the lease. It is not derived from an asset at all: it is simply a sum E which Parliament says shall be paid for expense and loss which are unavoidably incurred after the lease has gone.

Section 22(3) of the Finance Act 1965 delimits some particular matters which are liable to charge; paragraph (c) charges "capital sums received in return for forfeiture or surrender of rights, or for refraining from exercising rights." But in my judgment this section 34 compensation is not a capital sum received in return for the surrender of rights. It is not paid as a result of a bargain in which the tenant says, "If I get out, will you pay me £591?" It is a sum paid where a tenant is faced with a notice to quit and must get out. Parliament says that in these circumstances, which have F nothing to do with a surrender, the tenant is to have a sum by way of compensation for irretrievable loss. So, whether viewed by the light of nature or by the construction of the Finance Act 1965, I do not think that this sum is liable to capital gains tax, and the inevitable result is that the appeal must be dismissed with costs to be taxed.

CHANCERY DIVISION G

July 1 1976

(Before Mr Justice BRIGHTMAN)

VARTY (INSPECTOR OF TAXES) v. LYNES

Estates Gazette May 21 1977
(1976) 242 EG 625

Capital gains tax—Sale of house and part of garden followed about a year later by sale of rest of garden with benefit of outline planning permission—Claim for total exemption under section 29(2) of Finance Act 1965—Exemption not allowed in respect of part of garden sold subsequently H

This was an appeal by the Inland Revenue against a decision by the general commissioners that a taxpayer who sold his house and part of the garden first, and subsequently sold the remainder of the garden with outline planning permission for a house, was entitled to exemption under section 29(2) of the Finance Act 1965 on the second sale notwithstanding that it was sold at a later time than the house and part of the garden contained in the first sale.

Brian Davenport (instructed by the Solicitor of Inland Revenue) appeared for the appellant; the respondent appeared in person. J

Giving judgment BRIGHTMAN J said that in 1968, John Francis Edward Lynes bought the freehold house Dalesford, Hailsham, East Sussex, for £6,920. He sold the house and part of the garden in June 1971 for £10,000. The remainder of the garden, with outline planning approval, was sold in May 1972, also for £10,000. Mr Lynes was assessed to capital gains tax for 1972-3 in the sum of £6,930. He appealed and the general commissioners held that section 29(1) of the Finance Act 1965 identified the two items which could qualify for exemption from the tax — the house and the garden—and that section 29(2), which contained the exemption, K applied. The words of section 29(2) were "The gain shall not be a chargeable gain if the dwelling-house or part of a dwelling-house has been the individual's only or main residence throughout the period of ownership, or throughout the period of ownership except for all or any part of the last 12 months of that period." The question was a short one and was "Is the gain resulting from the disposal of the remainder of the garden a taxable gain?"

The contention of the taxpayer was that the exemption was equally applicable whether the house and garden were sold by one or two transactions. He relied on the words of section 29(1) which described in (a) the house and in (b) L "land which he has for his own occupation and enjoyment with that residence as its garden or grounds up to an area (inclusive of the site of the dwelling-house) of one acre . . ." The taxpayer contended that the phrase "land which he has" did not necessarily imply "at the date of disposal" and said it should mean land which he has at any time while owning the residence. The remainder of the garden was land which the taxpayer had for his own occupation and enjoyment with the dwelling-house as its garden during the whole period of his ownership of the dwelling-house. The taxpayer submitted, therefore, that there was much good sense in saying that it should fall within the exemption of section 29(2). The con- M tention of the Revenue was that the words "has for his own occupation and enjoyment" referred only to the moment of disposal. Accordingly the garden would not be exempt in any case where it was disposed of at a time when the taxpayer no longer owned the dwelling-house to which the garden had been attached. It was also contended by the Revenue that there was no provision in the section which would allow for the apportionment of the gain on some sort of time basis in relation to the garden comparable with the apportionment which was appropriate in certain circumstances in the case of the dwelling-house.

A The commissioners had relied on the last part of section 29(2) in reducing the assessment to nil. They had apparently reasoned that it was the intention of Parliament that the dwelling-house should be exempt notwithstanding that a few months before the sale the owner went out of occupation, but that if they did not accede to the taxpayer's argument in the present case they would, in effect, be saying that Parliament intended that a garden should not qualify for exemption in a case where a taxpayer went out of residence a month before he sold.

The cases put by each side produced anomalies and it was no part of the court's task to try to balance out the
B various anomalies and twist the language to produce what might seem to be the most sensible result. The difficulty in relating the expression "land which he has for his own occupation . . ." etc to any moment except the actual time of disposal was that section 29(1)(a) looked both to the present and past while section 29(1)(b) was looking only to the present, ie "land which he has . . ." In the face of that comparative wording it did not seem possible, without doing violence to the language of section 29(1)(b), to read it in the way desired by the taxpayer. In those circumstances the Revenue argument must be accepted and the appeal allowed.

C The appeal was allowed and the assessment restored. No application was made for costs.

CHANCERY DIVISION

November 18 1976

(Before Mr Justice FOSTER)

MAKINS v ELSON (INSPECTOR OF TAXES)

Estates Gazette May 21 1977

(1976) 242 EG 627

D **Whether jacked-up caravan a "dwelling-house" within the meaning of section 29(1) of the Finance Act 1965, for the purpose of exemption from capital gains tax—Electricity, water and telephone connected—Sale of caravan on site —"Only or main residence"—Exempt from tax—Appeal from general commissioners allowed**

This was an appeal by Hugh Broughton Makins by way of case stated from a decision of the commissioners for the general purposes of the income tax for the division of Ludlow, who had held by a majority of two to one that the appellant was chargeable to capital gains tax.

The appellant, an agricultural engineer, in 1970 purchased
E for £1,800 a plot of three-quarters of an acre at Upper Birchill, Chardstock, near Axminster. Before completing the purchase he had obtained detailed planning permission to erect a dwelling-house on the property. He took up residence with his family in a caravan and began to build the house, but little work on it took place. In 1973 he sold the property and was assessed to capital gains tax by the commissioners in the sum of £6,542, having appealed to the commissioners from a determination by the inspector of taxes.

The appellant appeared in person; Brian Davenport (instructed by the Solicitor of Inland Revenue) appeared for the Crown.

F Giving judgment, FOSTER J said that Mr Makins moved on to the site in a caravan in August 1970, and his house-building operation began in September 1970. He levelled the site of his future intended dwelling-house and in his caravan he provided essential services — water for the site and electricity for the caravan, which rested on some supports. A telephone was installed. As far as the dwelling-house was concerned no foundations were excavated and no concrete was poured in. Mr Makins sold the property for £8,800 in May 1973. The expenditure accruing at that date was £547 on the operations described. When the assessment to capital gains tax was made Mr Makins objected
G that he had nowhere else to live but in the caravan and that therefore it was a dwelling-house within the meaning of the legislation and so not subjected to capital gains tax. To this the inspector of taxes replied that the caravan was a "chattel" and not affixed to the land and that therefore it was not a dwelling-house within the meaning of section 29(1) of the Finance Act 1965. The commissioners had decided that the fact that Mr Makins had been living in the caravan was irrelevant. A further submission by Mr Makins was that he had to spend most of the money from the sale of the property for the purchase of another property to live in.

The main question which he (his Lordship) had to decide
H was whether the caravan was a "dwelling-house" and whether, in the relevant period, it was the taxpayer's "only or main residence." Rating cases had been cited, in particular *London County Council* v *Wilkins* [1957] AC 362 and *Field Place Caravan Park Ltd* v *Harding* [1966] 2 QB 484. In both those cases the court had been investigating the somewhat different question whether or not a caravan on a pitch was to be rateable. A different test had to be applied to decide whether a chattel could or could not become a rateable hereditament. In the *Field Place Caravan Park* case Lord Denning MR had said: "In parting with the case I would say that these caravans with their pitches are very
J like small bungalows with their gardens." Although a chattel was not a rateable hereditament it might become so if it was used in conjunction with land so as to form a unit with the land. Mr Makins had said he had his caravan on blocks and that the wheels did not touch the ground. Mr Davenport had submitted that section 29 was dealing with a dwelling-house, but that section 30 showed a contrast between movable property and a dwelling-house.

He (his Lordship) thought that the very particular facts of the present case would not apply to a great many others. When he looked at all the facts, the laying-on of electricity, water and a telephone and also the fact that the wheels of
K the caravan were off the ground, he could only come to the conclusion that the caravan was a dwelling-house occupied by Mr Makins. He thought the general commissioners had failed to ask the proper question. The appeal succeeded and there would be judgment for Mr Makins with costs.

CHANCERY DIVISION

November 16 1976

(Before Mr Justice FOSTER)

L
WYNNE-JONES (INSPECTOR OF TAXES) v BEDALE AUCTION LTD

Estates Gazette May 21 1977

(1976) 242 EG 627

Income and Corporation Taxes Act 1970—Expenditure on cattle market—Works to cattle sale ring, including new walls and roof—Whether improvements or repairs—Held to be capital expenditure on improvements—Not deductible for purposes of corporation tax—Nature of "repairs" discussed—Revenue's appeal allowed

M This was an appeal by the Crown from a decision of the General Commissioners for the Thirk division of Yorkshire, who held that expenditure on works to the sale ring of a livestock auction market at Bedale, owned by the respondents, Bedale Auction Ltd, was expenditure on repairs and therefore properly deductible in the computation of corporation tax.

Brian Davenport (instructed by the Solicitor of Inland Revenue) appeared for the appellants, and R M Bramwell (instructed by Hunton & Garget, of Richmond, Yorkshire) represented the respondents.

Giving judgment, FOSTER J said: This is an appeal by way of case stated under section 56 of the Taxes Management Act

A 1970 by the Crown against a decision of the General Commissioners of Thirsk, who held that a sum of £8,605 could be deducted from the profits of the respondents, Bedale Auction Ltd (to which I shall refer as "the company"), for the year ended December 31 1973, for the purpose of calculating corporation tax. It was common ground that the company was liable to corporation tax under section 238 of the Income and Corporation Taxes Act 1970 on the profits of the company, and those profits were to be assessed under Schedule D, Case I. The rules which apply to that case are to be found in section 130 of that Act. The question which the commissioners had to decide was shortly this: was the **B** expenditure of £8,605 to be regarded as a sum expended for repairs and therefore deductible under section 130(d) or was it capital employed in improvements to the premises under section 130(g), in which case it cannot be deducted in calculating the company's profits?

The rules in section 130 are in these terms: "Subject to the provisions of the Tax Acts, in computing the amount of the profits or gains to be charged under Case I or Case II of Schedule D, no sum shall be deducted in respect of . . . (d) any sum expended for repairs of premises occupied, or for the supply, repairs or alterations of any implements, utensils or articles employed, for the purposes of the trade, profession **C** or vocation, beyond the sum actually expended for those purposes, . . . (g) any capital employed in improvements of premises occupied for the purposes of the trade, profession or vocation."

The Appeal. It is clear that the answer to the question is largely, if not entirely, a question of fact. A court hearing an appeal is bound to follow the rules laid down by Lord Radcliffe in *Edwards v Bairstow & Harrison*, reported in (1954) 36 TC 207, where he says this at p 229:

> If the case contains anything *ex facie* which is bad law and which bears upon the determination, it is, obviously, erroneous in point of law. But, without any such misconception appearing **D** *ex facie*, it may be that the facts found are such that no person acting judicially and properly instructed as to the relevant law could have come to the determination under appeal. In those circumstances, too, the court must intervene. It has no option but to assume that there has been some misconception of the law and that this has been responsible for the determination. So there, too, there has been error in point of law. I do not think that it much matters whether this state of affairs is described as one in which there is no evidence to support the determination or as one in which the evidence is inconsistent with and contradictory of the determination or as one in which the true and only reasonable conclusion contradicts the determination.

E In this case the Crown submits that the decision reached is contrary to the only reasonable conclusion to which, on the facts, the commissioners could have come.

The Facts. The company was incorporated early in 1972 and it carried on business as auctioneers of cattle. It purchased as a going concern an auction mart (which I shall refer to as "the mart") at Bedale, in North Yorkshire, though I do not know the actual date of purchase. The purchase price paid was £16,000, of which £13,000 was paid for the freehold and £3,000 for the goodwill of the business. The mart has a roughly rectangular area of some 23,400 sq ft, and largely comprises livestock pens and stalls constructed from tubular **F** metal, some of which are covered by corrugated iron roofing supported on posts and some of which are open. In addition to the pens and stalls there is a large yard behind the cattle ring which is used by cattle lorries. Apart from the auction ring (which I shall call "the ring"), the only other building on the mart was a small office building of some 392 sq ft.

The Ring. Its area is 1,820 sq ft or about one-thirteenth of the whole. Livestock enter the ring by way of a weighbridge contained in a railed enclosure, and, after having paraded round the centre of the ring, the cattle are contained in a second railed enclosure known as "the crush" until the auction is over and they return to the pens.

The Works. The original ring building was roughly circular **G** in shape and was partly roofed, the centre of the ring being open. The walls were of timber clad with corrugated iron, and apart from the centre of the ring itself, which was concreted, the remainder of the floor was earth, with seating for the proposed buyers and an auctioneer's podium. During the year ended December 31 1973 the existing walls and the roof were demolished. The only parts which were left were the weighbridge, the crush and certain gates. New walls of externally-rendered breeze block were erected in a rectangular shape, and a new single-span roof which covered the whole area and incorporated a ventilation system. In addition the whole of the floor was also concreted and parts were raised **H** on three levels with seats upon them. This work cost £8,605 (the sum in question) and it is fair to say that almost nothing of the ring remained except the weighbridge, the crush and a few gates. On these facts the decision of the commissioners was in these terms: "We, the commissioners, . . . decided as follows: (a) that in our opinion the 'ring' is an integral but a subsidiary part of the auction mart, the latter being the entirety; (b) that the 'cattle crush' and the 'weighbridge' (*inter alia*) were retained, and accordingly the 'ring' was not replaced in its entirety; (c) that the expenditure of £8,605 was in the nature of a repair and was required to ensure the safety of people using the mart and was therefore an allowable **J** deduction; (d) that the assessment should accordingly be reduced to nil and the appeal determined accordingly."

The Law. Three cases were cited to the commissioners. The first was *O'Grady v Bullcroft Main Collieries Ltd*, reported in (1932) 17 TC 93, where Rowlatt J at p 101 said this:

> I am of opinion that the commissioners were right in both these cases. As regards the chimney, I think it is really very clear. Of course, every repair is a replacement. You repair a roof by putting on new slates instead of old ones, which you throw away. There is no doubt about that. But the critical matter is—as was pointed out in the passage read from Buckley LJ's judgment in **K** the case which has been referred to—what is the entirety? The slate is not the entirety in the roof. You are repairing the roof by putting in new slates. What is the entirety? If you replace in entirety, it is having a new one and it is not repairing an old one. I think it is very largely a question of degree, but it seems to me the commissioners have taken the only possible view here. What was this? This was a factory chimney to which the gases and fumes, and so on, were led by flues and then went up the chimney. It was unsafe and would not do any more. What they did was simply this: they built a new chimney at a little distance away in another place; they put flues to that chimney and then, when it was finished, they switched the gases from the old flues into the new flues and so up the new chimney. I do not think it is possible to regard that as repairing a subsidiary part of the factory. **L** I think it is simply having a new one. And they had them both. Perhaps they pulled down the old one; perhaps they kept it, because they thought it was an artistic thing to look at. There is no accounting for tastes in manufacturing circles. Anyhow, they simply built a new chimney and started to use that one instead of the old one. I think the chimney is the entirety here and they simply renewed it. I think they are right on this point.

The second case to which the commissioners were referred is the Scottish case of *Samuel Jones & Co (Devonvale) Ltd v Commissioners of Inland Revenue*, which is reported in (1952) 32 TC 513, where the Lord President, Lord Cooper, said this **M** at p 518:

> On these facts I take it from the way in which the case is stated that the decision by the Special Commissioners against the company is based, and solely based, upon the fact that they found themselves unable to distinguish the case before them from the case of *O'Grady v Bullcroft Main Collieries Ltd*. . . . The Special Commissioners agreed (what is plain from the report) that there is a distinction between the case now before us and the case of *O'Grady*, in respect that in the *O'Grady* case the new chimney was in many respects an improvement upon the old chimney; but they found it impossible to discover in that distinction any ground for not following Rowlatt J's judgment. Looking at the *O'Grady*

A case I find that it related to a chimney at the pithead of a colliery, and that in giving judgment Rowlatt J put the matter thus: "The critical matter is . . . what is the entirety? . . . I think it is very largely a question of degree." Then after describing the chimney with which he was concerned he says: "I think the chimney is the entirety here and they simply renewed it." It is no part of our duty to review the decision of Rowlatt J as applied to the facts in the *O'Grady* case, but so far as this case is concerned the facts seem to me to demonstrate beyond a doubt that the chimney with which we are concerned is physically, commercially and functionally an inseparable part of an "entirety," which is the factory. It is quite impossible to describe this chimney as being in the words of Rowlatt J the "entirety" with which we are concerned. It is B doubtless an indispensable part of the factory, doubtless an integral part; but none the less a subsidiary part, and one of many subsidiary parts, of a single industrial profit-earning undertaking.

In the third case cited to the General Commissioners—another Scottish case: *William P Lawrie v Commissioners of Inland Revenue*, reported in (1952) 34 TC 20—the Lord President, again Lord Cooper, said this at the top of p 25:

> In the first place, relying upon the decisions in the cases of the *Bullcroft Main Collieries* and *Margrett*, they express the opinion that the "entirety" to be considered is the roof, and that it was impossible to say on the facts before them that the roof was merely a subsidiary or subordinate part of the building as a C whole which was repaired. I dislike that ground of judgment, and refer to the opinions which we gave in a case decided subsequently to the disposal of the present appeal, now reported under the name *Samuel Jones & Company (Devonvale) Ltd v Commissioners of Inland Revenue*. In this case it seems to me that, if we have to look for an "entirety" in order to determine whether in a business sense the operation under consideration was a repair or a renewal or improvement within the meaning of the Income Tax Acts, the "entirety" to be considered is the building as a whole and not its roof, and the Inland Revenue before us were content to take the case on that footing.

Lord Carmont, at p 26, said this:

> D I agree. I think that the commissioners were wrong in thinking that the entirety that they had to consider was the roof. The entirety was the building as a whole, and I think if they had had the opinions which were subsequently delivered in the *Samuel Jones* case before them, they would have had no doubt as to what conclusion they should have reached in this matter. The importance of considering an entirety is, of course, with a view to determining the character of the work done because what might at first look like a renewal may, when applying the matter to a larger unit, be shown to be only a repair.

I had the benefit of having a number of other cases cited to me, in particular the well-known passage in *Lurcott v Wakely and Wheeler*, reported in [1911] 1 KB 905, from the E judgment of Buckley LJ, who was dealing with a covenant to repair in a lease. At p 923 he said this:

> "Repair" and "renew" are not words expressive of a clear contrast. Repair always involves renewal; renewal of a part; of a subordinate part. A skylight leaks; repair is effected by hacking out the putties, putting in new ones, and renewing the paint. A roof falls out of repair; the necessary work is to replace the decayed timbers by sound wood; to substitute sound tiles or slates for those which are cracked, broken or missing; to make good the flashings, and the like. Part of a garden wall tumbles down; repair is effected by building it up again with new mortar, and, so far as necessary, new bricks or stone. Repair is restoration by F renewal or replacement of subsidiary parts of a whole. Renewal, as distinguished from repair, is reconstruction of the entirety, meaning by the entirety not necessarily the whole but substantially the whole subject-matter under discussion.

In the first two cases the decisions rest, I think, on what the court considered was the entirety. Here I find Buckley LJ contrasting repairs with renewals. And in *Phillips v Whieldon Sanitary Potteries Ltd*, reported in (1952) 33 TC 213, Donovan J (as he then was) applied the test of what were the premises. He said this at p 219:

> In my judgment, the "premises" for the purpose of rule 3(d) may sometimes be the whole of the trader's business premises and G may sometimes be a specific building forming part of those premises. Thus, if a factory window were blown out and had to be repaired, it would be obviously wrong to argue that as the entirety of the window had been restored it was not a repair to the premises. In such a case the "premises" would be the entire factory, in relation to which the window would be a repair and nothing else. But if, for example, a retort house in a gasworks was destroyed and had to be rebuilt, one would hardly call that a repair to the gasworks. The size of the retort house would compel one to regard that as the premises for the purpose of rule 3(d); and since it had been replaced in full it could not be said to have been repaired. These examples illustrate what I think is the truth, that there is no one line of approach to the problem which is exclusively correct. In some cases it will be right to regard the H premises as the entire factory, and in others as some part of the factory. Whichever alternative is the right one to adopt will depend upon the facts of the particular case. Rowlatt J took the view in the *O'Grady* case that he must regard the chimney itself as the premises, or, as he described it, the "entirety," and I would respectfully agree with him.

However, in the Irish case of *Hodgins v Plunder & Pollak (Ireland) Ltd*, which is reported in [1957] IR 58, Kingsmill Moore J at p 70 said this:

> It was contended that the work done in connection with the weigh-house was not a repair in as much as it was a replacement. Repair and replacement are conceptions which are not mutually J exclusive. This is made clear in the judgments in *Lurcott v Wakely and Wheeler*.

Then he quotes the passage from Buckley LJ and goes on:

> *Lurcott v Wakely and Wheeler* was concerned with liability under a covenant to repair, and when Buckley LJ refers to the "entirety" he is referring to the "subject-matter of the covenant." The principles enunciated have been accepted as applying to the interpretation of the word "repair" in the income tax code, but the application of the principles is not always easy. If the question as to whether certain work is, or is not, a repair depends on the proportion between the work done and the "entirety," what is to be considered "the entirety"? There is no convenient indenture, K as there was in *Lurcott v Wakely and Wheeler*, to solve the problem. Is the "entirety" merely the building unit in respect of which the repair or replacement has been effected, or is it the congeries of buildings and premises occupied for the purposes of the trade—in this case the whole factory? Rule 3(d) speaks of "premises occupied . . . for the purposes of the trade." If the words were "the premises" then it would be clear that the totality of the premises was referred to, but in the absence of the definite article "premises" might mean "all or any of the premises." The difficulty may, I think, be eased by reminding ourselves that what is sought to be resolved is a commercial and business problem. What expenses are necessary to earn the profits? In a big undertaking there will be a necessity to do minor repairs every year L and also a recurrent necessity to replace units which have fallen into such a state of age or dilapidation that patching is no longer economical. The profit-earning entity is the whole of the premises taken together, and the replacement of a unit, even if such a unit be a separate building, must be viewed in relation to the profit-earning entity. If the work done is only "a renewal or replacement of a subsidiary part of the whole" it seems to me that it rightly ranks as a repair. It is a question of degree.

There the question posed was: What is the profit-making entity? In *Conn v Robins Bros Ltd*, reported in (1966) 43 TC 266—a case upon which the respondents greatly relied—Buckley J said this at p 274:

> M On the question of whether these works were or were not repairs on the one hand or improvements on the other, I have got, as I have said, an express finding by the commissioners, and it appears to me that that is a finding of fact which I ought not to disturb unless I am satisfied on the primary facts found in the case that the commissioners could not reasonably have arrived at that conclusion. Looking at those primary facts, it seems to me that they are certainly capable of supporting the conclusion at which the commissioners arrived, and that it would be impossible for me to say that the deduction from those primary facts and the consequential finding that the work was work of repair was so unreasonable that the court ought to displace the commissioners' finding. I proceed, therefore, on the footing that these works

A ought to be regarded as works of repair, and I will go on to consider whether in this case there are any special circumstances which would lead to the conclusion that the expenditure ought nevertheless to be regarded as expenditure of a capital nature.

Then later on that page he says:

> In the light of that circumstance it seems to me that this was expenditure incurred by the company with a view to enabling it to continue to earn profits from its business, not by acquiring some asset for that purpose but by putting the company's existing asset into a state of repair which would enable it to continue to use that asset. No doubt in the course of carrying out these works certain structural alterations were made, as one would
> B expect with any extensive repair of a building over 400 years old, when repairs were being carried out at a time when building techniques have completely altered. But the fact that there were alterations in the structural details of the building does not seem to me to be a good ground for proceeding upon the basis that the work produced something new. On the contrary, I think it is implicit in the commissioners' finding that the result of this work was not to produce something new but to repair something which had previously existed. Upon that basis it seems to me that there is no ground for regarding this expenditure as a capital expenditure. It was expenditure incurred for the purpose of enabling the company to continue to earn its profits, and was therefore in my judgment expenditure which would properly be chargeable to
> C income. For these reasons I think the commissioners were right in the conclusion at which they arrived and this appeal fails.

But in that case the property was very ancient, and though the work was extensive and expensive the learned judge came to the conclusion that no new asset was brought into existence.

Conclusion. I find it difficult, if not impossible, to spell any one test from these cases, and there are many others in the law reports. For the Crown it was submitted that for the purpose of the phrase "improvements of premises" in this case the premises can only be the ring itself. For the company, it was admitted, and rightly so, that if one looked
D only at the ring itself the expenditure could only be capital expenditure; but on the company's behalf it was submitted that one must look at the mart as a whole and, if one does that, the expenditure was not a substantial improvement to the whole mart. In my judgment each case must turn on its own facts and no exhaustive test can be found to assist the court. I find the idea that one can have premises within premises a difficult one, though it has behind it the weighty views of both Rowlatt J and Donovan J, judges of the highest repute, particularly in tax matters. I am not forced into saying in this case that the ring itself must be considered the premises, for if I take the mart as a whole the ring is the
E nerve centre where the whole of the business of buying and selling takes place and all the rest of the mart is only ancillary to the ring, permitting the transport and movement of animals to the ring and their exit and removal after the auction. I do not think that the conclusion of the commissioners that the ring, though an integral part, is a subsidiary part of the mart can be sustained. On any view as to what the test should be, their second decision—that because the crush and the weighbridge were retained the ring was not replaced in its entirety —is in my judgment irrelevant. The question is: was the expenditure capital employed in improvements of premises occupied for the purposes of the trade? The question was
F never discussed by the commissioners, nor answered. On the facts of this case, in my judgment, the only answer to that question can be "Yes."

Appeal allowed with costs and case remitted to the General Commissioners with a direction that the sum of £8,605 was not an allowable deduction in computing the company's profits, the exact terms of the order to be agreed between counsel.

COURT OF APPEAL G

February 11 1977

(Before Lord Justice BUCKLEY, Lord Justice ORR and Lord Justice GOFF)

EMMERSON (INSPECTOR OF TAXES) v COMPUTER TIME INTERNATIONAL LTD (IN LIQUIDATION)

Estates Gazette June 18 1977
(1977) 242 EG 963

**Finance Act 1965 Schedule 6—Computation of allowable deductions from capital gains on assignment of lease—
Question of rent arrears paid by liquidator—Held not expenditure qualifying for deduction—Taxpayer company's appeal dismissed** H

This was an appeal by Computer Time International Ltd (in liquidation) from a decision of Fox J on December 18 1975, [1976] 1 WLR 749, allowing an appeal by the Crown from a decision of special commissioners who allowed an appeal by the taxpayer company against an assessment to corporation tax in respect of chargeable gains on the disposal of assets, being the assignment of leasehold premises in Oxford Street, London W1.

Peter Whiteman (instructed by Kingsley, Napley & Co) J
represented the appellants, and Peter Gibson (instructed by the Solicitor of Inland Revenue) appeared for the Crown.

Giving the first judgment, ORR LJ said that the taxpayer company, incorporated in 1969, carried on business hiring out computer time from their premises in Oxford Street. The total annual rental for the premises was £35,000. The business was not a success and it went into voluntary liquidation in April 1970 at a time when it was in arrears with the rent. On condition that all arrears would be met, the landlord agreed to the liquidator assigning to new tenants the leasehold interests in the premises. The arrears were paid and the
liquidator received consideration of £93,155 from the new K
tenants. Appealing against an assessment to corporation tax for 1970-71 of £22,352, the company contended that £6,131 of that sum was paid to the landlord for arrears of rent due after the date of liquidation and qualified as an allowable deduction.

Paragraph 4(1) of Schedule 6 to the Finance Act 1965 provided that

> . . . the sums allowable as a deduction from the consideration in the computation under this Schedule of the gain accruing to a person on the disposal of an asset shall be restricted to . . . (b) the amount of any expenditure wholly and exclusively
> incurred . . . for the purpose of enhancing the value of the asset L
> being expenditure reflected in the state or nature of the asset at the time of the disposal, and any expenditure . . . in establishing, preserving or defending his title to, or a right over, the asset.

Paragraph 5(2) provided that

> . . . there shall be excluded from the sums allowable . . . as a deduction in the computation under this Schedule any expenditure which, if the assets . . . were . . . used as part of the fixed capital of a trade the profits or gains of which were . . . chargeable to income tax would be allowable as a deduction in computing the profits or gains or losses of the trade for the purposes of income tax.

The issue was whether the £6,131 was an allowable deduc- M
tion in computing the capital gains arising from the disposal of the taxpayer company's interests in the premises in Oxford Street. If the payment did not fall within paragraph 4(1)(b) of Schedule 6 it failed to qualify as a deduction. If it did fall within it, it was necessary to go on to decide whether it was disqualified as a deduction by paragraph 5(1) or (2). To qualify under 4(1)(b) the payment must be expenditure incurred either wholly and exclusively "for the purpose of enhancing the value of the asset" or "in establishing, preserving or defending" title to the asset. The special commissioners had been satisfied that the liquidator's purpose in

A entering into the agreement with the landlord was to obtain the right to assign the leases and as such the payment did "enhance the value." As to the second stage, paragraph 5, they held that the payment was a capital payment and not therefore disqualified. On the Crown's appeal from that decision, Fox J held that it was not apt to describe the payment as "enhancing the value," but he thought, without deciding the point, that it was possibly correct to say the payment "preserved" the title because, had it not been made, the landlord could have taken forfeiture proceedings. However, he did allow the appeal on the ground that the payment was rent coming within the express exclusion from
B allowable deductions in paragraph 5(2).

The first question in the company's appeal was whether the judge had been right in assuming that the payment qualified as a deduction under paragraph 4(1)(b). He (his Lordship) did not find it possible to fit the performance of obligations under a lease into either of the phrases used in the subparagraph. If a tenant in breach of covenants allowed premises to fall into disrepair, they became less valuable; if the tenant then effected repairs the value of the premises would be restored, but it would be a misuse of language to say that their value had been "enhanced." Similarly a payment of rent overdue under a lease did not "enhance" the
C value. Further, the phrase expenditure "incurred in establishing, preserving or defending" title to an asset, applied to such matters as evicting a squatter or registering a charge, but had no application to the performance of a tenant's obligations under a lease.

The conclusion having been arrived at that the payment failed to qualify for deduction under paragraph 4(1)(b), it was not strictly necessary to consider whether the judge had been right in holding that the payments were excluded by paragraph 5(2). However, the court had heard argument on the point. His (his Lordship's) view was that the judge had been right. The company's argument was that the D expenditure was not disqualified under paragraph 5(2), as the payments had been made with a view of disposing of the properties. But the subparagraph required the assumption to be made that the assets were to be treated as being at all times "held or used as part of the fixed capital of a trade the profits or gains of which were chargeable to income tax." Thus, the premises must be assumed to have been used as such fixed capital up to the time of the disposals. Mr Whiteman then, relying on *Littlewoods Mail Order Stores Ltd v IRC* [1969] 1 WLR 1241 and *IRC v Land Securities Investment Trust Ltd* [1969] 1 WLR 604, invited the court to treat the payment as being capital and not revenue expenditure. E These cases, however, had rightly been distinguished by Fox J on the ground that in those cases new capital assets had been acquired. That was not so in this case and no expenditure of a capital nature had been incurred in relation to the existing assets.

Finally, Mr Whiteman claimed that the taxpayer should be able to obtain relief either under the capital gains tax provisions or as a deduction under Schedule 6. He invited the court to construe paragraphs 4 and 5 so as to achieve that result. But the court's duty was to apply the terms of the statute and for the reasons given it was not possible to construe the provisions in the way suggested. F

The remaining issue related to paragraph 5(1) and involved matters of accountancy practice, as to which there was no evidence before the commissioners. It was sufficient to say that the Crown had accepted that the burden was on it to show that the subparagraph applied, and it had failed to discharge it. He (his Lordship) would dismiss the appeal.

Goff LJ delivered a concurring judgment and Buckley LJ agreed.

The appeal was dismissed. Leave to appeal to the House of Lords was refused.

TOWN AND COUNTRY PLANNING

A

QUEEN'S BENCH DIVISION

January 30 1976

(Before Mr Justice PHILLIPS)

KENT MESSENGER LTD v SECRETARY OF STATE FOR THE ENVIRONMENT AND ANOTHER

Estates Gazette January 1 1977

(1976) 241 EG 25

Listed building permission refused—Secretary of State fails to deal with substantial point canvassed before the inspector —Utility of restoration an important consideration, as well as its cost, but decision letter makes no mention of the B **matter—Order quashed, with comments on issues that ought to be taken into account in any further proceedings**

This was a motion by Kent Messenger Ltd, of 69 High Street, Ashford, Kent, for an order quashing a decision of the first respondent, the Secretary of State for the Environment, confirming the refusal by the second respondents, Kent County Council, of permission for the demolition of a listed building at that address.

A de Piro QC and D Keane (instructed by Nabarro Nathanson) appeared for the applicants, and H K Woolf (instructed by the Treasury Solicitor) represented the first C respondent. The second respondents took no part in the proceedings.

Giving judgment, PHILLIPS J said: Mr de Piro moves on behalf of the Kent Messenger Ltd for an order quashing a decision of the Secretary of State for the Environment given by a decision letter dated June 30 1975. The effect of that decision letter was to dismiss an appeal by the Kent Messenger Ltd against a decision of the Kent County Council. The effect of that decision was to refuse consent to the Kent Messenger Ltd for the demolition of a listed building at 69 High Street, Ashford, Kent. The building in question is to be found in the town centre of Ashford and was described as D being within an attractive group of old buildings on the south side of the High Street immediately to the north of St Mary's Church. For the purpose of this motion it has not been necessary to go into details of the nature of the building and its surroundings, but it can, I think, be generally said that the merit of the building lies not so much in the building itself but in its contribution to the group of buildings in which it is to be found.

The applicants' case, developed at length before the inspector who held an inquiry, raises a number of points. One of them was that the building was in a very dilapidated state, and that it would cost a very considerable sum of E money to put it into repair; a figure of something in the order of £45,000 or £50,000 was spoken of. Even then, it was said, it would not really meet modern requirements as to accommodation, in a number of respects. Many of the ceilings were too low, the fire precautions would be inadequate, and so on; with the consequence that even with the expenditure of that large sum of money the building could only be put to a fairly limited use, and quite large parts of it would be virtually unusable. That case was developed, of course, with regard to the use to which the Kent Messenger Ltd wanted to put it, but it is difficult to see whether the situation F would be much better in the case of any other occupier. That was only one of the matters urged on behalf of the Kent Messenger Ltd, but it was one of the matters which found acceptance by the inspector. He held his inquiry and made his report to the Secretary of State. He found that the condition of the building was not good, that it was decayed, that important parts of the timber frame required to be repaired and restored, that it was structurally unsound, and that the work of reparation would be extensive and costly. He then went on:

> Furthermore unless such repair work were to be accompanied G by relatively major internal alterations and improvements, the amount of the office accommodation provided by the restored building would be similar to that now available, and much of this would probably remain substandard in several respects, including daylighting and headroom. It seems to me, therefore, that there is some substance in the appellants' submission that the retention and repair of the building would be uneconomic, and that redevelopment with a new building providing more and better office accommodation would be a more practical and economic solution.

He went on to find that the building which the Kent Messenger Ltd proposed to put up in its place (if it got permission to demolish the existing building), although in part H a facsimile of the old building, was not in all respects suitable; so he recommended that the appeal against the refusal of planning permission, which he also had under consideration, should be rejected. It is clear, I think, without looking at it in detail, that in his opinion it was likely that some modification and adaptation of the plans for the proposed new building could result in a building which could take the place of the old building without much, if any, damage to the architectural amenity of the area. He went on to deal with other matters, and then in the last paragraph of his report he said this:

> J Having regard to these conclusions it seems to me that although it would be possible to repair the existing building, the cost of this would be high and the quality of the accommodation provided would be poor by modern standards, and therefore its repair would not be an economic proposition for the appellants. If the replacement building were to be of a suitable design and materials, the loss of the old building would not necessarily be detrimental to the character of the conservation area. Consequently I consider that it would be reasonable to grant listed building consent for the demolition of the existing building. However, I am of the opinion that the submitted plans for the proposed new building are not acceptable, because the design of the High Street facade is not entirely satisfactory for its setting, K for the reasons given in paragraph 53.

The matter in due course was considered by the Secretary of State. About the planning permission he agreed with the inspector. About the application for consent to demolish the listed building he disagreed with the inspector. In his decision letter the Secretary of State on June 30 1975 set out *in extenso* the whole of the part of the inspector's report under the heading of "Conclusions," including the parts which I have just summarised and the passage which I have just read.

A Having said that in his view the existing building was of high architectural and historic value (I am not sure what evidence there was of historic value), the Secretary of State in effect gives three reasons for disagreeing with the inspector. These are in paragraph 4 of the letter. They are not numbered, but it is convenient to isolate them in that way. First of all he says:

> He considers that the inspector gave insufficient weight to the importance of preserving the building which is both architecturally interesting in itself and in its setting.

In his submissions on this point Mr de Piro has complained B of that statement that if it is a fresh finding of fact no opportunity was given to the Kent Messenger Ltd to deal with it, and if it is not, it is not very clear what its importance is. I do not accept that criticism. I think there that all the Secretary of State is doing, in perhaps not very accurately drafted language, is saying that he considers that the building is of very considerable architectural interest. Then the decision letter goes on, and this is the second of the reasons:

> He cannot accept that a new building, only in part a facsimile of the original, would in any way compensate for its loss or that it would not have any very detrimental effect on the appearance of the row of old buildings, or the street scene, or in the C conservation area generally.

As to that Mr de Piro has said that the merits of the new building were irrelevant, and secondly, and in any event, it was an unfortunate way of expressing it, because the inspector had taken the view that planning permission should not be granted, but that a purely minor adaptation of the plans might yield a satisfactory result. So, says Mr de Piro, first of all he is wrong to consider this matter at all, and secondly, he is wrong to consider these particular plans rather than some other building, perhaps of a similar but adapted nature, which might be put up in its place. He draws attention, D among other things, to the view of the Secretary of State in the decision letter in a different case in relation to 25 Kensington Gore and 197-200 Queen's Gate, London SW7 [Reference APP/4434/A/73/12016 and APP/4434/E/73/66 dated November 20 1975 on appeals by the Royal College of Art against decisions of Westminster City Council]. Here was a similar problem where the Secretary of State in his decision letter says:

> The inspector's views as to the architectural merit of the proposed new building are noted, but in dealing with an application for listed building consent for the demolition of a listed E building the Secretary of State must have regard to the desirability of preserving the building, and the quality of the proposed replacement building is not material.

He suggested that there is there an inconsistency and that both approaches cannot be right. Now there is an appeal pending to this court in that case, in which that question is the substantial ground of appeal. In those circumstances, having regard to the view which I have formed of the third point, which I am coming to in a moment, I do not think it is necessary to decide the question in this case, particularly having regard to the fact that it was one of the contentions of the Kent Messenger Ltd that the building for which they F were seeking planning permission would in fact make a very adequate substitute for the building which they wished to pull down. Consequently the inspector dealt with it, and not surprisingly the Secretary of State dealt with it. All I would say in finality about these first two points is that because of my conclusion on the third point, which I intimated to Mr de Piro when he came to reply, he has not in his reply dealt with them. If, therefore, this case should go to appeal, he is not to be taken to have exhausted all that he wished to say about them. So I come to the third point, and I go back to paragraph 4 of the decision letter where the Secretary of State says:

> He is satisfied, in view of the intrinsic interest of the building G and its position, that despite its present condition, which is admittedly poor, and while accepting that it would be costly to restore, he would not be justified in giving consent to its demolition.

That is a reflection to some extent of the views expressed by the inspector in paragraphs 51 and 54 of the report. Paragraph 54 of the report seems to me to contain a very important part of the inspector's reasoning. He puts it right in the forefront of his reasoning. I will read part of it again: ". . . the cost of this would be high and the quality of the accommodation provided would be poor by modern standards, and therefore its repair would not be an economic H proposition for the appellants." If one sets that against what the Secretary of State says in paragraph 4, it is quite clear that the Secretary of State is dealing only with the cost of restoration, and makes no reference to the utility of restoration. He makes no mention of the fact that this very expensive restoration would, if the inspector was right, produce a result that would be of very limited value.

The question, first, is how that should be construed. Mr Woolf says that, particularly having regard to the fact that the Secretary of State has just set out in this decision letter the relevant part of the inspector's report, it cannot be J supposed that he was omitting from consideration what to the inspector was an important consideration, namely the value or the lack of value to be obtained by restoration. In effect what Mr Woolf suggests is that the form of words used, "it would be costly to restore," is shorthand that embraces the whole question of cost, the utility of what would be obtained as a result of the expenditure and the general economic equation of it. I cannot accept that. Cost is one thing, and saying it is costly to restore is one thing. The consequences of the expenditure, and the utility of the expenditure, are another. There seems to me to be a plain omission here. I think it is an omission which would immediately hit the eye K of anybody interested in the outcome of this appeal, in this case in particular the Kent Messenger Ltd. The intelligent reader would immediately say: "Well, the inspector was for me, and one of the reasons he was for me is because he thought the expenditure would be largely wasted. The Secretary of State is against me. What did he have to say about that? You can read this decision letter up and down and you will not find the answer." So I think that there was a substantial omission of an important point. The question then is what conclusions does one draw from that omission, and the answer, I think, is that without guessing it is not possible to draw any conclusion at all. It may be just a L mistake in draftsmanship; it may be that in fact the Secretary of State overlooked this consideration altogether, or it may be that the Secretary of State took this consideration into account but found that it was of so little consequence that he thought it not worth mentioning. Against that background I reach the conclusion that this is from the point of view of the Kent Messenger Ltd an unsatisfactory state of affairs. They have appealed, the inspector was for them, the Secretary of State was against them, and they do not know why.

The next question is, is there power to intervene, and if there is ought it to be exercised? It is necessary to turn to M section 245 of the Town and Country Planning Act 1971. Subsection (4) provides:

> "On any application under this section the High Court . . . (b) if satisfied that the order or action in question is not within the powers of this Act, or that the interests of the applicant have been substantially prejudiced by a failure to comply with any of the relevant requirements in relation thereto, may quash that order or action."

I will not take time to decide whether there would be power here to act on the grounds that the Secretary of State's order is not within the powers. That case has been argued, it being

A said that a material matter in this case was not only the cost of the restoration but the effect of restoration; whether it would be useful or not. The fair inference is that the Secretary of State ignored that, and if he did that then there was a failure by him to take into account matters which he ought to have taken into account, and if that is the case, then he acted outside the scope of his powers. I prefer not to decide that, because I would want to look a little more closely at the authorities, and because I am satisfied that I can decide it under the second part of paragraph (b); "failure to comply with any of the relevant requirements." The relevant rules are the Town and Country Planning (Inquiries B Procedure) Rules 1974 (SI 1974 no 419), where it is provided in rule 13 that "the Secretary of State shall notify his decision and his reasons therefor in writing to the applicant," and so on. In the case of *Lightfoot Court Properties Ltd v Devlin and others* (1970) 21 P & CR 689 the Divisional Court, presided over by Lord Parker CJ, when considering a different statute, expressly approved and adopted the words used by Megaw J in the case of *Re Poyser and Mills' Arbitration* [1964] 2 QB 467. That itself was a case dealing with a different statute and jurisdiction. There Megaw J said this in reference to the Tribunals and Inquiries Act 1958:

> C Parliament having provided that reasons shall be given, in my view that must be read as meaning that proper, adequate reasons must be given; the reasons that are set out, whether they are right or wrong, must be reasons which not only will be intelligible, but also can reasonably be said to deal with the substantial points that have been raised.

He goes on to say:

> I do not want it to be thought for a moment that I am saying that any minor or trivial error, or failure to give reasons in relation to every particular point that has been raised at the hearing, would be sufficient to invoke the jurisdiction of this court.

D That test, there approved, applies equally to the jurisdiction which I have to exercise in this case. If that be right, then the question is whether or not the reasons given by the Secretary of State here can be described as being proper and adequate. In my judgment they cannot, quite simply because they omit any reference whatever to what was an important material argument of the Kent Messenger Ltd and one, indeed, which had played a substantial and influential part in the mind of the inspector in leading him to the recommendation which he made. I accept, of course, two submissions of Mr Woolf: first, that it would be wrong for E this court to approach a decision letter with niggling pedantry (and I hope I have not done so in this case); and secondly, that a decision letter cannot be impugned merely because it does not refer to, or take account of, every submission that has been made to the inspector or every point that has been controversial. But it must, I think, if it is to be regarded as a satisfactory and sufficient statement of reasons, deal at least with the major important points. In my judgment this decision does not.

It is necessary, I think, however one phrases it, to go on and consider further whether that leads to the conclusion that there has been an error of law or (in the language used F in section 245) that "the applicant has been substantially prejudiced by a failure." Whichever way I approach it, the answer to me is yes. I think the applicant has been substantially prejudiced, because among other things he does not really know what view the Secretary of State took on this important point; and there is at least a potential error in law, because we do not know what view the Secretary of State formed about it. We do not know, because he omitted all reference to it. It may be that he regarded it as irrelevant. If he did I should consider that to be an error in law, making it clear that I am using the word "irrelevant" in this sense, that it was in my judgment a matter which ought to be taken into consideration, that is to say a matter G to which he had to apply his mind as being one of the relevant factors in reaching his decision, accepting of course that the weight which he chose to give to it would be entirely a matter for his consideration and nobody else's. For those reasons, in my judgment, the applicant is entitled to succeed, and entitled to have the decision of the Secretary of State quashed so that there will be a further consideration of the matter. What I would like to say in finality is this: although I have expressed no view, for the reasons given, on the second point argued by Mr de Piro, it would be right, I think, for me to say this, that when the matter is further considered, whether by the Secretary of State originally, or if there is a H further inquiry by the inspector or both of them, it ought to be borne in mind that if and in so far as it is material to have regard to the nature of a replacement building it ought not to be limited, I would have thought, to the building which the applicant has under contemplation, but ought to extend to and embrace any kind of building which might reasonably be considered to be likely to succeed in obtaining planning permission. In other words, it is less a question of a particular replacement building as such and more a question, I would have thought, of what, if any, replacement building is feasible. For the reasons given the application is allowed and J the order quashed.

The applicants were awarded costs.

QUEEN'S BENCH DIVISIONAL COURT

October 13 1976

(Before Lord WIDGERY CJ, Mr Justice FORBES and Mr Justice SLYNN)

TIDSWELL v SECRETARY OF STATE FOR THE ENVIRONMENT K

Estates Gazette January 8 1977

(1976) 241 EG 83

Enforcement notice—Planning authority has only to look at what has been done and what planning permissions have been granted—Authority need not ask itself whether the developer may be able to rely on some exemption, deemed permission etc—Developer who had so far operated a "Sunday market" on only nine Sundays held in any case to have been justifiably denied the protection of the general development order—Evidence of a permanent change of use before Secretary of State L

In these proceedings Harold Tidswell, of St Francis Way, Chadwell St Mary, Essex, moved (1) to set aside a decision of the respondent, the Secretary of State for the Environment, upholding an enforcement notice served upon him by Thurrock Borough Council, and (2) to quash the notice, which required him to discontinue using land at Tilbury Football Club as a Sunday market and to remove structures used for that purpose.

A C Mordern (instructed by Bates, Son & Braby, of Chelmsford) appeared for the applicant, and H K Woolf (instructed by the Treasury Solicitor) represented the respondent. M

Giving the first judgment, FORBES J said that the first point taken by the applicant was that before serving an enforcement notice, a planning authority had to be satisfied that there had been a breach of planning control or a failure to comply with conditions. Under the current general development order, he (the applicant) could hold a market on the site on 14 occasions without planning consent. He had in fact held markets on only nine separate days, and accordingly (he submitted) the enforcement notice was premature. Since the planning authority must have known this, it followed that it could not have appeared to them,

A before they served the notice, that there had been a breach of planning control. He (his Lordship) had been rather attracted to that submission, but it imported the premise that it was for the planning authority to investigate and satisfy themselves whether or not a developer could bring himself within some exemption or implied permission. If that were right, a local planning authority could not merely rely on the fact they had not given planning permission but would have to investigate each matter. The proposition could not be right. In his (Forbes J's) judgment, a planning authority needed only to look at the site to see what was happening, and then look at the records to see if planning B permission had been given, in order to see whether development had taken place. It was for the applicant to bring himself within an exemption or permission.

The applicant's second point was that the respondent had erred in ruling that the applicant was not entitled to claim exemption. It was, however, clear that at the time of an inquiry held into the applicant's appeal there was evidence that the use was permanent, and therefore the respondent was entitled to take the view that there had been a change of use and that the applicant could not avail himself of the exemption given in the general development order. The application should therefore be dismissed.

C LORD WIDGERY and SLYNN J agreed, and the motion was dismissed with costs.

QUEEN'S BENCH DIVISION

July 29 1976

(Before Sir Douglas FRANK QC, sitting as a deputy judge of the division)

KENT COUNTY COUNCIL v SECRETARY OF STATE FOR THE ENVIRONMENT AND ANOTHER

D

Estates Gazette January 8 1977

(1976) 241 EG 83

Planning permission for Kent oil refinery—Case called in by Secretary of State, who grants permission for all installations except a proposed access road and imposes a con-E dition that oil etc shall not be moved by road—Permission for the scheme minus the road held unobjectionable— "As a matter of common sense, the determining authority can grant as much of the development applied for as they think should be permitted"—Further points on the validity of the condition imposed and on the Secretary of State's reliance on a statement by the Secretary of State for Energy —Application to quash groundless, but court's discretion would in any case have been exercised in Secretary of State's favour

This was an application by the Kent County Council for an order quashing a decision of the first respondent, the Secretary of State for the Environment, granting the second respondents, Burmah-Total Refineries Trust Ltd, planning permission for construction of an oil refinery at Cliffe Marshes, Kent.

R Sears QC (instructed by Sharpe, Pritchard & Co, agents F for the solicitors to Kent County Council) appeared for the applicants; D Widdicombe QC and Lord Colville (instructed by the Treasury Solicitor) for the first respondent; and G Eyre QC and M Spence (instructed by Denton Hall & Burgin) for the second respondents.

Giving judgment, SIR DOUGLAS FRANK said: This is an application under section 245 of the Town and Country Planning Act 1971 for an order that a planning permission for an oil refinery granted by the first respondent to the second respondents be quashed. On September 23 1971 the second respondents made two applications for planning permission, one for development described as "oil refinery and jetty with ancillary activities" and the other as "road/rail terminal G facilities and ancillary pipelines and roadways." The purpose of the application was to obtain planning permission for a proposed new oil refinery at Cliffe Marshes alongside the Thames in Kent. It was explained that two applications were made, as two separate sites were affected, but that the local planning authority were asked to treat the two application forms as constituting one application. On November 29 1971 the first respondent gave notice that he had decided to exercise his powers under section 22 of the Town and Country Planning Act 1962 and to call in the applications for decision by himself. On May 2 1972 a local inquiry was held by one of the first respondent's inspectors, H M A H Stedham ARICS FRTPI, assisted by two assessors. It was completed on May 5 1972. Mr Stedham reported at length to the first respondent in September 1972. Having concluded that there were major objections to the development because of the effect it would have on the coastal area and because of the inadequacy of the road network to take the traffic involved, he recommended that the application be refused. However, the first respondent decided that it was in the national interest that additional refinery capacity should be made available to meet the expected requirements in south-east England, and for that reason was unable to accept his inspector's recommendation. J Nevertheless, he recognised that the road system was inadequate, and said that he had it in mind (1) to refuse permission for the construction of the access road as proposed in the application, and (2) to grant outline planning permission for the oil refinery, the jetty, the rail terminal facilities and the pipelines between the refinery and terminal, subject to a condition, *inter alia*, that save as should be agreed from time to time with the local planning authority (or in default of agreement, as should be determined by the Secretary of State) no deliveries of oil, oil products or by-products should be made to or from the oil refinery or terminal except by sea, pipeline or rail transport. K

The first respondent also stated that "since a planning permission along the lines indicated above would result in development significantly different from that for which planning permission was originally sought, the Secretary of State thinks that he should afford all the parties concerned an opportunity to make further representations to him on the matter." Further representations were made, particularly by the applicants, and accordingly on December 10 1974 the first respondent gave notice that he intended to reopen the inquiry, but limited to hearing representations relating to his proposed intention to grant permission in the way he had stated and to material considerations which had arisen L or come to light since the original inquiry. The reopened inquiry was held by Mr Stedham from April 8 to 16 1975, and he reported in the following month. In his report he said that in his opinion there had been sufficient changes since March 1974, certainly since 1972, to require a fresh assessment as to whether the public interest in securing extra refining capacity in the region was still sufficiently strong to outweigh the siting of a proposed refinery on the Cliffe Marshes, especially with the prospect of either restricted distribution arrangements for a lengthy period, or the imposition of tanker traffic on an unsuitable road network. He felt unable to make a recommendation. On December 31 M 1975 the first respondent granted planning permission on the lines previously stated and subject to a number of other conditions, of which only one will have to be mentioned. The applicants challenged the first respondent's decision on a number of grounds, and it will be convenient for me to deal with these under separate headings.

(1) *Permission for part only of the development*. The contention is that the first respondent erred in law, and was acting *ultra vires*, in purporting to grant planning permission for part only of the development which was the subject of the application. Mr Sears, for the applicants, submitted that

A the first respondent is in the same position with a called-in application as a local planning authority in dealing with an application, and that his powers are not greater than those set out in section 29 (1) of the Act of 1971. Thus he may grant permission either conditionally or subject to conditions, or may refuse permission, but not both. It is clear from the language of the section that the grant or refusal can relate only to the development proposed in the application and not to a different development, for that, Mr Sears said, is the scheme of the Act. He asked how it would be possible to appeal under section 36 or section 37 where an application had been granted in part and refused in part, or how sections B 41 and 45 could apply to a part-only permission. He contrasted the provisions of section 183, and claimed that had it been intended that part-only permission could be granted under section 29, then words similar to those in section 183 would have been used. He emphasised that the reason why the Secretary of State had reopened the inquiry was because he had said that a planning permission along the lines indicated would result in a development significantly different from that for which planning permission was originally sought. He referred to *Federal Steam Navigation Co Ltd v Department of Trade and Industry* [1974] 1 WLR 505 for C authority for the proposition that the word " or " in section 29 should be used disjunctively—that is to say, that planning permission could not be granted and refused on the same application.

Mr Widdicombe conceded that the first respondent had no greater power than the local planning authority in dealing with an application, but submitted (1) that where an application contains a number of separate and divisible elements it is lawful for them to be separately dealt with, and (2) alternatively, that if the elements are not divisible there is power to modify the application provided (a) the scope of the development is not enlarged, (b) the essential nature of D the development is not altered, and (c) any persons affected are given a chance to make representations. He (counsel) did not think that the agreement of the applicant for permission to the modification was necessary, but that question did not arise in this case, as the second respondents had agreed to the modification. As to his first submission, Mr Widdicombe said that the fact that there is only one application form is not conclusive, for there can be several developments in one application. Here there were four elements expressed in the application and by the Secretary of State, namely (1) the construction of the refinery, (2) the provision of road and rail terminals, (3) the construction of a E pipeline, and (4) the construction of an access road. Each part involved separate land, and a separate and distinct use of any one part could have been carried out without the carrying-out of the others, save that the construction of the refinery was a development to which all the other elements were ancillary. He denied that any problems could arise under the sections mentioned by Mr Sears, for in each case the provision concerned would apply to the part of the development granted or the part refused, as the case might be. He said that a permission related not to the application but to the development applied for, because by section 23 planning permission is required for the development. Mr F Eyre, for the second respondents, adopted Mr Widdicombe's arguments, and conceded that the first respondent had no jurisdiction to grant planning permission for development to which the application did not relate, or for development which was different in nature; nevertheless, he had power to grant permission for any development included in the application. He (counsel) argued that section 38 of the Act of 1971 clearly envisaged that the first respondent had power to grant permission in respect of part of the land to which the application related. As to the meaning of the word " or," counsel contended that it was not to be construed in the sense of mutual exclusivity and that it was not an inviolate disjunctive where the clear intention of the statute was G otherwise: *Mersey Docks & Harbour Board v Henderson Brothers* (1888) 13 App Cas 595 at 603. Dealing with other objections raised by Mr Sears, he said that the applicants' consent to the amendment was not required, as upon the application being called in they ceased to have any jurisdiction and had merely a right to be heard. He said that the form of a decision could not affect the subject-matter of the grant.

In my judgment, the correct approach to this matter is to ascertain the powers under section 29 of the Act by reference to the purposes of Part III, in which it appears. It seems to me that everything in Part III flows from and is conse- H quential upon the provision in section 23 that planning permission is required for the carrying-out of any development of land. Hence when the matters come before the determining authority, in this case the first respondent, what that authority has to do is to decide whether, having regard to the provisions of the development plan and to any other material considerations (that is, planning considerations), permission ought to be granted, and if so, what if any conditions should be imposed. It further seems to me that as a matter of commonsense the determining authority can grant as much of the development applied for as they think should be permitted. J It may be that the applicants' consent should first be obtained, but no question as to that arises in the instant case, as on April 2 1974 the solicitors for the second respondents told the first respondent, " We confirm the condition proposed by paragraph 5 of your letter is acceptable to our clients." That was the condition eliminating the use of the road. Indeed, as Mr Sears admitted, it is common practice for applications to be dealt with in this way, for example, where 50 houses are applied for and the local planning authority grants permission for 40 houses. Mr Sears found this unobjectionable, but said that a formal amendment to the application was required and that where K the application was called in the consent of the local planning authority must have been given. On the facts of this case the first respondent and the second respondents by their conduct agreed to an amendment. I see no reason why the local planning authority had to be a party to that amendment. Accordingly, I agree with Mr Widdicombe that where an application consists of a number of separate and divisible elements it is lawful for them to be separately dealt with, as was done in this case. It seems to me to matter not in this case whether the decision of the first respondent was upon an amended application or whether it was a part refusal, save that in the latter case it is necessary to construe L the word " or " conjunctively. But I think that may be done, having regard to the context of the part of the statute in which the word appears. In the *Mersey Docks & Harbour Board* case Lord Halsbury said:

> I know of no authority for such a proceeding unless the context makes the necessary meaning of " or " " and," as in some instances it does: but I believe it is wholly unexampled so to read it when doing so will upon one construction entirely alter the meaning of the sentence, unless some other part of the same statute or the clear intention of it requires that to be done.

In my judgment, Part III of the Act of 1961 does require M " or " to be read as " and," and apart from the general purpose of Part III, section 38 impliedly recognises that as the correct construction. There is one further way of looking at the first respondent's decision. Not only did he expressly refuse permission for the construction of the access road, but he also imposed a condition which had the same effect, in that in the absence of the agreement of the local planning authority deliveries were to be made only by sea, pipeline or rail transport. Thus if the refusal was *ultra vires* then it is arguable that that part of the decision should be disregarded. Of course in truth what the first respondent decided was to grant planning permission for the terminal proposed subject to a condition which would restrict the way in which the development was carried out. There would be nothing unusual in that, but whether this particular condition was *ultra vires* I will consider when I come to the last ground of appeal set out in the notice of motion.

(2) *Natural justice.* It is said under this head that the first respondent erred in finding, as he did, that the proposed oil refinery at Cliffe was needed to meet demand in the south-east of England and that " there is evidence of a deficit in refining capacity " in the south-east. The error alleged is that he had regard to the views of the Secretary of State for Energy when there was no evidence given by that Secretary of State as to the need for the refinery, and that the applicants had no opportunity of testing the assertion of that Secretary of State that the proposed refinery was needed to meet demand in the south-east or of the factual basis of the statement made by him and relied upon by the first respondent. This allegation arises from paragraph 15 of the decision letter, which sets out the respondent's conclusions in these terms:

> So far as the south-east region is concerned, there is evidence (as indicated in paragraph 13 above) of a deficit in refining capacity, and the Secretary of State concludes that this deficit will be increased by the extent to which any North Sea oil refined in the region is exported. Given that it will take a number of years to build the proposed refinery, the Secretary of State does not agree with the argument that reliance for oil supplies can or should be placed on other projects, the implementation of which is not guaranteed. He sees no reason to doubt that additional capacity should be provided in the region, and he also considers that the need to provide for exports strengthens the case for such additional refining capacity. In all the circumstances he has concluded that a need for the refinery proposed in the present application has been established.

The evidence indicated in paragraph 13 was expressed in these words:

> At the reopened inquiry forecasts were submitted by the county council that, assuming no new capacity is created, the regional deficit would be about 5 mty, whereas national capacity would be in surplus. Documentary evidence was submitted on behalf of your clients that the regional deficit could be in the order of 10 or 12 mty. In his statement of December 6 1974 (referred to in paragraph 8 above) the Secretary of State for Energy has expressed the view that the proposed refinery at Cliffe is needed to meet the demand in the south-east.

In his statement to the House of Commons, the Secretary of State for Energy had said:

> If projects now under consideration go ahead, including those at Canvey Island and Cliffe which are needed to meet demand in the south-east, I expect our refinery capacity in the early 1980s to approach 150 million tons a year. This should be enough overall to meet our own needs and provide for some exports.

Mr Sears, while conceding that the first respondent was entitled to have regard to his own policy and that of any other Government department, said that the applicants were entitled to be provided with the information on which the Secretary of State for Energy's judgment was made, and were entitled to cross-examine to test the credibility of that information. He further complained that the decision in this case was made not by the first respondent but by the Secretary of State for Energy, and he referred to *H Lavender & Son Ltd v Minister of Housing and Local Government* [1970] 1 WLR 1231. Dealing first with that last point, the *ratio decidendi* in the *Lavender* case was that the respondent had delegated the making of his decision to the Minister of Agriculture and had not properly or at all exercised his duty of discretion. In the instant case, although the first respondent had regard to the view expressed by another Minister, there is no evidence whatsoever that the decision was other than his own. Accordingly, I do not find this complaint well founded, and indeed it was abandoned by Mr Sears. As for the applicants' first point, I think the short answer to it is that the Secretary of State for Energy's statement in the House of Commons that two more refineries were required in the south-east to meet the expected demand was a statement of policy. He did not say that it was his policy that those refineries should be at Canvey Island and Cliffe, but merely that if those projects went ahead they would meet that particular need. He was thus confining himself to a policy peculiar to his own department, leaving the decision for the siting with the appropriate authority. If that is right, I think it is clear from the Town and Country Planning (Inquiries Procedure) Rules 1974 that far from the Secretary of State being required to justify that policy in any way, it was a matter which could not have been questioned at the inquiry (see rule 8 (5)), in that the first respondent could have taken it into account even if it had not been raised at the inquiry (see rule 12 (1) (b)). Apart from the rules, there is ample authority for the proposition that the first respondent was entitled to take into account the views of another Government department even without giving the parties an opportunity to comment on them (see, for example, *Miller v Minister of Health* [1946] KB 626; *Robinson v Minister of Town and Country Planning* [1947] KB 702; *Summers v Minister of Health* [1947] 1 All ER 184; *Johnson v Minister of Health* [1947] 2 All ER 395; and *Darlassis v Minister of Education* (1954) 4 P & CR 281). However, it happens that in this case the statement in the House of Commons was made before the 1975 inquiry, so that the parties were able to, and did, comment on it at the inquiry. Moreover, the applicants themselves stated that there would be a regional deficit, and this was part of the evidence relied on by the first respondent for his decision. I would add that in any event there is no rule of natural justice which could have required the Secretary of State for Energy to give evidence at the inquiry. It follows that the applicants' contention under this heading fails.

(3) *Invalidity of condition.* The applicants say that condition 5 (that is, the condition which requires that the delivery shall be made only by sea, pipeline or rail transport) is invalid, first because it is unworkable, and secondly because it takes away a substantial part of the benefit of the planning permission. Mr Sears said, however, that this is not the most important part of his case. In my judgment there are short answers to these two points. First, if the second respondents carried out deliveries by road (say as stated in the condition) then they would be liable to enforcement procedure under the Act. The fact that there might be practical difficulties as alleged in an affidavit, namely in distinguishing the second respondents' traffic from other traffic, does not go to the validity of the condition. As to the second point, a conditional permission is almost invariably less beneficial than an unconditional permission. It must always be a question of fact and degree whether a particular condition is such as to take away the substance of the permission, in which event that condition may be invalid. In this case, however, the development sought is the construction of an oil refinery, and all else is ancillary to that purpose. Of course if the condition had been such as to render the oil refinery unworkable, that would be a different case; but the second respondents' acceptance of the condition is evidence that it is certainly not this case.

There is one final matter with which I must deal. Mr Widdicombe submitted that if I concluded that the decision of the first respondent was right in principle and wrong in form, then I had a discretion which I should exercise in favour of the respondent. He pointed out that under section 245 (4) (b) the court, if satisfied that the order or action in question is not within the powers of the Act, *may* quash the order or action. In *Miller v Weymouth & Mel-*

A *combe Regis Corporation* (1974) 27 P & CR 468 at 478 Kerr J said: "In my judgment, Mr Slynn is correct in his submission that both as a matter of history and on the basis of such authority as there is the word "may" is here used in its ordinary permissive sense, though it would no doubt only be rarely and in very unusual cases that a court would not exercise its discretion to quash a ministerial order or action which was not within the relevant statutory powers, and it would never do so if a refusal to exercise the discretion to quash would or might be unjust in the circumstances." Thus if I am wrong in my view that the first respondent's decision is within the powers of the Act,
B the question is then whether the applicants would be entitled to have the order quashed *ex debito justitiae*. The most that can be said against the first respondent is that he failed to state formally that the application was amended to exclude the construction of the road. The parties at the second inquiry were well aware that construction of the road was to be refused. That refusal or amendment, far from being prejudicial to the applicants, came about in order to meet one of their major objections. It cannot therefore be said that they are entitled to the order which they seek *ex debito justitiae*. Accordingly, even if I had found the first respondent had no power to refuse the
C application in part, or that in truth he dealt with it as an amended application, then I would decline to exercise my jurisdiction to grant the order sought by the applicants. In all the circumstances I accordingly dismiss this action.

QUEEN'S BENCH DIVISION

October 19 1976

(Before Mr Justice WILLIS)

D FRENCH KIER DEVELOPMENTS LTD v SECRETARY OF STATE FOR THE ENVIRONMENT AND ANOTHER

Estates Gazette January 8 1977

(1976) 241 EG 87

Planning permission for houses in metropolitan green belt—Inspector recommends grant of permission—Secretary of State refuses it, but fails properly to explain his decision—References to differences of "emphasis" not adequate to justify departure from inspector's carefully-reasoned conclusion—"No one reading decision letter could understand why inspector's recommendation was rejected"—Decision
E **quashed**

This was an appeal by French Kier Developments Ltd, of 50 Epping New Road, Buckhurst Hill, Essex, against a decision of the first respondent, the Secretary of State for the Environment, dated September 24 1975, dismissing an appeal against the refusal by the second respondents, Thurrock Borough Council, of planning permission for the residential development of some 40 acres of land known as Grey Goose Farm, at Long Lane, Fairfield Avenue and Brookmans Avenue, Stifford Clays, Grays, Essex.

A de Piro QC and D Prebble (instructed by Bates, Son & Braby, of Chelmsford) appeared for the appellants, and H K
F Woolf (instructed by the Treasury Solicitor) represented the first respondent. The second respondents took no part in the proceedings.

Giving judgment, WILLIS J said that a number of grounds were advanced in the notice of motion, but the nub of Mr de Piro's argument centred upon the alleged failure of the Secretary of State to comply with rules 12 and 13 of the Town and Country Planning (Inquiries Procedure) Rules 1974, thus involving substantial prejudice to the appellants. French Kier were well-known developers, and they embarked on the notoriously difficult task of obtaining planning permission to develop a substantial piece of land in the approved metropolitan green belt. They were no doubt G
encouraged to do so by some shift in Government policy, as evidenced by documents which came before the inspector and before the court, indicative that some land in the metropolitan green belt should be found to meet urgent housing needs. The council's reasons for refusal were set out in the inspector's report. At the inquiry held on April 22 1975 the single issue was whether in all the circumstances, including the important fact that the land was in a major growth area where Government policy encouraged a five-years' supply of residential land at any time, the site could properly be released for development despite the undoubted contribution it made, if undeveloped, to preventing the eastward expansion H
of Grays and to inhibiting any coalescence of Grays and communities to the east. The inspector's conclusions plainly included a number of findings of fact, and to the extent that they did Mr Woolf accepted that the Secretary of State agreed with them. The inspector recommended that the appeal should be allowed. In his letter of decision the Secretary of State set out his approach to the matter as follows:

> No reason is seen to dissent from the inspector on any fact but—in a situation where the arguments for and against the appeal proposal are thought to be in many respects fairly finely balanced—there is some disagreement with him about the degree J
> of emphasis which he put on certain of the factors discussed in his conclusions. In particular it is thought that more importance than he allows should be attached to the positive contribution which this site of more than 40 acres makes to the metropolitan green belt in that it forms part of an open area which checks the growth of Grays and inhibits its coalescence with neighbouring towns. As regards the availability of housing land, while the evidence indicates that there is only just over three years' supply in the Thurrock district, it is apparent that the situation in South Essex as a whole (where, although the matter was not argued in detail, there was said to be a seven years' supply) is better. There is also the consideration that the general search K
> for more land for housing has to be set alongside the importance which the Government attach, and have from time to time reaffirmed, to supporting local authorities in their resistance to unacceptable encroachment in the green belt. When these factors are taken together it is considered that the loss of the quite large appeal site from the metropolitan green belt, to which it makes a positive contribution, would not in all the circumstances be justified, and it has therefore been decided that the council's decision refusing permission should be upheld.

Mr de Piro submitted, and Mr Woolf did not dissent, that there appeared to be three so-called "factors" identified by the Secretary of State where there was "some disagreement" L
as to the "degree of emphasis" expressed by the inspector. The "factors" were (1) the sentence beginning, "In particular it is thought that more importance than he allows . . ."; (2) the passage beginning, "As regards the availability of housing land," and ending, ". . . is better"; (3) the passage from "There is also the consideration" down to "resistance to unacceptable encroachment in the green belt." Mr de Piro's first submission was that there had been a breach of either rule 12 (2) (a) or (b) in relation to factor (2). Despite the inspector's observations as to the absence of any evidence as to the contents of a document emanating from the Essex County Council entitled "South Essex Interim Development M
Control Policy," which referred to seven years' supply of land in SE Essex, the Secretary of State had accepted that statement as correct, said Mr de Piro, notwithstanding that at the hearing the local planning authority could not explain where the document came from, called no evidence about it, and conceded that the figure of seven years was in conflict with a figure of four years given on behalf of the county council at a recent appeal to the Secretary of State. Reference was made to the document in the report, and it was clear that the inspector regarded it as of no evidential value in the appeal. Mr Woolf submitted that the Secretary of

A State was only saying that the land supply position in SE Essex as a whole was better than in the Grays-Thurrock area, and he was entitled (said counsel) to look at the county council document and give it such weight as he thought fit.

The inspector had made a finding as to land supply in the Grays-Thurrock area, but none as to SE Essex, since he did not consider he had the evidence to do so. The decision letter, in his (Willis J's) view, in factor (2), if not by reference to the document itself then at least by reference to its recorded contents, concluded that the land supply position in SE Essex was better than in the Thurrock district. He (his Lordship) confessed to a failure to understand the relevance B of this conclusion of the Secretary of State or how it represented a disagreement on emphasis from the inspector. But he did not think it could possibly be said to represent a differing from any fact found by the inspector, and thus a breach of rule 12 (2) (a), because the inspector had made no finding of fact on this matter. Faced with this point, Mr de Piro said in the alternative that the Secretary of State should have ignored the document or any references to its contents, as the inspector did, since it was not produced by any witness, its provenance was unexplained, and it could not be tested by cross-examination. Certainly, said counsel, it should not have been treated as providing any evidence for C disagreeing with the inspector on anything he said with regard to land supply.

It hardly needed to be said that legal rules of evidence were not applied at local inquiries, and both oral and documentary evidence was freely admitted in circumstances where even the more relaxed rules of evidence at the present time would not allow of its admission in a court of law. Nonetheless some limit must surely be imposed in fairness to an appellant on the scope of so-called "evidence" which by no stretch of the imagination could be said to have the slightest evidential value; this must particularly be so when such "evidence" was used to support a conclusion unfavourable D to the appellant. He (his Lordship) thought that the inspector was right to ignore this document and the Secretary of State was wrong, in the particular circumstances, to attach any weight to it or its contents. If he (the Secretary of State) wanted to explore the matter further, the right course, in his (Willis J's) view, would have been to follow the procedure of rule 12 (2) (b). He did not do so, and had therefore acted in breach of the rule.

Factor (3), said Mr de Piro, begged the whole question, and he (his Lordship) agreed with counsel. Mr Woolf, with his usual candour, conceded that factors (2) and (3) added E little, if anything, to factor (1) and that factor (3) was a "mirror," as he put it, of factor (1). He (Mr Woolf) did not feel quite able to accept the description of "window-dressing" as applied to factors (2) and (3), but if that was not an unfair description it was perhaps pertinent to wonder why the author of the decision letter found it necessary or thought it appropriate to dress the window of factor (1) in the way he did. Mr de Piro had a short answer: that the author of the letter was purporting to find some sort of reasons for disagreeing with the inspector, and had only succeeded in showing that there was no valid reason for doing so. It was clear, therefore, that the real issue centred F upon factor (1), although in considering whether the Secretary of State had complied with rule 13 it seemed to him (his Lordship) important that the passage from the decision letter should be looked at as a whole, including the so-called factors (2) and (3). Mr Woolf submitted that the Secretary of State had had to perform the balancing act which must always precede a decision on whether or not to release green belt land for development, that his decision was based on his policy in relation to any release of green belt land, and that he had only differed from his inspector in respect of a planning judgment, as he was entitled to do. It would have been quite sufficient, said Mr Woolf, for the Secretary of State to have said: "I can find no reason for departing G from my green belt policy."

It was, of course, established law that the Secretary of State was entitled to have a policy and to apply it, where appropriate, to any case which fell for his decision. The decision letter made no reference to policy, but he (his Lordship) naturally assumed that those who decided the matter had in mind current Government policy in relation to the need to consider the release of some green belt land for housing. He did not think there was any indication that the Secretary of State differed from his inspector on any ground of policy, whatever that policy might be. It seemed plain H from the report that the inquiry was conducted against the background of Government policy as it had developed since Circular 42/55, and there was no suggestion in the decision letter that the inspector had misdirected himself at any stage in relation to policy. To him (Willis J) it appeared that the Secretary of State, far from differing from his inspector on any matter affecting policy, was on the face of the decision letter differing only on questions of emphasis on three so-called factors in the conclusions, and that it was necessary to consider whether he had set out his "reasons" for dismissing the appeal as required by rule 13. If reasons had been given, they could be found only in what was said under J factor (1). If he (his Lordship) understood what was there being said, it was that the inspector had underemphasised the importance of the contribution made by the appeal land in containing the eastward expansion of Grays. Yet it was clear that the inspector accepted that the land made such a positive contribution. His conclusions as a whole and in detail seemed a model of careful reasoning and weighing of all the facts before giving his opinion that the site made little positive contribution and its loss would result in little, if any, harm to the metropolitan green belt strategy.

In these circumstances, where all the facts found in the conclusions were accepted, did factor (1) and the intro- K ductory remarks provide adequate reasons for disagreeing with the inspector and dismissing the appeal? He (his Lordship) thought not. Where had the inspector gone wrong? What was the nature of the disagreement? Where should the emphasis have been put? Nothing could be vaguer than the words used in the decision letter. Of course, a decision letter was not to be construed like a statute or philosophical treatise. But the appellants, and indeed the local planning authority, were entitled to a clear and intelligible statement of the reasons for the decision, and he (Willis J) did not think that such a statement was to be found in factor (1), either alone or together with its appendices. It was true L that there were cases where words used in a decision letter could properly be regarded as surplusage which did not vitiate or detract from the effect of preceding words, but this was not such a case. The fact that two of the three "factors" which together were said to have led to final disagreement were plainly mere verbiage and of no substance must inevitably invite the question whether, if only factor (1) remained, the decision would have been the same. No one reading the decision letter could answer that question. If the decision letter gave any reasons at all—which he (his Lordship) was inclined to think it did not—he regarded them M as so vague, inadequate and unintelligible as to leave anyone reading the letter unable to understand why, when so much of the factual conclusions were accepted, the inspector's decision was rejected. In his judgment there had been a breach of both rule 12 (2) (b) and rule 13, the appellants had suffered substantial prejudice, and the Secretary of State's decision should be quashed.

The appellants were awarded costs.

A **QUEEN'S BENCH DIVISION**

October 19 1976

(Before Mr Justice WILLIS)

CLYDE & CO v SECRETARY OF STATE FOR THE ENVIRONMENT AND ANOTHER

Estates Gazette January 15 1977

(1976) 241 EG 155

**Planning permission cannot be refused merely to encourage retention of an existing use—If from a planning point of view flats might just as well be offices, permission for offices
B must be granted—Council's wish to foster housing accommodation not a "material consideration"—Secretary of State's decision quashed**

This was an appeal by Clyde & Co, solicitors, of Dunster House, Mincing Lane, London EC3, from a decision of the respondent, the Secretary of State for the Environment, upholding the refusal by the Guildford Borough Council of planning permission for use of 9,500 sq ft of accommodation at Miller House, Bedford Road, Guildford, as offices.

R Phillips (instructed by Wedlake Bell) appeared for the appellants, and H K Woolf (instructed by the Treasury
C Solicitor) represented the respondent. Guildford Borough Council were joined as second respondents but took no part in the proceedings.

Giving judgment, WILLIS J said that the case raised a short but important point which had not hitherto come up for decision. The facts were a little unusual. Miller House was originally intended by its owners, Miller & Co, as their principal office. Because they needed space urgently, Millers applied for permission for offices in the western half of the proposed building and eight flats with garages in the eastern half. This was in point of fact a device to avoid the need for an office development permit for the whole block, and the
D delay that that would involve. The western half had been built and occupied, and the structure of the eastern half had been put up at the same time because that seemed the sensible course, though it had become clear that the western half was sufficient for Millers' requirements. The eastern half had not been completed as flats, and could just as well be completed as offices. Clyde & Co, a very-well-known City firm of solicitors specialising in marine insurance work, moved a number of partners and staff to Guildford in 1969 with the assistance of the Location of Offices Bureau. They came to need more and better accommodation, and finally were able to obtain a lease of the eastern half of Miller
E House subject to grant of an office development permit and planning permission. An office development permit was issued, but Guildford Borough Council refused planning permission, and Clyde & Co appealed to the respondent Secretary of State. An inquiry was held by an inspector. In his report the inspector noted that "there appeared to be considerable weight in the arguments advanced on behalf of the firm in respect of its need for accommodation," but he went on to state: "There is a shortage of housing accommodation in Guildford, at least in the public sector, and hence the loss of residential accommodation is to be resisted save in the most exceptional circumstances. I am not convinced,
F therefore, despite the strength of the appellants' case, that the loss of the eight flats that would be provided by the appeal premises can be justified." The Secretary of State accepted the inspector's recommendation that the appeal be dismissed. Clyde & Co now challenged the Secretary of State's decision, submitting that he had erred in law in taking into account, or basing his decision upon, irrelevant considerations, namely that a refusal of planning permission would ensure that the eastern half of Miller House would be used for residential purposes, and that the proposed change of use would result in an unjustified loss of residential accommodation.

Mr Phillips, in his very able argument, had gone back to
first principles. He referred to section 29 (1) of the Town and G
Country Planning Act 1971, which provided that "where an application is made to a local planning authority for planning permission that authority, in dealing with the application, shall have regard to the provisions of the development plan so far as material to the application and to any other material considerations." He (Mr Phillips) submitted that the word "material" when used in relation to "other considerations" had the same meaning as in relation to the development plan, namely "material to the application." He (his Lordship) thought that that was right, and he did not understand Mr Woolf to disagree. Further, by section 36 (3) the Secretary
of State was required to deal with an appeal on the same H
basis as did a local planning authority in dealing with an application. Then Mr Phillips said that there would appear to be nothing in the development plan relative to the desirability of retaining housing uses save in exceptional circumstances. The proposed use, counsel continued, clearly complied with the zoning for the area. The office policy objection had been decided in Clyde & Co's favour, and the only ground for dismissing the appeal, namely the desirability of retaining the permitted housing use, was not a consideration material to the application.

Mr Phillips was not without some support for his submis-
sions. Indeed, it was upon the case of *Granada Theatres Ltd* J
v Secretary of State for the Environment [1976] JPEL 96 that the notice of motion was based. That case concerned a proposal to change the use of a building from a cinema to that of a bingo hall and social club. Permission was refused, and the refusal upheld on appeal, not, it was said, on the ground that the proposed use was objectionable on any planning ground but because it was desirable to continue the cinema use. The case was not argued before Judge Stabb, since the Secretary of State agreed to the quashing of the decision on two grounds, the material ground being in identical terms to
the ground in this case. In agreeing to the quashing on this K
ground, counsel for the Secretary of State conceded that planning control could not be used to force the continuation of a use for any particular purpose and that the decision letter was open to that interpretation. In the present dispute, Mr Phillips summarised his submission in the words of the editorial comment on the *Granada* case: "The local planning authority cannot take into consideration the desirability of retaining the existing use by refusing permission for another use if there is no other reason for refusing that use." That, counsel said, was exactly what the Secretary of State had done here. On behalf of the Secretary of State, Mr Woolf
naturally accepted the principle contained in the concession L
made in the *Granada* case, but he pointed to the very different circumstances in which the concession was made and submitted that the decision letter in the present case was not open to the same interpretation. He said that in deciding whether or not to allow a change of use it was permissible to take into account the merits of the existing use from a planning point of view, and that any consideration of a proposed change of use would be incomplete without taking into account the appropriateness of the existing use. He said that the real issue here was a choice between retaining a housing use and allowing an office use, that the local planning authority
could have a policy not to allow a change from a residential M
use unless the circumstances were exceptional, and that the choice made by the Secretary of State was within the powers of section 29.

Although it was clear that the local planning authority concerned was anxious to retain the permitted use of the building, no doubt in the hope that if office permission were refused the owners would complete the flats or with the possibility of compulsory purchase in mind, though not expressed, it did not appear to him (his Lordship) that the authority had any policy to the effect referred to in the report. On the contrary, the only matter of policy relied on

A was the "office" policy, and that argument was rejected. In other words, any "regard to the provisions of the development plan" (see section 29) was favourable to the appellants, and the question was whether the Secretary of State had had regard to "other material considerations" or to something which lay beyond their scope. No doubt cases did arise where it was legitimate to consider two or more competitive uses within the ambit of section 29. If it were right that "other considerations" to be "material" must be referable to the application, as he (his Lordship) thought it was, it seemed to him to be clear on the facts of this case that all the "considerations" referable to the application were con-
B cluded in favour of the appellants. It was common ground that save in the most oblique and improbable circumstances, the Secretary of State could not compel the implementation of the permitted use. It therefore followed that on what he (his Lordship) would think were planning grounds properly so called the proposed change was unobjectionable, and the planning appeal had failed only because the Secretary of State thought it desirable that the permitted use should remain unaltered. In his (Willis J's) judgment, that was not a proper consideration on which to found the decision. It did not fall within section 29, and in particular it was equally open to the same interpretation as in the *Granada* case, namely that
C the Secretary of State took into consideration what he regarded as the desirability of retaining the existing housing use by refusing permission for office use when there was no other reason for doing so. Accordingly the appeal would be allowed and the Secretary of State's decision quashed.

The appellants were awarded costs.

D

QUEEN'S BENCH DIVISION

October 19 1976

(Before Mr Justice WILLIS)

SPACKMAN v WILTSHIRE COUNTY COUNCIL; SAME v SECRETARY OF STATE FOR THE ENVIRONMENT

Estates Gazette January 15 1977

(1976) 241 EG 157

Nature of "specified operations" necessary to preserve an unimplemented planning permission of 1948—A soakaway will do, even if the contractor has got it out of line, and so will a length of drive laid out with a hardcore
E **foundation**

This was a claim by Mrs Barbara Yvonne Mary Spackman, of Bath Road, Swindon, against Wiltshire County Council for a declaration that valid planning permission existed for construction of a house on land owned by her at Hodson, Chisledon, Wiltshire. By a further application, Mrs Spackman sought to have quashed a decision of an inspector appointed by the respondent, the Secretary of State for the Environment, whereby permission was refused for a house on the same site.

P Mottershead (instructed by Hewitt, Woollacott & Chown,
F agents for Lemon & Co, of Swindon) appeared for Mrs Spackman; C Cochrane (instructed by Collyer-Bristow & Co, agents for the solicitors to the council) for the planning authority; and H K Woolf (instructed by the Treasury Solicitor) for the Secretary of State.

Giving judgment, WILLIS J said that the plaintiff and her late husband had been very anxious to make their home on the appeal site, and on July 21 1948 they obtained detailed planning permission for a house there. Unhappily, the husband developed multiple sclerosis, which meant that he had to live his life in a wheelchair until his death in 1965. That sad fact, and the problems the plaintiff faced with her
G two children after her husband's death, provided a readily-understandable explanation of why the 1948 permission was not acted on. Eventually, in 1972 or 1973, the plaintiff decided that she wanted to make her home at Hodson, but on a scale appropriate to one person rather than four. That involved modification of the 1948 plans, and the planning authority required submission of a fresh application. In the course of negotiations with the council, moreover, the plaintiff discovered that there was a deadline which applied to the 1948 permission. By paragraph 19 of schedule 24 to the Town and Country Planning Act 1971, it was provided that every planning permission granted before April 1 1969
H should, if the development to which it related had not been begun before the beginning of 1968, be deemed to have been granted subject to a condition that the development must be begun not later than the expiry of five years from April 1 1969. No development relative to the 1948 permission had been begun before the beginning of 1968, and therefore the deadline, if that permission was to be preserved, was April 1 1974. Section 43 of the 1971 Act set out what constituted the beginning of development, and provided, so far as material, that development should be taken to have been begun on the earliest date on which any specified operations comprised in the development had been carried
J out. The section went on to define "specified operations." On February 8 1974 the planning authority refused permission for the modified development. The plaintiff appealed against that decision and at the same time carried out what she was advised could be regarded as "specified operations" with a view to keeping the 1948 permission alive. One question for him (his Lordship) was whether the works carried out, or any of them, were in fact "specified operations." The appeal against refusal of permission for the plaintiff's modified plans was subsequently dismissed by an inspector given power by the Secretary of State for the
K Environment to hear and determine it, and the other question for the court was whether the inspector's order was valid.

He (his Lordship) would deal first with the question of the "specified operations." Two operations were relied on. The first was a partially-constructed soakaway 6 ft deep, partially filled with hardcore but not covered over. Drainage trenches were dug, 3 ft deep and 20 ft long, and in each trench a 4-ft length of pipe had been laid leading to the soakaway. The second operation was the laying-out with a hardcore foundation of a short section of the access drive. It was conceded by the plaintiff that the soakaway was about 35 ft from the nearest soakaway shown on the 1948
L plans, and that there was some deviation in the section of access drive constructed from the line approved in 1948. It was clear, however, that these deviations were immaterial. Mr Cochrane's point on the soakaway was that if the trenches dug in March 1974 were projected, two of them would miss the notional 1948 building altogether. He (counsel) suggested that the works were so wide of the mark that they could not be regarded as operations relative to implementing the 1948 permission. Mr Mottershead conceded that the drive was, as it were, off course, and that the contractor made a mistake in the location of the soakaway and the direction of at least one trench, but said that the
M other two could have been used or adapted to connect the soakaway with the 1948 house had it been built. He (his Lordship) agreed with counsel that if the 1948 house had been built, a substantial part, at any rate, of the soakaway works could have been used. If, moreover, the test to be applied was a subjective one, he (Willis J) was satisfied that the plaintiff intended the works to be relative to the 1948 plans, and that she should not suffer by reason of any error due to the failure to comply with her instructions. He was therefore of the opinion, though he expressed it with diffidence, that the soakaway and drainage trenches were "specified operations."

There was no question but that a section of the drive, with the associated hedge-grubbing, was laid out in a preliminary way before April 1 1974. But was a laid-out vehicular access, 60 yds long, from a highway to a dwelling-house with garage, a "road"? It need occasion no surprise that neither in the 1971 Act itself nor in its long definition section was there any assistance as to what was in the draftsman's mind when he used this word. Mr Cochrane submitted that "road" involved the concept of access by the public, even though an access road might not be a repairable highway. Counsel said that he would concede that an estate road serving a number of houses, and thus a road to which in one form or another the public had access, was a "road" within section 43, but he contended that a drive serving only one house was not. He (his Lordship) had come to the opposite conclusion. It should not be forgotten that the Act's provisions were in a sense penal, taking away from the subject, albeit in a fairly leisurely way, a vested entitlement. In his view they should be construed benevolently towards, and not strictly against, the subject. But whether or not that was the right approach, he found no difficulty in holding that the access drive to the 1948 house was a "road" within section 43 (1) (d). It seemed a ludicrous situation that if the plaintiff had chosen to construct the whole of her access drive to the house before April 1 1974, much in the same way that estate developers built their estate roads before building houses, it would be said not be a "specified operation" because it was not a "road," whereas if she had chosen to build one soakaway in the exact 1948 position, with trenches pointing in the correct direction, that would be accepted as a "specified operation." In his judgment, the plaintiff had carried out "specified operations" prior to April 1 1974, and so preserved a valid planning permission to carry out the development approved in the 1948 plans subject to immaterial variations. He accordingly granted the declaration sought by the summons against the council.

It must follow that the refusal of the Department of the Environment to grant planning permission for what Mr Mottershead called the "revised version (1973)" must be quashed. The inspector plenipotentiary was plainly in a difficulty in deciding the question whether or not the "authorised version" was extant when he was considering the "revised version." It was agreed that it was within his power to decide this question had he thought it right to do so. Instead, it seemed clear from his decision that he "noted" the submission on the 1948 permission, but then put it on one side and decided the appeal solely on the merits of the 1973 application, unaffected in any way by the earlier planning history. It went without saying that if at that time the plaintiff had a valid planning permission it was a vitally material consideration for the inspector to take into account. This involved no criticism of the inspector, but must result in the quashing of his decision.

The plaintiff was awarded her costs against the local planning authority in respect of the summons, and against the Secretary of State in respect of the application.

QUEEN'S BENCH DIVISIONAL COURT

February 3 1976

(Before Lord WIDGERY CJ, Mr Justice KILNER BROWN and Mr Justice WATKINS)

WILD AND ANOTHER v SECRETARY OF STATE FOR THE ENVIRONMENT AND ANOTHER

Estates Gazette January 22 1977

(1976) 241 EG 235

Enforcement notice—Sole issue whether development occurred before January 1 1964—Evidence each way before inspector—No error of law on part of inspector or Secretary of State—Owner's appeal dismissed

This was an appeal by David Arthur Wild and AGM Car Hire against an enforcement notice served by the second respondents, Hillingdon Borough Council, in May 1974 and approved with minor amendments by the first respondent, the Secretary of State for the Environment, requiring them to discontinue the use of land at Sipson Farm, Sipson Road, West Drayton, Middlesex, for the purposes of a car-hire business.

B Payton and D Di Mambro (instructed by Wood, Nash & Winters) appeared for the appellants, and H K Woolf (instructed by the Treasury Solicitor) represented the first respondent. The second respondents took no part in the proceedings.

Giving judgment, LORD WIDGERY said that the sole issue between the parties was whether the development of which complaint was made in the enforcement notice, namely that of change of use to a car-hire business, had taken place before January 1 1964. The appellants' contention was that it had, and the respondents claimed the contrary. At the inquiry before a ministry inspector, there was a conflict between the oral evidence given on behalf of the appellants, to the effect that use of the site had begun before the end of 1963, and written statements of the appellants produced by the respondents which tended to show that the use had begun only in the spring of 1964 at the earliest. The inspector found in the light of all the evidence that the appellants had not discharged the onus on them of proving that the development had taken place before January 1 1964, and his finding, with his recommendation that the appeal should be dismissed, had been accepted by the Secretary of State. It was clear that there was evidence before the inspector tending each way on the relevant issue, and clear that it was for the inspector, in the first instance, to resolve the conflict of evidence that resulted. He had done so, and it could not be said that there had been any error of law, either on his part or on the part of the Secretary of State. In particular, it could not be said that the use of the phrase "no uncorroborated evidence" meant that the Secretary of State had adopted a criminal standard of proof; he (his Lordship) thought that these words meant only that there was no disinterested witness who was in a position to support the appellants' case. In these circumstances there was no ground upon which the court could interfere with the decision, and the appeal accordingly failed.

KILNER BROWN and WATKINS JJ agreed, and the appeal was dismissed with costs.

QUEEN'S BENCH DIVISIONAL COURT

October 12 1976

(Before Lord WIDGERY CJ, Mr Justice FORBES and Mr Justice SLYNN)

PARRY v FOREST OF DEAN DISTRICT COUNCIL

Estates Gazette February 5 1977

(1976) 241 EG 387

Enforcement notice—Information charges breach of notice "since January 8 1972" by use of land for storage of equipment—Equivalent to a charge of a breach every day since that date—Information bad for duplicity—Conviction quashed

This was an appeal by Maurice Dudley Parry from his conviction by justices sitting at Newent, Gloucestershire, on November 19 1975 on an information laid by the respondents, the Forest of Dean District Council, alleging that since January 8 1972 he had used certain land at North Farm, Kilcot, Newent, for the storage of motor vehicles, trailers and

A equipment in contravention of an enforcement notice dated November 8 1971.

B Galpin (instructed by Hewitt, Woolacott & Chown, agents for Leslie J Slade & Co, of Newent) appeared for the appellant, and R Evans (instructed by the solicitor to the council) represented the respondents.

Giving judgment, LORD WIDGERY said that the appellant submitted that the respondents' information was bad for duplicity in that it charged continuing offences. One had to look at the information, and on doing that one saw that it alleged that the appellant had since January 8 1972 used the land in contravention of the enforcement notice. It therefore
B alleged a breach every day since January 8 1972. That was a breach of the principle that informations must not be bad for duplicity, and the appeal should accordingly be allowed.

FORBES and SLYNN JJ agreed, and the appeal was allowed with costs.

CHANCERY DIVISION

November 15 1976

(Before Mr Justice FOX)

C

UNITED REFINERIES LTD v ESSEX COUNTY COUNCIL

Estates Gazette February 5 1977

(1976) 241 EG 389

Condition in planning permission for oil refinery requiring works to be commenced by a specified date satisfied by soil stripping and by construction of an access road, albeit one in part at least of a temporary nature—Expenditure of £30,000-odd out of a contemplated contract cost of
D **perhaps £15,000,000 not to be brushed aside as de minimis**

This was a claim by United Refineries Ltd against the Essex County Council for a declaration that construction of an oil refinery at Canvey Island, Essex, had been commenced by January 1 1969 in fulfilment of a condition imposed by the Minister of Housing and Local Government in a letter of December 14 1965 allowing a planning appeal and granting permission for the development.

G Dobry QC, G Lightman and R Carnwath (instructed by Linklaters & Paines) appeared for the plaintiffs, and J Harman QC and Miss E Appleby (instructed by Sharpe,
E Pritchard & Co) represented the defendants.

Giving judgment, Fox J said that by an application dated July 29 1964 the plaintiffs applied to Canvey Island District Council for permission under the Town and Country Planning Act 1962 for the development of 262 acres of land on Canvey Island. The purpose of the application was for "use of land for an oil refinery, including railway and pipeline approaches." The plaintiffs appealed, and on December 14 1965 the Minister of Housing and Local Government issued a decision letter allowing the appeal and granting permission for construction of the refinery, including ancillary offices and the railway and pipeline approaches. The per-
F mission was made subject to six conditions, of which two were presently relevant. By condition (b), the siting, design and external appearance of the buildings and the means of access were to be such "as may be agreed with the local authority (account being taken of the need to keep land across the north part of the appeal site free of buildings so as not to prejudice proposals for the construction of a proposed new road between A13 and Canvey Road), or, in default of agreement, as shall be determined by the minister." Condition (f) read: "The building and other operations hereby permitted shall be commenced not later than January 1 1969." By an application of May 17 1966 the plaintiffs submitted to the district council a number of G
detailed proposals in pursuance of condition (b), showing among other matters a connection between the external refinery road and a proposed roundabout to be constructed at the junction of the proposed Canvey approach road and classified road B1014. The proposals were approved by the district council on October 4 1966 with certain conditions, one of which was that access to the site was to be taken from the roundabout, except that until the roundabout was constructed and made available for use access might be obtained at another point shown on the plans.

After discussion with officers of the Essex County Council, the plaintiffs' contractors in March 1967 began con- H
struction of an access road and the stripping of topsoil in preparation for the erection of buildings. About this time, too, plans were submitted showing the siting of the buildings, their design and external appearance. By August 1967 the access road had been constructed along the line proposed to a length of 3,400 ft and to an average width of 24 ft, the cost being £27,000. Furthermore, the topsoil had been stripped from areas shown on the plans as the location of administrative buildings, measuring in aggregate 5.8 acres, and this part of the work had cost £3,200. The defendants, Essex County Council, now said, nevertheless, J
that none of this work had to do in any substantial sense with the development of the land; it was all entirely preliminary, and the "building and other operations . . . permitted" therefore could not be said to have been "commenced" by January 1 1969. Alternatively the defendants said that what work had by then been commenced was *de minimis* and that on its true construction the condition in the letter of December 14 1965 required a start to be made on the main building work and other substantial aspects of the proposed development. He (his Lordship) rejected both these arguments. Work costing a total of £30,000-odd might be only a small part of a project which K
on completion could cost in the region of £15,000,000, but was by no means *de minimis*. As for the argument that what had been done was merely preliminary, both the items of work concerned were items for which permission had been given, and in his (Fox J's) judgment they could not be dismissed as preliminary only. The road might have a temporary quality, but it was in accordance with plans that had been approved, and the final development would occur when the roundabout was completed. As for the stripping of the topsoil, that seemed to him to be a building operation, work that would normally be done in erecting a building, and it had been commenced by the specified date. The L
word "commence" must be taken in its ordinary meaning, and there was a distinction between starting before a particular date and completing by such date; here, the only condition attached was that development should commence before the date in question, and development had so commenced. Accordingly there would be a declaration in the terms sought.

The plaintiffs were awarded costs.

M

QUEEN'S BENCH DIVISIONAL COURT

July 22 1976

(Before Lord WIDGERY CJ, Mr Justice Melford STEVENSON and Mr Justice CAULFIELD)

BROOKS & BURTON LTD v SECRETARY OF STATE FOR THE ENVIRONMENT AND ANOTHER

Estates Gazette February 12 1977

(1976) 241 EG 457

Enforcement notice—Established use for general industrial purposes held to cover only a building on owner's land

and not the open ground surrounding this—Notice based on intensification of light industrial use outside building accordingly upheld—Further point on estoppel by statements of council employees dismissed on the facts

This was an appeal by Brooks & Burton Ltd, owner-occupiers of land at Holtwood, Wimborne, Dorset, from a decision of the first respondent, the Secretary of State for the Environment, upholding with amendments an enforcement notice served on them by the second respondents, the former Dorset County Council, requiring them to cease using the land for manufacture of concrete blocks or other similar industrial purposes.

Miss M Viner (instructed by Riders, agents for Derek T Wilkinson & Co, of Bournemouth) appeared for the appellants, and P Boydell QC and H K Woolf (instructed by the Treasury Solicitor) represented the first respondent. The second respondents took no part in the proceedings.

Giving judgment, LORD WIDGERY said that in 1959 planning permission was granted for use of at any rate some of the land in question for light industrial purposes within Class III of the Town and Country Planning (Use Classes) Order 1950, now replaced by Class III of the Town and Country Planning (Use Classes) Order 1972. From 1963 to 1972 the land was occupied by two brothers who carried on business as makers of concrete blocks, installing in a building which stood on part of the land some fairly basic equipment involving a minimum of mechanical power. In 1972 the land was occupied by two brothers who carried orally inquired of officers of the local planning authority whether planning permission was needed to carry on the business of concrete-block making on the land, and were told no. They bought the land, and then set up a larger business than existed previously. In addition to retaining the old equipment they installed more modern machinery in the open; and they increased the number of employees from a maximum of six to about 19, and the production from under 300,000 blocks a year to over 1,200,000. The appearance of the land was altered, and the activities carried on upon it became much more intense. On September 13 1973 the planning authority served an enforcement notice on the appellants reciting a breach of planning control, namely a material change of use to a use for the purpose of the manufacture of concrete blocks and other industrial purposes distinct from those indicated under Class III of the use classes order. The notice required the appellants to discontinue the use and remove the plant and machinery connected with the manufacture.

It was obvious that what the planning authority was complaining about was the more intense activities taking place on the land. Nevertheless it was clearly wrong, having regard to the existing use within Class III, for the authority to require removal of all the appellants' plant and machinery. There was an appeal to the minister, who appointed an inspector to hold an inquiry. The inspector came to the conclusion that a use within Class IV of the order as "a general industrial building for any purpose" had become established on the land by 1972, and that accordingly the concrete-block-making business did not constitute development for which planning permission was required. The minister accepted this conclusion, but went on to hold that the intensification of the activities constituted a material change of use. He therefore upheld the enforcement notice, having made the amendments necessary to permit the established use to continue. The question which arose on that decision was how, since both the pre-1972 and the present use of the land were within Class IV, a material change of use could have taken place. The answer lay in article 2 (3), by which references to a building might include land occupied therewith and "used for the same purposes." By that article land was excluded from a building unless it was attached to and ancillary to the building. The minister took the view that the open land did not acquire the capacity for use as a general industrial building before 1972, since the machinery was then installed indoors and the blocks were made indoors. If that view were right the enforcement notice could not be set aside.

The remaining question of substance was whether the planning authority were estopped from serving the enforcement notice because of the views expressed by their employees to the appellants about the need for planning permission. Any attempt to expand the doctrine of estoppel expressed in *Lever Finance Ltd v Westminster (City) London Borough Council* [1971] 1 QB 222 was to be deprecated, because it was extremely important that planning authority officers should feel free to help applicants without having the shadow of estoppel hanging over their heads and the possibility of their immobilising the authority by some careless remark. In the present case it had not been shown that the appellants gave the officers sufficient information for them to give a precise answer so as to bind the authority. The appeal failed on this ground also, and should be rejected.

MELFORD STEVENSON and CAULFIELD JJ agreed, and the appeal was accordingly dismissed.

QUEEN'S BENCH DIVISIONAL COURT

June 18 1976

(Before Lord WIDGERY CJ, Mr Justice CROOM-JOHNSON and Mr Justice MAY)

REDBRIDGE LONDON BOROUGH v PERRY

Estates Gazette February 12 1977

(1976) 241 EG 457

Enforcement notice—House converted without permission to multiple paying occupation—56 days allowed by notice to rectify change of use—Magistrates in error in holding this period unreasonable as too short to enable landlord to get occupiers out—Enforcement notices are served on occupiers as well as owners—Correct test is what time is reasonably required to clear premises—One period of compliance only for all affected

This was an appeal by case stated from a decision of North East London Area Magistrates dismissing an information laid by the London Borough of Redbridge against Mrs H Perry, the owner of premises known as 23 De Vere Gardens, Ilford, for alleged failure to comply with an enforcement notice concerning the use of the premises.

M B Horton (instructed by L C Alexander, Borough Secretary to the London Borough of Redbridge) appeared for the appellants, and P H Collins (instructed by E Edwards, Son & Noice) represented the respondent.

Giving judgment, LORD WIDGERY said: The information alleged that Mrs Perry was the owner of premises known as 23 De Vere Gardens, Ilford, that an enforcement notice under section 87 of the Town and Country Planning Act 1971 had been served upon her to the effect that within a period of 56 days from the date on which the notice took effect the use of the said premises as a house let in lodgings or for multiple paying occupation should be discontinued, and the information went on to say that on and after February 18 1975 the said Mrs Perry had permitted the use of the said premises as a house let in lodgings or for multiple paying accommodation. That sounds a great deal to swallow. The essence of the charge is easy enough to understand. You have Mrs Perry as the owner of a house. You have the London Borough of Redbridge (the prosecutors below) as the local planning authority. The complaint that they make against Mrs Perry is that she has made a material change in the use of the house 23 De Vere Gardens,

A Ilford, by using it or permitting it to be used as a house let in lodgings. She had broken it up into lodgings whereas previously it was in single occupation.

The facts found in the court below consequent upon that information being laid were that at all material times Mrs Perry was the owner of the house. In March of 1973 work was being carried out in that house to convert it into seven bed-sitting-room units. The local authority, hearing of this, on March 5 1973 sent her a request to stop this work because, as was pointed out by the planning authority, it was in breach of planning control. Mrs Perry does not seem to have been much impressed by letters from the local
B authority because, notwithstanding the timely warning which she received in March 1973, in March 1974 she was still at it and she got another letter from the planning officer saying that he was still awaiting an application for planning permission in respect of this conversion which was going on and saying that the facts would be reported to his committee.

In August 1974 the conversion of the house into seven bed-sitting-room units was complete, and Mrs Perry, having got to that point, as it were, ahead of the council, lost no time in providing occupiers who took over and occupied the seven bed-sitting-room units. Then came the service of the enforcement notice to which I have already briefly
C referred. Mrs Perry, having completed the conversion operation and having let the premises, an enforcement notice was served by the Borough of Redbridge, and it is always important in these cases to see exactly what is complained of. The notice was sent altogether to eight people; seven of them were occupiers of these seven bed-sitting-room units and the other was Mrs Perry herself. The reason why the notice was sent to all those people was no doubt the requirement of section 87 (4) of the Act of 1971 that an enforcement notice shall be served on the owner and on the occupier of the land to which it relates. So
D service appears to be in accordance with the Act. The notice recited that a material change of use had been made by the house being used as a house let in lodgings, and the notice required that that use should be discontinued. In fact the use was not discontinued. Use as seven separate bed-sitting-room units continued and proceedings were ultimately brought by the London Borough of Redbridge against Mrs Perry.

The basis of the proceedings was section 89 of the same Act, which provides in subsection (5) that a continuance of the use forbidden by an enforcement notice would amount to a criminal offence. I will read section 89 (5) because it is
E important in one or two places:

> "Where, by virtue of an enforcement notice, a use of land is required to be discontinued, or any conditions or limitations are required to be complied with in respect of a use of land or in respect of the carrying out of operations thereon, then if any person uses the land or causes or permits it to be used, or carries out those operations or causes or permits them to be carried out, in contravention of the notice, he shall be guilty of an offence. . . ."

So once the enforcement notice has required a use to be discontinued any person who uses or permits the land to be
F used in contravention of the notice commits the offence, and at this stage in the proceedings the Borough of Redbridge are saying to Mrs Perry, "You have permitted this breach. You have received an enforcement notice. You have done nothing about it and we therefore prosecute you for an offence under section 89 (5)."

When the case came before the magistrates I get the impression—I hope I am not doing an injustice to counsel by saying so—that counsel for Mrs Perry was somewhat short of inspiration as to exactly what the defence ought to be. He might have considered an argument that Mrs Perry had not permitted the continuance of the use complained of and I must return to that possibility again later G
on. But he has frankly told us that he did not take that point and he frankly and properly accepts that, that being so, he cannot in this court take that point for the first time. What point did he take? An ingenious one. He reminded himself of section 87 of the Act, which in subsection (6) specifies what an enforcement notice shall say. Among other things the enforcement notice has to specify the period for compliance with the notice, that is to say the period within which the steps to rectify the change of use are required to be taken. In our enforcement notice that period was stated to be 56 days.

There is authority, surprising as it may seem, for the H
proposition that if a case comes before magistrates on a prosecution for failure to comply with an enforcement notice it is open to those magistrates to inquire into whether the period for compliance was a reasonable period or not. I say that is a somewhat surprising situation because one might well have thought this was essentially a matter for the administrative authorities and not for the court, but it was so decided in *Smith v King* (1970) 21 P & CR 560 DC and therefore one cannot now in this court question the propriety of counsel for Mrs Perry raising the issue of whether 56 days was a reasonable period or not. Of course on the face of it
56 days would seem to be more than reasonable. If all that J
had to be done was discontinue the use of these seven bed-sitting units, one would not have thought that any period like 56 days would have been required. But more ingenuity again is shown here by the argument put forward by counsel, because evidence was called before the magistrates to tell them how long it might take to get the occupants of these seven bed-sitting-room units out of possession if they took advantage of every possible statutory right which a tenant now has under the various Landlord and Tenant and Rent Acts. The evidence given before the magistrates was that at a conservative estimate it might take 16 weeks to get
the occupants out of the bed-sitting rooms, and so the K
magistrates said, well, a reasonable period for the compliance with this enforcement notice must be more than 56 days. It requires more than that to get the occupiers out and, clearly recognising that 112 days would have been more to the point, the magistrates dismissed the information and Mrs Perry walked out a free woman.

Where are we now? The matter comes to this court and the first question which has to be considered is whether the justices were right in tailoring the period of compliance to the position of Mrs Perry or whether they should not
have taken a more objective view. In my judgment they L
should have taken a more objective view, and the authority for that proposition is the case of *Johnston and Another v Secretary of State for the Environment and Another* (1974) 28 P & CR 424. This was a case in which a somewhat similar situation arose, namely, that an enforcement notice had been served in respect of a material change of use and the landlord was, as in the instant case, out of possession by virtue of having granted a tenancy. So it was argued in *Johnston's* case that the period of compliance must refer to time which the landlord would take to get the tenant out —exactly the same sort of argument which was used in the
present case. That point is dealt with in the judgment of M
the court at page 429 in a judgment of my own in these terms:

> The third and last point taken by Mr Hames is concerned with the period specified in the enforcement notice for compliance. What the notice says is that the recipient—and for present purposes I am regarding the recipient as Mr Hames' clients—is required within 28 days after the date on which the notice takes effect to discontinue the use of the garages for motor vehicle repairs. The point taken is this: that since the garages in question are let, and, indeed, let to tenants who have a measure of security under the Landlord and Tenant Act 1954,

it is not possible for Mr and Mrs Johnston to comply with the notice and obtain the discontinuance of this use of the garages within 28 days, or indeed within any foreseeable time.

Then, leaving out the next paragraph, I go on:

> I think that this is a misunderstanding of the structure of the enforcement provisions of the Act of 1971 because it should be remembered that enforcement notices are not served only on the owners of the land. They are also served on the occupiers. So far as the occupier is concerned, the notice to which I have referred is intelligible and a direction with which he can comply. All that he has to do is discontinue the use of the garage for the purpose of repairs. When one comes to the landlord and asks oneself what the consequences will be if the use is not discontinued, one finds the answer in section 89 (5) of the Act. . . .

and I have read that section and will not read it again. I stated this conclusion at the bottom of page 429:

> So it is only a person who is in a position to determine the unlawful use and who, notwithstanding the notice, either continues to use the land in defiance of it or causes or permits some other person to do so who is at risk at all.

The points of importance going beyond the scope of this case which arise out of these considerations, I think, are as follows. First of all, there is to be one period of compliance and only one. It has been suggested, although not very seriously, in this case that an enforcement notice can give different periods of compliance to different owners and occupiers to match the difficulty or lack of difficulty which they will have in complying with the notice. That, I am confident, is wrong. The notice must state a single period of compliance and that single period of compliance following *Johnston's* case must be related to the time which would be taken by those in active occupation to correct the particular breach of planning law which is in contemplation. If 56 days is enough for the seven tenants in this case to pack their bags and go, then 56 days is sufficient as a period of compliance, and it does not matter in my judgment that 56 days might be too short a period for Mrs Perry to initiate proceedings to obtain possession, because the compliance period is related not so much to people in possession who have additional problems of that kind but to people who are in possession who are in a position to comply with the notice. So the justices were misled here. They should not have asked themselves how long it would take Mrs Perry to get possession. They should have asked how long was reasonably required to clear the premises. The Act says quite clearly that it must be carried out by the present occupiers and for that operation a period of 56 days in my judgment would be ample. Although it is not a question of law but a question primarily of fact, I would think it right in this case to go thus far as to say that the matter should be sent back to the justices with a direction that for the reasons which I have given a period of 56 days was ample and that the ground upon which they dismissed this information has thereby proved to be unsure, and that they should receive the case back with a direction to convict.

That is really an end of the matter, but I think it necessary for avoidance of confusion in the future if I add one or two words about Mrs Perry's position in regard to "permitting." It will be remembered that she was barred from any argument on "permitting" in this court because she had not taken the point below. Questions may arise hereafter where the point is taken in the court below and where some slight further investigation of the section may be worthwhile. I come back again to section 89 (5), which is really the basis of the decision in *Johnston's* case and look at it again. The offence is committed if these conditions are satisfied: that an enforcement notice has been served, the use has been required to be discontinued and thereafter a person has used or permitted the land to be used in contravention of the notice. In the ordinary way one cannot use land in contravention of a notice before the notice is served, and consequently when considering whether an offence has been committed or not one has to have regard to the fact that section 89 (5) is concerned with activity following the service of the notice, activity in contravention of the notice. A person in Mrs Perry's position, who has been served with the notice and therefore on the face of it is at risk of prosecution, can in my judgment plead that she did not use the land for that purpose or permit the land to be used for that purpose if she had not used it since the enforcement notice was served, nor had she since that notice was served granted a permission to some person to use it in breach of the terms of the notice.

This has proved to be a somewhat difficult and complex case, although it has some important points embedded in it, but at the end of the day the proper order is that the appeal should be allowed and the case should be sent back to the justices with a direction to convict of the offence charged.

CROOM-JOHNSON and MAY JJ agreed, and the appeal was allowed with costs.

QUEEN'S BENCH DIVISIONAL COURT

June 17 1976

(Before Lord WIDGERY CJ, Mr Justice CROOM-JOHNSON and Mr Justice MAY)

DUFFY AND ANOTHER v PILLING

Estates Gazette February 26 1977

(1976) 241 EG 607

Enforcement notice—Alleged conversion of house to multiple paying occupation—Meaning of "multiple paying occupation"—Not merely two or more people living in house—Must be two or more people occupying parts in the sense of "exercising control"—Convictions quashed

This was an appeal by case stated from a decision of Newham Justices on October 8 1975 convicting Lawrence Ruire Duffy and Mrs Doris Banks of the offence of failing to comply with an enforcement notice served by the London Borough of Havering requiring the discontinuance of alleged multiple paying occupation of a house in 165 Hornchurch Road, Hornchurch, Essex.

N B Primost (instructed by Graham Harvey & Co) appeared for the appellants, and B A Payton (instructed by J Head, Borough Secretary to the London Borough of Havering) represented the respondent.

Giving judgment, LORD WIDGERY said: This history of the case takes us back a few years. It concerns a two-storey semi-detached house in a mixed residential area of Hornchurch. At the date of the enforcement notice to which I will come in a moment the premises were owned by Mrs Banks, who was the owner at all material times, and had been let out in bed-sitting rooms. By implication rather than by express provision the bed-sitting rooms had been well equipped with sink and refrigerator and the like, and Mrs Banks was running the house so adapted and which was 165 Hornchurch Road. The local authority were dissatisfied with this use and so they served an enforcement notice on the footing that it had been undertaken without consent and was a material change of use. The language of the enforcement notice is important, at all events so far as we are concerned, for the recital of the infringement of the planning code. It runs thus: "Whereas it appears to the council that the premises known as 165 Hornchurch Road, Hornchurch, shown coloured red on the attached plan marked 'A' within the said London Borough are being used for multiple paying occupation without the grant of consent required in that

A behalf under the Town and Country Planning Acts 1962 to 1968." That notice was served on September 27 1971. Of course it means that there has been a change of use, and that the change of use is from something to multiple paying occupation, and it contains the allegation that that change was a material change requiring consent.

For some time events marked time because Mrs Banks, who was the recipient of the notice, appealed against it to the Secretary of State, and, although her appeal was unsuccessful, the notice was in suspense until the appeal was disposed of and the notice became effective, I think, on May 3 1973 or some date very close to that. The notice B having become effective, the planning authority had to turn their minds to the question of enforcement action and the control of this, as they saw it, unauthorised use. We find Mrs Banks also showing very natural concern for the situation and seeking advice from the local planning experts on planning law. No one would wish to discourage the councils from helping their citizens as much as they can, and I certainly do not wish anything I say in this case to discourage the officers of the council from giving the best service which they can. Unfortunately the officers consulted by Mrs Banks on the meaning of the phrase "multiple paying occupation" C gave her, as though it were the law of the Medes and Persians, what in fact seems to have been no more than a set of guidelines used by council officials concerned with planning. However that may be, she was given advice which was perhaps not of the most useful kind, but again events in this case had taken a turn which renders it unnecessary to take that further. I would pick up the story again at the summer of 1974 when, it will be remembered, the enforcement notice is now effective again and Mrs Banks is concerned about what she is doing in that regard.

By July 25 1974, which is the date chosen in the information as the date of the offence, the cookers and the D refrigerators and the meters which had graced the bed-sitting rooms had all gone. By that time or a little later, and a week or two matters not for this purpose, Mrs Banks had been in touch with a Mr Duffy (the appellant in these proceedings) and, following the advice that she received, she granted Mr Duffy a weekly tenancy of the whole of this house at a rack rent. The intention quite clearly was that Mr Duffy should then rent it and that it should be occupied by a number of people, whom one might perhaps call lodgers if the term is not too accurately construed, but Mr Duffy was to rent the house for the benefit of a number of people who were going to live there. When we come E to July 25 in this connection we find living on the premises Mr Duffy (the appellant) and also a bus driver and four other people, all working men in that part of Hornchurch, and the arrangement under which they worked was that Mr Duffy was paid £10.50 a week for board and accommodation. They had their bedrooms, although sometimes a bedroom was shared, and Mr Duffy had to pay Mrs Banks, the owner, for gas and electricity and various matters of that kind, and also, of course, for the rent which was charged under the lease. At the vital moment, therefore, one gets the picture of the situation in this form. The house is let to Mr Duffy. Mr Duffy is maintaining four other F men—five of them living in the house—and they eat Mr Duffy's food and Mr Duffy is responsible for the organisation of the premises. I have deliberately tried to describe all that in as neutral language as possible.

The point that arose in the court below and arises in this court can be put this way. The planning authority (the respondent in these proceedings) sought to enforce the enforcement notice which had been served as long ago as 1971 and they issued informations against Mrs Banks and Mr Duffy alleging a breach of the requirements of the notice. The defendants in those proceedings, that is Mr Duffy and G Mrs Banks, could not challenge the correctness of the statement contained in the notice that there had been a material change of use. The appeal to the Secretary of State had put an end to any argument on that point. What the prosecution had to show, and what the defendants Mr Duffy and Mrs Banks would themselves endeavour to challenge, was that their conduct on July 25 1974—the date chosen—was in breach of the requirement of the enforcement notice.

So one goes back to the enforcement notice to see what that requirement is and one then asks oneself whether that requirement has been complied with or not. The questions may not be as easy to answer as that, but they are posed as H simply as that and we are not here concerned with construing the Town and Country Planning Act or anything of that kind. We are simply concerned with looking at the enforcement notice, looking at the language used and saying: was the conduct of these people on July 25 a breach of that requirement? What was the requirement? The complaint is that the house is used for multiple paying occupation. We have been told that this a phrase much used in town and country planning circles nowadays, and no doubt there are many versions of its meaning in circulation. We have got to adopt the meaning which we think it ought to have in this J context having regard to the actual language used and the natural meaning of those words. It seems to me it is a phrase which is not by any means easy to give a meaning to. However, one thing is clear and that is that it envisages the use of the house for multiple occupation. I do not know that the word "paying" in the context of this case makes very much difference and I do not propose to say anything about it. But one thing which one can say with confidence is that the condition prohibited involved, among other things no doubt, multiple occupation and so one asks oneself what is the meaning of "multiple occupation" in the K context of this case? What is the meaning which the recipient of the notice should give to it having regard to the meaning of the word?

I again am in no doubt that what is contemplated here is not that there should be two or more people living in the house but that there should be two or more people occupying parts of the house in the common law sense of occupation meaning "exercising control." I think what one has to look for to give this phrase its proper meaning is some kind of physical division of a house so that there are parts in separate occupation. Then I think you can speak of multiple occupation, and if you have such a separation you L can go on and consider what "paying" means should you be unlucky enough to have to reach a conclusion on that as well. I do not say that the phrase will necessarily have this meaning in another enforcement notice, although I suspect it probably will. But as far as this notice is concerned it seems to me that multiple occupation in the sense in which I have defined it is essentially to be proved before the planning authority can say that the enforcement notice has been breached.

Accordingly, since the justices quite obviously did not approach the matter on this basis, I think we have to look at it again, and when one looks at the situation prevailing M on July 25 it is demonstrable, as I hope I have demonstrated, that there were not then two separate occupations of the kind to which I have referred. There was a single occupation and Mr Duffy was the occupier. There seems to be no reason to send this case back for further elucidation, because that, as I see it, is the only possible conclusion consistent with the findings of fact in the case. I think therefore that these convictions were wrongly entered and I would allow the appeal and quash them both.

CROOM-JOHNSON and MAY JJ agreed, and the appeal was allowed with costs.

A

QUEEN'S BENCH DIVISION

May 13 1976

(Before DOUGLAS FRANK QC, sitting as a deputy judge of the division)

TARMAC PROPERTIES LTD v SECRETARY OF STATE FOR WALES AND ANOTHER

Estates Gazette March 12 1977

(1976) 241 EG 765

Application to quash Secretary of State's decision upholding refusal of outline permission for warehouse and distribution centre—No evidence that Secretary of State had
B **disregarded material considerations such as alternative powers to impose conditions or to purchase compulsorily —No failure to take account of a section 52 agreement as no such agreement completed—Application dismissed**

This was an application by Tarmac Properties Ltd under Section 245 of the Town and Country Planning Act 1971, for an order quashing a decision by the Secretary of State for Wales given on November 28 1975 dismissing an appeal against the refusal by the predecessors of the second respondents, Gwent County Council, of planning permission for the development of 90 acres of land near Chepstow.

K C Bagnall QC and Miss K L Megarry (instructed by
C Church Adams Tatham & Co) appeared for the applicant, and H K Woolf (instructed by the Treasury Solicitor) represented the first respondent. The second respondents took no part in the proceedings.

Giving judgment, DOUGLAS FRANK said: The matter arises in the following way. The applicants are the owners of some 90 acres of disused farmland lying immediately to the south of Junction 22 on the M4 motorway just on the Welsh side of the Severn/Wye Road Bridge. In September 1973 the applicants applied for planning permission for "the development of land as a warehouse and distribution centre by the layout of roads, services and landscaping." The application
D was expressed to be an outline application. On November 16 1973 permission was refused by the local planning authority, at that time the Monmouthshire County Council, on 18 grounds. Accordingly, the applicants appealed to the Secretary of State. A three-week-long public inquiry commenced on September 29 1973, held by K G Robbins, an inspector appointed by the Secretary of State. One of the matters which arose at the inquiry concerned the reservation of land so as to provide a possible means of access in the future from the elevated roundabout at Junction 22 through the application site to land to the west which might
E be released for development. During the inquiry the applicants had stated that they were willing to enter into an agreement under section 52 of the Act of 1971. The inspector in his report of the inquiry summarised the applicant's case in this respect in these words:

"186. It was entirely proper that the authority should be concerned to keep growth strategies open, and it was recognised that any development on the appeal site should not be allowed to preclude access from Junction 22 to other land which might, at some future date, be released for development. The appellants were quite prepared to allow access to other land and, if the authority were not satisfied that this could be secured through normal powers of development control, the best way to deal
F with this would be by way of an agreement under section 52 of the Act."

Section 52 of the Act of 1971 provides for agreements which regulate the development or use of land. Under Section 52 (1):

"A local planning authority may enter into an agreement with any person interested in land in their area for the purpose of restricting or regulating the development or use of the land, either permanently or during such period as may be prescribed by the agreement; and any such agreement may contain such incidental and consequential provisions (including provisions of a financial character) as appear to the local planning authority to be necessary or expedient for the purposes of the agreement. G
(2) An agreement made under this section with any person interested in land may be enforced by the local planning authority against persons deriving title under that person in respect of that land. . . ."

In his report the inspector set out 26 findings of fact of which the last was in these terms: "In the event of the development at Rogiet/Caldicot/Portskewett. And although the 22 the range of structural plan options would be reduced. The appellants have suggested that such potential access could be secured by a section 52 agreement." The reference there to the range of structure plan options was to the possible future development of the land to the west which I have H
already mentioned. The inspector in accordance with the usual procedure then went on to state his conclusions of which the following are relevant for the purpose of this case, paragraphs 271 and 272.

"271. As to the fear that the recommendations of the South Gwent Study (and hence, possibly, the policies of the structure plan) would be seriously pre-empted were the appeal to succeed, in my view the present site could be considered independently. It falls within an area divorced from the main Caldicot levels by development at Rogiet/Caldicot/Portskewett. And although the proposed estate is very substantial (compared with a run-of-the-mill application) in the context of the whole study area and the potential Uskmouth port and maritime industry its scale seems J
less impressive. Nevertheless the site is critical because of its relationship to a motorway access point, and in my view it is essential that the range of structure plan options should not be restricted by the non-availability of Junction 22.

"272. In my judgment a section 52 agreement between the parties would be necessary to secure such potential right of access beyond peradventure. Had such a document been forthcoming (as appeared to be the intention at the inquiry) I would have felt able to recommend that permission be granted for development in the terms defined by the appellants, subject to conditions to the following effect,"

and he there sets out the conditions. Then followed his K
recommendation: "In the absence of a section 52 agreement which would safeguard potential access via the site to Junction 22, I recommend that the appeal be dismissed."

Following the inquiry, discussions and correspondence took place between the applicants and the Gwent County Council (who had become the local planning authority) with a view to entering into an agreement under section 52. The inspector was aware of these discussions and caused inquiries to be made on the progress of them, and had delayed the submission of his report until September 29 1975. On November 28 1975 the decision of the Secretary of State was conveyed in a letter to the parties. The letter L
first set out the inspector's conclusions and then went on:

"The inspector recommended that, in the absence of a section 52 agreement, the appeal be dismissed. The Secretary of State agrees generally with the inspector's conclusions and accepts his recommendation. He makes no comment on the question of a section 52 agreement since this, and the details of any conditions which may be imposed on any subsequent issue of planning consent, would be matters for consideration by the local planning authority in the first instance. For the reasons given the Secretary of State hereby dismisses your client's appeal."

I must now return to the discussions which took place concerning the section 52 agreement. The correspondence was M
initiated by letters headed "Without Prejudice" and began with one of June 16 from the applicants to the county council's chief executive officer. I think I should quote some of this correspondence as the applicants attach great importance to it. In the letter of June 16 the applicants said:

"At the planning appeal relating to the above site, it was indicated on behalf of Tarmac Properties Limited that they would be prepared to enter into an agreement under section 52 of the Town and Country Planning Act 1971 to accommodate, if possible, the reasonable requirements of the county council to have access to the land adjoining the appeal site in the event

A of that land being required for development. The requirements were somewhat vague at the inquiry and we would be grateful if you would now submit to us a draft agreement so that we can have this approved on behalf of our clients and then submit it to the inspector and to the Welsh Office."

Then they went on to discuss some matters concerning a possible agreement. The chief executive officer replied on June 19. In his letter he said: "I have asked the county planning officer to let me have details of his requirements for such an agreement, but of course consultations will also have to be carried out with the county surveyor and also Professor Goss. While, therefore, I will do my best to let B you have a draft agreement as soon as possible, I anticipate it may be some weeks before I am able to do so." Then followed some unimportant correspondence. Then on September 5 the chief executive officer wrote enclosing a draft agreement and saying "I must stress, however, that the draft merely represents the personal views of the writer as to what was suggested at the recent planning inquiry. The comments of the county planning officer have yet to be received and further amendments may be required. Also of course, in the event of your client company being successful in its appeal, the approval of the council will be necessary prior to entering into such an agreement." There is then C further correspondence. On November 6 the chief executive officer wrote again and said: "I am taking final instructions as to the draft section 52 agreement and plan, kindly submitted, and hope to be able to confirm that the same is approved in the course of the next few days." On October 10 the applicants had written to the Welsh Office reporting that they had received a draft agreement and were arranging a meeting to resolve matters of detail and said that they would keep the Welsh Office advised of developments. On November 4 the applicants told the chief executive officer that the draft agreement was acceptable to them, and then there was a reply as I have mentioned.

D In the event the draft agreement never got beyond that stage and was not executed. Two points are taken against the Secretary of State, and I will deal with them in the order in which they were argued. First it is said that the Secretary of State had a number of powers which would have enabled him to ensure that part of the application site was reserved so as to be available as an access to the area to the west. It is said that he could have imposed conditions by reasons of the provisions of sections 29 and 36 of the Act of 1971; alternatively that there are powers of compulsory purchase for the provision of roads contained E in section 112 of that Act and section 215 of the Highways Act 1959. Mr Bagnall argues that these powers are material considerations within the meaning of section 29 (1) of the Act of 1971 and that the Secretary of State was bound to have regard to them. He cited a number of authorities in support of the general proposition that the Secretary of State has power to impose conditions reserving land for a future road, and as Mr Woolf did not challenge that proposition it is unnecessary for me to refer to them. I think the right approach to this matter is to ask myself whether there was any evidence that the Secretary of State failed to take a material consideration into account, and if F so whether in the circumstances of this case he should have granted permission. There is no positive evidence that the Secretary of State disregarded his power to impose a condition or overlooked the powers of compulsory purchase. However, it seems to me that these questions were never realistic and are not realistic today. In order for the Secretary of State to reserve part of the application site for a future road he would have needed to have had evidence before him upon which he could determine with precision the boundary of the land to be reserved. I have searched in vain for any such evidence. Mr Bagnall's answer to this difficulty was that the Secretary of State could have made inquiries of the applicants and of the county council which G would have enabled him to delineate the land needed for access. He pointed to the plan attached to the draft agreement showing such land. Whether or not the Secretary of State could have obtained adequate information for that purpose seems to me to be irrelevant, because although he has power to receive further evidence after the close of the inquiry, in my judgment he has no duty to do so. If he has no duty to make inquiries and need apply his mind only to the evidence given before the inspector, then it seems to me that it was impracticable or indeed impossible to impose a suitable condition. In truth it was the applicants' failure to lead evidence necessary for the imposition of a H suitable condition which has made it impossible for the Secretary of State to have imposed one. It seems to me, therefore, that there is no material consideration before the Secretary of State which he failed to take into account.

As to the exercise of compulsory purchase powers, I do not think that this was a matter which he was bound to take into account. He was being asked to give planning permission for a distribution depot, and that permission (which would necessarily have extended to the whole of the application site) would be inconsistent with a concurrent intention that at some future time compulsory powers would J be exercised for the carrying out of some development, namely the construction of a road, which as to part would frustrate the very permission which he purported to grant. Moreover the applicants would be entitled to develop the whole site, which in the event of the road being constructed would commit the highway authority to the payment of compensation appropriate for the taking of land which by then would have formed part of the distribution centre. In any event I do not think that any party contended before the inspector that this matter or problem concerning the reservation of land for access should be dealt with other than by a section 52 agreement. Mr Bagnall says that K paragraph 186 of the report shows that he had submitted that a condition could be imposed. I do not read that paragraph in that way. On the contrary, I think he was putting aside the question of a condition, or indeed, had he had a condition in mind he would have expressed it with sufficient precision to enable the minister to have drafted it for inclusion in his decision. It follows from what I have said that the Secretary of State did not fail to take into account a material consideration, if for no other reason, because there was no way in which he could have taken it into account. Moreover, I think the Secretary of State's approach to this matter was that contended for by the applicants L themselves.

I turn then to the second ground on which the decision is challenged, which is that the Secretary of State acted unreasonably in not taking into account the section 52 agreement. Had the section 52 agreement been executed and transmitted to the Secretary of State before he gave his decision, there might be something in this point. However, as no agreement had then been executed—nor has been yet—I cannot see how it could possibly be said that the Secretary of State acted unreasonably in not taking into account a non-existent agreement. Accordingly, I consider M this point misconceived.

In the circumstances the application is dismissed.

The application was dismissed with costs.

A **QUEEN'S BENCH DIVISION**

January 28 1977

(Before Mr Justice SLYNN)

GENERAL ACCIDENT FIRE & LIFE ASSURANCE CORPORATION LTD v SECRETARY OF STATE FOR THE ENVIRONMENT AND ANOTHER

Estates Gazette March 19 1977

(1977) 241 EG 842

**Listed building—Refusal to consent to demolition—Public inquiry—Appeal to Secretary of State—Criticisms of his
B decision rejected—Natural justice—No failure to take into account material considerations or to disregard immaterial ones—Appeal dismissed**

This was an application under Order 94, Rules of the Supreme Court, by General Accident Fire & Life Assurance Corporation Ltd, who were aggrieved by a decision of the Secretary of State for the Environment dated February 2 1976 on an appeal by the applicants under paragraph 8 of Schedule 11 to the Town and Country Planning Act 1971 against a refusal of listed building consent by Cheltenham Borough Council. The applicants sought an order that the decision of the Secretary of State dismissing the appeal be
C quashed, contending that he had failed to take into account material considerations and had had regard to immaterial ones, had acted contrary to the rules of natural justice in taking into account matters not raised at the inquiry, and had given no clear or adequate reasons for his decision.

George Dobry QC and R J A Carnwath (instructed by Iliffe & Edwards) appeared for the applicants. J Newey QC and D Latham (instructed by the Treasury Solicitor) represented the respondent. Cheltenham Borough Council took no part in the proceedings.

Giving judgment, SLYNN J said the appeal concerned Oriel
D Lodge, Oriel Road, Cheltenham, a building owned by the applicants. In 1973 the building, in the Gothic revival style of the early 19th century, was listed under section 54 of the Town and Country Planning Act 1971. The applicants applied under section 55 for listed building consent to demolish the building. The local planning authority refused, stating that the buildings was of architectural merit and was one of the few remaining examples of its kind in the town. It was within the Cheltenham conservation area. In 1963 the applicants had obtained outline planning permission to erect an office building on an adjoining site to the east and to demolish Oriel Lodge and erect a four-storey office block
E on its site. The first phase, on the land to the east of Oriel Lodge, was completed in 1967, and the applicants occupied that new office block. The applicants now wished to proceed with the second phase, which would provide about 9,300 sq ft of additional floor space. The inspector who conducted the public inquiry reported that Oriel Lodge was a well-preserved produce of is period and worthy of listing under section 54.

Mr Dobry complained that the Secretary of State had adopted conclusions in the inspector's report based on matters which were not explored at the trial, so that the applicants had no opportunity of dealing with them. These
F were the possibility, which in the inspector's view remained, of effecting some "first aid" in architectural or landscaping terms, so that the building could more effectively be integrated into its surroundings, which consisted of more modern buildings, and the further point that, although the office block to the east was designed as the first phase of a larger proposal, it was so designed as to stand as an individual design on its own. Mr Dobry relied on *Charlton Sand & Ballast Co Ltd v Minister of Housing and Local Government* [1964] EGD 386 and *Fairmount Investments Ltd v Secretary of State for the Environment* [1976] 1 WLR 1255, per Viscount Dilhorne at p 1260.

It had long been the law that a person exercising functions G
like those exercised by the Secretary of State must not take into account matters prejudicial to the applicant without giving the latter a chance to deal with them. But his Lordship did not read Viscount Dilhorne's speech as meaning that an inspector could make no comment or express no view without first putting it to the applicant and giving him the opportunity of dealing with it. One must look at the circumstances of the case and see from where the inference had been drawn. There was here no serious challenge to the contention that the building was of historic interest, though the degree of such an interest might well
have been a matter for debate. A major question, if not *the* H
major question, at the inquiry was whether this particular building was so incongruous in relation to other buildings that it would be wrong to preserve it. The inspector came to a number of conclusions relative to the question of incongruity. He clearly thought it a well-preserved product of the period and that it contained features rendering it worthy of preserving in its own right. The question of whether the building really "fitted in" was dealt with by the applicants in their contentions. The applicants had the opportunity to comment on, and lead evidence on, the question of incongruity. The inspector's view was that
there was sufficient open space around both Oriel Lodge and J
two of its near neighbours to enable visual incongruity to be avoided. His Lordship saw no risk of prejudice, since the Secretary of State accepted the inspector's conclusions in general, and he did not believe that there was any real risk that the comments played any real part in the Secretary of State's decision, or were capable of having prejudiced the applicant in any real sense.

Mr Dobry also complained that the assertion that the building to the east of Oriel Lodge was designed intentionally to stand alone was a conclusion of fact by the inspector. But his Lordship did not so read it. It seemed to him that
the inspector recognised that the office block was planned K
as the first phase of a larger proposal, and was clearly, on the appellants' own case, intended to stand alone for some time. His Lordship did not consider that, if the inspector made a comment of this kind on material before him on which everybody had had an opportunity to comment, that would involve any breach of the rules of natural justice.

In his Lordship's view, the inspector was not bound to set out all the facts laid before him, but he was not satisfied that the Secretary of State had left out of account any matter which was material. The inspector's report and the
documents had been considered by the Secretary of State, and L
his Lordship was unable to say that any matter was introduced which might have affected the Secretary of State's mind, but with which the applicants had no chance to deal.

The appeal was dismissed with costs.

QUEEN'S BENCH DIVISIONAL COURT

February 10 1977

(Before Lord WIDGERY CJ, Mr Justice MICHAEL DAVIES and Mr Justice GOFF) M

TYNAN v MELTON MOWBRAY DISTRICT COUNCIL AND ANOTHER

Estates Gazette April 16 1977

(1977) 242 EG 205

Town and Country Planning Act 1971—Motion for certiorari to quash conviction by justices for breach of enforcement notice—Notice of appeal to Secretary of State under section 88 lodged one day out of time—Application refused—Conviction stands

A In these proceedings Richard Tynan, company secretary, Tynan's Cash and Carry, Saxby Road Industrial Estate, Melton Mowbray, moved for an order of certiorari to quash a decision of justices sitting at Melton Mowbray on February 10 1976 whereby they convicted him of breach of an enforcement notice served upon him by the Melton Mowbray District Council and fined him £200.

The applicant appeared in person; Igor Judge (instructed by P J G Herrick, solicitor to the council) appeared for the first respondents, the Melton Mowbray District Council; and H K Woolf (instructed by the Treasury Solicitor) represented the second respondent, the Department of the Environment.

B Giving judgment, LORD WIDGERY said that the enforcement notice, for breach of which the applicant was convicted, complained that there had been a change of use without planning consent in warehouse buildings in Saxby Road and required him to cease using such warehouse premises for the purpose of wholesale and retail sales. The notice having been served, the applicant lodged notice of appeal under section 88 of the Town and Country Planning Act 1971. In due course he received a stereotyped letter from the second respondent informing him that his appeal had been "accepted."

Section 88 of the Act provided that once an appeal had been lodged there was an automatic stay on any enforcement C notice pending outcome of the appeal or until it was withdrawn. The appeal notice was lodged on March 14 1975. The enforcement notice was dated February 12 1975 and took effect on March 13. A perusal of those dates showed that the applicant was one day out of time in lodging his notice of appeal.

In due course the first respondents, who were no doubt ignorant of the appeal application, commenced proceedings alleging breach of the order. When the matter first reached the justices on June 9 they quite properly adjourned the proceedings *sine die* after the applicant had told them of the D appeal application. On August 21 the applicant was told in a letter from the second respondent that his appeal was one day out of time and therefore of no effect. After further adjournments of the proceedings, the justices finally convicted on February 10 1976.

The applicant's contention was that since he was told on June 9 that his appeal had been "accepted" section 88 applied and still applied because there had been no determination or withdrawal of the appeal. He (his lordship) could see how the reference to the appeal being "accepted" was being construed by the applicant, but there was nothing in the Act which required the Secretary of State to reach a E preliminary view on an appeal when lodged. The Secretary of State was acting properly if he stated that an appeal was hopeless and he could then give a final decision. On the other hand he could not give an interlocutory decision. He (his lordship) did not think the letter of June 9 made any difference to the legal position. The position was that if the appeal was valid, then section 88 protection was not taken away. The essential defect in the applicant's notice of appeal was that it was one day out of time during the whole time and deprived the applicant of the protection of section 88. That state of affairs continued until the justices determined the matter as they were entitled to do. The application failed.

F MICHAEL DAVIES and GOFF JJ agreed and the application was dismissed.

G **QUEEN'S BENCH DIVISION**

July 30 1976

(Before Sir DOUGLAS FRANK QC, sitting as a deputy judge of the division)

HERON CORPORATION LTD AND ANOTHER v MANCHESTER CITY COUNCIL

Estates Gazette April 16 1977

(1976) 242 EG 209

Outline permission for large Manchester development granted subject to conditions—Approval of matters reserved under outline grant—Subsequent listing of buildings in the complex under section 54 of the Town and Country Planning Act 1971—New application made for approval of revised plan designed to avoid demolition of listed buildings—Planning authority in error in rejecting new application on grounds of invalidity H

By this summons Heron Corporation Ltd and an associated company, Gerard Securities Ltd, sought a declaration that a second application for approval of matters reserved under an outline planning permission was valid notwithstanding that approval had been given on the first application.

M H Spence (instructed by Saunders, Sobell, Leigh & Dobin) appeared for the plaintiffs, and D G Nowell (instructed by Leslie Boardman, of Manchester, represented the defendants). J

Giving judgment, SIR DOUGLAS FRANK said that on September 20 1972 the plaintiffs made an application for outline planning permission for what, in effect, was to be a comprehensive development comprising offices, shops, hotel, public house, restaurant, car-parking, discotheque, residential accommodation, filling station and entertainment buildings. The land was a site of about 8 acres in the centre of Manchester and there could be no doubt that the scheme was to be one of great importance. There was sent with the application a number of drawings on which were written the words "for illustrative purposes only." There was also sent K a location plan which was substantially varied in part to meet the wishes of the defendants. On September 5 1973 outline planning permission was granted subject to 13 conditions including the following:

(3) Before the development is commenced, the applicant shall submit to the council and obtain their approval under the Town and Country Planning Acts of detailed plans and elevations of the proposals and detailed particulars of the materials to be used in external elevations.

(4) The permission hereby granted relates solely to the erection of buildings for the uses specified in the application and does not L imply approval of the submitted layout which has been prepared for illustrative purposes only. . . .

(5) The permission relates only to the approval in principle of the development of the whole of the application site and in no way implies approval of the development of any part of the site other than as part of an overall phased programme to be approved by the council. The detailed plans shall include the phasing proposed together with details of the timing of each phase.

(6) Before the development is commenced the applicants shall submit to the council and obtain their approval . . . of detailed plans and elevation of the proposals and particulars of materials to be used in the external elevation.

(7) The total gross floor space of all uses in the development M shall not exceed 92,905 sq metres (1,000,000 sq ft).

On December 7 1973 the plaintiffs submitted a form entitled "application for permission to develop land," but which was by common consent an application for the approval of matters reserved under the outline permission, wherein the development for which approval was sought was described as "demolition of existing premises and erection of hotel, etc." The total gross floor area was stated to be about 230,000 sq ft, being the hotel plus car park. Although that application was for that part of the site which was to be used for the hotel, there went with it a phased programme for the development of the whole site showing commencement in September 1974 and completion in June 1980. By a notice dated February 6 1974 the council approved the plans subject to 11 conditions. Soon after that two of the buildings in the area were listed as buildings of special architectural or historic interest in accordance with section 54 of the Town and Country Planning Act 1971. Accordingly, in the absence of written consent by the defendants or the Secretary of State for the Environment, it was an offence to execute any works for their demolition. The construction of the hotel would have required the demolition of one of the buildings. As a result the plaintiffs revised their layout so as to put the hotel on a different part of the site. On August 11 1975 they made a fresh application for approval with details showing the retention of the listed buildings.

The city planning officer refused to entertain the application on three grounds: (1) the area exceeded 1m sq ft; (2) the outline permission involved the demolition of all other buildings; (3) once reserved matters had been approved they could not be revised or varied. These in effect were also the submissions of Mr Nowell for the defendants, although he relied mainly on ground (3). Mr Spence for the plaintiffs submitted that there was nothing in the Act nor in the General Development Order 1973, which regulated applications and grants of planning permission, to suggest that a second application could not be made. Indeed, section 42 (2)(b) of the Act referred to approval on different dates.

In his Lordship's judgment not only could more than one application for approval be made under the same outline planning permission, but in many cases this was inevitable and the present case was one such example. It must have been clearly contemplated, having regard to the fact that it was a common aim that the development should be carried out in phases, that there would be a number of applications. It was true that the local planning authority was *functus officio* when it had given a decision on a particular application and it was also *functus officio* in the present case in respect of the approval dated February 6 1974 in the sense that it could not now change its mind and revoke the approval which it had given or attach further and different conditions to it. This did not, however, prevent more than one application for approval being made under the same outline permission. His Lordship found it impossible to draw a distinction on grounds of validity between an application for one component of a scheme and an application for a different component. It also offended common sense to say that a planning permission in effect became valueless because the listing of a building prevented the implementation of a previously approved scheme. Under the circumstances it followed that the plaintiffs were entitled to the declaration sought.

COURT OF APPEAL

October 6 1976

(Before Lord DENNING MR, Lord Justice STEPHENSON and Lord Justice SHAW)

KENT COUNTY COUNCIL AND ANOTHER v BATCHELOR

Estates Gazette April 23 1977

(1976) 242 EG 287

Action against farmer in respect of tree preservation order—Breach of interlocutory injunction—Judge's sentence of one month's imprisonment for contempt—Bail granted by Court of Appeal and appeal subsequently allowed—Judge's order quashed

This was an appeal by Hubert Dorrington Batchelor, of Thurnham Court, Bearsted, Kent, from an order of Caulfield J, finding him guilty of contempt of court for breach of an interlocutory injunction restraining him from cutting down or damaging any tree on land owned or occupied by him which was the subject of a tree preservation order, and sentencing him to one month's imprisonment. The interlocutory injunction had been granted by Chapman J on the application of the plaintiffs, Kent County Council and Maidstone Borough Council, who had issued a writ claiming an injunction and damages. Caulfield J's order was made on June 25 1976 and on the same day the Court of Appeal granted Mr Batchelor bail on his own recognisance of £500.

George Newman (instructed by Hallett & Co, of Ashford) appeared for Mr Batchelor; Anthony Dinkin (instructed by W G Hopkin, County Hall, Maidstone) represented the local authorities.

Giving judgment, LORD DENNING said that in 1975 Mr Batchelor, a farmer, bought a large mansion house and some land which had been derelict for 15 years. It was in an area of natural beauty between Maidstone and Sittingbourne where there were about 200 acres of woodland. At that time Kent County Council made a tree preservation order under sections 60 and 61 of the Town and Country Planning Act 1971 covering the 200 acres and a great deal of the land bought by Mr Batchelor. He did not seem to have taken much notice of the order, and was said to have cut down a lot of trees in part of the area. He had been served with 54 summonses, which had not yet been heard by the magistrates. Civil proceedings had also been brought in the High Court and Chapman J had made an order for an interlocutory injunction restraining Mr Batchelor from cutting down, uprooting or in any way damaging any trees of named species on his land which were the subject of a tree preservation order.

The court was concerned with a tiny corner in old meadow land. On May 19 1976 operations on the land were observed by a Mr Holt from his garden about 150 yds away. He saw that scrub was being taken away by a bulldozer and that a large pine tree was on the fire. He thought that trees were being uprooted within the area of the tree preservation order. The county council took proceedings to commit Mr Batchelor, and Caulfield J ordered him to be imprisoned for one month for contempt consisting of disobedience to the injunction.

On a review of all the evidence it appeared that Mr Holt, who was clearly an impressive and honest witness, did not have the opportunities of judging whether or not there was a breach of the order. He was looking from 150 yds away and without a map. The only safe way of ascertaining what had occurred was for someone to go to the prohibited area and see whether trees were being grubbed up there. But no one went to the prohibited area for a few days. It was very unsafe on Mr Holt's evidence to say that there had been a breach of the injunction. The bulldozer driver had been given instructions to keep clear of the prohibited area. Much of the cleared matter had come from an area not within the order.

There was a further point. There was an ambiguity in the Act and order, because they did not say what was a "tree" as distinct from saplings, shrubs and bushes, which were not trees. There was no definition of "tree."

His Lordship sympathised a great deal with Kent County Council. There was nothing so important as that, in areas of natural beauty, the trees and the woodland should be preserved. It was only because it seemed to him that an offence against the tree preservation order had not been proved that he would allow the appeal and quash the order of the judge.

STEPHENSON and SHAW LJJ agreed and the order committing Mr Batchelor to prison for contempt was quashed.

QUEEN'S BENCH DIVISIONAL COURT

February 18 1977

(Before Lord WIDGERY CJ, Mr Justice MICHAEL DAVIES and Mr Justice ROBERT GOFF)

NEWBURY DISTRICT COUNCIL v SECRETARY OF STATE FOR THE ENVIRONMENT AND ANOTHER

Estates Gazette April 30 1977

(1977) 242 EG 377

Town and Country Planning Act 1971—Appeal against decision of Secretary of State for the Environment quashing enforcement notices—Alleged failure to comply with conditions of planning consent—Conditions ultra vires—Demolition of existing building required—Planning powers not to be used for an ulterior object, however desirable—Secretary of State's decision correct in law

This was an appeal under section 246 of the Town and Country Planning Act 1971 by the Newbury District Council against a decision of the Secretary of State for the Environment on July 24 1975 quashing two enforcement notices served by the former Hungerford Rural District Council on the ground of a failure by the second respondents, the International Synthetic Rubber Co Ltd, to comply with conditions subject to which planning permission had been granted for the use of buildings at Membury Airfield, near Newbury.

P Boydell QC and R M K Gray (instructed by Sharpe, Pritchard & Co, agents for W J Turner, Newbury District Council) appeared for the appellants; H K Woolf (instructed by the Treasury Solicitor) represented the first respondent, the Secretary of State; and David Widdicombe QC and A Anderson (instructed by Herbert Smith & Co) appeared for the second respondents.

Giving the first judgment, MICHAEL DAVIES J said that the appeal was primarily concerned with two hangars at the former wartime RAF station at Membury. The hangars were described by the inspector who held a local inquiry following the International Synthetic Rubber Co's appeal against the enforcement notices as "large, prominent and ugly." Until 1959 they had been used by the Home Office for the storage of some 162 "mothballed" civil defence vehicles. In October 1959 planning permission was given to a Mr Gilbey for a change of use "to allow for storage of agricultural products." There was a condition that "the buildings shall be removed at the expiration of the period ending December 31 1964 and all works necessary for the reinstatement of the land to its former state and condition shall then be carried out to the satisfaction of the council." Mr Gilbey protested as to the time allowed and the requirement to reinstate the land, and on December 2 1959 a fresh permission was given with the relevant condition reduced to "the buildings shall be removed at the expiration of the period ending December 31 1970."

In 1961 the International Synthetic Rubber Co Ltd (hereinafter called "ISR") became interested in the hangars and they applied for planning permission for "use of the two hangars at present on Membury Airfield as warehouses for the storage of synthetic rubber." In May 1962 planning permission was duly given with the condition that "the buildings shall be removed at the expiration of the period ending December 31 1972." In July 1962 ISR bought the hangars and lease of the land and had used them ever since for the purpose stated above. They did not remove the hangars in December 1972 or at all. In November 1973 enforcement notice A was served. The principal question argued before the court was whether the Secretary of State was right in quashing notice A on the grounds that the condition as to the removal of the buildings was *ultra vires,* invalid and unreasonable.

The power of a planning authority to grant planning permission subject to a condition or conditions was contained in sections 29 and 30 of the Town and Country Planning Act 1971. Certain possible conditions were specifically referred to in section 30, but quite apart from those the planning authority's discretion was virtually unlimited. It was, however, well established that that discretion was subject to some limitations. The case most often looked at was *Pyx Granite Co Ltd v Minister of Housing and Local Government* [1958] 1 QB 554 and the observations of Lord Denning MR at p 572 where he said:

> The principles to be applied are not, I think, in doubt. Although the planning authorities are given very wide powers to impose "such conditions as they think fit," nevertheless the law says that those conditions, to be valid, must fairly and reasonably relate to the permitted development. The planning authority are not at liberty to use their powers for an ulterior object, however desirable that object may seem to them to be in the public interest. If they mistake or misuse their powers, however bona fide, the court can interfere by declaration and injunction.

That approach had been endorsed by the House of Lords in *Fawcett Properties Ltd v Buckinghamshire County Council* [1961] AC 636 in which case Lord Denning in effect pointed out at p 678 that the existence of some limitation upon what would otherwise be the unfettered discretion of a public authority was recognised and explained before the *Pyx Granite* case in *Associated Provincial Picture Houses Ltd v Wednesbury Corporation* [1948] 1 KB 223.

There were a number of factors in the present case which had tempted him (his Lordship) to strain to conclude that the condition was valid and notice A was good. It was plain that, from a planning point of view, the sooner the hangars were removed the better. ISR had not appealed to the Secretary of State when the condition was imposed. ISR had acquired the hangars with their eyes open. But, on reflection, he (his Lordship) was satisfied that the removal of the buildings would involve some expense, would prevent reversion to the previous use and might cause injustice to the freeholder. So he was constrained to conclude that the Secretary of State was right in concluding that the condition, however well intentioned, went too far and was bad. He did not say that a condition as to the removal of a pre-existing building could never be good—for example a permitted development might be likely to be going to cause the building to become a health hazard—and he was content to base his judgment upon "the circumstances of this case," as the Secretary of State put it in his decision letter. In the circumstances of this case the local planning authority did "use their powers for an ulterior object, however desirable."

A second point taken on behalf of ISR was that no planning permission was needed in 1962 for their use of the hangars because they had been used, certainly since 1955, for purposes within Class X of the Town and Country Planning (Use Classes) Order 1950 (now the order of 1972). There was no discernible error of law in the Secretary of State's conclusion that planning permission was needed in 1962 on the ground that the Home Office use was not within Class X. It was submitted that if the Secretary of State were wrong about that then ISR by their conduct in applying for permission in 1962 and subsequently enjoying its benefits gave up any existing use rights. It was said that ISR could not "blow hot and cold" and reliance had been placed on *Prossor v Minister of Housing and Local Government* (1968) 67 LGR 109. On his Lordship's view of the appeal that argument did not arise. The appeal should be dismissed.

Agreeing that the appeal should be dismissed ROBERT GOFF J said that the Secretary of State allowed the appeals by ISR against the enforcement notices on the ground specified in section 88(1)(b) of the Act, namely, that the matters alleged in the notices did not constitute breaches of planning control. So far as enforcement notice A was concerned, the first condition in the 1962 permission was held to be invalid because the Secretary of State considered:

> . . . in the circumstances of this case where planning permission was sought merely for a change of use of existing substantial build-

ings, that a condition requiring the removal of the buildings after the expiration of a specified number of years was not sufficiently related to the change of use in respect of which the planning permission was granted and was unreasonable.

However, the Secretary of State took the view that severance of the condition from the permission would have the effect of altering the character of the permission by making it a permanent permission and he therefore concluded that the 1962 planning permission must be regarded as void. So far as enforcement notice B was concerned, the condition in the 1959 permission was held to be invalid because it was not reasonable, since the building to be extended, a Nissen hut used as an office, was not subject to a comparable condition and no useful purpose would have been served by requiring demolition of part of an office building. The Secretary of State therefore exercised his power to discharge that condition. In the result the Secretary of State directed that the two enforcement notices be quashed. In the appeal before the court it had been recognised that the appeal in respect of enforcement notice B was insignificant and argument had been concentrated on enforcement notice A.

Under section 29(1) of the Act a local planning authority had the power to grant planning permission either unconditionally or subject to such conditions as it thought fit. Section 30(1) provided:

> Without prejudice to the generality of section 29(1) of this Act, conditions may be imposed on the grant of planning permission thereunder: (a) for regulating development or use of any land under the control of the applicant (whether or not it is land in respect of which the application was made) or requiring the carrying out of works on any such land, so far as appears to the local planning authority to be expedient for the purposes of or in connection with the development authorised by the permission. . . .

These provisions were in wide terms, but, as Lord Widgery CJ said in *Kingston upon Thames Royal London Borough Council v Secretary of State for the Environment* [1973] WLR 1549 at 1553:

> It has been recognised for a very long time that they are subject to certain restrictions. The two principal restrictions which the courts have placed upon those words are first that the condition is invalid as being contrary to law unless it is reasonably related to the development in the planning permission which has been granted. It must not be used for an ulterior purpose and must, in the well-known words of Lord Denning MR in *Pyx Granite Co Ltd v Minister of Housing and Local Government* [1958] 1 QB 554 at 572, "fairly and reasonably relate to the permitted development." The second restriction on those words which the courts have adopted in recent years is that a condition which is so clearly unreasonable that no reasonable planning authority could have imposed it may be regarded as *ultra vires* and contrary to law and treated as such in proceedings in this court.

In the case of the 1962 permission the Secretary of State appeared to have regarded the condition as invalid by reason of the first of these two restrictions—that it "was not sufficiently related to the change of use in respect of which the planning permission was granted" (para 8 of the decision letter). However, the first restriction was clearly related to the second. They both derived from the general principle upon which the court would interfere with the exercise of executive discretion, which is at bottom a principle that discretion must be exercised reasonably: *Associated Provincial Picture Houses v Wednesbury Corporation* [1948] 1 KB 223 at 229, per Lord Greene MR, and *Fawcett Properties Ltd v Buckinghamshire County Council* [1961] AC 636 at 679, per Lord Denning. It would be wrong, therefore, to treat Lord Denning's words in *Pyx Granite* as though they were the words of a statute; they were to be applied broadly as embodying a particular aspect of a general principle of reasonableness.

The present case was concerned with the imposition, upon a planning permission for a temporary change of use, of a condition which required demolition of an existing building at the end of the period in question. He (his Lordship) was not prepared to say that any such condition would necessarily contravene Lord Denning's principle. It was possible to conceive of circumstances, for example where the temporary use involved a health hazard, in which it might be justifiable to impose a condition requiring demolition of an existing building after the expiry of the temporary period of use. But such a case would be rare. The aspect of reasonableness which Lord Denning's principle embodied was that the imposition of conditions should be reasonably related to the permitted development which was the subject-matter of the exercise of the discretion; it was not enough that they merely related to the land which was the subject of the development. There might be many objectives which, from a planning point of view, a local planning authority considered desirable, but it was not a proper exercise of its discretion to achieve such objectives by attaching such a condition to a planning permission if the condition did not fairly and reasonably relate to the permitted development. In such circumstances, although acting from no doubt laudable motives, the local planning authority was using its powers for an ulterior purpose, and that was not allowed. A local authority had to use other powers to achieve its objective, for example its powers under section 52 of the 1971 Act to regulate the development or use of land by agreement, though the exercise of such power might cost money.

Applying these principles to the present case he (his Lordship) did not consider that the Secretary of State misdirected himself in law in holding that the condition was *ultra vires*. He (his Lordship) did not see how a condition that the buildings be removed related to the permitted development which was the use of the building as a warehouse. No doubt, from the broad planning point of view, it was desirable that the buildings should be removed, but the local planning authority was not entitled to achieve that result in the present way. In concluding, in the circumstances of the case, that the condition was invalid because it was not sufficiently related to the change of use in respect of which the planning permission was granted, the Secretary of State reached a conclusion which was in law entirely justified. The appeal in respect of notice A should be dismissed.

Although the issue on ISR's alternative submission did not arise he (Robert Goff J) proposed to deal with it in view of the full argument the court had heard upon it. The question on that issue was whether the Secretary of State misdirected himself in law in holding that the Home Office use from 1955 to 1959 was not within Class X of the Use Classes Order 1950, the submission of ISR being that, if it was within Class X, then no planning permission was required by them in 1962. In his Lordship's judgment the Secretary of State did not misdirect himself in that respect. Class X was concerned with "use as a wholesale warehouse or repository for any purpose." Clearly the Home Office use was not use as a wholesale warehouse; the only question was whether it was use as a repository. Having regard to the natural and ordinary meaning of the word and the context in which it was found, a repository did not just mean a store. It meant "a place where goods are stored away, to be kept for the sake of keeping them safe, as part of the storage business"; *G Percy Trentham v Gloucestershire County Council* [1966] 1 WLR 506 at 512, per Lord Denning MR; and *Horwitz v Rowson* [1960] 1 WLR 803 at 810, per Havers J, and *Calcaria Construction Co (York) Ltd v Secretary of State for the Environment* (1974) 72 LGR 398 at 402, per O'Connor J. That was precisely the test which the Secretary of State applied in the present case. Whether the Home Office use was in fact a use as part of the storage business was a question of fact for the Secretary of State to decide; it could not be said that he misdirected himself in law in reaching his conclusion.

Finally, it had been argued on behalf of the appellants that, even if the Home office use was within Class X, nevertheless since the use was Crown use and no planning permission had been granted ISR could not claim that they did not require planning permission in 1962. His Lordship could not accept that submission. Section 266(3) of the 1971 Act provided that "No enforcement notice shall be served under section 87 of this Act in respect of development carried out by or on behalf of the Crown after the appointed day on land which was Crown land at the time when the development was carried out." The subsection provided that no enforcement notice should be served in respect of such a development. It did not provide, as it easily could have done if such had been the intention, that no enforcement notice should be served on the Crown in respect of such development. It followed that if a private citizen subsequently acquired any such land from the Crown he would not have to apply for planning permission in respect of development within the subsection. He would be protected from the service of an enforcement notice by the terms of the subsection, which were explicit. No argument had been addressed to the court why the Secretary of State's decision in respect of enforcement notice B should not stand. The appeal should be dismissed.

LORD WIDGERY CJ agreed and said that he had nothing to add.

The appeal was dismissed and the appellants were ordered to pay costs of the second respondents, the International Synthetic Rubber Co Ltd. Leave to appeal to the Court of Appeal was granted.

COURT OF APPEAL

February 22 1977

(Before Lord Justice CAIRNS, Lord Justice STEPHENSON and Lord Justice WALLER)

LINK HOMES LTD AND ANOTHER v SECRETARY OF STATE FOR THE ENVIRONMENT AND ANOTHER

Estates Gazette May 14 1977

(1977) 242 EG 545

Planning appeal—Refusal to permit residential use of farmland—Proposed development alleged to threaten extension of green belt and to risk prejudicing structure and local plans—Court of Appeal, agreeing with Willis J, upholds dismissal of developers' appeal

This was an appeal by Link Homes Ltd and Ernest Baker from the decision of Willis J on February 26 1976, upholding the Secretary of State's dismissal of appeals from the refusal of Basildon District Council to permit the residential development of portions of agricultural land adjacent to the eastern end of the developed area of Wickford, Essex.

G Dobry QC and C J Lockhart-Mummery (instructed by Norton, Rose, Botterell & Roche) appeared for the appellants, and Michael Mann QC and H K Woolf (instructed by the Treasury Solicitor) represented the respondents.

Giving judgment, CAIRNS LJ said: This is an appeal from a judgment of Willis J refusing to quash a decision of the Secretary of State for the Environment, who had dismissed four appeals from the refusal of the Basildon District Council to grant planning permission for residential development of four areas of land at Shot Farm, Wickford, in the County of Essex. The areas were referred to as (a), (b), (c) and (d). (a) and (d) were of 62 and 77 acres respectively, and (b) and (c) only 7.5 and 2.75 acres respectively. Areas (a) and (b) overlapped, and area (a) included (b), which in its turn included (c). It was all agricultural land adjoining the east side of the developed area of Wickford, and between that township and Rayleigh.

Planning permission was refused by the council on four grounds, of which only two were sustained before the inspector. These were:

(i) The site lies within an area of no notation as shown on the approved county development plan and is to be included within the extensions to the metropolitan green belt, as shown in the review map of the county development plan proposals now with the Secretary of State for consideration. Within these areas new development is not permitted unless required for agricultural purposes or for purposes appropriate to a green belt; (ii) the proposal represents a substantial extension of the Wickford development area into open countryside separating the townships of Wickford and Rayleigh which ought to be protected from further development in order to prevent the eventual coalescence of the townships.

Planning permission for areas (b) and (c) was refused on the same ground numbered (i). On area (d) the council had failed to give a decision in time and the appeal was against such failure. However, out of time, they gave five grounds, of which only one was alive at the date of the inquiry. That was a ground similar to ground (i) to which I have already referred, but with slightly different wording:

The site lies outside any area allocated for residential development in the first review of the county development plan and is within the proposed extension to the metropolitan green belt where a new development is permitted only in the most exceptional circumstances and when required for agricultural or allied purposes.

In substance, therefore, the inspector had only to consider the green belt grounds of refusal. His report set out as usual very fully the cases of the parties, his findings of fact and his conclusions. The conclusions are set out in four paragraphs which I will read:

44. Bearing these facts in mind I am of the opinion that, as the council have withdrawn their objections to the appellants' proposals on both sewage disposal and highway grounds, the principal issue in the determination of the appeals is whether the development of the appeal sites with dwellings would be acceptable in principle having regard to their location outside the development area of Wickford and within the proposed extension to the metropolitan green belt. I have considered whether the appeal sites have been appropriately included in this proposed extension to the green belt, but I find no reason to question their inclusion for the time being, pending decisions on the green belt proposals as a whole and on other planning considerations affecting this part of Essex.

45. It seems to me that this determining issue must be considered in the light of current government policy relating to the determination of residential applications, as expressed in annex A to circular 122/73 and subsequent guidelines. Annex A requires that proposals to develop "white land" within a growth area defined by strategic plans must be considered against the background of a strong general presumption in favour of housing in those parts of the growth area which would clearly fall to be considered for inclusion in any specific growth area plan. This presumption should only be overriden if there are exceptionally compelling planning objections which outweigh the need to make the land available for housing, such as the location of the site within a green belt where it would be unlikely to form part of the growth area when this is further defined. As the sites are within a proposed extension to the metropolitan green belt, and as there do not appear to be any specific plans for the major growth area in which they lie which might require the exclusion from the proposed green belt of the land including the sites, it seems reasonable to assume that the general presumption in favour of housing on land within a growth area does not apply to the appeal sites.

46. If the sites were not within a proposed extension to the green belt, then their development might be considered to be a natural extension of the existing development on the east side of Wickford. Furthermore, as the sites are physically suitable for housing and are immediately available for development, their development could make a material contribution to the demand for houses at Wickford, where the amount of suitable housing land now available is sufficient only to provide about four years' supply.

47. Although the appellants' proposals would only marginally reduce the width of the narrowest part of the gap between Wickford and Rayleigh, the development of the appeal sites and especially

the two larger ones would nevertheless result in a major extension of the urban area of Wickford into the open land to the east, to the detriment of the principal function of the proposed extension to the metropolitan green belt of preventing the expansion of the towns beyond their urban boundaries and their eventual coalescence. Consequently, I am of the opinion that because of their detrimental effects in relation to the function of the proposed green belt extension the appellants' proposals would not be acceptable in principle. Moreover, having regard to the apparent conflict arising from the inclusion of the sites within both a proposed extension to the green belt and a major growth area, the grant of planning permission for the proposals could prejudice the wider future planning of the Wickford area when the proposed structure and local plans are prepared, and so would be premature at the present time.

The Secretary of State in his letter of decision set out those four paragraphs, and concluded, in paragraph 3 of the letter, as follows:

No reason is seen to dissent from the inspector's conclusions on the merits of the appeal proposals. In view of the green belt considerations to which he refers and of the importance which is attached to avoiding any risk of prejudicing the wider future planning of the Wickford area in the context of the preparation of structure and local plans, it is considered that to decide the future of the appeal land at the present time would be premature. The Secretary of State therefore accepts the inspector's recommendation, dismisses all four appeals, and hereby refuses to grant planning permission for the residential development of 77 acres at Shot Farm.

The applicants applied to the court under section 245(1) of the Town and Country Planning Act 1971 to quash the decision, and the application was heard by Willis J, when three points were taken: first, that the inspector and the Secretary of State failed to deal separately with areas (b) and (c); secondly, that paragraphs 46 and 47 of the report were mutually contradictory; and thirdly, that there were no firm facts on which the decision as to prematurity was founded. The learned judge decided against the applicants on all three points. Only the first of them has been pursued in this court. The case for the applicants so far as recorded in the inspector's report did not differentiate between areas (b) and (c), on the one hand, and areas (a) and (d), on the other, except in two sentences in paragraphs 19 and 20 of the report. In the middle of paragraph 19 there are the words: "The development of the two smaller appeal sites would not extend the urban area beyond the present eastern limit of Wickford formed by Fanton Walk." And in paragraph 20: "The development of the two smaller sites would be a form of rounding-off and infilling of the existing Alicia Avenue estate." There is no reason to suppose that either of those statements was in issue. The first was a statement of the obvious if one looked at a plan—and a number of plans were annexed to the report. The second statement really adds nothing to the first. It remained the fact that the sites (b) and (c) were on land proposed for extension of the green belt, and Mr Dobry concedes that there is no distinction to be drawn between such land and land which is already part of the green belt. That being so, and the same considerations applying to all four areas, there was no need for areas (b) and (c) to be dealt with separately in the inspector's conclusions or in the Secretary of State's decision.

Mr Dobry complains that in his judgment the learned judge directed his attention to the question of whether there were special circumstances affecting the sites, and, in particular, affecting areas (b) and (c). That conflict arises in this way. There is a circular no 42/55, which is set out on page 4192 of the *Encyclopedia of Planning*, paragraph 5 of which says: "Inside a green belt, approval should not be given, except in very special circumstances, for the construction of new buildings . . . for purposes other than agriculture, sport, cemeteries, institutions standing in extensive grounds, or other uses appropriate to a rural area." There is another circular, 122/73, paragraph 2 of which says: "Annex A to this circular sets out the new guidelines and should be regarded as supplementing and amplifying the advice given in circular 102/72. The Secretary of State expects authorities to apply the new guidelines in determining planning applications and he will follow them himself in deciding applications which come before him on appeal or at his direction." In annex A, paragraph 4 provides: "In growth areas planning applications must be considered against the background of a strong general presumption in favour of housing. But as a growth area is a general location only, this strong general presumption applies only to those parts of the area that clearly would fall to be considered for inclusion in any specific growth area plan. . . . For those parts the presumption should only be overridden if there are exceptionally compelling planning objections to a development which, in the circumstances of the case, outweigh the need to make the land available for housing. Such objections would, for example, include the fact that the development would seriously prejudice the design of a satisfactory plan for the area as a whole; that the development was proposed on land situated in green belt"—then the words that follow are not relevant. Paragraph 5 provides: "Outside growth areas, white land can be regarded as suitable for housing, providing its development represents a natural extension of existing development and is compatible with its size, character, location and setting, and provided there are no further compelling planning objections (which would of course include those mentioned in paragraph 4 above) . . .", one of which is land situated in green belt. Then the last sentence of that paragraph reads: "The question of need is relevant only to the extent that other objections arise and it has to be decided whether the evidence of need outweighs them."

Paragraph 9 says:

Where it is reasonable for the development to take the form of an expansion of the town, extension of development into the countryside will be justified. But such an extension is objectionable on planning grounds if it entails creating ribbons or isolated pockets of development or reversing accepted policies to separate villages or towns, or if it conflicts with national policies for the protection of the environment such as those for safeguarding green belts. . . . Such objection would normally rule out development unless the circumstances of the case are such that there is an exceptional need to make land available for housing.

The 1973 circular and the annex were undoubtedly relevant and were taken into account in the report and the decision. The learned judge said that both parties and the inspector recognised that both circulars were relevant. But, in fact, says Mr Dobry, he never recognised them as relevant, and though it was mentioned as part of the respondents' case and was referred to as such in the inspector's report, it was not made the basis of the conclusions of the report nor of the decision of the Secretary of State. Since the relevant principle was that a green belt objection would normally rule out development unless there was exceptional need to make land available for housing, this seems to me to be in effect the same as saying that there must be either an exceptional need for housing or some other special circumstances to make it proper to permit building development on green belt land.

Then it is said that the first sentence of paragraph 47 of the report quoted in the decision letter is open to strong criticism. I will read that sentence again. "Although the appellants' proposals would only marginally reduce the width of the narrowest part of the gap between Wickford and Rayleigh, the development of the appeal sites and especially the two larger ones would nevertheless result in a major extension of the urban area of Wickford into the open land to the east, to the detriment of the principal function of the proposed extension to the metropolitan green belt of preventing the expansion of the towns beyond their urban boundaries and

A their eventual coalescence." This, it is contended, leaves the applicants in doubt whether the development of sites (b) and (c) would in view of the inspector and the Secretary of State constitute a major extension of the urban area. I do not consider that any sensible person reading the sentence could be in any doubt of its meaning, namely, that the development of any of the sites would constitute a major extension.

The Town and Country Planning (Inquiries Procedure) Rules 1974, which are to be found at page 3880 of the *Encyclopedia of Planning*, provide in rule 13(1): "The Secretary of State shall notify his decision, and his reasons therefor,
B in writing to the applicant, the local planning authority and the section 29 parties and to any person who, having appeared at the inquiry, has asked to be notified of the decision."

Mr Dobry referred to a number of authorities for the propositions that the reasons must be clear, must leave no substantial doubt in the mind of the applicant, must be in sufficient detail to enable him to know the conclusions on the principal issues, and must be reasonable conclusions on the evidence adduced. I do not find it necessary to refer to the authorities, because I fully accept all those propositions. In my judgment, the reasons that were given passed all those
C tests, and for the reasons that I have given, and for the reasons contained in the judgment of Willis J, with which I wholly agree, I would dismiss this appeal.

STEPHENSON and WALLER LJJ agreed and the appeal was dismissed with costs.

QUEEN'S BENCH DIVISIONAL COURT

March 31 1977

D (Before Lord WIDGERY CJ, Mr Justice EVELEIGH and Mr Justice WIEN)

R v COLCHESTER JUSTICES EX PARTE NORTH ESSEX BUILDING CO LTD

Estates Gazette May 28 1977

(1977) 242 EG 696

Town and Country Planning Act 1971—Alleged unlawful demolition of listed building—Committal to Crown Court —Magistrates' Courts Act 1952—Bench told of previous conviction—Committal order quashed—Mandamus to justices to hear and determine

E In these proceedings the North Essex Building Co Ltd, estate developers and building contractors, of Anglian House, Victoria Place, Brightlingsea, Essex, applied for an order of certiorari to quash an order of Colchester Justices dated January 29 1976. By that order the applicants were committed for trial at the Crown Court on an information preferred against them by Tendring District Council under section 55 of the Town and Country Planning Act 1971 alleging that without lawful authority from the Secretary of State for the Environment they demolished a listed property at Victoria Place, Brightlingsea. The applicants also sought an order of mandamus directed to the justices requiring them to hear
F and determine the information.

Anthony Scrivener QC and P Surrey (instructed by John Fowler, Oldman & Co, of Colchester) appeared for the applicants; Raymond Sears QC and J Holt (instructed by the legal adviser to Tendring District Council) represented the respondent local authority.

Giving the first judgment EVELEIGH J said that the ground for the application was that the justices, who had begun to inquire into the information under section 18(1) of the Magistrates' Courts Act 1952, considered, at the invitation of the defence, whether or not to proceed to summary trial. In the course of that consideration, the justices heard the prosecution give evidence of a previous conviction of the defendants. The justices went on to hear further evidence and then committed the respondents to the Crown Court for trial. G

It was the general policy that in cases before justices no information concerning previous convictions should be disclosed in any form before trial. In the present case the previous conviction was disclosed. The reason for that, according to an affidavit filed on behalf of the justices, was that under the provisions of Schedule 2 to the Magistrates' Courts Act, paragraph 7, the provisions relating to committal for sentence did not apply to corporations. In the case of individuals justices had power to try summarily and, after conviction, to commit for sentence under section 29 if they H
felt a greater sentence should be imposed than they had power to impose. When exercising their powers under section 29 they could hear about character and antecedents. Section 29 did not apply to corporations, but the justices had taken the view that it was right for them to be informed of any previous conviction which they could consider in deciding whether or not to commit a corporation for trial.

The court was concerned with the general principle. Justices, when asked to consider committing for trial, should not be told of any previous conviction. The Magistrates' Courts Act firmly stated that section 29 did not apply to corporations and there was no reason why the general J
principle preventing disclosure of previous convictions before trial should be ignored. The applications should be allowed.

LORD WIDGERY and WIEN J agreed that the applications should be allowed. Accordingly, it was ordered that certiorari should go to quash the committal and that mandamus should be granted to order the justices to hear and determine the information according to law. The applicants were awarded costs out of central funds.

K

QUEEN'S BENCH DIVISIONAL COURT

February 8 1977

(Before Lord WIDGERY CJ, Mr Justice MICHAEL DAVIES and Mr Justice GOFF)

ROBERTS AND ANOTHER v VALE ROYAL DISTRICT COUNCIL AND ANOTHER

Estates Gazette June 4 1977

(1977) 242 EG 811

Town and Country Planning Act 1971—Condition attached to permission to tip waste materials—Materials to be approved before tipping—Whether condition void—Tipping a "material change of use"—"Reserved matters" unusual in permissions for change of use—Condition nevertheless valid —Appeal from Secretary of State's decision upholding enforcement notice dismissed L

This was an appeal by R A Roberts and H T Roberts, partners in a quarry firm who worked Manley Quarry, Cheshire, from a decision by the second respondent, the Secretary of State for the Environment, upholding an enforcement notice served by the first respondents, the Vale Royal District Council, requiring the appellants to remove materials which, it was alleged, had been tipped in contravention of a M
condition attached to a grant of planning permission.

Alistair Dawson QC and R J Furber (instructed by Thompson, Quarrell & Megaw, agents for Oliver & Co, of Ellesmere Port, Cheshire) appeared for the appellants; H K Woolf (instructed by the Treasury Solicitor) represented the second respondent; the first respondents took no part in the proceedings.

Giving judgment, LORD WIDGERY said: This is an appeal under section 246 of the Town and Country Planning Act 1971 by Robert Alfred Roberts and Harold Thomas Roberts against an enforcement notice under the Act served by the

A first respondents, the Vale Royal District Council, upon the appellants as partners in a firm controlling a quarry known as Manley Quarry at Manley in Cheshire.

Permission under the planning Acts for the tipping of refuse into this quarry was given by a permission, dated March 2 1960. The permission extends to the proposed tipping of industrial and builders' waste material at Manley Quarry and grants permission subject to conditions, the second condition being: "The materials to be tipped in this area shall be approved by the local planning authority before tipping commences." As required by law, the council give their reason for imposing the condition, and they put it in these terms: B "To prevent nuisance arising from the tipping operations." All seems to have gone on happily under that permission for a number of years, but the matter has now become the subject of challenge because the enforcement notice alleges that materials have been tipped into the quarry which are of a kind other than those approved by the local planning authority. On September 10 1974 we get the enforcement notice which gives particulars of the condition not complied with, and goes on to specify the acts required to restore the position as it was prior to the unauthorised development taking place.

C It is a peculiarity of modern town planning legislation that, although the deposit of refuse or waste materials on land can be development with all the important consequences that flow from the Town and Country Planning Act 1971, the nature of the development which is created by the deposit of refuse or waste materials is not described as an operation, be it an engineering operation or a waste disposal operation or whatever. Instead it is described in section 22(3) of the Act as a material change of use. Thus one has the conception that if waste material is to be tipped, permission from the planning authority is required on the footing that the land affected will be subject to a material change of use.

D There cannot be many instances in which a material change of use gives rise to what in planning jargon are called "reserved matters," in other words gives rise to decisions which have to be postponed and given at a future time. By contrast, the type of development which comes under the description of building development and engineering operations quite frequently leaves for the future certain of the detailed decisions which are involved in that planning application. If one wants particular examples of that, there is section 42 which deals with outline applications and section 24(5) which deals with conditions affecting the exterior of buildings. These and many other provisions are examples of E the Act accepting that part of the decision giving permission to planning development must and should be postponed and dealt with after the main outlines have been settled.

Mr Dawson, appearing for the appellants today, has drawn our attention to the fact that, although for other types of development, namely, the operations type, there are these provisions whereby matters can be reserved and dealt with later, there are no such provisions in regard to development springing from a material change of use. It may be, as I have already said, that the explanation of that is that development consisting in a material change of use does not attract this kind of reserved decision. But it is the case that so far as that F type of development is concerned no provision has been made for the detailed administration of a reserved matter. One finds, as Mr Dawson says, that, if having been given a planning permission with reserved matters, you want to put in an application to have permission granted for the reserved matter, or if you want to have a decision by a particular time, or if you want to appeal against the failure of the local authority to give you an answer on a reserved matter, you may well find you will be in difficulty because of the absence of these provisions. From that he asks us to infer that, as a matter of construction of the Act, it cannot have been intended that a condition of this kind with reserved matters should apply to change of use developments. G Accordingly he asks us to say that this condition was void because it was postponing the final decision until a date when the materials were tipped, and he says that if he can persuade us that the permission is void, that means that the enforcement notice is bad. If we get to that point, of course we must send the case back to the Secretary of State to think again.

For my part I do not draw the fundamental conclusion which is basic to Mr Dawson's argument. I do not infer as a matter of construction of the Act or otherwise that change of use developments are all that different from operation developments. Certainly I do not find any ground for saying that the condition was void on the basis relied upon in H argument. If that condition is not void, the rest of Mr Dawson's argument tumbles to the ground. I would dismiss this appeal.

MICHAEL DAVIES and GOFF JJ agreed and the appeal was dismissed with costs. Leave to appeal to the Court of Appeal was refused.

QUEEN'S BENCH DIVISIONAL COURT J

May 2 1977

(Before Lord WIDGERY CJ, Mr Justice PARK and Mr Justice PETER PAIN)

MONOMART (WAREHOUSES) LTD AND OTHERS v SECRETARY OF STATE FOR THE ENVIRONMENT AND ANOTHER

Estates Gazette June 11 1977

(1977) 242 EG 881

Town and Country Planning Act 1971—Appeal from decision of Secretary of State upholding enforcement notice—Meaning of "builders' merchants' warehouse"—Use of K **warehouse as retail "do-it-yourself" supermarket not permitted—Appeal dismissed**

This was an appeal by Monomart (Warehouses) Ltd, Wetton Bros Ltd, Clapton Building Supplies Ltd and Bosskon Ltd under section 246 of the Town and Country Planning Act 1971 from a decision of the Secretary of State for the Environment, upholding with minor amendments an enforcement notice served upon them by the Bristol District Council requiring them to discontinue the use of a warehouse in Channons Hill, Fishponds, for the purpose of retail sales other than such sales as were incidental to the use of the building as a builders' merchants' warehouse. L

W J Glover QC and C S Fay (instructed by Stanleys & Simpson, North, agents for Stanley, Wasbrough & Co, of Bristol) appeared for the appellants; H K Woolf (instructed by the Treasury Solicitor) represented the first respondent; the second respondents, Bristol District Council, were not represented and took no part in the appeal.

Giving judgment LORD WIDGERY said that by virtue of planning consent the appellants erected four warehouses on the appeal site, one warehouse being described in the planning consent as a "builders' merchants' warehouse." In no time at all after the completion of the warehouses, the building described as a "builders' merchants' warehouse" began M being used as a "do-it-yourself" supermarket with 95 per cent of the trade being retail to the general public. The local authority did not want that. The development generated an enormous amount of traffic, as people used their cars to visit the building, and on April 2 1973 the local authority served an enforcement notice. From that notice the appellants appealed to the Secretary of State, who dismissed it.

The appellants now contended that, on a proper construction of the original planning consent and in all the circumstances, the consent covered the activities now complained of. Put in another way, the argument was that "builders'

A merchants' warehouse" had acquired a special meaning and covered the retail sale of "do-it-yourself" materials. The Secretary of State had been right to reject that contention. The word "warehouse" had a well-known meaning. It meant a place where goods were stored before being taken elsewhere for sale or disposal. It was not a place where retail sales were carried out. The addition of the words "builders' merchants" to the word "warehouse" did nothing more than describe the type of goods to be warehoused.

The Secretary of State, in dismissing the appeal, had amended the notice so as to allow retail sale of materials ancillary to the use of the premises as a builders' merchants'
B warehouse. It was not a notice which required amendment. The notice, without any reference to retail sales ancillary to the use of the building as a builders' merchants' warehouse, was a valid notice. The original notice banned the use of the building for retail sales in excess of retail sales permissible as being incidental to the main use. The appeal should be dismissed.

Park J and Peter Pain J agreed that the appeal should be dismissed.

The appeal was dismissed with costs. The appellants were given leave to appeal to the Court of Appeal.

C

QUEEN'S BENCH DIVISIONAL COURT

November 1 1976

(Before Lord WIDGERY CJ, Mr Justice EVELEIGH and Mr Justice SLYNN)

LIPSON AND ANOTHER v SECRETARY OF STATE FOR THE ENVIRONMENT AND ANOTHER

Estates Gazette June 25 1977

(1976) 242 EG 1049

D **Town and Country Planning Act 1971—Appeal from enforcement notices—Houses alleged to be in multiple paying occupation—Apt description of the situation—Condition attached to notices requiring restoration to original condition—Secretary of State's decision valid—Appeal dismissed —Observations on meaning of multiple paying occupation**

This was an appeal by A Lipson and E Lipson from a decision of the Secretary of State for the Environment dated January 17 1976, whereby he upheld nine enforcement notices served by the Salford City Council upon the appellants requiring discontinuance of the use of the appellants'
E properties for multiple paying occupation and the restoration of the properties to their original condition.

A Rumbelow (instructed by Betesh & Co, of Manchester) appeared for the appellants; Harry Woolf (instructed by the Treasury Solicitor) represented the Secretary of State. The second respondents, the Salford City Council, were not represented and took no part in the appeal.

Giving judgment, Lord Widgery CJ said: This is an appeal under section 246 of the Town and Country Planning Act 1971, brought by Mr Lipson against the decision of the Secretary of State for the Environment in respect of an appeal against certain enforcement notices served under the
F same Act in respect of a variety of properties in Salford. There were originally 10 enforcement notices. Of those, no 10 has gone out of consideration as far as we are concerned because the parties, we are told, have made an arrangement about it, so we are left, for all practical purposes, with the remaining nine notices.

I go straight to the Secretary of State's decision letter for the relevant facts upon which his conclusions are based. Paragraph 5 says this: "The inspector found as facts, which are accepted, that each of the appeal properties"—that is the nine remaining—"contained two or more bed-sitting rooms and that at present there were seven separate lettings,
all bed-sitting rooms, in 240 Great Clowes Street, with one bathroom and wc; seven lettings in 250 Great Clowes Street, with two bathrooms and three wcs; five lettings on the first and second floors of 273 Great Clowes Street, with two bathrooms and wcs; and three lettings, all bed-sitting rooms, on the ground floor (15 The Priory), with one bathroom and wc; five lettings in 275 Great Clowes Street, with one bathroom and wc; five lettings in 285 Great Clowes Street, including one self-contained unit, and one other bathroom and wc; five lettings in 287 Great Clowes Street, with one bathroom and wc"—and so it goes on. I do not think I need burden the judgment with greater detail, save to say G
that in the view of both the planning authority and the Secretary of State that user of these premises justifies their description as premises in multiple paying occupation. H

These houses, as the Secretary of State said, had two or three storeys when originally built. No doubt they were originally single family occupations, possibly passing through the stage of self-contained flats, but the situation when they first came to the notice of the planning authority was as I described. The planning authority acknowledge that all this had happened since 1963, and served the enforcement notices to which I have referred. I look at one just to see the relevance of it. I take one which was served on Mr
Middleton in respect of 250 Great Clowes Street. It recites that he is the occupier of the land at that address, it recites that the council of the City of Salford are the local planning authority, and then comes the important recital: "It appears to the council that after December 31 1963 there has been a breach of planning control in that the said land has been developed by the making of a material change in the use (thereof) to a use for the purpose of multiple paying occupation." Then one goes to the operative part of the notice where the council, in exercise of their power contained in the Act, require the recipient of the notice, within the period specified, "to discontinue the use of the said land for the J
purpose of multiple paying occupation and to restore the use thereof to that of a single dwelling-house or single household." That enforcement notice is, of course, based on sections 87 and 88 of the Act of 1971, and of particular relevance is subsection (1)(d) of section 88, dealing with the time within which enforcement action should be taken. This prevents enforcement where it is shown "in the case of a notice not falling within paragraph (c) of this subsection, that the breach of planning control alleged by the notice occurred before the beginning of 1964." K

However, the first question which we have to decide, and
I think we have to decide on the facts before the law was amended, is whether the enforcement notice, which I have read out, aptly describes the factual situation found to exist at the time it was served and I am bound to say that I cannot imagine language more suitable to explain that situation than the phrase "multiple paying occupation." That does not mean it is the only possible way of describing such a situation. That does not mean the phrase has anything inherent in it, except it appeals to me as being a very good set of words to describe the picture which is found. Some support for that is to be obtained from the case of *Duffy and Banks v Pilling* which was heard in this court on June L
17 1976 [(1977) 241 EG 607] and was also concerned with the phrase "multiple paying occupation." In *Duffy's* case there were five men living together in, as it were, a sort of mess. Duffy paid the rent and did the buying and I think he did the cooking as well. The others were lorry drivers, and working men of that kind, and Duffy charged them so much per week and they all had, I have no doubt, a happy time in the house. In the course of argument in this court the question arose as to whether or not that was properly to be described as multiple paying occupation. We took the view, and the relevant judgment is my own, that it was not a multiple paying occupation. The contrast which I sought to M

A draw, at page 6 of the transcript, is this. I said: "I again am in no doubt that what is contemplated here is not that there should be two or more people living in the house but that there should be two or more people occupying parts of the house in the common law sense of occupation meaning 'exercising control.'"

Now that exactly covered what I was seeking to say in *Duffy*, but I wish at once to say that that is not intended, in every similar case hereafter, as an authoritative proposition that there must be a situation of common law occupancy in order to meet the meaning of those words. But the phrase "multiple paying occupation" is not a statutory phrase. It
B is a phrase which has come into acceptance in the practice of town and country planning, and it does not necessarily have one single inexorable meaning in all the contexts in which it is used. Every time it is used you must look at it in its context and discover what its meaning is. Notwithstanding anything in *Duffy* I have no hesitation in saying that the phrase "multiple paying occupation" in this case was totally apt to deal with a situation which, as I read from the decision letter, the inspector found. It will be apparent that the question of dates is very important. If all these changes had occurred before 1964 nothing could have been done about
C it. If the changes had occurred after 1963, then the planning authority was in a position to take action. In such cases the onus is upon an occupier to show, if he can, that the change occurred before 1964.

The way in which the Secretary of State dealt with this point to be found in paragraph 7 of his letter, where he says this:

> It appears from the evidence that, in 1963 [*a number of houses which he lists*] were rated as three flats, two flats and three flats respectively. These flats appear to have been self-contained ones and consequently it is not considered that their use constituted multiple paying occupation of the respective premises. While the
> D evidence of the rating valuation lists is not necessarily regarded as conclusive it is considered that your clients have not discharged the onus of proof which lies with an appellant that any of the appeal properties were in multiple paying occupation before the beginning of 1964. The appeals therefore fail on ground (d).

I do not agree, for the reasons I have just given, with the proposition that self-contained flats necessarily exclude multiple paying occupation. The point is not really open to analysis in this case any longer because the Secretary of State, I think, in the end has decided this point on the onus of proof, and in any case, for reasons which I have made clear, it seems to me only too obvious that in this case the phrase "multiple
E paying occupation" is a perfectly good phrase to use. Therefore, in so far as an attack is made on the description of the development area, I would reject it. The real attack, I think, comes in the next stage. The Secretary of State took the view—and I think quite rightly—that assuming all this happened after 1963 it was still not appropriate for the planning authority to seek to require the occupier of the land to resume a particular use of the land. The point is quite important because section 87 makes it quite plain that the enforcement notice can require the landowner to restore the land to its F
condition before the development took place. There is nothing in the section which says that the landowner shall apply a positive use to it. The notice, it will be remembered—or one of them at all events—had such a requirement. The Secretary of State dealt with that by removing from the notice any requirement which went beyond the restoration of the condition of the premises. The point taken by Mr Rumbelow is that it is not possible to carry out this requirement because he says nobody knows what the condition of the premises was in 1963. I do not quite know why he says that. I asked him if there was a finding of fact to that effect, and he said, No. In the ordinary way, if you ask, and if you G
are painstaking, you can find out something about the condition of a building five, six or 10 years ago. In any case, it does not seem to me that the requirement of the notice to restore the condition of the premises to that which it was before the development took place is bad, simply because, if it be the fact, the landowner says, "I do not know what this involved." If he does not know what it involved, presumably nor does anyone else, and if nobody else knows what it involved he cannot be expected to observe the requirements of the notice. In other words, he will have an effective defence in any proceedings brought against him.

H
There is a whisper of a similar consideration in the case of *Trevors Warehouse v Secretary of State for the Environment* (1972) 23 P & CR 215. That was a case where the local planning authority tried to preserve, on service of an enforcement notice, a certain limited use that they regarded as not being capable of restraint. Criticism was made of the language used to describe this, and it was said it was not effective, it was ambiguous and vague, and unenforceable. In the short judgment I delivered I pointed out that in this kind of situation a local authority cannot do better than its best, and if the language is somewhat obscure, somewhat ambiguous, then to some extent that must be accepted. It is J
only when it is quite evident that better alternatives were available that one could make any kind of progress in this kind of argument.

For those reasons it seems to me that this case, complicated as it has become, really resolves itself into two simple points, the first being whether the notice effectively described the development, and I think it did: and the second, whether the requirement of the amended notice was in any way a breach of the law, and I think it was not. For these reasons I would dismiss this appeal.

EVELEIGH J: I agree. The difficulty in this case seems to me to have arisen by attempting to argue that *Duffy's* case K
gives an authoritative and exclusive meaning to the phrase "multiple paying occupation." Those words are no more than ordinary words to afford particulars of the breach alleged. I agree, for the reasons stated by my Lord, that they cover this particular case and that for the other reasons given the appeal must be dismissed.

SLYNN J also agreed and the appeal was dismissed with costs.

VENDOR AND PURCHASER

COURT OF APPEAL

February 13 1976

(Before Lord Justice BUCKLEY, Lord Justice ORR and Sir JOHN PENNYCUICK)

CAPITAL & SUBURBAN PROPERTIES LTD v SWYCHER AND OTHERS

Estates Gazette January 15 1977

(1976) 241 EG 153

Vendor and purchaser—A vendor who elects to pursue specific performance cannot, on continued default by the purchaser, claim on the basis of non-compliance with a preceding notice to complete

This was an appeal by Capital & Suburban Properties Ltd from an order of Foster J in the Chancery Division on November 18 1975 in so far as it declared that the respondents, Sonia Swycher, Dennis David Swycher, Frank Joseph Kershaw and Adrienne Shaw Kershaw, owners of Savile House, 74-90 Savile Street, Sheffield, were entitled, should the property be resold before November 12 1976 for less than the price contracted to be paid for it by the appellants, to receive the difference (after giving credit for the deposit) as well as the expenses of resale.

M Miller QC and M Reynolds (instructed by Stowe & Co) appeared for the appellants, and D J Nicholls QC and M Hart (instructed by Bower, Cotton & Bower, agents for Kershaw, Tudor & Co, of Sheffield) represented the respondents.

Giving a reserved judgment, Buckley LJ said that on March 25 1974 the respondent vendors agreed to sell Savile House to the appellants, the sale to be subject to the Law Society's General Conditions of Sale (1973). The purchase price was £150,000 with certain additional sums as and when the property should be substantially let. The date fixed for completion was July 25 1974. On August 16 1974 the vendors served notice to complete expiring on September 17 1974. The purchasers did not complete by that date, but on September 12 issued the writ in the action. On March 6 1974, not long before the contract of sale, Sheffield Metropolitan District Council passed a resolution under section 17 of the General Rate Act 1967 taking power to impose a general rate on unoccupied property. The vendors did not know of this at the date of the agreement, the resolution not being gazetted until April 1 1974. The purchasers' writ alleged that the vendors had by an agent falsely represented that the local authority had passed no such resolution, and that the purchasers had contracted in reliance on that representation. They claimed specific performance with an abatement of the price by £27,000. The vendors counterclaimed specific performance and other relief, and applied for summary judgment. An order was made for completion on November 12 1975 on terms as to the amount in dispute. When the purchasers failed to comply with that order, Foster J directed that the deposit be forfeited, that the vendors be at liberty to resell, and that should the property be resold before November 12 1976 for less than the contract price the purchasers pay the difference, after crediting the deposit, together with the costs of the resale. The purchasers appealed from the order as to payment of any difference arising on resale.

It was unfortunate that the judge had given no reasons for his decision. That was a most unsatisfactory practice. The order evidently treated general condition 19 (4) of the Law Society's form as applicable to the circumstances of the case as if the "date fixed for completion" for the condition's purposes were November 12 1975, the date fixed under the court's order of November 5 1975. The condition, so far as relevant, provided that on non-compliance by the purchaser with the terms of an effective notice to complete, the vendor might forfeit the deposit and resell the property, and if on resale within one year from the date fixed for completion the vendor incurred a loss, the purchaser should be liable for that loss as liquidated damages, including costs and expenses of the resale but subject to credit for the amount of the deposit. The argument for the purchasers was that by counterclaiming specific performance and electing to pursue that remedy, the vendors had abandoned their right to proceed on the basis of a repudiation. He (his Lordship) thought it would be useful for him to state explicitly his own view of the principles involved. Where time had been made of the essence of a contract and the purchaser failed to complete within that time, the vendor might (a) treat the purchaser as having repudiated the contract and sue for damages, or (b) seek specific performance. He might seek both remedies in one writ, but must elect at the trial which road he proposed to travel. If he chose specific performance, and the purchaser still failed to complete, he could forfeit the deposit and claim any other monetary relief that might be available to him under the terms of the contract, and it was not in dispute that he could, if he chose, seek an order of the court rescinding the contract at this stage. Here the word "rescind" did not import treating the contract as though it had never been made; the rescission order would mean that the vendor was entitled to retain the land in consequence of the purchaser's repudiation. If the vendor took this course, he could not also seek damages. Rescission was useful to a vendor where the land had risen in value, but could not benefit him otherwise; there would therefore be no virtue in allowing him both to rescind and to claim damages. Foster J's order should be amended by deletion of the declaration relating to resale on or before November 12 1976 for less than the purchase-price.

Sir John Pennycuick gave a concurring judgment, and Orr LJ agreed with both the judgments delivered. An order was made in the terms proposed by Buckley LJ, and the purchasers were awarded the costs of the appeal. Leave to appeal to the House of Lords was refused.

CHANCERY DIVISION

June 9 1976

(Before Mr Justice SLADE)

JANMOHAMED v HASSAM

Estates Gazette February 26 1977

(1976) 241 EG 609

Oral agreement for sale of land—Term that purchaser should obtain mortgage "satisfactory to himself" within one month held not to be void for uncertainty—Purchaser entitled to return of deposit and damages for breach of contract

This was a motion for judgment in default of defence, the defence having been struck out, in an action by Bahadurali Janmohamed claiming the return of his deposit from Tajdih Hassam pursuant to an oral agreement for the sale of a freehold property in Sutton Court Road, Chiswick, London.

R Reid (instructed by L Silkin & Partners) appeared for the plaintiff; Mr Hassam did not appear and was not represented.

Giving judgment, SLADE J said that the defence had been struck out on order of the master made on January 6 1976, the defendant, Mr Hassam, having failed to comply with an order for discovery. The statement of claim asserted an oral agreement made on or about June 16 1971, by which Mr Hassam agreed to sell to Mr Janmohamed a freehold property in Sutton Court Road, Chiswick, for £11,000. The agreement was subject to the condition that if Mr Janmohamed did not obtain an offer of a mortgage on terms satisfactory to himself within one month the contract should thereupon be rescinded. On June 27 1971 pursuant to the contract Mr Janmohamed paid by cheque a deposit of £1,000 to Mr Hassam. Mr Janmohamed obtained a mortgage on terms satisfactory to himself within one month. He alleged that in breach of the agreement Mr Hassam had sold the property elsewhere for approximately £13,500. He claimed that he was entitled to the return of the deposit and to damages for loss of bargain, amounting to £2,500. Alternatively, it was said, the agreement was too vague to amount in law to a binding contract, so that Mr Janmohamed was entitled to the return of his deposit.

The first relief sought was the return of the deposit with interest at 9 per cent. He (his Lordship) thought that the appropriate rate was 8 per cent, but subject to that Mr Janmohamed was entitled to that relief whether or not the contract was void and unenforceable.

The second relief sought was an inquiry into damages suffered because of the breach, certification of the amount, and payment of the amount certified. That claim raised in a direct form the question whether the plaintiff had established the existence of a binding contract, or whether the existence of a binding contract was negatived by the condition that if Mr Janmohamed did not obtain an offer of a mortgage on terms satisfactory to himself within one month the contract should thereupon be rescinded. As the statement of claim itself recognised, it was perhaps arguable that the introduction of such a condition might make the agreement too vague in law to amount to a binding contract.

Mr Reid for the plaintiff referred to *Lee-Parker v Izzet (No 2)* [1972] 1 WLR 775, where Goulding J had to consider a special condition embodied in the National Conditions of Sale (17th edition) that the sale was "subject to the purchaser obtaining a satisfactory mortgage." Goulding J had pointed out that these words made the vendor's obligation to sell and the purchaser's obligation to buy dependent on a satisfactory mortgage being obtained and that until that condition was fulfilled there was no binding contract. He concluded that the condition was void for uncertainty on the ground that in the circumstances the concept of "a satisfactory mortgage" was too indefinite for the court to give it a practical meaning.

In the present case his Lordship agreed that it made all the difference that the satisfaction was expressed to be the *plaintiff's* satisfaction. The courts should be slow to reject an agreement for uncertainty in circumstances like the present unless obliged to do so. The case did not fall within that category. He therefore held that the contract was not void for uncertainty and granted an inquiry as to damages and the other relief sought.

CHANCERY DIVISION

April 2 1976

(Before Mr Justice WALTON)

BEECH PROPERTIES LTD v G E WALLIS & SONS LTD

Estates Gazette March 5 1977

(1976) 241 EG 685

Specific performance—Contract conditional on vendor obtaining deeds of grant of easements over neighbouring lands to enable services to be connected up to public installations—Condition not satisfied, in respect of surface-water drainage, by a grant in mistaken form which coincidentally permitted a junction with a 9-in pipe nearby at present used for limited disposal of foul sewage—Specific performance refused, deposit ordered to be returned—Judgment involving numerous points of potential importance to developers etc

This was a claim by Beech Properties Ltd, of Mulcaster Street, St Helier, Jersey, against G E Wallis & Sons Ltd, of Cosmos House, Homesdale Road, Bromley, Kent, for specific performance of a contract for purchase by the defendants of Pennington Place, Southborough, Kent for £540,000. The defendants counterclaimed the return of the deposit of £54,000 and interest thereon.

A L Price QC and J M Henty (instructed by Theodore Goddard & Co) appeared for the plaintiffs, and F M Drake QC and T Coghlan (instructed by Vernor, Miles & Noble) represented the defendants.

Giving judgment, WALTON J said: This is a vendor's action seeking specific performance of a contract dated March 26 1974 for the sale of property called Pennington Place, Southborough, Kent, against the purchaser. The purchase price was £540,000, and the usual 10 per cent deposit was paid to the vendor's solicitors, Theodore Goddard & Co, as stakeholders. There is the usual counterclaim for the usual declaration in relation to this deposit by the purchaser.

The contract was conditional, and the conditions had to be performed by September 26 1974. Mr Price, for the plaintiff vendor, has conceded that time was of the essence of this date, as it plainly was. The sole question for decision is whether, by that date, there had been, in the words of clause 14 (c) of the contract:

> . . . the execution of a deed or deeds of grant of easement giving to the purchaser the right in perpetuity to run from the development site all such foul sewers surface water drains electricity gas and water conducting media over so much of the land adjacent to the development site as may be necessary to bring all such services from the development site to the public supplies or outlets such deed or deeds of grant of easement to be executed by the freeholder or respective freeholders of the land in question.

This is a case where it is necessary to describe the *locus in quo* in some little detail, and I shall now proceed to do so by reference to document D1, a plan of the entire drainage systems of the whole area, which was put in by the defendant and not challenged by the vendor in any way. To the south of the relevant site there is another batch of houses, whose precise date I do not know, but which were

A originally on cesspool drainage. In the year 1966 a foul sewer was put in serving these properties, and that foul sewer proceeds first of all to the east of the development site, and then runs from east to west along a portion of the development site, and finally turns north in a sweep to a pumping station which was erected in 1966 or 1967. There run into the pumping station combined drains coming from land to the west of that station, and the sewage coming from the site to the south of the development site is there combined with the water coming in the combined sewers and is pumped under pressure away out to the west. Proceeding north from the pumping station, there was originally a line
B of sewers which connected with the combined drains and were originally, I think, a combined series of sewers. When the pumping station was erected, the link with the combined sewers was disconnected and the line of sewers was led into the pumping station itself. However, this was only, as it were, a spur into the pumping station, designed for the occasion when the pumping station might be overloaded or something might go wrong with the electric pumps therein, as a result of which the pressure would fail, and therefore this spur was put on so as to enable, under those circumstances, all the sewage and surface water at the pumping station to flow along the
C old system. The old system proceeded in a northerly and slightly easterly direction to an area by now well to the north of the development site, in which the effluent was disposed of. With the building of the new pumping station, this system became largely, but not wholly, disused, and as a matter of pure fact, after the system was built it was used only for the drainage of a large property known as The Glen, taking foul drainage only, and subsequently in 1968 another property called Beverley was built which was presumably difficult to connect to the existing line of combined drains, and that was also connected to this system.
D So when our story begins, this system, which it will be convenient to call the 9-in system because that is the diameter of the pipes, was taking the sewage from two places only, The Glen and Beverley, and taking nothing else; although it was capable of taking an overflow from the pumping station, the evidence is that no such overflow had in fact yet occurred.

From the description I have given, it will be seen at once that from the point of view of development of the site there was no difficulty so far as foul sewage was concerned. That could quite easily be connected up to the existing foul sewer starting to the south of the development
E site, and we are in this case throughout concerned only with the drainage of surface water from the development site. In order to procure the fulfilment of the contract condition in this respect, the vendor obtained two grants of easements over the properties intermediate between the development site and the disposal area. The first property belongs to a Mr and Mrs Nicholls, and as to the grant of the easement over their land, no question arises; it proceeds from one point on their boundary to another point on it. The second piece of land belongs to Mrs Dunmall, and to this I shall have to return. Both grants are dated September 6 1974. No question arises in either case to the ability of the
F grantors to grant the easement. Now the grant from Mrs Dunmall aimed at proceeding from the second point on the boundary of Mr and Mrs Nicholls land to the boundaries of the substantially disused disposal area to which I have already referred at a point where there was a chamber, or only just short of this chamber (which was a slight distance inside the water authority's land), basically a settling tank in 1974, although it may well have started life as a septic tank, shedding the necessary pipes for fully effecting this purpose as it grew older and became more disused. Unfortunately, the terms of this grant were such that the wayleave thereby granted was to a range of pigsties some little way to the west of this chamber. As, however, G
there was in the same locality a chamber which could improperly but very plausibly be described as a "local authority chamber," and as the line of the wayleave on the plan attached to the deed of grant was the document by reference to which the wayleave itself was granted, and it was shown as clearly proceeding to the pigsties and not to any other possible feature, Mr Price judged (correctly) that he would have but little hope of persuading me that the wayleave was intended to run to a wholly different point only just short of the water authority's chamber. I need therefore spend no further time on the construction of this deed, or upon dissecting the attempted correction of this H
mistake (which I judge it to have been) by a subsequent deed which was, nevertheless, executed after September 26 1974. However, it will be appreciated from the general description of the area which I have already given that on its way to a point west of the water authority chamber the wayleave crossed the line of 9-in sewer pipes. I shall call this "the point of intersection." Mr Price's final refined submission was that the combined effect of the two deeds of grant was to entitle the purchasers to construct a sewer for their surface water to the point of intersection, and that they then had, under section 34 of the Public Health Act 1936, a right to connect this sewer into the 9-in sewer at J
this point. Accordingly the conditions of clause 14 (c) of the contract were satisfied, and had been satisfied by September 26 1974—indeed, had at all times been satisfied since the grant of the two easements on September 6 1974.

These are truly remarkable submissions. Firstly, no such suggestion to the effect that the condition had been fulfilled in this manner was ever made by the vendor to the purchaser at any time prior to September 26; indeed, although such a claim is indeed foreshadowed in the statement of claim served in this action on February 20 1975, no such suggestion was ever made at all in the correspondence, so K
far as I can trace it, prior to February 20 1976—17 months after the date by which the condition should have been fulfilled and is in fact contended was fulfilled. Secondly, I am by no means happy that what the parties had in mind in clause 14 (c) was anything other than an "outlet" in the popular sense of the term. I am fortified in this view by observing the steps which the vendors took to satisfy the condition, namely, although they got it, as a matter of pure geography, wrong, the obtaining of a deed of grant for an easement which led to an outlet in this sense. So that if the purchaser is indeed to be held to its contract, it will be because notwithstanding that it was not informed at the L
time precisely how the condition had been fulfilled, and notwithstanding that the documents tendered to prove fulfilment were clearly and obviously based upon the premise that an outlet of the nature I have described had been provided for the purchaser, yet totally unknown to both purchaser and vendor, the condition was indeed fulfilled. I should not regard a binding agreement of this nature as one of which I ought properly to decree specific performance. The purchaser might well have to pay damages for breach of the agreement at law, but it appears to me that it would be most inequitable to force specific performance of it upon him. Thus refusing specific performance, M
I would be entitled under the Law of Property Act 1925, section 49 (2), to order a return of the deposit: but I apprehend that this would not invalidate the contract or deprive the vendor of his right to damages, following my own decision in *Node Estates Ltd* v *Omertec Engineering Ltd* (1976).

I think, however, that in spite of this lame conclusion it would be right to summarise the course of the negotiations over the disposal of surface water as contained in the documents shortly as follows. The first scheme for the run-off of surface water was that it was going to the River Medway, starting from a pond, or dry pond, at the site, and continuing along the course of a stream which eventually discharged into that river. This idea was pursued at least from the letter of May 6 1974 from Henry Osborne Associates, chartered architects and planners, who had been concerned with the site for some time, doubtless originally on behalf of the vendor, but had later been engaged as architects by the purchaser. It was pursued from that date, with variations (such as the reconstruction of the existing pond) until at least July 30 1974, when at a meeting between Henry Osborne Associates and the planning officers of the Tunbridge Wells District Council a second possible solution was outlined, namely the laying of a new surface water drain, following the line of the existing foul sewer and continuing further to the water authority's old septic tanks and from thence along a further pipe direct to the major stream, thereby taking advantage of the water authority's statutory powers. The suggestion at this stage, it must be clearly borne in mind, was that there would be a new surface water pipe all the way to the water authority's land, and indeed one on from that land to the main stream. It was realised that it would be necessary to drain the two houses, The Glen and Beverley, which the existing system served, into fresh septic tanks, but this would be a comparatively trifling problem. On August 6 1974 the Tunbridge Wells District Council (as agents for the Southern Water Authority) wrote to say that they had no objection to the principle of this scheme, but that it would require "detailed agreement." In the light of what subsequently happened, it made the doubtless well-meant but in the event unfortunate suggestion that on investigation it might be found that the length of sewer from the site to the water authority's land could be materially reduced: that is to say, by connecting up with the existing 9-in sewer at some point south of that land. However that may be, it was realised that if easements could be provided across Mr and Mrs Leslie's land and Mrs Dunmall's land to the water authority land, the condition would have been satisfied, and on August 30 1974 Theodore Goddard wrote to the purchaser's solicitors as follows:

> The local authority have confirmed that they accept this alternative method of disposing of surface water from the site, and by producing easements to enable the water to be passed through the land of the Leslies and Mrs Dunmall our client has satisfied the condition in the contract.

Of course, the vendor had done no such thing, for the reasons already noticed, certainly not in the way intended. On September 4 1974 Mr Scott-White, of Scott-White & Hookins, consulting engineers, who had been retained by the purchaser, wrote to the Tunbridge Wells District Council in the following terms:

> The alternative proposal to the surface water is to construct a new sewer to connect into an existing local authority storage tank which is situated below the pumping station. The capacity of this tank, together with its outfall to the river, is unknown and therefore a survey will have to be made. Who would be responsible for carrying out the survey and who would be responsible for the construction of the new drainage work and any up-grading of the existing system?

On September 12 the Tunbridge Wells District Council replied as follows:

> The alternative proposal may well be preferable, but it will demand further survey and investigation. The points which, *inter alia,* require attention are (a) the structural and hydraulic adequacy of the tanks and outfall; (b) whether the route of the proposal can be shortened by linking into the pipeline north of the pumping station; (c) arrangements for the disposal of the drainage from the adjoining dwellings. The council would expect your clients to carry out any necessary investigation and bear the cost of all necessary works. I would, however, be prepared to assist in the practical aspects of the survey, and subject to my being satisfied with the proposals, I would recommend that an agreement under section 187 of the Public Health Act be entered into for the eventual taking over of the sewers by the council.

The next move was that the vendor instructed BMMK & Partners, a firm of consulting engineers, to act on its behalf, and that firm, acting with commendable speed, had a report on the discharge of surface water from the site ready the day after they had been instructed. However, acting with that speed, they made mistakes in the terms of that report. Among other matters, they misapprehended the course of the 9-in sewer, and produced a plan showing it as continuing past the pumping station to at least a point south thereof, which it is convenient to call point A. From this moment on, for a considerable period of time, both BMMK and Mr Spashett, an architect in the office of Geering & Colyer, the estate agents acting on behalf of the vendor, were under the impression that it was possible to connect with the 9-in sewer at this mythical point. Mr Spashett lost no time in writing to the Tunbridge Wells District Council, which he did on September 25 1974. The nub of that letter is as follows:

> I would, therefore, be most grateful if you would please confirm (a) that the existing surface water disposal system is a local authority sewer, and (b) that the connection as shown on BMMK Drawing 4746/01 for the drainage of the development site is acceptable to your authority.

Where Mr Spashett got the idea that the 9-in system was a surface water disposal system, he was totally unable to explain in the box. The "connection as shown" was, unfortunately, to the mythical point A. On October 1 1974 Mr Spashett wrote a second letter to the Tunbridge Wells District Council in the following terms:

> Following our discussions regarding the above, we understand that you are of the opinion that the existing council sewer between the points A and B shown on the drawing 4746/01 prepared by BMMK & Partners is not adequate to take the outfall from the development site. You will recall that BMMK investigated this at the council's suggestion that this might be a more economical way of arranging the outfall. However, we accept what you say, and we therefore propose to revert to the attached plan. The whole of the new sewer would be run in 300 mm pipe. The gradients have been checked and are more than adequate.

This letter, and the plan enclosed with it, suggested of course a further alternative, which had not previously been discussed, and for which there were certainly no easements available to the vendor. This fresh scheme was, however, approved by the council by a letter of October 2. It is, incidentally, to be remarked that Mr Spashett was wholly unable to give any coherent explanation of how it came about that he wrote the letter of October 1 in apparent complete contradiction of his letter of September 25 without there having been, according to him, any contact between himself and the council in the meantime. This must remain an intriguing puzzle, as neither of the two council representatives concerned could throw any light on it. By now, of course, one is well past the September 26 1974 deadline. Misconceptions, however, still lingered on. Theodore Goddard wrote on October 9 to the purchaser's solicitors reiterating that there was a foul sewer which discharged at the mythical point A, and more surprisingly, "The whole of this part of the old installation is designated for surface water disposal." However, when the purchaser's solicitors pressed the point that point A was mythical, they were sharply told by Theodore Goddard that this assertion:

> . . . clearly indicates a profound misunderstanding of the facts. Our client has obtained easements to lay a new surface water drain which will not involve any junction with an existing sewer.

So even as late as October 22 1974, the way in which it was alleged that condition 14 (c) had been satisfied was by means of easements which enabled a new surface water drain to be laid which would "not involve any junction with an existing sewer." The action was in fact started on that very day. I think it was about this time that the Southern Water Authority

A came on to the scene. The former sewage works were in fact vested in them, and the Tunbridge Wells District Council were merely acting as their agents in negotiating terms upon which the surface water arising from the site could be discharged over the Southern Water Authority land. But on February 28 1975 the Southern Water Authority made it clear to the Tunbridge Wells District Council that they were not happy about the possibility of flooding if the water was in fact discharged on their land, and that there was a special point as regards surface water originating on the highway:

> Further to the recent meetings between members of our staff, B I write to confirm that with regard to the proposed development I raise no objections to the roof drainage in the properties being allowed to connect to the disused section of the foul sewer, such as the surface water discharges over the existing 2-acre land treatment area, but as mentioned by yourself, I am concerned that properties or land downstream of the erection area may be flooded if we permit the road drainage to discharge to this system as well. As there is no agreement with the highway authority in accordance with section 21 of the 1936 Public Health Act enabling them to connect to this system, I would be grateful if you would communicate with the developers and tell them that they should negotiate with the highway authority for your area so as to enable satisfactory arrangements for the disposal of road drainage from the development.

C All this left the purchaser more than somewhat mystified as to what was actually happening, and on March 4 1975 Mr Scott-White again wrote to the Tunbridge Wells District Council asking, in substance, whether the council's attitude was still accurately reflected in the letter of September 12. Before the council replied to that letter, in view of the Southern Water Authority's letter, they wrote on March 21 1975 to Mr Spashett saying that in view of the attitude of the Southern Water Authority they were no longer able to support the previously-agreed solution in so far as it related to highway drainage. The reply to Mr Scott-White's letter was dated March 21 D 1975, and the vital paragraph read:

> Reverting to your letter and copies of previous correspondence attached, other than for the specific detail outlined in the third paragraph of my letter dated September 12 1974 and listed in subparas (a), (b) and (c), the facts stated in that letter still apply. The reason for deleting the specific detail is that depending on the ultimate solution achieved, these facts may or may not be relevant. The basic problem at present is highway drainage and the fact that the Southern Water Authority are not prepared to accept highway drainage into the public sewers without an agreement with the highway authority.

E So that even at that date, the position was the simple one that so far as the council were concerned they had approved the scheme in principle, but it had not been finalised. Moreover, the Southern Water Authority having raised the point about surface water originating from the highway, if the consent of the Southern Water Authority to the connection was required, it was not possible at all in respect of all the surface water in question.

To complete the history very briefly, on May 2 1975 Geering & Colyer wrote to the Southern Water Authority putting forward an amended scheme prepared by BMMK & Partners, together with certain calculations relating to the flow F of surface water in the whole of the system then proposed, which were stated to have been "agreed" by the Tunbridge Wells District Council. This scheme involved a connection at a point actually on the 9-in sewer to the north-east of the pumping station. The calculations purported to show that the 9-in sewer could take all the surface water under the conditions that the rainfall was not worse than what is called a one-year storm. Consent to these proposals was given by the Tunbridge Wells Borough Council on behalf of the Southern Water Authority on May 9 1975. I gather that the Southern Water Authority were not particularly happy with the scheme, but felt that their freedom of action had been somewhat curtailed by reason of the approvals in principle given by G their agents the Tunbridge Wells Borough Council on their behalf. Finally, on February 20 1976 Geering & Colyer submitted a fresh proposal, namely that the connection should be made at the point of intersection of the wayleave and the 9-in pipe—the proposal which is now relied upon in this action—the connection being from a 12-in pipe into the 9-in pipe. This was agreed in principle by the Tunbridge Wells Borough Council on February 25 1976. It is therefore plain as a pikestaff that the conditions of clause 14 (c) of the contract had not been met by September 26 1974, unless the local authority could have been made to accept a solution which was never put to them until February 20 1976 (nearly 17 H months later) and which, as matters stood on September 26 1974, involved a proposal—namely the use of a part of the 9-in drain—which they had up to that point declined to accept otherwise than "in principle."

I come back now to the precise terms of clause 14 (c) of the agreement. Picking out the relevant words, the agreement was conditional upon "the execution of a deed of grant of easement giving to the purchaser the right in perpetuity to run from the development site all such . . . surface water drains . . . over so much of the land adjacent to the development site as may be necessary to bring . . . such service(s) from the development site to the public . . . outlet." What is J meant here, in connection with this condition, by (i) "surface water" and (ii) "public outlet"? I do not think that there is any difference between the parties over the scope of "surface water": it includes all water arising on the site and running off therefrom in the course of nature, including drainage from roofs, impermeable areas, and roadways. Indeed, the calculations of the experts on both sides of surface water run-off have been made on this basis, and no word was directed to either expert to suggest that this was not perfectly proper or otherwise than in strict accordance with the terms of the agreement. As regards "public outlet," not much time was K devoted to discussing precisely what this meant, but I think that it must have meant at least a public outlet of size sufficient to take all the surface water which was going to be generated by the development site, and which the purchaser was to be entitled to use as of right. I do not think that any more restricted definition will meet the obvious requirements of the case. It would be no use offering the purchaser an outlet which was too small, or which he might or might not be able to use.

Now, in my judgment, neither of these two requirements was met. On the question of capacity, there is a regrettable conflict of evidence between Mr Maggs of BMMK & Partners L on the one hand and Mr Scott-White of Scott-White & Hookins on the other. I have no doubt that I must prefer the evidence of Mr Scott-White, which was not only much more convincing in itself, but was founded upon observations actually made by him (admittedly together with an assistant, but the kind of measurements which he made could not have been effected single-handed) and was checked by taking levels back to the original bench mark, to which he closed within 0.003 of a metre. The observations upon which Mr Maggs relied were not made by him but by members of his staff, and while I give them all credit for being careful and conscientious, M it is quite impossible to understand how they could have missed the obvious feature of a wall, giving a steep and pronounced drop along the line of the 9-in pipe; nor have I any evidence as to how they fared when (if they did) they closed their measurements. The result is that, in my opinion, not only is it unsound engineering practice to lead a 12-in pipe into a 9-in one, which was freely conceded; it is established that the 9-in pipe would not have been sufficient to carry the volume of surface water expected at peak periods from the development site. In other words, the outlet was not an adequate outlet. Mr Price sought to meet this possible difficulty in his way by saying that this might be very unfortunate, but it could be simply solved by the Southern Water Authority using their statutory powers under section 22 of the Public Health Act 1936 to enlarge the diameter of the 9-in pipe. Doubtless; but I do not think that this is what was contemplated as a satisfaction of the condition. A developer has to live with a water authority, and for him to commence their relationship over a site by presenting the water authority with a sizeable bill for a totally unnecessary piece of engineering is hardly the best way to achieve a good rapport. But this is peripheral; I interpret "outlet" as a "sufficient" outlet, and the outlet here in my judgment was not, on the evidence, going to be sufficient.

Secondly, was the connection at this particular point one which the purchaser was entitled as of right to make? I answer this question also in the negative. One may start a consideration of this problem, as Mr Price did, with a consideration of the Public Health Act 1956, section 14, but I do not think that that carries one very much further. It confers no individual right on any individual. Moreover, it has been repealed by section 14 (1) of the Water Act 1973, although this in substance only substitutes the water authority as the body liable. I think one has to go straight to section 34. I read subsection (1):

> Subject to the provisions of this section, the owner or occupier of any premises, or the owner of any private sewer, within the district of a local authority shall be entitled to have his drains or sewer made to communicate with the public sewers of that authority, and thereby to discharge foul water and surface water from those premises or that private sewer: Provided that nothing in this subsection shall entitle any person . . . (b) where separate public sewers are provided for foul water and for surface water, to discharge, directly or indirectly, (i) foul water into a sewer provided for surface water; or (ii) except with the approval of the local authority, surface water into a sewer provided for foul water. . . .

Now it is obvious that the right thereby given is not an absolute, but a qualified, right: "Subject to the provisions of this section." That refers principally to subsection (3). *In limine*, however, it is very difficult to say what is meant by the words "where separate public sewers are provided for foul water and for surface water" in proviso (b). Earlier, the section has referred to sewers "within the district of a local authority"; now this is to be read as a reference to a water authority and its area (Water Act 1973, section 14 (2)), and these areas are considerably larger in all cases than the local authority areas. Can it be that because there is a surface water sewer in one part of the area, and a foul sewer in an entirely different part of the area, the proviso comes into effect? Fortunately, I do not, I think, need to discuss this problem. It appears to me that the 9-in sewer, although at the present time only used for foul sewage, was originally provided for both foul and surface water, and that, although it is now only as a matter of fact used for foul sewage, is still provided for both, in that it is possible for water from the pumping station, both foul and surface, to be discharged into it when required. I cannot therefore regard it as a sewer "provided" for foul water alone. It is in my judgment a sewer provided for both foul and surface water, but at the moment actually only used for foul, although at any moment it may be required to take a combined mixture. Accordingly, in my judgment there is no point on proviso (b) which would assist the defendant.

However, it does appear to me that, wide as the words of subsection (1) may be, and for the moment ignoring the opening qualification, they do not confer upon an individual the right to connect his sewer to the water authority's sewer at any point which he may choose. In most cases, of course, the matter will be quite academic. There will be the water authority's sewer, going along the road; a new house is built in the road; and quite obviously and clearly the owner will expect to have a right to drain into that sewer, and it would be very difficult, assuming that there are no problems under the proviso to subsection (1), to imagine a set of circumstances where the water authority would be entitled to say that he must not connect to that sewer but to some other sewer. Even so, if the new house was built at a crossroads and there were available sewers in both roads, I can see no reason why the owner should be entitled to drain into the sewer of his choice if the water authority required him to drain into the other, which might, for example, well be a relief sewer expressly provided for the district because the other sewer was approaching capacity. Similarly, I see no reason why the owner is entitled to connect at point X rather than an adjacent point Y, if the water authority requires him to connect at Y. I think that I am only saying here in less felicitous language what has already been said in very much more forthright terms by Romer and Stirling LJJ, particularly the latter, in *Wilkinson v Llandaff & Dinas Powis Rural District Council* [1903] 2 Ch 695; see also *Halsbury's Laws of England*, 3rd ed, vol 31, p 211, n (1), where all the relevant cases are collected.

Of course, it may be said that in the present case the water authority, through its agent the Tunbridge Wells District Council, raised no objection in principle, and finally no objection *tout court*, to a connection with the 9-in sewer. This is perfectly true: the local authority, through its officers, was trying to be as helpful as it possibly could be to assist the proposed development. But I do not think that at the end of the day this counts for anything. What is equally clear is that the water authority, the principal, was far from happy about the proposed connection. Again, it is true to say that, so far as one can see, its unhappiness had its roots in financial, rather than in engineering, matters. But it was definitely unhappy, and only finally agreed to the proposal—one which it made perfectly clear was not to be taken as a precedent, hence by implication asserting its rights to deny connection with the 9-in sewer to *any* developer—because it felt that its position had to some extent been compromised by the actions of its agents. The whole of this factual history has, however, to be read against the background that the calculations which were presented to the local authority, and which the local authority checked only in the sense of redoing the calculations and not in the sense of checking the assumed data, were themselves inaccurate. As they led to the conclusion that the 9-in pipe was adequate to take the projected flow, naturally the local authority never gave any consideration to what the position would have been if the reverse had been the case, as I find that it undoubtedly was. In so far as it is a matter of inference from the surrounding circumstances of the case, I am prepared to infer that the local authority would certainly not have agreed (if only for the reason that it would obviously have involved their principal, the Southern Water Authority, in additional expense), and that beyond all conceivable question the Southern Water Authority itself, if presented with the correct data, would never have agreed. But, be that as it may, neither principal nor agent ever did agree, for the simple reason that the correct facts were never presented to either.

Therefore it appears to me that on this ground the claim of the plaintiff that the conditions of clause 14 (c) of the agreement of March 26 1974 were satisfied by September 26 1974—or at all—must fail. I now proceed, however, to consider the implications of subsection (3) of section 34. This subsection of course requires notice of the intended connection to be given to the water authority. Mr Drake sought to submit that for this reason alone the suggested outlet was not sufficient to satisfy the conditions of clause 14 (c), since no such notice had been given by September 5 1974, or indeed at all. Of course, case after case in the books stresses the importance of the giving of such notice; but if this had been the only point at issue, I should have been against Mr Drake's submission. An outlet would nevertheless have satisfied the conditions of that subsection even though

A a notice would have to be given, or indeed even more elaborate machinery gone through, before it could actually be used, if there could be no conceivable doubt as to the outcome of the procedure. The subsection provides, however, that the water authority may refuse to permit the connection to be made. It is, of course, perfectly true that such refusal can only be on a limited number of grounds, and that if such refusal is not based on reasonable grounds it may be overruled by a court of summary jurisdiction. In the present case, the grounds include the ground that the mode of construction of the sewer proposed to be connected to the water authority sewer is such that the making of the B communication would be prejudicial to the water authority's sewer. Now I do not know what the water authority would have done if application had been made to them to connect a 12-in sewer to their existing 9-in sewer and they had at the same time been in possession of the fact that what was proposed was of the same general natúre as pouring a quart into a pint pot. I have the shrewdest possible suspicion that they would have objected violently, and none the less because the quart into pint pot could have been solved by the expenditure of money by them in enlarging the existing 9-in sewer into a 12-in. Since the question was never posed to the water authority, it never answered it. But whether C I am right or wrong in my deduction as to what their attitude would have been, it is quite impossible for anybody to predicate with certainty that it would not have been such as I think it would. Nor is it possible to predicate with any certainty that if the water authority had taken that view, any court of summary jurisdiction would have overruled their determination. The conclusion is that in spite of Mr Price's silken oratory, it is quite impossible for it to be predicated even now—and still less was this possible on September 26 1974—that beyond all question the purchaser would have been entitled to connect its sewer to the water authority's existing 9-in drain at the point of intersection, or indeed D at all. So for this complementary reason also, the conditions of clause 14 (c) were not satisfied on September 26 1974, or E indeed subsequently.

Mr Drake also had a third point, based on the question of highway drainage. It will be recalled that some of the surface water drainage arising from the development site arises from water draining off the highways. One of the points—indeed, the main point—taken by the water authority when refusing in its letter of February 28 1975 to permit the discharge was that this in part related to surface water arising from the highway, and that there was no agreement under section 21 of the Public Health Act 1936 in existence. Subsection (1) of that section reads as follows:

> Subject to the provisions of this section, a county council and F a local authority may agree that (a) any drain or sewer which is vested in the county council in their capacity of highway authority may, upon such terms as may be agreed, be used by the local authority for the purpose of conveying surface water from premises or streets; (b) any public sewer vested in the local authority may, upon such terms as may be agreed, be used by the county council for conveying surface water from roads repairable by the county council.

It does not, however, appear to me that this point was a good one. The term "surface water" is not really defined in the Act (though see section 90 (1) for a partial definition), but I see nothing in any of the provisions of the Act to G limit the width of that term, or to restrict the rights of any person under section 34 so as to exclude the drainage of surface water from roadways forming part of his premises. I can well appreciate why the Southern Water Authority might have wished to have the matter dealt with by means of such an agreement as is specified in section 21, because doubtless any such agreement would contain financial provisions in its favour; but I cannot myself so construe the provision of section 21 as to exclude from the ambit of section 34 highway surface water. However, for the reasons already given, I propose to dismiss the action, and allow H the counterclaim.

LANDS TRIBUNAL

COMPENSATION

WILSON AND ANOTHER v SOUTH OXFORDSHIRE DISTRICT COUNCIL

(Ref/272/1975)

(W H REES FRICS)

December 15 1976

Estates Gazette February 12 1977
(1976) 241 EG 466

Purchase notice for land in " white " area and area of great landscape value at Sonning Common, Oxon—Certificate of appropriate alternative development for " agriculture, educational, cultural or recreational facilities and certain types of open development," where grounds large in relation to built-up area—Cedar buildings erected under Camps Act 1939, formerly used as school and since severely damaged by vandals, held to have no additional value—Hope value rejected—Consistent refusals for residential development in area—December 1973 valuation date in period of stagnation for building land sales—Land valued overall with allowance for some deferment

A J Butcher (instructed by Arthur F Clark & Son, of Reading) appeared for the claimants; M Rich (instructed by the solicitor to South Oxfordshire District Council) for the acquiring authority.

Giving his decision, MR REES said: This is a reference to determine the compensation to be paid in respect of 22.5 acres of land at Sonning Common, Oxfordshire: the matter arises from a purchase notice served by the claimants on Henley Rural District Council on September 21 1968.

The following are taken from an agreed statement of facts:

The date of the deemed notice to treat is April 30 1969 and the date of entry December 7 1973, the date of valuation. The land is on the western side of the village of Sonning Common, which is some five miles north of Reading. It has a frontage of 430 ft to a class III road, Cane End to Peppard Common. The boundaries are hedges for the most part, but part of the land is woodland (" Old Copse ") and where the boundary of the subject land runs through this it is undefined. The claimants retained some land out of their original holding: the boundaries of this portion are also undefined. The retained land also has a frontage to the class III road. On the east of the subject land there is a field gate which gives access on to Woodlands Road (which was made up and adopted on May 17 1974, when it became a public highway maintainable by the highway authority) and a pedestrian gateway giving access to a public footpath into " Old Copse." There is a wayleave in favour of the electricity board, but it is proposed to dismantle this overhead line.

The land has a level plateau on the west and slopes gently down to the east: most of it is under grass. There are water mains under both the class III road and under Woodlands Road. There is a gas supply in Woodlands Road and another at a point just off the class III road, 730 yds from the site. Electricity is to be supplied by underground cable to a new school on part of the subject land on the west; this was in course of construction at the date of the hearing before me.

Timber Buildings Constructed in 1939

The foul drainage for some buildings erected on the subject land in 1939 was to a private disposal system constructed on the land retained, but this installation is now derelict. The nearest point for main drainage at the material date was about 267 yds north-east from the site under the class III road. The buildings constructed in 1939 were erected under the Camps Act 1939 and were on the plateau: they were of Canadian cedar wood with shingle roofs. They originally comprised: assembly hall; dining hall; boiler house; four classrooms; two store buildings; workshop; two ablutions (male and female); seven dormitories; hospital block and a small bungalow. In addition, there were a small swimming pool, a single garage, double hard tennis court and air-raid shelters dispersed around the site. All the buildings were connected to each other by tarmacadam paths, the site was landscaped around the buildings and in front of the dining hall was a grassed or garden area. Pipes for hot and cold water both for washing and for central-heating purposes were laid in trenches across the site.

It was not disputed that the buildings were in use as a school until 1954, but in 1955 a receiver was appointed to sell the land. The buildings were used to house refugees after 1955. It was also not disputed that, at the relevant date, the buildings had been damaged by vandals. The claimants purchased the freehold just before auction, the advertised date of which was December 18 1959. I was told that they paid £9,500 for it and that their object in so doing was to sell it at its full value.

The following facts are taken from a bundle of agreed documents. In January 1960, one of the claimants applied for planning permission for the establishment of a caravan site and siting of caravans for residential and holiday purposes on the whole of Bishopswood Camp Site, ie the subject land and on the land retained. Consent was refused. In May 1960, the claimants applied for a determination under section 17 of the Town and Country Planning Act 1947 as to whether the use of existing buildings for the purpose of holiday and/or permanent residence required planning permission. In reply to this application, the county council said that this would constitute development and would require an application for planning permission. The claimants lodged appeals to the minister against the refusal of planning permission and against the decision of the county council with regard to the section 17 application. After a local inquiry, both appeals were dismissed.

In June 1962, the claimants applied for planning permission to erect 10 dwelling-houses in part of " Old Copse ": this application was refused. The claimants appealed to the minister, and after an inquiry the appeal was dismissed. In February 1965, an enforcement notice was served on the claimants requiring the discontinuance of the use of the subject site for the purpose of a private dwelling-house. The claimants appealed: after another inquiry the appeal was dismissed.

Purchase Notice Served

In June 1968, one of the claimants applied for planning permission to adapt the existing cedar buildings to residential

A bungalows for old people and to develop some of the land for residential purposes at 10 houses to the acre. This application was refused. On September 21 1968, the claimants served a purchase notice under section 129 of the Town and Country Planning Act 1962 in respect of the 22.5 acres. Henley Rural District Council (who were succeeded by the respondent authority) served notice of refusal to accept this purchase notice. An inquiry was held and the minister confirmed the purchase notice (a letter written by the minister before this inquiry is referred to later).

In December 1971, Henley RDC applied to the county council for a certificate of appropriate alternative develop-
B ment for recreational purposes. In the same month the claimants also applied for such a certificate but for residential development subject to conditions. In February 1972, the county council issued a certificate to Henley RDC to the effect that planning permission might reasonably have been expected to be granted in respect of the subject land for (1) the use of that part of the land occupied by buildings for sporting or cultural purposes and the use of the remainder as amenity open-space or playing fields, and (2) the use of that part of the land occupied by buildings for the purpose of a residential or boarding school or residential college and the use of the remainder as playing fields ancillary thereto.
C Two days later the county council issued a certificate in similar terms to the claimants.

The claimants appealed to the minister against both certificates. Again an inquiry was held, following which the minister cancelled both certificates issued by the county council and substituted one to the effect "that planning permission directly related to agriculture, educational, cultural or recreational facilities and certain types of open development, such as hospitals, sanatoria or other institutions where the grounds would be large in relation to the built-up area (subject to no building being erected on any part of the land other than" a specified area of 8.5 acres; to the effect that
D 3.9 acres should remain in existing use (woodland) except for such development which is proposed to be carried out by the respondent authority. The remainder of the land comprises 10.1 acres.

I note here two further matters in connection with town planning: (1) In January 1973, planning permission was granted subject to conditions in respect of 3.6 acres of the subject site (on the east) together with some adjoining land for the erection of a primary school with access from Woodlands Road. (There were two consents and one was granted on the understanding that a further application would be
E made adding another strip of land, so providing a wider access road.)

(2) In 1975 (the date is immaterial) planning permission was granted in respect of an area of about 2.1 acres of the subject site on the west, with frontage to the class III road, for the erection of a school for the educationally sub-normal (the "ESN" site).

Counsel for the claimants called D J Morton FRICS, who has practised for about 27 years in Reading. Counsel for the respondent authority called: A P Twigger, senior assistant education officer, Oxfordshire; G T Barrow MRTPI, control officer, South Oxfordshire County Council;
F and F W Read ARICS, a senior value of the Inland Revenue Valuation Department, who has had 33 years' professional experience, deputy district valuer, Oxford.

The respective valuations of the two valuers are as follows:

For the claimants—D J Morton FRICS

(a) Area developed as a residential school and available for redevelopment for institutional use	£
8.5 acres @ £5,000 per acre	42,500
Existing buildings	
25,000 sq ft @ £1 per sq ft	25,000
G (b) Area of open land available for playing fields and sports purposes or as grounds of institution.	
6.5 acres @ £2,000 per acre	13,000
Area subject to planning consent for primary school.	
3.6 acres @ £5,000 per acre	18,000
(c) Woodland	
3.9 acres @ £1,000 per acre	3,900
22.5 acres	£102,400

Alternatively

Taking the entire site at "hope" value for residential development. H	
22.5 acres @ £4,500	£101,250

Note

In each valuation a deduction is necessary to allow for compulsory redemption of tithe by the purchasers.

£3.82 pa @ 9.73 YP = £37.17	
Value adopted	£100,000

For the authority—F W Read ARICS

18.6 acres @ £2,000 per acre	37,200
3.9 acres @ £1,000 per acre	3,900 J
22.5 acres	41,100
Less Tithe Redemption Annuity	
£3.82 x 9.73	37.17
	41,062.83
Say	£41,000
Surveyor's fees baser on scale 5 (a) of the scales of professional charges of the Royal Institution of Chartered Surveyors	457.80
	£41,457.80 K

NB—Corn rent	62½p
Quit or fee farm rent	2.27½p
	£2,90 per annum

These have not been paid for many years and have been ignored in the valuation.

It will be seen that the value of 3.9 acres of woodland and the cost of redemption of the tithe redemption annuity are agreed. Both valuers ignore the corn and quit rents. There
L is apparent agreement in respect of 6.5 acres of land at £2,000 per acre, but Mr Read's is an overall figure and Mr Morton's one for playing fields without consent to erect buildings.

The areas of disagreement fall under three headings:

(1) Whether an amount should be added for the buildings.

(2) Whether there is any "hope" value.

(3) The existing use value of the land without buildings taking into account the certificate of appropriate alternative development and the planning permission for the primary school.

M I consider them in that order.

(1) *The buildings*

Mr Morton excluded 9,000 sq ft of ancillary buildings and valued the remaining 25,000 sq ft at £1 per sq ft having regard to the availability of services and to the fact that the buildings, when repaired, could be used as a residential school or for purposes indicated in the certificate. He had considered the *cost* of building them from scratch at the relevant date and the *cost* of repairing them as they then existed: there was a difference well in excess of £1 per sq ft between the two. Mr Read considered that the buildings in their damaged state, at the relevant date were beyond

A redemption: the cost of repair would be high and there would be no demand for them when repaired for the purposes of the last use or for those uses indicated in the certificate. Other somewhat similar buildings on sites quoted by both sides as comparables were referred to, all of which together with the subject site and buildings I inspected.

Having considered the evidence in the light of my inspection, I have little difficulty in coming to the conclusion that no purchaser of the land would give more because of the presence of these severely damaged structures.

(2) *Hope value*

B Mr Morton said that he would value the subject land at the relevant date at £45,000 per acre if it had had planning permission for residential development. He supported this figure by reference to the sale by auction by his firm of a site, also at Sonning Common, in 1972. The subject land was on the edge of the village, which had grown rapidly, and there was a keen demand in that area. There was no land allocated for residential development which had not been taken up and pressure on the local planning authority to release more land for building. He considered possible purchasers at the relevant date and came to a figure of 10 per cent of £45,000. He referred to a site at Burghfield C Common where figures just in excess of £4,500 per acre had been paid for a school site: he considered this to be very comparable to the subject site.

He took the view that the door would be opened (ie planning permission for residential development on the subject site would be granted) as the knocking on it had been so persistent: he agreed that there were few examples of sales at "hope" value in the open market.

Mr Barrow said that he considered that there was no hope of obtaining planning permission for residential development on the subject site at the relevant date. It was in a "white area" and in an area of great landscape value, but 200 m D from an area of outstanding natural beauty, not in it. He referred to planning policy documents, to the village plan for Sonning Common, to the planning history of the subject site and to the results of surveys to support this view and compared and contrasted the planning position of Sonning Common with Burghfield Common.

Mr Read said that the planning history of the site left no room for hope: there had been consistent refusals of permissions for residential development. Burghfield Common was a case where hope value was reflected, but he considered that this was not similar to Sonning Common. December E 1973 was in a period of complete stagnation as far as sales of building land were concerned. The only faint light in the gloom of the history of the subject land was given by the minister in a letter written before the inquiry held just before the purchase notice was confirmed: the minister then considered that the land could be used for housing subject to conditions, but the lamp which gave this light was shattered when the minister subsequently accepted the inspector's recommendation that there was no evidence of an overriding housing need in the area to justify granting planning permission for residential development. Thereafter the light in the dust lay dead. Quite different planning F considerations and policies apply to Burghfield Common.

Having regard to the planning policies for the area and to the planning history of the subject land, I come to the conclusion that no one would have purchased it at a figure which reflected hope value at the relevant date, which was during a period of stagnation as far as sales of land with planning permission for residential development were concerned. I therefore reject this approach. It is perhaps not without significance that the respondent authority offered to agree to the claimants' withdrawing the purchase notice so that they (the claimants) could market the land: the claimants did not accept this proposal.

(3) *The value of the land taking account of the certificate* G

Leaving aside the woodland, the value of which is agreed, Mr Morton takes a total of 12.1 acres of land at £5,000 per acre because buildings can be put on it and 6.5 acres at £2,000 per acre as the value of land for playing-fields etc on which no buildings can be erected. Mr Read values 18.6 acres overall at £2,000 per acre, this figure allowing for some deferment, since he considered that there was no immediate demand for it at the relevant date. When comparisons of land for such uses as are envisaged here are made, there should be no difference in results whichever of these approaches is used, provided that the ratio of "building" land to "playing field" land is constant. I come H to the conclusion that there is no evidence other than Mr Morton's opinion, which is barely supported, to warrant his approach. On Mr Read's overall approach, Mr Morton's total figure of £73,500 equals about £3,950 per acre.

There is some difference of view between the valuers as to the range of uses: Mr Morton includes such purchasers as those requiring land for such a use as a laboratory. He pointed to similar uses in the area. Mr Read excludes such uses, as he considered that there was no demand for the subject land for such a purpose, an opinion which was supported by the results of searches which he had had carried out of registers of planning applications in the area. J

In support of his figures, Mr Morton relied on five transactions which he said were comparable; one of these was not concluded, another was land at Burghfield Common to which I have already referred. Another was £9,000 paid by the county council to the district council for the "ESN" site, 2.1 acres. This transaction was complicated by requirements that the purchasers should construct an entrance from the class III road on adjoining land and by a provision for adjustment of price in the sale agreement. The sale of this site reduces the frontage which the subject land formerly had to the class III road by over half. There may be some injurious affection to the remainder. I therefore place no great reliance on it. Mr K Read put forward details of 14 transactions, some of which he relied on to support his figure of £2,000 and some of which he used as illustrations to criticise Mr Morton's opinion.

"Some Deferment" Taken into Account

Before coming to my conclusion, there is one matter I should touch on. In approaching his valuation, Mr Read said that the only bidder must be the actual purchaser under the purchase notice, that the district council were not willing purchasers and that the law does not import willingness to purchase. Nevertheless he put a figure on the land such as L would be arrived at in friendly negotiations. Counsel for the claimants submitted that the buyer is to be considered willing and consequently that to defer the value as Mr Read had done was wrong in principle. Having regard to section 186 (1) of the Town and Country Planning Act 1971, this purchase must be considered as being on the same footing as if it followed a compulsory purchase order and notice to treat in the usual way. Having regard to the decision in *Inland Revenue v Clay and Buchanan* [1914] 3 KB 466, all likely buyers must be considered. But I do not see that deferment is precluded; it is a matter of valuation evidence and I have accepted the view that it is right to take some deferment into M account in this instance.

Taking the evidence as a whole in the light of my inspections, I come to the conclusion that Mr Read has taken too gloomy a view when considering deferment. He would, he said, have taken £2,500 per acre if he had not deferred. The planning permission for a primary school site on part of the subject land approached from Woodlands Road (an access would probably have to be widened and this would involve negotiations with an adjoining owner) must add a little to the value, although it was conceded by counsel for the claimants that by the relevant date, the education authority had abandoned its plans for a school on this site. I heard evidence about this and its history and about the "ESN" site from Mr Twigger. However, the planning permission for the primary school site at least indicates that subdivision of the subject site might be favourably considered. On the other hand, Mr Morton's figure of £3,950 overall seems to me to be far too high. I accept Mr Read's view that there was no demand for the subject land for such purposes as laboratories, etc.

I come to the conclusion that the figure should be £2,250 per acre overall to give £41,850, to which must be added the agreed figure for the woodland, £3,900, a total of £45,750 less cost of redemption of tithe redemption annuity £37.17, but, say, £45,750 plus surveyor's fee based on scale 5 (1) of the scales of professional charges of the Royal Institution of Chartered Surveyors, £507.675 plus VAT at 8 per cent £40.614, total £46,298.29, and the acquiring authority will pay to the claimants their proper legal costs (if any) from the date of the deemed notice to treat up to the date of the notice of reference.

I heard submissions as to costs. The acquiring authority will pay to the claimant their costs of this reference, such costs unless agreed to be taxed by the Registrar of the Lands Tribunal on the High Court scale.

MOSS v BIRMINGHAM CITY DISTRICT COUNCIL

(REF/125/1976)

(J R LAIRD FRICS)

December 7 1976

Estates Gazette February 26 1977

(1976) 241 EG 617

Leasehold interest in shop and dwelling-house at Small Heath, Birmingham—Unexpired term of 10 years—Shop disused—Council valuer's valuation, given orally, on basis of sale with vacant possession accepted—No expert evidence for claimant—Tribunal says that valuations should be tendered in writing and recommends expert witnesses "to read and follow advice given in" John Watson's "Nothing but the Truth"

The claimant appeared in person; Mrs N E Field, assistant solicitor of City of Birmingham District Council, for the acquiring authority.

Giving his decision, MR LAIRD said: This reference by E L Moss dated June 28 1976 requires the Lands Tribunal to determine the compensation payable to him by the City of Birmingham District Council (hereinafter called "the council") upon the compulsory acquisition of his leasehold interest in a shop and dwelling-house known as 127 Golden Hillock Road, Small Heath, Birmingham 10, where Mr Moss now resides but has not carried on business for some years, as I will explain later.

The acquisition arises under the Housing Act 1957 and the Birmingham (Cooksey Road) Golden Hillock Road, Small Heath Housing Compulsory Purchase Order No 15, 1974, which was confirmed with certain modifications by the Secretary of State for the Environment on July 8 1975 and became operative on October 2 1975; under it 127 Golden Hillock Road was reclassified as "grey" land situated outside the clearance area. Notice to treat was served on March 11 1975 and a notice of entry was served to take effect on December 15 1975, which is thus the material date for the valuation of Mr Moss' interest, which is underleasehold for a term of 99 years (less three days) from December 25 1886 (expiring December 22 1985) at a ground rent of £6.25 per annum. It is understood that this interest is not subject to any easements, restrictions, charges or liabilities of any kind.

Sale of Claimant's Business

The history of events is as follows:

About 1955. Mr Moss acquired the leasehold interest and set up business as an ironmonger in the ground-floor premises and occupied the two floors above as living accommodation.

About 1960. On doctor's advice, Mr Moss sold the ironmongery business and let the shop and dwelling accommodation to a Peter Gannon on lease until about 1974.

About 1974. When the lease expired in 1974 Mr Gannon vacated no 127 because, Mr Moss said, Mr Gannon was unable to make a living from the shop because the area had been "blighted" by the compulsory purchase order made by the council on August 22 1974; the business was closed down and the shop became disused.

About June 1974. Mr and Mrs Moss returned to no 127 and lived, as before, on the two upstairs floors, but the ground-floor shop remained closed.

November 11 1975. The council served notice to treat and notice of entry, the latter to take effect on December 15 1975.

November 21 1975. A form of claim was submitted on behalf of Mr Moss by Edwards, Bigwood & Bewlay claiming £8,000 for his leasehold interest in the property. No figure was stated for business disturbance for the simple reason that Mr Moss was not trading.

December 15 1975. The council took possession.

Mr Moss told me in evidence that he did not carry on the ironmonger's business when his tenant gave up because, on inquiry of the Department of the Environment in July 1974, Mr Moss had been told that the property would be demolished "in six months' time"; had he known the acquisition was going to drag on for so long he would have reopened the business, even if only for a short period. He sought legal advice, but this, he said, did not help him. Mr Moss contends that a basic cause of his troubles had been because initially the council had classified no 127 as unfit for human habitation. Following his expenditure of a mere £87.30 on the property, the minister's inspector had excluded no 127 from the "pink," "unfit" clearance area and had included the subject property in Part II (" grey " area) of the schedule to the order.

At the time when the property was said by the council to be "unfit," a council valuer had told Mr Moss that his compensation was likely to be a nominal £10 site value for his short leasehold interest.

I interpose here to say that it seems to me there are other reasons why Mr Moss is affected adversely when it comes to compensation, namely: (a) the fact that, sadly, owing to ill health, he was obliged to cease trading as an ironmonger and left the premises about 1960, never to resume. Thus he has no claim for compensation for disturbance of business. He has my sympathy, for he has been unemployed since July 1974; (b) the remaining term of the lease of only 10 years.

The subject property is situated on the west side of Golden Hillock Road approximately 2 miles south east of the city centre. It comprises a medium-sized three-storey combined shop and dwelling-house of brick and slate construction with an open forecourt, small rear garden and large garage. The total site area is 300 sq yds or thereabouts with a frontage of 20 ft and a depth of 136 ft. There is a vehicular right of way to the rear approached from Bolton Road. At the material date the property was assessed at £830 gross value and rateable value £663. The property was built about 1886 and the council agrees that it is in both good structural and internal decorative condition. The accommodation comprises:

Ground floor

Shop 19 ft 7 in (max) x 31 ft 3 in (max) with office/lobby off. Store 10 ft 3 in x 10 ft 6 in. Store 20 ft 6 in (max) x 20 ft (max). Workshop 7 ft x 8 ft. Store 20 ft (max) x 13 ft 6 in (max).

A *First floor*
Living room, bedroom, office, bathroom, kitchen.
On the second floor
Two attic bedrooms.

P J A Taylor FRICS, of the city estates office, called as expert witness by Mrs N E Field, assistant solicitor for the council, told me in evidence that the subject property is in a "mixed" area, zoned for shopping use in the current approved development plan. His approach to the valuation had been on the basis of a sale with the benefit of vacant possession on December 15 1975; he had estimated the net annual rental value of the whole property and then capitalised the resulting B net annual figure for the 10-year unexpired term of the lease. He gave his valuation orally; my note was as follows:

Shop	
Zone "A"—(first 20 ft depth)—380 sq ft at £1.60 psf	£608
Zone "B"—remainder (15 ft deep)—152 sq ft at 80p	£122
Stores—662 sq ft at 30p	£199
Living accommodation	£312
Garage	£ 50
Total rack rental value	£1,291 pa
Less ground rent	£ 6
C Net profit rental	£1,285
YP 10 years at 14% and 3% (adjusted for tax)	3.647
Total vacant possession value	£4,686
Say	£4,700

It will be seen that whereas Mr Taylor estimated the total annual rack rental value of no 127 at December 15 1975 at £1,291 clear, after repairs and insurance had been paid, the valuation officer, Inland Revenue, had valued it as at April 1 1973 for the purpose of the new valuation list at £830 gross value on the assumption that the hypothetical landlord would D do all the repairs and pay the insurance; the latter figure would therefore fall to be reduced by the amount of the repairs and insurance paid by the hypothetical landlord before it could be compared with the former figure of £1,291. As an approximation, having in mind inflation between 1973 and 1975, the figures bear comparison. In support of his valuation, Mr Taylor adduced evidence of four transactions concerning roughly similar sized and type of properties, 129, 143, 147 and 139 Golden Hillock Road, which I find are of some assistance; these were the only transactions known to him in the whole area, he said.

E Mr Moss did not call any expert evidence as to value, relying instead upon five offers to purchase that he had received for his property, some as recently as March 1976, following a "For Sale or To Let" notice that he had placed in his shop window when it became empty following the closure of the ironmongery business. These offers ranged from £8,500 in August 1975 to £12,000 (cash) in March 1976; he had also received an offer to rent of £30 per week for the whole property if a seven years' lease could be granted. However, as none of the offerers, who were mainly immigrants, were professionally advised, and were not called to give evidence and to be tested under cross-examination, I find that F the weight of evidence is in favour of the council and I accept it following my inspection of the subject property and the locality.

Accordingly I award the sum of £4,700 as compensation for Mr Moss's leasehold interest in 127 Golden Hillock Road to which will be added a surveyor's fee based on scale 5 (a) of the scales of professional charges of the Royal Institution of Chartered Surveyors and proper legal costs incurred from the date of notice to treat up to the date of this reference. Interest on the amount of this award is payable from December 15 1975 in accordance with statute.

One final point. As I said at the hearing, valuations by G expert witnesses should be tendered to the tribunal in writing and not given verbatim. I recommend expert witnesses to read and follow the advice given in the book written by John Watson CBE PPRICS, lately a member of the Lands Tribunal, entitled *Nothing But the Truth*. The foreword to the book was written by the Rt Hon Lord Widgery, Lord Chief Justice of England, who said, *inter alia*, "This book is full of suggestions for making the presentation of expert evidence more positively helpful to the tribunal which is to receive it."

Having read the decision in this matter and having then caused to be opened the sealed offer of compensation lodged by the acquiring authority, I find that the amount thereof H exceeds the amount of my award; accordingly the acquiring authority will pay the claimant his costs of this reference up to the date of the sealed offer and the claimant will pay the sum of £50 towards the costs of the acquiring authority as from the date of the sealed offer.

HALL v HORSHAM DISTRICT COUNCIL
(REF/246/1975)
(W H REES FRICS)
J October 29 1976

Estates Gazette March 5 1977
(1976) 241 EG 695

Coal merchant's freehold premises in Horsham—Acquired for new highway but for seven years included in "area of indecision," which depressed values—Best comparable a delicensed public house bought by claimant on borrowed money—Claimant's failure to seek further loan to provide money needed for conversion to replace subject premises—Tribunal says money "could have been raised somehow"— K Offers of alternative accommodation refused—Claim for total extinction of business not sustainable

Christopher Beaumont (instructed by Rawlison & Butler, of Horsham) appeared for the claimants; Jeremy Sullivan (instructed by the solicitor to Horsham District Council) for the acquiring authority.

Giving his decision, MR REES said: This is a reference to determine the amount of compensation to be paid on the acquisition of premises at 33 Park Street, Horsham, used as a dwelling-house and as offices in connection with a coal merchant's business. Leonard P Hall was in partnership with his two brothers, Lester S Hall and Gordon I Hall, L and they, together, are the claimants, but it was agreed between the parties that I should make my decision in terms of a lump sum. Since L P Hall gave evidence and appeared to be the one most concerned with running the business and with the compulsory acquisition, I shall refer hereinafter to the claimants as "Mr Hall" or "the claimant."

Counsel for the claimant called G E Freemantle, a coal merchant of Cranleigh, L P Hall and L A Tatham FRICS FRVA. Counsel for the acquiring authority called its solicitor, F C Kraus, its financial planning officer, C J Field, and V C Turner ARICS, district valuer, Chichester.

M Residential and Business Use

The business was started in 1947 by Mr Hall's father; other coal businesses were purchased from time to time and amalgamated with it. In 1954 the subject premises were purchased and the business was thereafter carried on from there; they were also Mr Hall's home until he was evicted by the acquiring authority at the beginning of November 1973. Plans for the redevelopment of parts of the centre of Horsham had been prepared in the 1950s and were submitted to the minister concerned in 1963 for confirmation. The plans showed that the subject premises were to be in an area of comprehensive development and included in

A the area of a compulsory purchase order. In April 1966 the minister indicated that he intended to modify the scheme, and in his decision, given in June 1966, the subject premises were omitted from both areas and were left in one of four "areas of indecision"; that was still the position at the date of the hearing before me. A proposal for a new highway was made and the subject premises were acquired under the Horsham Urban District (Park Street) Compulsory Purchase Order, which was confirmed on February 25 1972 by the Secretary of State for the Environment. Notice to treat was served on August 18 1972 and the acquiring authority went into possession on either November B 1 or 2 1973. I will return to that matter later.

Certainly before the plans were prepared in the 1950s, the centre of Horsham was Carfax. Park Street runs north/south, a short distance east of it, and they are connected by a number of alleys. The subject property was towards the northern end on the west side. The buildings, which were of brick and slate, had been demolished by the date of the hearing before me, but I was supplied with photographs and an agreed schedule of condition which contained plans which showed that the accommodation comprised: on the ground floor: shop, office at rear, two living rooms, kitchen, bathroom and yard; on the first floor: four bed- C rooms and two rooms used for storage, one of these being in the roof space. The site area was given by Mr Turner at 185.81 m^2 and the floor area at 113.34 m^2 plus the two storage rooms. Only very small quantities of fuel were kept on the premises, for example fire lighters and 28-lb bags of charcoal; the main coal yard was at Faygate, a few miles east of Horsham. The freehold of this yard is also owned by the claimant; I will refer to it later.

Before coming to the claim spoken to by Mr Tatham, I note that counsel for the acquiring authority did not contest the submission made by counsel for the claimants D that the "scheme" for the purpose of this reference comprised the whole of the town centre redevelopment proposals, including the areas of indecision, and was not confined solely to the road proposal. I also note here that I heard evidence, mainly from Mr Hall, but also from Mr Kraus and from Mr Field, as to the events which took place when the acquiring authority took possession. It seems that an attempt was made to evict Mr Hall by the sheriff's officers on October 31 1973. Mr Hall resisted and boarded up wall openings for this purpose: a little damage was done as a result. Mr Hall moved out on November 1 1973, but it seems that the acquiring authority may not E have entered until the next day and some damage was done by vandals in the short interim. I have come to the conclusion that the total damage due to both these causes is of no great significance and I propose to disregard it. In this connection, I note that Mr Turner valued the freehold as at November 1 1973, taking the agreed schedule of condition as a basis: I do the same.

I consider the claim under two headings:

1. The value of the freehold interest in the subject premises (no question of injurious affection to the yard at Faygate arises). The claimant's figure was £30,000, the acquiring authority's £16,300.
F 2. Disturbance. The claimant's total figure was £32,991 (the total of 10 individual items was in fact £33,399) and the acquiring authority's £1,500.

Freehold Interest

Mr Tatham, who now practises at Lancing and Arundel on his own account, joined West Sussex County Council as a valuer in 1950, was appointed county valuer and land agent in 1963 and retired from that appointment in 1973. He spoke to the following valuation:

G Two-storey property containing in all 11 rooms; occupied by claimant and used

(a) as to two ground-floor rooms as order office for coal merchant's business.

(b) as to remainder as residence for family in form of 4 bedrooms, 2 reception rooms, kitchen, bathroom and store. say **£30,000**

Mr Turner submitted the following valuations because of the partnership but agreed that if the occupation was treated as being that of a freeholder in possession, there would be no difference in the total amount:

H *Freehold interest:*		
Residential portion as adjunct to business premises		£10,000
Offices and use of yard—estimated rental worth, say, net	£500	
YP in perpetuity—8%	12.5	
		6,250
Tenant's interest. In accordance with section 20 of Compulsory Purchase Act 1965		
Value of unexpired term purely nominal, say ...		50
	Total	**£16,300**

J Mr Tatham relied solely on one comparison—the purchase by the claimant at auction early in 1973 of the Alexandra Inn (the "Alexandra"), a delicensed public house in Queen Street, Horsham, for £19,250. It was not disputed that some money would have to be spent on it to make it suitable as a replacement for the subject premises, but no estimate had been obtained from a builder and no specification or measurements were submitted to price the necessary work. Opinions as to the probable cost varied from £3,000 to £6,000, but it was agreed that £19,250 was a fair value at the beginning of 1973 and that by November this had fallen to £18,000 or £18,500.

K Mr Turner relied on four purchases by the council of three houses and a shop, all close to the subject premises, to support his "spot" valuation of £10,000. A rental value of £500 per annum net for the offices was agreed. Mr Turner also referred to the Alexandra and to other premises. The valuers' opinions differed as to the value of the Alexandra compared with that of the subject premises and as to the effect of the notation on the plan, "area of indecision." Counsel for the acquiring authority submitted that as this notation had been in operation for seven years (1966-73) it was of no effect, and this was Mr Turner's opinion also, but on the evidence I prefer Mr Tatham's view that it must L have depressed values. I consider that Mr Turner's figure of £16,300 is the better supported, but something must be added for the effect of the scheme, particularly for this rare notation and also for the advantage of the combined uses, residential and commercial. I agree with Mr Tatham that the Alexandra is the best comparable and I think that Mr Turner may have compared it too unfavourably with the subject premises. Supported by my view of the other properties referred to and of the area generally, in the evidence before me, I arrive at the figure of £22,000 as the value of the freehold interest in the subject property.

M

Disturbance

The claim spoken to by Mr Tatham for disturbance is as follows:

(a) Loss of profits on total annihilation of business On basis of accounts for year to March 31 1972 —net profit	£5,142		
Less interest on capital of £6,920 at 6% ..	415		
	£4,727		
Add 25% for depreciation as consequence of scheme	1,182		
		£5,909	
At 3 YP		3	
			£17,727

(b)	Interest incurred on loan for purchase of possible alternative accommodation, at last available figure (to be adjusted at date of Lands Tribunal)				9,200
(c)	Loss on sale of plant, vehicles and equipment, value from accounts to October 30 1973		£3,515		
	Less recovered on sales	£400			
	Estimated present value of items not sold	100			
			500		
				say	3,000
(d)	Loss on stationery, order books, etc				220
(e)	Redundancy payments				
	2 at £230		£460		
	1 at £154		154		
					614
(f)	Provision of temporary residential accommodation				
	Caravan purchase		£506		
	Less estimated current value		100		
				say	400
(g)	Removal and storage of furniture, etc from November 1 1973 to June 26 1976				
	137 weeks at £6.05		£829		
	VAT at 8%		66		
	Removal costs		50		
	Insurance surcharge during storage at 7p per £		58		
				say	1,000
(h)	Legal and surveyors' fees incurred in connection with advice on claim and purchase of the Alexandra				
	L M Doffman		£471		
	R Maile		35		
	Henry Smith		132		
					638
(i)	Costs (estimated) to be incurred in connection with acquisition of alternative housing accommodation				
	Legal and surveyors' fees, say		£350		
	Removal and incidental costs		150		
					500
(j)	Incidental costs in connection with formulation of claim, provision of plans, etc				100
			Total		£32,991

(Note by tribunal—the total of the individual items is £33,399)

At the hearing some items were agreed and some reduced.

Mr Turner, following the form of his valuations of the interests in the land, submitted the following:

Freehold interest . . . disturbance, removal costs and other incidental expenses occasioned by disposition including legal costs, say ...	£1,000
Tenant's interest. In accordance with section 20 of the Compulsory Purchase Act 1965.	
brought forward	£1,000
Disturbance	
(a) Just allowance by incoming tenant, having regard to nature of interest ...	Nil
(b) Loss sustained. Bearing in mind possibility of removal to former Alexandra Inn, say for temporary disturbance to trade and incidental expenses ...	500
Total	£1,500

As with the interest in the land, Mr Turner said that his figure, taking the interest as that of a freeholder in possession, would be £1,500.

Much of this very large difference is due to the approaches adopted: the claim is on the footing of total extinction of the business, but the respondent authority took the view that other premises, the Alexandra, were available.

As to which of these approaches is the correct one the relative evidence was as follows: Mr Freemantle considered a central position to be essential for a coal merchant's office; customers, who tended to remain with their current merchants, preferred to call rather than to write or telephone. A move away from a central position would entail some loss of customers. The tonnage of Mr Freemantle's business had increased over the last eight years.

Mr Hall said that at one time he ran his business from 70 East Street (" no 70 "). His business was at its peak between about 1961 and 1964/5 but declined when the first redevelopment plan was published. Later, Corralls and the " Co-op " were the only two coal merchants in Horsham except for one small supplier. Mr Hall married in 1955: his wife ran the office and continued to do so after about 1964 when his brothers took over Mr Hall's work; thereafter Mr Hall just supervised. His brothers became partners in the 1960s. He felt the effect of the redevelopment plan before 1964, trade going down thereafter until a loss was made in 1971. From semi-retirement, he went back to the business and thereafter a profit was made, although in 1972 the effect of the redevelopment plan was still felt. The effect of his return to the business was to cut the outgoings rather than to increase the tonnage sold. He started looking for other premises in the 1960s through estate agents. He was offered a tenancy of no 70, where the business had been started, by the acquiring authority in the 1970s (counsel for the acquiring authority conceded that it was reasonable of Mr Hall to refuse this offer—and another in respect of 4 Denne Parade (" no 4 ")—*at that time,* ie before the compulsory purchase order was made).

Attempts to Find Alternative Premises

Mr Hall considered that no 4 was off the beaten track: it was a freehold dwelling but there was no written confirmation that a change of use to business would be granted. In the 1970s he asked the acquiring authority about 9 Brighton Road (" no 9 "), no great distance from the Alexandra and in the same road (between these two properties the name of the road changes from Queen Street to Brighton Road). He said he would have moved there but the acquiring authority were not successful in purchasing it: they did not offer enough, Mr Hall said. Under cross-examination he agreed that in April 1973 he inspected it with Mr Field and from a question put to Mr Field when under cross-examination, it appears that when the acquiring authority were negotiating with the owners of no 9 at £13,000-£15,000, Mr Hall, unknown to them, had offered the owner £20,000.

Mr Hall said that two council houses a mile-and-a-half away at Roffey which were mentioned to him were not suitable nor was a house, 8 Wellington Road (" no 8 ") close to the subject premises. No 8 was not in a commercial position: there was no passing trade and no written confirmation that an application for a change of use would be granted. A central position convenient for customers to call was essential.

When he purchased the Alexandra he did not tell the acquiring authority of his intention to buy it—he was not then on speaking terms with them—later in 1973 they asked him if he had purchased it. At that time he was advised by two firms of surveyors that he could expect £30,000 for his freehold interest in the subject premises (he did not ask the acquiring authority or the district valuer what they would pay) and on the strength of this he borrowed the purchase money for the Alexandra, £19,250, from his bank, but when, in the middle of 1973, he learned from the district valuer that only £13,000 (not £30,000 as he had been advised) was in his (the district valuer's) mind, he told his bank, who requested payment of the outstanding balance. He did not ask his bank specifically for the amount required to convert the Alexandra for use for his business and as his home (he put it at £3,000-£4,000 but £6,000 was also mentioned). He did not consider that his bank would advance the money to convert the Alexandra. If he had known that he would only get £13,000 for his freehold interest in the subject premises, he would never have purchased the Alexandra.

Purchase of Caravan

In November 1973, when he was evicted from the subject premises, he had nowhere to go: he feared that his compensation would be reduced in amount if he took a council house although he was not offered one. So two days after eviction he purchased a caravan and has since lived in it at Faygate. He began to close his business when he was evicted and ran down his stocks. In September 1973 he had advertised his business for sale at £450,000 to include Faygate but did not find a buyer. In November 1973 he had conditional offers of £265,000 and £400,000 for the site at

A Faygate. Nothing came of these offers. At the beginning of 1974 he put the Alexandra on the market without success: it has not yet been sold and some of the £19,250 purchase money which he had borrowed from the bank continues to attract interest.

In December 1973, he received £13,116 from the council, being a home-loss payment of three times the rateable value (ie £1,416), plus 90 per cent of £13,000, and this went to reduce his liability to the bank. At no time did he ask the bank specifically if any part of this sum could be used to adapt the Alexandra although he had several discussions with the manager. He did not lodge a claim until October 2
B 1973 (the notice to treat was dated August 18 1972), so no payment on account could be made before then. He did not consider that it was practicable to use the Alexandra for offices only and to live elsewhere. His wife fell ill in January 1973, went into hospital in February 1973, and is still not, at least fully, recovered. He considered the Alexandra to be suitable alternative accommodation although not in as good a position as the subject premises. He could not live in the Alexandra without forming a kitchen in it. He wanted freehold residential and business premises but would have accepted leasehold if suitable and, as a last resort, business premises only.

C Mr Tatham said that he did not see why Mr Hall should have asked the acquiring authority what they had in mind as regards the amount of compensation: it was reasonable of him to have relied, as he did, on the advice given him by his then professional advisers.

Mr Field told me that he was responsible for the management of Horsham Urban District Council's (predecessors of the acquiring authority) properties in the town centre from 1966 onwards. He was one of a team concerned with the redevelopment of the town centre including the matter of alternative accommodation for displaced traders. Some 50
D businesses were involved together with seven or eight voluntary associations and 95 occupiers of dwellings. Alternative accommodation had been provided by the acquiring authority for just under half of the 50: of the others some had found their own, others had moved out of the town. He had first met Mr Hall (to discuss alternative accommodation) in July 1968, when Mr Hall indicated his requirements. Mr Field then wrote to Mr Hall offering him no 70: Mr Hall did not tell Mr Field that the business had been started in no 70. An annoyed Mr Hall, said Mr Field, declined this offer in a telephone conversation.

Mr Field produced copies of the correspondence which
E followed between the acquiring authority, the claimant and R A Allen & Co, solicitors, of Horsham, who were then acting for the claimant; a letter of June 3 1971 addressed to Horsham Urban District Council and signed A Y Mika pp L P Hall denied that Allen & Co had ever been instructed to act for the claimant, but Mr Hall denied knowledge of this letter. This correspondence, which consisted of about a dozen letters, related to no 4, to the Alexandra and to no 9. Only one reply was received from Mr Hall: this gave the only reason for rejecting suggestions as to alternative premises made by the acquiring authority until the hearing before me. Mr Field said that he was not asked to give any written
F undertaking regarding planning permission for no 4; Mr Hall himself could have applied for planning permission.

Mr Turner said that he knew no reason why Mr Hall could not have established his business elsewhere. It was sensible of him to have purchased the Alexandra and he considered that Mr Hall could have borrowed the money required for adaptation. He would think in terms of £60,000 as the value of Faygate in November 1973. He considered £450,000 to be unreasonable as an asking price. Apart from the two firms of surveyors referred to by Mr Hall, a third had also been employed by him to negotiate terms for the subject premises.

I was referred by counsel for the acquiring authority to G
the decision of this tribunal in *Bede Distributors Ltd v Newcastle upon Tyne Corporation* (1973) 227 EG 1663 and by counsel for the claimants to three decisions, also of this tribunal: *Knott Mill Carpets Ltd v Stretford Borough Council* (1973) 227 EG 336; *John Line & Sons v Newcastle upon Tyne Corporation* (1956) 6 P & CR 466; *Koch v GLC* (1968) 209 EG 1503. It seems to me that *Knott* and *John Line* turned on the balance between the terms asked for the alternative premises and the level of the business likely to be achieved therein, and that *Koch* was concerned with the position of a claimant holding under a lease which had only a very short term unexpired and with the multiplier to be H
used to capitalise the net profits from the business. I find the principle in *Bede* to be of more assistance. At p 320, this principle is summed up:

> "Where a loss has been incurred . . . the issue is not whether anyone else might have avoided it, the issue is whether the acquiring authority have proved that the claimant in incurring his loss and failing to take mitigating steps, can be said to have acted 'unreasonably' ie in a way that no reasonable person would have acted."

The question I have to answer, therefore, is: "Has the acquiring authority shown that Mr Hall acted unreasonably?" J

Counsel for the acquiring authority submitted that Mr Hall did act unreasonably in that he failed to take any of the premises offered to him by the acquiring authority which could have saved his business; he failed to use the Alexandra after he had purchased it. Furthermore, his general course of conduct was unreasonable. Counsel for the claimant submitted that there were good reasons why in each instance the claimant had not accepted the alternative premises suggested by the acquiring authority: the Alexandra could not have been used solely for business purposes because the interest charged on the purchase money could not have been carried by the coal merchant's business. To raise money on K
Faygate would have made it difficult to sell that property.

General Conduct "Fatal" to Total Loss Claim

I consider that if counsel for the acquiring authority had had to rely solely on the alternative premises offered by the authority, it might have been difficult for him to succeed in his argument as to alternative accommodation. But it seems to me that the failure of the claimant to proceed with the Alexandra coupled with his general conduct is fatal to his claim for total loss of his business. The money required to adapt the Alexandra could have been raised somehow having L
regard to the value of Faygate, and/or an advance payment could have been obtained earlier from the authority if the claimant had made the claim before October 1973 (notice to treat requiring "a claim within six weeks" had been served in August 1972). I do not see why a mortgage on Faygate would make it difficult to sell.

Mr Hall's attempt to purchase no 9 without informing the authority when he knew they were negotiating to buy it so that it could be offered to him (he had expressed interest in it) is an example of his general conduct. I note also that Mr Hall twice criticised alternative properties suggested to him by the acquiring authority on the grounds of the lack of M
planning permission, yet he purchased the Alexandra without planning permission to use it for his business and without ensuring that he had the money to adapt it. The only explanation put forward by Mr Hall as to why he did not convert the Alexandra for his business and his dwelling accommodation was that he considered that he could not borrow the necessary money from a bank.

There is no evidence that he put to the acquiring authority his difficulties as they were in the middle of 1973, ie after he had purchased the Alexandra and had himself come to the conclusion, without inquiry, that he could not

A raise the money to adapt it; nor as to what advice he was given by his various professional advisers comprising four different surveyors and three firms of solicitors, besides another adviser. I take into account that Mr Hall's wife was ill as from the time he contracted to purchase the Alexandra, and that reading and writing are not his *forte*. But there is no evidence that the acquiring authority told him specifically that if he failed to take other suitable alternative premises he might not be able to claim disturbance on the basis of total extinction. On the other hand, there is no doubt that he was a successful and competent coal B merchant.

Intention to be "Awkward"

In a letter, dated May 24 1971, sent by Mr Hall to the Department of the Environment objecting to the compulsory purchase order, he wrote that he intended "to be as awkward as I can." I agree with counsel for the respondent authority that he did just that. He had half retired in about 1964 and it may be that he thought that he could advantageously take the opportunity afforded by the compulsory acquisition to do so completely, but this is no more than speculation on my part. I have, therefore, come to the conclusion on the evidence before me that a claim for C compensation on the footing of total extinction of the business cannot be sustained.

On the basis of partial extinction, I have only the evidence of Mr Turner: counsel for the acquiring authority said that the amount was for me to decide on impression and by judgment. Mr Tatham did not comment at all on this part of Mr Turner's evidence; few questions on it were put to Mr Turner in cross-examination. Having given the matter due thought, I am satisfied on the facts before me that Mr Turner's figure of £1,500 is, if anything, on the generous side. The total compensation payable to the claimant, therefore, is £23,500 (freehold £22,000, disturbance £1,500), to D which falls to be added a surveyor's fee based upon scale 5 (a) of the scales of professional charges of the Royal Institution of Chartered Surveyors and proper legal costs from the date of the notice to treat up to the date of the notice of reference.

In conclusion I refer to one other matter: it was said that Mr Field was "devious and not to be trusted"; I see absolutely no evidence that this was so. I found him to be a straightforward and honest witness.

Having opened the sealed offer of compensation lodged by the acquiring authority, I find that the amount thereof is less than the amount of my award; accordingly the acquiring E authority will pay to the claimant his costs of this reference, such costs unless agreed to be taxed by the Registrar of the Lands Tribunal on the High Court Scale.

WOOD MITCHELL & CO LTD v STOKE-ON-TRENT CITY COUNCIL

(REF/33/1976)

(V G WELLINGS QC)

F February 8 1977

Estates Gazette March 19 1977

(1977) 241 EG 856

Printers' office and warehouse premises at Hanley sold to, and part bought back from, council by agreement—Disturbance compensation for temporary loss of profits agreed —Question of adjustment of amount to take account of corporation tax—Application of para 11, Schedule 19, Finance Act 1969—"Represents a crucial development of the law"—Held amount to be regarded as in the nature of income and to be treated as a trading receipt taxable in the hands of the claimants—Alternatively, matter to be left to Inland Revenue to assess corporation gains tax in claimants' hands G

C Whalley, of Breton, Deacon & Co, of Longton, Stoke-on-Trent, appeared for the claimants; R M Bramwell (instructed by the solicitor of the Stoke-on-Trent City Council) for the acquiring authority.

Giving his decision, MR WELLINGS said:

This reference relates to the amount of compensation to which the claimants are entitled for temporary loss of profits and other matters under Rule (6) of section 5 of the Land Compensation Act 1961. In effect, the only dispute is as to what adjustment (if any) on account of tax should be made H to agreed figures for compensation. By virtue of the City of Stoke-on-Trent (Lindop Street, Bucknall New Road, St Ann Street and Bucknall Old Road, Hanley) Housing Compulsory Purchase Order 1965, which became operative on March 31 1966, the acquiring authority became entitled to acquire from the claimants certain office and warehouse premises at 5, 7 and 9 Lindop Street, Stoke-on-Trent, and a yard and garage adjoining those premises. In all those land and premises the claimants carried on the business of printers. Notice to treat was served on the claimants on May 27 1966. Thereafter the parties, so far as possible, resolved their differences by agreement. J

By an agreement in writing dated March 13 1969 the claimants agreed to sell and the acquiring authority to buy the above-mentioned land and premises at the price of £6,000. In the same agreement the acquiring authority agreed to sell and the claimants to buy back part of the land and premises acquired by the acquiring authority and at the price of £3,000. All the land retained by the acquiring authority after March 13 1969 was in its possession by November 1968. The transaction provided for in the agreement dated March 13 1969 was completed on July 21 1970, when conveyances putting the agreement into effect were executed. K The purpose of the claimants' buying back part of the land and premises was to enable them to re-establish their offices and warehouse. Their claim to compensation under Rule (6) relates to the temporary losses which they suffered in the years 1969 to 1971 while their offices and warehouse were being re-established.

Reference by Consent

In the agreement dated March 13 1969 it was acknowledged by the acquiring authority that the property acquired by it under the agreement did not take into account any compensation for disturbance consequent upon the making of the L compulsory purchase order; it was thus agreed that the acquiring authority should pay to the claimants such sum as might be agreed in respect of compensation for disturbance together with any interest properly payable thereon and in default of agreement such sum as might be fixed in respect thereof by the Lands Tribunal. This reference is therefore to be regarded as a reference by consent.

No evidence was called at the hearing because all the facts had been agreed. Indeed the parties were in agreement as to the gross amount of compensation under Rule (6) to which the claimants are entitled. It was agreed that that gross M amount is £12,633.77. It was further agreed that of that such items totalling £405.50 are of a capital nature, leaving a balance of £12,228.27. The only question which the tribunal was asked to resolve is whether, and if so to what extent, that balance should be adjusted to take account of corporation tax.

Corporation tax is imposed on the profits and chargeable gains of companies: see Income and Corporation Taxes Act 1970, section 238. The profits are computed in accordance with income tax principles (this aspect of corporation tax I will call "corporation income tax"): see Act of 1970, section

250. The chargeable gains are computed in accordance with capitals gains tax principles (this aspect of corporation tax I will call "corporation gains tax"): see Act of 1970, section 265.

Mr Bramwell, for the acquiring authority, contended that the sum of £12,228.27 should be reduced by applying to it corporation income tax at the rate of 42.5 per cent. He said that the principle of *British Transport Commission v Gourley* [1956] AC 185 and *West Suffolk County Council v Rought* [1957] AC 403 required such a deduction:

(1) because the sum of £12,228.27 would not be subject to corporation income tax in the hands of the claimants as recipients because, as a matter of law, it is to be regarded as part of the price paid by the acquiring authority for the land acquired, and

(2) in order to place the claimants in the same position as that in which they would have been if the losses had not occurred.

Grossing Up

Mr Whalley, for the claimants, agreed that if any deduction on account of corporation income tax were to be made, the appropriate rate of deduction was 42.5 per cent. He, however, did not agree that any such deduction ought to be made. His main argument was that if the sum of £12,228.27 were to be reduced by 42.5 per cent on account of corporation income tax, then it would be correct to gross up the resultant figure of £7,031.26 for corporation gains tax. He would apply to that figure for that purpose the multiplier of 10/7 (that is to say, an effective rate of 30 per cent). The application of that multiplier would produce £10,044.65. On this basis, adding back £405.50, the total compensation would be £10,450.15.

Mr Bramwell agreed that if it were right to gross up for corporation gains tax 10/7 was the correct multiplier, but he said that there should be no grossing-up.

Mr Whalley supported his contention that grossing-up should take place by reference to the decision of the tribunal (Sir Douglas Frank QC, President, and J D Russell-Davis FRICS) in *Hobbs (Quarries) v Somerset County Council* (1975) 234 EG 829. In that case the tribunal awarded to the company compensation for losses arising out of a revocation of a planning permission for the winning and working of limestone in a quarry owned by the company. The compensation was assessed by reference to the profits which the company might have expected to earn under a motorway contract which they might have expected to obtain (by reason of the proximity of the quarry to a section of a motorway). From the gross amounts of compensation, the tribunal deducted corporation income tax and then grossed up the resultant figure for corporation gains tax.

Mr Bramwell said that the tribunal in the *Hobbs* case was correct to gross up for corporation gains tax because in that case no land was taken. That was the distinguishing feature. In the present case, land having been taken, the situation was governed by paragraph 11 of Schedule 19 to the Finance Act 1969. That provision was crucial. It provides:

"Where land or an interest in or right over land is acquired after April 29 1969 and the acquisition is, or could have been, made under compulsory powers, then in considering whether, under paragraph 21 (4) of Schedule 6 to the Finance Act 1965, the purchase price or compensation or other consideration for the acquisition should be apportioned and treated in part as a capital sum within section 22 (3) (a) of the said Act, whether as compensation for loss of goodwill or for disturbance or otherwise, or should be apportioned in any other way, the fact that the acquisition is or could have been made compulsorily, and any statutory provision treating the purchase price or compensation or other consideration as exclusively paid in respect of the land itself, shall be disregarded."

Paragraph 21 (4) of Schedule 6 to the Act of 1965 says:

"For the purposes of any computation under this Schedule any necessary apportionments shall be made of any consideration or of any expenditure and the method of apportionment adopted shall, subject to the express provisions of this Schedule, be such method as appears to the inspector or on appeal the Commissioners concerned to be just and reasonable."

Paragraph 2 (1) of the same schedule says:

"There shall be excluded from the consideration for a disposal of assets taken into account in the computation under this Schedule of the gain accruing on that disposal any money or money's worth charged to income tax as income of, or taken into account as a receipt in computing income or profits or gains or losses of, the person making the disposal for the purposes of the Income Tax Acts."

Mr Bramwell continued that it was right in the present case to make the apportionment referred to in paragraph 21 (4) of Schedule 6 to the Act of 1965 and that the effect of paragraph 11 of Schedule 19 to the Act of 1969 was to enable the statutory fiction that all compensation was treated as being part of the price paid for the land taken to be disregarded in respect of compensation which was attributable to loss of profit. When once the apportionment was made and the statutory fiction was disregarded, he said, no part of the sum of £12,288.27 would be subject to corporation gains tax because it was not derived from any form of property known to the law; see section 22 (3) (a) of the Finance Act 1965 and *Inland Revenue Commissioners v Montgomery* [1975] Ch 266 at 271 *per* Walton J.

Mr Whalley replied that if a severance took place under paragraph 11, then the sum of £12,228.27 ought to be regarded as trading receipts chargeable to corporation income tax in the hands of the claimants in the year in which that compensation were paid. In support of that submission he relied upon the decision of the Court of Appeal in *London & Thames Haven Oil Wharves v Attwooll (Inspector of Taxes)* [1967] Ch 772. In that case the taxpayer company owned a jetty at its oil installation at which tankers bringing oil could berth in order to discharge the oil. In April 1953 one tanker was so negligently navigated that it struck the jetty and caused serious damage to it. Ultimately the owners of the tanker paid to the taxpayer company *inter alia* a sum of £21,404 in respect of the loss of profit incurred by the taxpayer company while the jetty was being repaired. The company was assessed to corporation income tax under Case I of Schedule D on that sum for the fiscal year 1955-1956. It was held by the Court of Appeal that the sum was taxable as a trading receipt and that the assessment was correct.

Three Assumptions

I have no doubt that it is right to make three assumptions:

(1) That in the present case there has been a chargeable capital gain. This follows from the agreements made between the advocates in the present case. The amount of the gain has not been expressly quantified, but, I conceive, no one would quarrel with the suggestion that it amounts to at least the amount of my award.

(2) That the principle of the *Gourley* and *Rought* cases is of no application if the compensation to be paid to the claimants will be taxed in their hands when they receive it: see *Julien Praet et Cie v H G Poland* [1962] 1 Lloyd's Rep 566; *Parsons v BNM Laboratories* [1964] 1 QB 95.

(3) That it is a rule of law in respect of compensation for compulsory purchase of land that compensation for loss of goodwill, disturbance or any other matter is treated as part of the price payable for the land taken: see *Commissioners of Inland Revenue v Glasgow and South Western Railway Company* (1887) 12 App Cas 315; *Horn v Sunderland Corporation* [1941] 2 KB 26.

None of the above assumptions is in dispute in the present case. In relation to the second assumption, the advent of capital gains tax and corporation gains tax is important and it may well be that the decision in the *Rought* case would have been different if it had been given after the passing of the Finance Act 1965. I doubt, however, whether *Gourley's* case would, in that event, have been decided differently, because damages for loss of earnings by a private individual, though a capital sum, would not seem to be derived from any asset known to the law, so that in respect of such damages there can be no chargeable gain.

I agree with Mr Bramwell that paragraph 11 of Schedule 19 to the Finance Act 1969 represents a crucial development in the law. It is not free from difficulty of interpretation, more particularly because it requires *any statutory provision* (my italics) treating the compensation as exclusively paid in respect of the land itself to be disregarded, whereas the third of the assumptions which I have made would seem to be not so much a statutory provision (that is to say a specific express provision in a statute) but a rule of law which the courts have inferred from a number of statutory provisions. Nevertheless, paragraph 11 cannot be given effect to unless that rule be regarded as a statutory provision and I intend so to regard it.

I accordingly agree with Mr Bramwell that paragraph 11 applies in the present case, the result being that the sum of £12,228.27 is to be regarded as in the nature of income, but my agreement with Mr Bramwell's argument stops here. I agree with Mr Whalley that, the severance having been made, the sum of £12,228.27 ought to be treated as a trading receipt taxable in the hands of the claimants in the same way that the sum of £21,404 was taxable in the hands of the plaintiffs in *London & Thames Haven Oil Wharves v Attwooll, supra.* After the severance has been made, it appears to me that the present case is quite indistinguishable from that case. The effect of an apportionment to which paragraph 11 refers, though it be made for the purposes of a computation of a capital gain, is to free the compensation for loss of temporary profits of its capital nature and enable it to be treated for what it in truth is, namely, a trading receipt.

For these reasons I award to the claimants the sum of £12,633.77 together with their surveyors' fees (if any) on scale 5 (a) of the scales of the Royal Institution of Chartered Surveyors. If I were wrong in the decision to which I have come, rather than adjust the gross amount of compensation by reference to corporation income tax and then gross up for corporation gains tax (which course may be contrary to the views, albeit *obiter*, of the Court of Appeal in *Parsons'* case) I would still prefer to make no adjustment for corporation income tax, but (following the *obiter* views of the Court of Appeal in *Parsons'* case) to leave matters to the Inland Revenue with a view to their making an assessment to corporation gains tax in respect of the compensation in the hands of the claimants when they receive it.

The acquiring authority will pay the claimants' costs of this reference, such costs if not agreed to be taxed by the Registrar of the Lands Tribunal on the High Court Scale.

WAKERLEY AND ANOTHER v ST EDMUNDSBURY BOROUGH COUNCIL

(REF/267/1975)

(R C WALMSLEY FRICS)

January 28 1977

Estates Gazette March 26 1977 and April 2 1977

(1977) 241 EG 921; (1977) 242 EG 49

First reference under s 48 of Land Compensation Act 1973—Compensation payable on surrender of annual tenancy of farmland at Bury St Edmunds required for town development scheme—Gratuitous licence granted to claimants for 3 harvest years on surrender—Development delayed—On expiry of licence, claimants continue in caretaker capacity and council offer further 3-year licence—Capitalisation on life-expectancy-of-profit basis at 20 per cent risk rate—Claim for addition of year's rent for disturbance rejected—Tribunal accepts set-off against loss of profit of value to claimants of first 3-year licence but not of offered licence—Set-off might have been more simply approached on "before-and-after" basis

J A Denbin and J V Davies (instructed by Ellis & Fairbairn) appeared for the claimants; A P Fletcher (instructed by the solicitor to the St Edmundsbury Borough Council) for the authority.

Giving his decision, Mr Walmsley said: This reference is the first to reach the tribunal under section 48 of the Land Compensation Act 1973. It concerns the compensation payable on the surrender of an annual tenancy of farmland in Suffolk required for a town development scheme. The subject land comprises part of Eldo House Farm on the eastern outskirts of the town of Bury St Edmunds. The claimants are (a) the personal representatives of Arthur Cecil Wakerley deceased and (b) Joseph Arthur Wakerley. The compensating authority is the council of the Borough of Bury St Edmundsbury.

The late A C Wakerley first took a tenancy of Eldo House Farm, on the Orttewell Estate, in 1941. J A Wakerley (to whom I shall refer as "Mr Wakerley") joined his father as a partner in 1953, on completion of his national service, and the current tenancy agreement of Eldo House Farm was drawn up at that time, naming Mr Wakerley and his father as joint tenants. The agreement is in a customary form and it includes a clause 34:

> "That it shall be lawful for the landlords at any time or times upon giving four calendar months' notice in writing . . . to enter upon and resume possession of . . . any part or parts of the farm . . . for purposes of building mining or other industrial purpose or for any other purpose not being the use of the land for agriculture."

Mr Wakerley and his father farmed together from 1953 until 1958, when his father retired and the partnership was dissolved. The farm tenancy, however, continued to run in their joint names, and Mr Wakerley senior died only very recently, on May 26 1976. I propose, for convenience, to use the term "the claimant" or "the tenant" as referring to J A Wakerley and/or to the joint claimants, as the context may require.

Conveyance of Freehold

On the town map of Bury St Edmunds a portion of the Orttewell Estate, about 354 acres, was shown as within a town development area, and in 1971 the council (at that time the borough council of Bury St Edmunds) decided to buy this land, and certain other land that was needed, for the town development scheme. Terms were negotiated with the Orttewell Trustees in 1972, and the freehold of the 354 acres was conveyed to the council on January 5 1973. Part of Eldo House Farm, about 155 acres, was included in the conveyance, and this land was of course sold subject to the Wakerley tenancy.

At that time, at the beginning of 1973, the council were anticipating being able to make an early start on phase 1 of their town development scheme. In order to do this they required possession of the Wakerley land. An arrangement to this end was made with the claimant on the basis of a surrender of the tenancy of the 155 acres as at July 14 1973, this surrender being linked with two grants by the council to Mr Wakerley viz: (a) the grant of a gratuitous licence to farm 129 acres (later adjusted to 126.6 acres), being land included in phase 1 of their scheme, and (b) the grant of an annual tenancy in respect of about 26 acres not

A required for phase 1. This present reference is concerned only with the compensation payable in respect of the 126.6 acres of licence land, and I need not mention again the land let on annual tenancy.

At the hearing, J A Denbin and J V Davies appeared for the claimants, and Alan Fletcher appeared for the council. Valuation evidence was given by F A Laws FRICS, senior partner in the firm of Rutters, for the claimants, and by H O Wilkin MBE FRICS, senior partner in the firm of Cruso & Wilkin, for the authority. Mr Laws and Mr Wilkin are both diploma holders, being two of the very few chartered surveyors who hold the special diploma of the Chartered B Auctioneers' and Estate Agents' Institute in the subject of agricultural valuations, and both are gold medallists.

The valuers had been able, prior to the hearing, to agree figures in respect of a number of valuation factors; and during the five-day hearing agreement was reached on certain other figures. By the end of the hearing the compensation payable was being calculated by Mr Laws at £35,295 and by Mr Wilkin at £10,455.

A summary of their respective adjusted valuations, slightly rearranged for ease of comparison, is as follows:

C

	Mr Laws for the claimants (per acre, per annum)		Mr Wilkin for the authority (per acre, per annum)	
	£	£	£	£
Gross maintainable output		103.50		103.50
Less Variable costs	16.50		16.50	
Acts of husbandry	14.00		14.00	
Fluctuation risk	15.00		15.00	
Interest on capital	Nil	45.50	4.50	50.00
Profit prior to deduction for rent		58.00		53.50
Notional rent		15.50		15.50
Net profit		42.50		38.00
Y.P.		6.404		4.989
Capitalised net profit per acre		272.17		189.60
Acres surrendered		126.6		126.6
D Capitalised loss of profit		34,457		24,003
Add One year's rent for disturbance		838		Nil
		35,295		24,003
Deduct for the licence granted and for a further licence offered		Nil		13,548
Compensation payable		£35,295		£10,455

Four Matters of Dispute

It will be seen that the factors on which the valuers had been unable to reach agreement are four in number, *viz*:

(a) Whether, in the estimation of a net profit per acre, an E allowance should be made for interest on capital.

(b) The years' purchase at which the ascertained net profit per acre should be capitalised.

(c) Whether one year's rent for disturbance is properly included.

(d) Whether a set-off falls to be made for the value of licences.

Before dealing with these differences it is necessary to describe in rather more detail the train of events covering the period from the autumn of 1972 up to the date of the hearing before me in November 1976.

F By September 1972 Mr Laws, acting on behalf of the Orttewell Estate, had completed his negotiations for the sale of the trustees' freehold interest in the 354 acres being purchased by the council. Thereafter Mr Laws embarked on negotiations, acting now on behalf of the present claimant, in respect of the tenant's interest. These negotiations were conducted by Mr Laws with Mr Wilkin and, as might be expected with two such experienced agricultural valuers, agreement was soon reached on the arrangements which I have briefly outlined. This was at the point of time when the council had contracted to purchase the freehold and were hoping to start shortly on actual development. But just at that same time there intervened an enforced and G lengthy delay on planning grounds. The land comprised in the town development area lies, as it happens, on rising ground in a locality referred to as "Moreton Hall," this particular locality being of such prominence as to be locally dubbed "the backcloth to the Abbey." By early 1973 considerable local opposition to the Moreton Hall development was being evinced; in the outcome and several months later, the council's scheme was recast on environmental grounds to include *inter alia* a greater area of public open space. One result of the delay involved in all this was to defer the date fixed for the surrender of the claimant's tenancy; that date had been originally contemplated as for January 1973, but H it first was delayed to April 1973, and in the event it became deferred to July 14 1973. By that later date the crops on the 126.6 acres of licence land were almost ready to harvest.

"Special Arrangements" for Continued Farming

To meet this situation the valuers agreed that, as a condition of the arranged surrender on July 14 1973, the claimants should harvest their growing crops and, further, that if the council should require to enter any of the subject land before September 14 1973 then compensation would be paid for the crop that would normally have been grown in the 1974 season as well as for any crops lost in the 1973 season. J The council were also to pay compensation for fertiliser and lime residues on all the land surrendered. It was appreciated, looking beyond the 1973 harvest, that the council would not be requiring the whole of the 126.6 acres for development all at once; and an arrangement was made for the claimants to be granted a gratuitous licence covering the ensuing harvest years 1974, 1975 and 1976. If the council took possession of any cropped land between September 14 1973 and September 14 1974 they were to pay for loss of crop, but in respect of the next two harvest years, 1975 and 1976, the claimants would crop the land at their own risk, knowing K that there would be no compensation payable for crop loss.

These arrangements were practical, sensible and economically sound. Indeed they suited everyone. They enabled the claimants to continue farming what had been their "own" land, pending its development; they saved the council having to take over, and pay for, undamaged crops; they took care of the council's civic responsibility for seeing that good farming land did not fall unnecessarily out of cultivation; and they obviated the valuers having to bring into account the difficult question of entitlement under section 15 (2) of the Agriculture (Miscellaneous Provisions) Act 1968.

Common ground having been reached on the above L arrangements, a single solicitor was instructed jointly by both parties to draw up the requisite documents. But on February 23 1973, before these documents had become finalised, a Bill was introduced in the House of Commons which included a clause providing improved compensation to tenants of agricultural holdings dispossessed by an acquiring authority. The valuers for the parties thereupon agreed that if this provision as enacted were found to throw up a higher figure of compensation than that payable under existing legislation, then the claimant should receive that higher figure. (The clause in question took its final form as section 48 of the Land Compensation Act 1973.) It was M further agreed that an interim payment of compensation should be made, at the rate of £60 per acre. This £60 per acre was related to the legislation existing up to May 1973 and it represented an amount equal to five years' rent under section 12 of the Agriculture (Miscellaneous Provisions) Act 1968 (equating to four years' rent under section 9 of that Act and importing the equivalent of one year's rent from section 34 of the Agricultural Holdings Act 1948). For the purpose of this interim payment the annual rent was taken to be £12 per acre, albeit the actual rent being paid was apparently only about £6.50 per acre.

A Four deeds incorporating all the agreed matters (and which I shall now call "the special arrangement") were duly drawn up and these deeds were all executed on March 26 1974, a few days in advance of the date on which the powers and duties of the mayor, aldermen and burgesses of the Borough of Bury St Edmunds devolved under local government reorganisation on the council of the Borough of Bury St Edmundsbury.

The four deeds comprised:

(1) A *deed of surrender* reciting that:

1. In consideration of the covenants herein contained and
B on the part of the council to be observed and performed the tenants hereby acknowledge to the council as follows:

(i) That they will surrender to the council their tenancy described in the First and Second Schedules hereto with effect from July 14 1973 [the First Schedule is of the "licence land" and the Second Schedule is of the "tenancy land"]

(ii) That all claims to compensation which might otherwise have been made by either the council or themselves are replaced by this deed and agreements marked "A" and "B" attached and that the tenants have no further claim against the council in respect of any statutory or common law provisions arising out of their surrender of the land
C set out in the First and Second Schedules hereto or otherwise.

2. In consideration of the tenants surrendering their tenancy of the said land conveyed to the council as described in the First and Second Schedules hereto the council covenant that they will:

(a) Forthwith pay to the tenants the compensation to which they may be entitled under the provisions of section 48 of the Land Compensation Act 1973 or the sum of sixty pounds (£60) per acre whichever shall be the greater in respect of the land hereinafter described in the First Schedule hereto together with interest at the rate of 15 per cent per annum from July 14 1973 until paid . . .

(2) A *supplemental deed* reciting that:

D 1. . . . the council hereby covenants

(a) . . . .

(b) that the tenants shall be entitled to be paid compensation under section 48 of the Land Compensation Act 1973 in respect of any matters arising in respect of [the deed of surrender] . . .

2. It is hereby agreed and declared between the parties that . . . any disagreement as to compensation under section 48 of the Land Compensation Act 1973 shall be referable by either party to the Lands Tribunal.

(3) A *licence agreement* (*Agreement "A"*) in respect of the subject land incorporating a nominal yearly payment of
E £1; the licence to run from July 14 1973 to October 11 1976.

(4) A *tenancy agreement* (*Agreement "B"*) in respect of the 26 acres not required for the council's phase 1 development.

No Impetus Left in Market

By March 1974, when the above deeds came to be signed, Mr Wakerley had not only harvested his 1973 crops on the licence land (and had done so without loss), he had also put in hand the cropping of the whole of the 126.6 acres
F for the 1974 season. This may seem surprising in the light of the council's known intention to make an early start on developing the licence land, but what had happened was that the boom conditions prevailing in the property world in 1972 and in the first half of 1973 had, by March 1974, changed dramatically to conditions of slump. The whole of 1973 and the early part of 1974 had been spent by the council in disposing of the planning obstacles to their Moreton Hall development, and by the time they were ready to proceed with their scheme, there was no impetus left in the market. Indeed the first substantial take of land by the council was not until May 1974, this being essentially for the laying of roads and the provision of sewers and other G
services. During the course of the harvest year 1974, Mr Wakerley lost about 18 acres of crops, compensation for which has been duly settled and paid. In the harvest year 1975 Mr Wakerley cropped 63 acres, without loss. And in the harvest year 1976 just completed, he has cropped an area of 73.5 acres—again without loss.

As has been noted, the licence in respect of the subject land ran only to October 11 1976. Since that date things have been in suspense. As a commonsense temporary arrangement, Mr Wakerley has agreed to husband, in a caretaker capacity on behalf of the council, what remains available of the 126.6 acres; the understanding being that he shall H
be reimbursed for his cultivations, seeds and fertilisers unless in the event he should decide later to take a fresh licence. On this basis some 93 acres have been drilled with winter corn for the harvest year 1977. The council have offered Mr Wakerley a licence for a three-year period from October 11 1976 to October 10 1979 on a payment at the rate of £15.50 per acre per annum. The terms of the offered licence, a draft of which was prepared just prior to the hearing before me, would include *inter alia* a clause that:

> "the licensors may at any time or times after the expiration of J
> not less than ten days written notice to that effect resume possession of any part or parts of the land for the purpose of laying roads and services or carrying out any development for which planning permission has been or shall hereafter be granted *provided always but not further or otherwise* that in the event of the said development causing major interference with the access to or use of any land the subject matter of this agreement the licensors will pay to the licensee compensation for the loss (if any) which the licensee shall sustain for the estimated "gross value" of any crops actually grown on the said land *and provided also* that for the purposes of this agreement . . . "gross value" shall mean the estimated gross output (including any subsidy) from adjacent or similar crops by comparison with actual K
> realisation after harvest."

This, then, is the background against which I return to the valuations. Both valuers adopt the same premise: that in the absence of the scheme Mr Wakerley could have expected to enjoy a life expectancy of profit from the subject land, and that as a result of the scheme this expectancy has been extinguished. Mr Laws, for the claimant, takes the view that that is the end of the calculation, save for the addition of one year's rent for disturbance. Mr Wilkin, for the authority, takes the view on the other hand that the special arrangement must now be brought into account, and that the L
(admittedly considerable) value to the claimant of the 1974-76 licence and also the value of the offered 1977-79 licence must be credited as a set-off against the loss suffered as shown at the first stage of the calculation.

I propose to deal with the differences between the parties under four heads:

(1) Calculation of a net profit per acre.

(2) Capitalisation of the ascertained net profit.

(3) One year's rent for disturbance.

(4) Set-off. M

There have already been references in this decision to certain statutory provisions, and references to further such provisions will be made hereafter. Rather than burden the body of the decision I am attaching as an appendix, the wording of the relevant parts of the relevant sections of the relevant statutes, *viz*:

Agricultural Holdings Act 1948 ("the Act of 1948"), sections 24 and 34.

Land Compensation Act 1961 ("the Act of 1961"), section 5.

A Compulsory Purchase Act 1965 ("the Act of 1965"), section 20.

Agriculture (Miscellaneous Provisions) Act 1968 ("the Act of 1968"), sections 9, 10, 12 and 15.

Land Compensation Act 1973 ("the Act of 1973"), sections 48 and 89. [Appendix not reproduced].

Calculation of Net Profit per Acre

Eldo House Farm is a 600-acre holding, of light, mixed, free-draining soil. For many years it has been farmed on a system of barley, wheat and sugar beet; and with about 50 B acres down to grass. Each year 250-300 Border Leicester/Cheviot ewes are bought in Scotland as lambs and are sold a year later as breeding sheep.

It will be recollected that Mr Wilkin's estimate of a maintainable net profit was £38 per acre, arrived at thus:

Gross maintainable output		£103.50
Less variable costs	£16.50	
Acts of husbandry	14.00	
Fluctuation risk	15.00	
Interest on capital	4.50	
		50.00
		53.50
C Notional rent		15.50
		£38.00

Mr Laws' figure was £4.50 per acre higher, because he included nothing for interest on capital. In this type of calculation, however, the tribunal recognises it to be right to include a proper allowance for interest on the claimant's capital employed in his business—as for instance in *Pearce v Bristol Corporation* (1950) 1 P & C R 367. At the hearing before me some considerable time was spent on the quantum D of tenant's capital, both as to peak requirements and as to average requirements. On the basis of that evidence I find for a figure of £30 per acre as the average amount of capital that would be required. Regarding the cost of providing that capital, both valuers concurred in the adoption of an interest rate of 15 per cent—on the assumption that the item should appear in the calculation at all. Accordingly I accept Mr Wilkin's allowance of £4.50 per acre for interest on capital. This finding brings with it an acceptance of his figure of £38 per acre net profit.

In addition to the above open disagreement as to interest on capital, there was some latent disagreement between the E valuers as to the concept underlying the mutually adopted figures of £103.50 per acre for "gross maintainable output," and of £15 per acre for "fluctuation risk." Mr Laws, it appeared, had regarded "gross output" as being related essentially to the barley crop. His £103.50 had been derived from an estimated yield of 35 cwt of barley per acre at £60 per ton (rounded down) and his "fluctuation risk" represented accordingly an allowance for possible changes in the price of barley. Mr Wilkin had taken a broader view. He had looked at the cropping of the farm as a whole, of which the barley shift represented only about 40 per cent in acreage terms. Mr Wilkin's "fluctuation risk" was accor- F dingly intended to take into account all the year-to-year risks of farming. I propose to adopt Mr Wilkin's wider view as the concept deemed to underly the ascertained net profit figure of £38 per acre.

Capitalisation of Ascertained Net Profit

At what years' purchase should the ascertained net profit per acre be capitalised?

For over a quarter of a century the agricultural code has given tenant farmers considerable security of tenure. This security is less complete if the tenancy agreement contains a clause (such as clause 34 in the agreement for Eldo House G Farm) which enables the landlord to resume possession of land for building purposes. This the landlord can do under the aegis of section 24 (2) (b) of the Act of 1948. For completeness it should be mentioned that an effective notice to quit can also be served if the tenant commits one or other of the "deadly sins" specified in later paragraphs of section 24 (2), such as not farming in accordance with the rules of good husbandry, or becoming bankrupt. It is common ground in the present reference, however, that (in the absence of the authority's scheme) Mr Wakerley—a highly respected farmer—may be assumed to have been impregnable against any risk of dispossession during his lifetime, other than H whatever risk might attach to the aforesaid clause 34. And, as regards Mr Wakerley's present claim for compensation, even that risk must be deemed to have been removed, because section 48 (3) of the Act of 1973 now provides, in effect (and for my present purpose), that in assessing the compensation payable by the acquiring authority to the tenant in connection with the taking of possession of part of an agricultural holding there shall be disregarded any right of the landlord to serve a notice to quit under the aegis of section 24 (2) (b) of the Act of 1948.

It was for this reason—that of an assumed complete J security of tenure—that both valuers adopted a life expectancy of profit, when calculating a loss on extinguishment, despite the claimant's interest being no greater than as tenant from year to year.

At the relevant date in 1973 Mr Wakerley was aged 38, with a life expectancy of 33.5 years. Accordingly Mr Laws looked to p 442 in *Parry's Valuation Tables* (9th edition), which shows the years' purchase for a single life (male) aged 38, at 15 per cent, to be 6.404—and this is the multiplier that Mr Laws uses. Mr Wilkin would have used the same table were it not for the circumstance that he considers the appropriate risk rate to be 20 per cent and this particular K table in *Parry* does not go beyond 16 per cent. For convenience Mr Wilkin looked instead to p 38 in the same volume which shows the years' purchase for an income for 33.5 years, at 20 per cent, to be 4.989—and this is the multiplier that Mr Wilkin uses.

The reason why Mr Laws adopts 15 per cent is that this is the interest rate to be found in clause 2 (a) of the deed of surrender, and it is also the agreed risk rate to be applied to tenant's capital. Mr Wilkin, in preferring 20 per cent, has in mind that factors beyond year-to-year farming risks require bringing into account when capitalising an assumed net income for a life expectancy of over 30 years, such as: L

(a) that Mr Wakerley might not have continued farming until he dropped. In a no-scheme world he might have retired, or he might have become incapacitated some years before death.

(b) that, in as much as the subject land lies on the outskirts of the town of Bury St Edmunds, there was always the risk in a no-scheme world of the land, or part of it, being required for development. In this event, the landlord could resume possession of it and the profit therefrom would cease—albeit the claimant M would receive compensation under section 9 of the Act of 1968.

(c) that over such a long period abnormal factors could intervene, as, for instance, freak weather conditions or EEC directions.

I think Mr Wilkin is right and that a higher risk rate than 15 per cent attaches at this stage of the calculation. I adopt his 20 per cent. As to the consequent years' purchase, the life table certainly seems the appropriate table to use. Mr Laws calculated, at my request, that at 20 per cent this table (if extended) would show a years' purchase of 4.868. While

A noting this, I have nevertheless decided to adopt Mr Wilkin's years' purchase of 4.989; partly because this higher multiplier is in favour of the claimant, but also because some weight should be given to an additional factor—that when a tenant farmer dies his tenancy devolves on his personal representatives, who may be deemed to continue earning the same net profit for a further year or so.

The effect of my findings up to this stage is to produce a capitalised loss of profit of £24,003 as calculated by Mr Wilkin.

One Year's Rent for Disturbance

B At this stage Mr Laws adds one year's rent for disturbance. His cited authority for so doing is section 34 of the Act of 1948. This section, however, only comes into play when the tenancy of an agricultural holding terminates by reason of a notice to quit given by the landlord and in consequence of the notice to quit the tenant quits the holding. The one year's rent under section 34 is a minimum payment relating to loss or expense incurred by the tenant on the sale or removal of goods and implements. But this provision applies, as can be seen, on the resumption of possession by a landlord; it does not apply on the taking of possession by an acquiring C authority. In 1968 the minimum payment by a landlord on resuming possession was increased (by section 9 of the Act of 1968) to five years' rent, the additional four years' rent being intended to assist in the reorganisation of the tenant's affairs. And by section 12 of the Act of 1968 an acquiring authority, on the taking of possession of an agricultural holding, is required to pay the tenant the same amount as if the acquiring authority were the landlord. This amount, equal to five years' rent, payable by an acquiring authority, becomes merged into any larger sum properly claimable by the dispossessed tenant under the general compensation code.

As a separate item of claim in the present reference, the D "one year's rent for disturbance" has no place.

Set-off

Having disposed of the foregoing preliminary matters I come to the substantive issue which separates the parties. Should the value to the claimant of the 1974-76 licence, and also the value of the offered licence for 1977-79, be brought into credit as a set-off against the loss of profit, the amount of which has now been ascertained?

The claimant strongly challenges this proposition. Mr Laws' evidence (which I fully accept) was that in the latter E part of 1972 and during the early part of 1973, when the special arrangement was being negotiated, the council were evincing an anxiety to go ahead quickly with the implementation of their town development scheme; that in the boom conditions then prevailing, Mr Laws had expected the major part of the 126.6 acres to be developed within two years so that by 1975 and 1976 there would be only "bits and pieces" left to farm; that the agreed £1 nominal rent was what he considered in 1973 to be an adequate payment for the 1974-76 licence; that at no time, until very recently, had any question of a set-off been mentioned by the council; that he regarded the council's present proposal, to make such F a set-off, as constituting a major departure from the basis of negotiation; and that, if he had known of this proposal at the time, he would never have allowed the licence for 1974-76 to be taken up.

Mr Denbin, for the claimant, contends that the tribunal is dealing here with a "contract situation," which has to be construed, and applied, as at the particular date of July 14 1973; that nowhere in the documents embodying the agreed arrangements is there to be found any mention of a set-off; that, as a matter of basic principle, a deed must be construed against the grantor; that the attitude of the parties at the time had been described by Mr Laws, and that the council were now seeking to vary their grant. Counsel cited G the dictum of Lord Hodson in *White & Carter (Councils) Ltd v McGregor* [1962] AC 413, 445:

> There is no duty laid upon a party to a subsisting contract to vary it at the behest of the other party so as to deprive himself of the benefit given to him by the contract.

Also the dictum of Lord Reid in *McCutcheon v David MacBrayne Ltd* [1964] 1 All ER 430, 433:

> The judicial task is not to discover the actual intentions of each party: it is to decide what each was reasonably entitled to conclude from the attitude of the other.

While recognising the persuasiveness of the above argu- H ment, I accept from Mr Fletcher, for the authority, that the council's proposal of a set-off is in no way inconsistent with the terms of the special arrangement.

"Contract Situation" within a "Compulsory Purchase Situation"

I agree with Mr Denbin that this is a "contract situation," but I find it be a contract situation within a "compulsory purchase situation." The covenant by the council in the supplemental deed of March 26 1974 that "the tenants shall be entitled to be paid compensation under section 48" of the J Act of 1973 does not provide a self-contained code for assessing the compensation payable. Indeed section 48 by itself is meaningless; it stands as no more than a small part of the general compensation code, being doubtless designed to reverse the rules established by *Rugby Joint Water Board v Foottit* (1972) 24 P & CR 26 HL and *Minister of Transport v Pettit* (1969) 20 P & CR 344 CA.

Under the general compensation code the compensation in the present case falls, as I understand it, under three heads, *viz:*

1. An amount equal to five years' rent, under section 12 of the Act of 1968, equating to four years' rent under section 9 of the Act of 1968 and importing one year's rent from section 34 of the Act of 1948. K
2. The value of the "unexpired term or interest in the land," under section 20 of the Compulsory Purchase Act 1965, assessed in accordance with rule (2) of section 5 of the Land Compensation Act 1961, at open market value.
3. "Disturbance or any other matter not directly based on the value of land," assessed as referred to in rule (6) of the Act of 1961.

As to the first head above, section 48 (5) of the Act of L 1968 requires the amount payable to be deducted from the total amount payable under the last two heads, in order to avoid a double payment.

As to the second head, the claimant here is not putting forward any figure as representing the open-market value of the claimant's interest in the land. The present case is on this account distinguishable from *Watson v Secretary of State for Air* (1954) 5 P & CR 13 and (1956) 6 P & CR 51. Nevertheless an interest in land *is* being acquired here, and one with a profit rental element. It is not for the tribunal to attach a value of its own to that interest, in the absence of evidence, instead I propose to adopt, as the value of "interest M in land," a token figure of £100.

The whole of the claim as presented falls to be looked at under the heading of disturbance. And in dealing with a claim for disturbance, the tribunal must properly have regard to what has actually happened during the interval between July 14 1973 and the date of hearing. Thus Wynn-Parry J in *London County Council v Tobin* (1959) 173 EG 533, 537:

> Now the duty of the tribunal was to estimate the capital value of the business in the new premises as at the point of time immediately after the move into those premises. Had they been hearing the case at that time any figure at which they arrived

would necessarily have been wholly an estimated figure. In fact, they were sitting nearly two years after the move. In those circumstances they were bound to have regard to what had actually happened during the interval. That this is so appears clearly from the opinions of Lord Macnaghten and Lord Robertson in *Bwllfa and Merthyr Dare Steam Collieries (1891) Ltd v Pontypridd Waterworks Co* [1903] AC 426. Lord Macnaghten said this at p 431: "If the question goes to arbitration, the arbitrator's duty is to determine the amount of compensation payable. In order to enable him to come to a just and true conclusion it is his duty, I think, to avail himself of all information at hand at the time of making his award which may be laid before him. Why should he listen to conjecture on a matter which has become an accomplished fact? Why should he guess when he can calculate? With the light before him, why should he shut his eyes and grope in the dark?" Lord Robertson said on p 432: "Any estimate of this profit necessarily involves the question, 'How long would the coal in question take to work out?' if for no other reason than in order to compute the cost of working. If the question thus presented has to be solved *ab ante*, the arbitrator, or whoever has to deal with it, must form the best conjecture he can of probable prices and probable wages during the period required to work out the area in question. But these calculations are resorted to merely in the absence of facts, and not because the time to be considered is that moment of time when the notice of October 15 1898, was presented. And if, owing to the course of the procedure, the period required for the working out of the coal in question has come to be matter of history, then estimate and conjecture are superseded by facts as the proper *media concludendi*. I do not mean that the proper course in awarding compensation is to wait for the expiry of the period required for working out the coal. On the contrary, it is natural that the compensation should be assessed once for all and by estimate. But the point is that as in this instance facts are available, they are not to be shut out."

The nature of the compensation claimable on an acquisition was defined by Lord Denning MR in *Bailey v Derby Corporation* [1965] 1 WLR 213:

> You must first ascertain the value of the land. . . . Next you must ascertain the compensation for disturbance, as it is called. That must be ascertained by looking at what has in fact happened since the notice to treat. The claimant can get compensation for his loss due to the acquisition; provided, in the words of Romer LJ in *Harvey v Crawley Development Corporation* [1957] 1 QB 485, "first that it is not too remote, secondly, that it is the natural and reasonable consequence of the dispossession of the owner." . . . All that is acquired is the land. The compensation is given for the value of the land, not for the value of the business. The business still remains the claimant's. The only thing to which he is entitled is compensation for his loss by reason of the acquisition of the land.

The question to be answered, therefore, is: What has been the claimants' "loss by reason of the acquisition of the land"? Mr Wilkin approaches this question by estimating the value to Mr Wakerley of the licence that was granted for 1974-76 and of the licence that is now offered for 1977-79; the total value of these licences being then set against the estimated loss on the basis of a total extinguishment of profit on July 14 1973. The set-off, as calculated by Mr Wilkin at £13,548, is made up as shown in the calculation on p 52.

I propose to deal with that calculation in two stages; taking first the harvest years 1974-76 and then the harvest years 1977-79.

Set-off for 1974-76

In Mr Wilkin's calculation for 1974-76 the profit figure of £53.50 per acre is the same as that used in his estimate when arriving at total loss and can be accepted without further comment. His figures of acreages cropped in each of the three years are non-contentious, as is also the risk rate of 15 per cent applied to the profit for the harvest year 1974. The two other factors in the calculation are not agreed, viz (a) a deduction of 20 per cent for injurious affection and severance and (b) a risk rate of 20 per cent for 1975 and 1976.

		£	£	£
Harvest year 1974 (rent free) profit per acre		53.50		
PV of £1 deferred 1 year @ 15%		0.870		
		46.54		
Acres cropped		126.6		5,892
Harvest year 1975 (rent free) profit per acre		53.50		
PV of £1 deferred 2 years @ 20%		0.694		
		37.13		
Acres cropped		63.0	2,339	
Harvest year 1976 (rent free) profit per acre		53.50		
PV of £1 deferred 3 years @ 20%		0.578		
		30.92		
Acres cropped		73.5	2,272	
			4,611	
Deduct for injurious affection and severance 20%			922	3,689
Value of licence for 1974–76				9,581
Harvest years 1977–79 (rent payable £15.50 per acre)				
Profit per acre		38.00		
PV of £1 deferred 4 years @ 15%	0.5717			
do. 5 years @ 15%	0.4971			
do. 6 years @ 15%	0.4323			
say		1.500		
		57.00		
Acreage available for cropping:				
1977 93 acres				
1978 87 acres				
1979 81 acres *average*		87.0		
			4,959	
Deduct for injurious affection and severance 20%			992	3,967
Total set-off				£13,548

As to the "injurious affection and severance" factor, I was told that this could be more accurately termed the "fragmentation" factor, being intended to bring into account the circumstances that the crops grown in 1975 and 1976 had produced less profit per acre because parts of the cropped areas were in smaller parcels and irregular in shape, thus making it less economical to carry out cultivations and to apply sprays; furthermore on the licence land the cultivations and sprays had been kept to a minimum. Mr Laws, while not readily accepting the adequacy of a 20 per cent allowance for these matters, did not seriously challenge that figure and I accept it.

What Mr Laws *did* seriously challenge was the use of a risk rate of only 20 per cent when capitalising the net profit per acre for 1975 and 1976. In his opinion the risk rate was nearer 100 per cent, because a farmer who was invited to crop land within a town development area and on a catch-as-catch-can basis with no compensation for crop loss might not feel inclined to accept such an invitation, even if he were offered the licence rent free. I can envisage circumstances in which such a comment would be wholly warranted. But in the circumstances of the present case I propose to adopt Mr Wilkin's 20 per cent. The evidence was that there has been the fullest co-operation throughout this difficult period, both between the two valuers, and between the council and the claimant; that the borough surveyor has kept Mr Wakerley informed about any potential risk of entry being required on any particular area during a harvest year; and that more extensive acreages had been available for cropping in 1975 and 1976 than were actually planted. In confirmation of this minimising of the risk factor it is noteworthy that the crops on the whole of the 63 acres and 73.5 acres which Mr Wakerley planted in those two years were harvested without loss. The adoption of an increased risk rate, from 15 per cent on the licence conditions for 1974 to 20 per cent on the licence conditions for 1975 and 1976, seems to me fair and reasonable.

These findings mean that I accept the set-off of £9,581, as calculated by Mr Wilkin, for the harvest years 1974-76. The claimable compensation for disturbance becomes thereby reduced from £24,003 to £14,422 or, say, £14,425.

A **Set-off for 1977-79**

Passing now to the proposed set-off for the next three harvest years 1977-79, the figure of £3,967 which Mr Wilkin seeks to bring into credit represents, of course, capitalised amounts of profit which the claimant has not yet received. Mr Fletcher, while recognising this fact, nevertheless contends that such a set-off is wholly justified, because it is well recognised that there is a duty on every claimant to mitigate his loss; accordingly the offer to Mr Wakerley of a new licence on reasonable terms to continue as from the expiration of the previous licence should be brought into account whether the claimant decides to accept it or not.

B I should mention here that the area available for cropping in the harvest year 1977 is agreed to be 93 acres, but the areas assumed by Mr Wilkin to be available for 1978 and 1979 (of 87 acres and 81 acres) are estimates made by Alan Reed MIMunE CEng, chief assistant engineer in the technical services department of the council, who gave detailed evidence in support of those figures.

What are the circumstances to be considered? Here we have part of an agricultural holding of which possession is deemed to have been taken compulsorily in July 1973 by an acquiring authority. No alternative land was offered to
C the tenant and none was found by him. The tenant's expectation of future profits, in so far as these had been previously derived from the subject land, were thus extinguished. But the tenant's total loss on this extinguishment became substantially abated as a result of sensible arrangements ("the special arrangement") negotiated and entered into between the authority and the claimant, whereby that dispossession became linked with the grant and acceptance of a licence to reoccupy the land on specified terms for a period of three years. In this way the claimant's loss became reduced, as has been seen, by not far short of £10,000, from about £24,000 to £14,425.

D This special arrangement, however, ran only to October 11 1976 and has now expired. Since October 11 1976 the claimant has had no interest in the land at all, albeit he is currently—and helpfully—acting in a caretaker capacity for it on behalf of the authority. The authority contends that the claimant is under a duty to further mitigate his loss by returning to farm such parts of the land from which he was dispossessed as the authority may find to be still available for cropping; and to do so not only for the harvest year 1977 but also for the two following years; and furthermore to do so on the footing that all the fruits of his labours over those three years are to be taken away from
E him.

The proposition put forward is, in effect, that of expecting a dispossessed trader whose business has been extinguished to mitigate his loss by starting to trade again from part of the very same premises from which an acquiring authority has just expropriated him; and to pay that authority not just a rent but also the whole of any profit which may accrue to the trader as a result of such reoccupation. I know of nothing in the compensation code that would warrant such a contention.

In my opinion no further set-off falls to be made against the figure of £14,425 already arrived at. Certainly it is open
F to the claimant to take a fresh licence or licences to farm his former land, or part of it, on terms mutually acceptable to the authority and himself, but whether he does so or not seems to me irrelevant to the present reference.

Throughout this decision I have been using the approach and the form of valuation that was used by both parties. The issue referred to as "set-off" might, however, have been more simply approached on a "before-and-after" basis, by looking first at the claimant's financial position in the absence of the authority's scheme and then looking at his position as a result of the scheme. Such an approach serves to demonstrate more clearly that the claimant is receiving "a money payment not less than the loss imposed G on him in the public interest, but, on the other hand, no greater," and that he is being thereby put "so far as money can do, in the same position as if his land had not been taken from him" (Scott LJ in *Horn v Sunderland Corporation* [1941] 2 KB 26 at 42). A reworking of the disturbance claim in this way would produce the figures set out below.

	£	£
Claimant's financial position in the absence of the scheme		
For the 1974 harvest year and thereafter		
126.6 acres cropped, at a profit per acre of £38.00 after debiting a notional rent of £15.50 per acre		H
126.6 acres @ £38.00 profit per acre		4,810.8
YP for a life expectancy, as ascertained		4.989
		£24,003
Claimant's financial position as a result of the scheme		
For the 1974 harvest year		
126.6 acres cropped, rent free and with compensation payable for any crop loss (*Note* about 18 acres were repossessed during 1974 and compensation has been paid for that crop loss)		
126.6 acres @ £53.50 profit per acre	6,773	
Deferred 1 year @ 15%	0.870	
		5,892 J
For the 1975 harvest year		
63.0 acres cropped, rent free but with no compensation payable for any crop loss due to the scheme (and accordingly at an increased risk rate), and with a fragmentation of field areas resulting in a reduced profit per acre, of only £42.80 instead of £53.50		
63 acres @ £42.80 profit per acre	2,696	
Deferred 2 years @ 20%	0.694	
		1,871
For the 1976 harvest year		
73.5 acres cropped under the same conditions as in 1975		
73.5 acres @ £42.80 profit per acre	3,146	K
Deferred 3 years @ 20%	0.578	
		1,818
		£9,581

Loss sustained as a result of the scheme:		
Financial position—before	£24,003	
Financial position—after	9,581	
Loss	£14,422	say £14,425

L I find the compensation payable to be £14,525 made up of:

Interest in land:	Token figure, as adopted	£100
Disturbance:	Loss of profit, as ascertained	14,425
		£14,525

The authority will pay the claimant the sum of £14,525, to which will be added a surveyor's fee based on scale 5 (a) of the scale of professional charges of the Royal Institution of Chartered Surveyors plus VAT, also any legal costs properly incurred from the date of the deed of surrender M up to the date of the notice of reference.

I was informed that the amount of compensation paid over on March 26 1974 in accordance with clause 2 (a) of the deed of surrender was in the sum of £7,605.90. There thus remains a balance of £6,919.10 now due to be paid as the outcome of this award. It will be recollected that interest payable on the compensation awarded is provided for in clause 2 (1) of the deed of surrender.

Having opened the sealed offer lodged by the authority, I find that the amount thereof exceeds the amount of my award. This being so, Mr Fletcher asks that the claimant

A should pay the costs of the authority as from the date of the sealed offer, while conceding that the authority should pay the costs of the claimant up to that date. Against this, Mr Denbin submits that the authority are responsible for the whole of the costs of the reference, and that the question of costs is taken out of the discretion of the tribunal by the wording of clause 4 in the deed of surrender executed on March 26 1974. This clause reads:

> The council hereby further covenant to pay the costs of the professional advisers to the tenants in relation to the negotiations relating to this deed, the gratuitous licence and the tenancy agreement and to the compensation agreed as herein mentioned
> B and all other costs arising in respect of this matter on the basis set out in the correspondence between the parties' valuers.

I find the above wording insufficient in itself to support Mr Denbin's argument; and none of "the correspondence between the parties' valuers," which might have thrown some further light on the matter, was produced.

The authority will pay the claimant his costs of this reference up to the date of the sealed offer, and the claimant will pay the costs of the authority as from the date of the sealed offer, such costs unless agreed to be taxed by the
C Registrar of the Lands Tribunal on the High Court Scale.

WILLIAMSON AND STEVENS (EXECUTORS OF WALTER WILLIAMSON (DEC'D) v CAMBRIDGESHIRE COUNTY COUNCIL (REF/143/1976)

(Sir DOUGLAS FRANK QC, President)

January 12 1977

D Estates Gazette April 30 1977

(1977) 242 EG 369

Small area of agricultural land at Peterborough acquired as caravan site for gipsies—Interim decision on points of law —Tribunal rejects claimants' submissions (1) that account be taken under s 14(12) of 1961 Act of deemed planning permission for caravan site "occupied by gipsies" in force at date of notice to treat, (2) that deemed permission was a "personal permission"—Held that "occupied by gipsies" amounts to a limitation of permission and to use land as site for "general caravans" would be a material change
E of use—BSH Holdings v West Ashford RDC "wrongly decided" by tribunal where inconsistent with present decision—President's "protest" that no application made for s 17 certificate—"Would have determined case one way or the other"—Not for tribunal to take on mantle of Secretary of State for "essentially administrative decision" whether grant of permission could reasonably have been expected

B Marder (instructed by Mellows & Sons, of Peterborough) appeared for the claimants; J Grove (instructed by the solicitor to the Cambridgeshire County Council) for the acquiring authority.

F Giving his decision, THE PRESIDENT said:

This is a reference to determine the compensation payable for the compulsory acquisition of 2.1 acres of land situated in Peterborough at the junction of Oxley Road and Edgerly Drain Road, which immediately adjoins the designated area of Peterborough new town. The interest to be valued is the unencumbered freehold with vacant possession as at the date of entry, which was March 11 1974. A compulsory purchase order was made by the county council in 1973 and confirmed in that year. The land was then used for agriculture. In the statement of reasons for the acquisition it was made clear that the purpose of acquiring the land was to provide a caravan site for gipsies. Moreover, on February 2 G
1973, the county council registered under the Town and Country Planning General Regulations 1969 a deemed planning permission in respect of development thus described:

> The use of land as part OS field no 226, Oxley Road, Peterborough, as a site for caravans occupied by gipsies (as defined in section 16 of the Caravan Sites Act 1968) for the Huntingdon and Peterborough County Council.

The permission was subject to conditions which are not relevant to these proceedings.

The parties agree that if the value is to be ascertained on the assumption that no development is likely to be permitted other than as a caravan site for gipsies or for development H
related to agriculture, then the value is £2,100. There is what purports to be an alternative agreed valuation to which I will return later.

Two Submissions in Law

Mr Marder for the claimants makes two submissions in law and I will deal with each in turn.

The first arises under section 14(12) of the Land Compensation Act 1961, which provides that there shall be taken into account any planning permission in force at the date of the service of the notice to treat. He says shortly that the planning permission deemed to be granted (to which I have J
referred) was in force at the date of the notice to treat. It is immaterial that, having regard to subsection (4), it was deemed and not an actual planning permission and that although that deemed permission is expressed to be for caravans occupied by gipsies, there would be no material change of use in using that land as a site for what I will call a "general caravan" site, that is to say as a site for caravans occupied by anybody whether gipsies or not. I think that there are two reasons why that argument cannot succeed. The first is that, in my judgment, the words "occupied by gipsies" amount to a limitation on the per- K
mission, a limitation of a kind to be found in the case of *Wilson v West Sussex County Council* [1963] 2 QB 764. I think that the words "occupied by gipsies" have a functional significance and are to be construed as limiting the proposed use to one as to occupation by gipsies.

Enforceability of Limitation

Mr Marder argues that such a limitation is not capable of enforcement. He refers to the definition of gipsies as in section 16 of the Caravan Sites Act 1968, namely:

> "Gipsies" means persons of nomadic habit of life, whatever their L
> race or origin, but does not include members of an organised group of travelling showmen, or of persons engaged in travelling circuses, travelling together as such.

and says that great difficulties could be encountered on deciding who are "persons of nomadic habit". What is a site owner to do if a person comes along asking for a site and he says he is of a nomadic habit and he is not? He gave other demonstrations of the difficulty of enforcing that limitation. As I listened I heard echoes of the illustrations given in the case of *Fawcett Properties Ltd v Buckingham County Council* [1961] AC 636, where great play was made of the difficulty in enforcing a condition restricting a house M
to occupation by agricultural workers.

But whether the limitation would be difficult to enforce is not the question before me. When there is a limitation, the question is whether it is a valid limitation. If there is difficulty then either the planning authority overcome it or they fail to enforce the limitation; that does not invalidate the limitation as such, nor do I think, to deal with another argument, that there is no power to grant a permission subject to a limitation. He said that the references in section 87(2) of the Town and Country Planning Act 1971 and in section 243 to limitations apply only to limitations on a grant under

A a General Development Order. That submission seems to me to fly in the face of the case of *Wilson* v *West Sussex County Council* [1963] 2 QB 764, where it was held in terms that the phrase "agricultural cottage" in a grant of planning permission for the erection of an agricultural cottage was a limitation. On p 777 Willmott LJ said:

> On either view there is no doubt that the phrase must be construed as limiting the use of the building that is proposed to be erected.

On p 779 Danckwerts LJ said:

> I think it is accepted that as the result of the argument in this B case first permission given, in this case by the planning authority, did not create any implied condition and in my view did amount only to a limited permission. The owner of the land asked for and received a permission for exactly what she had asked for and that was a permission to build an agricultural cottage.

So there was a case where it was held that in an expressed permission granted by the planning authority the words in dispute were a limitation.

Returning to the matter of the difficulty of enforceability, of course whether there has been a breach of a condition of limitation becomes a question for the planning authority (or an appeal to the Secretary of State), and whether occupa- C tion is by gipsies as defined would have to be determined on the particular facts at the time. In any event, even assuming in Mr Marder's favour that the words concerned are not a limitation, the question arises whether it would be a material change of use to use the land as a site for "general caravans". In my judgment there can be no doubt that it would be a material change of use. The county council has gone out of its way to make specific provision for fulfilling a duty in relation to sites for gipsies. It is perhaps worth contrasting the definition of caravan site in section 1(4) of the Caravan Sites and Control of Development Act 1960, namely:

> D A "caravan site" means land on which a caravan is stationed *for the purposes of human habitation* and land which is used in conjunction with land on which a caravan is so stationed.

So that a caravan site *simplicita* is one on which a caravan can be stationed for the purpose of occupation by anyone, whereas one which is to be occupied by gipsies is one to be occupied by gipsies as defined in the Act of 1968, and I think that is a different form of occupation. It is a matter of fact and degree, but I think that the fact and degree here point conclusively to the change being a material one.

I now turn to Mr Marder's second submission on law. It E arises under sections 15(1) and (2) of the 1961 Act. He says that the deemed permission here was one which did not accrue to the benefit of all persons for the time being interested in the land and he relies in the deemed planning permission on the words "for the Huntingdon and Peterborough County Council." I do not think that that has the effect of making the deemed permission what is called a "personal permission." If it had been intended that the permission was to be personal to the county council, then I would have expected to have found an express condition to that effect and there would seem to me to be very little point in restricting the permission in that way, but be that F as it may it is a matter of judgment, and my impression of this document is that it was not intended to be a personal permission.

"Far-reaching Propositions"

But let me again assume in favour of Mr Marder that it is a personal permission so as to bring into consideration paragraph (a) of subsection (1), namely that it shall be assumed that planning permission would be granted such as would permit the development of the site in accordance with the proposals of the acquiring authority. The argument there is a short one. G He says that the proposal of the acquiring authority was for the use of the site for caravans and that was all the permission was required for — a change of use of the land from agriculture to caravans. That, he said, exhausts the description of the development. He found himself at the end up against some rather far-reaching propositions in putting that submission to its test. For example, he conceded that if a permission were granted for the erection of an industrial building for the uses as described in the definition of light industry in the Use Classes Order, then that description would be otiose and the building could be used, for example, for general industrial purposes.

I find myself unable to accept such a far-reaching pro- H position. I think that the proposals of the acquiring authority in this case were for the change of the use of the land for caravans to be occupied by gipsies. That I think is the full description of the development. I think, indeed, that every description of a development is in itself a limitation of the development permitted. If one gets permission to build a house, that does not entitle one to build a factory. A permission to use land for a caravan site would not permit it to be used as a scrapyard. In my judgment the development permitted here is, as it was stated in the reasons for the compulsory purchase order and in the deemed planning permission, as the site of caravans occupied by gipsies and not J for a general caravan site.

Mr Marder referred to *BSH Holdings Ltd* v *West Ashford Rural District Council* (1972) 223 EG 1355. That was a case very similar to the instant case and in which Mr Marder appeared and advanced similar arguments and which apparently found favour with the member of the tribunal who heard that case. I am bound to say, however, that on reading his reasons at p 1363 I find some difficulties, because it seems to me he did not distinguish between the question of law and the question of whether permission would be expected to be granted. I find it very difficult to follow his K finding in law, still less his reasons for that finding. In any event, in so far as anything done in that case is inconsistent with my judgment, then in my judgment that case was wrongly decided.

So I turn to section 14(3), under which nothing in the provisions referred to in the previous subsections shall be construed as requiring it to be assumed that planning permission would necessarily be refused for any development which is not development for which in accordance with those provisions the granting of a planning permission is to be assumed. I protest that no application was made under section 17 for a certificate of appropriate alternative develop- L ment. It seems to me that where a claimant comes before this tribunal saying, as is said here, that permission could reasonably have been expected to have been granted, on the date of notice of entry for a development, then the appropriate course is to apply for a certificate. Otherwise, the tribunal is being asked to take on the mantle of the Secretary of State and make what essentially is an administrative decision. Of course there is no doubt on the facts of this case that a nil certificate would have been issued by the planning authority, and what I really have to assume is an appeal to the Secretary of State. What I am being asked to do is to say what the Secretary of State would have done had an appeal been M made to him in 1974 against a refusal of permission for use as a general caravan site.

Now the obvious person to answer that question is the Secretary of State himself and not this tribunal, and a certificate not having been applied for, the assumption must be that the claimant had considerable doubt as to whether an application would have been successful. A certificate would have determined the case one way or the other and indeed almost certainly would have avoided reference to this tribunal. Section 14(3) is not intended to require the Lands Tribunal to undertake the sort of exercise which has been

A imposed on me in this case. On the contrary it is directed to the hope of getting a planning permission some time in the future and not the realisation of that hope by the date of the notice of entry. However, the matter has been argued, evidence has been given on both sides, and I have no evidence on "hope" value as such, and therefore the decision I give must be an interim decision.

However, I have been asked to give some guidance to the parties,

Notional Purchaser's Views

What would a notional purchaser in 1974 have taken into
B account? First, the certainty of being refused planning permission and having to appeal to the Secretary of State with all the delay and uncertainty inherent in that course; secondly, that this is a small site; thirdly, that £17,000 would have to be spent on drainage; and fourthly, the arguments which have been advanced by the planning officer who has given evidence here. What, on the other hand, would be in favour of a planning permission? Of course, in the forefront of the purchaser's mind would be the deemed permission, and one might ask, if permission can be given for a gipsy caravan site, then why not for a general caravan
C site? Secondly, the need for a caravan site, and on the evidence I have heard, it would seem that a case could have been made, on the planning authority's own documents, for a site for an additional 130-140 caravans and moreover in this general area of Peterborough. Then again he would take into account the characteristics of this site, which in general conform with the guide-lines which have been set down by the Secretary of State himself. So, perhaps the chances of a buyer getting planning permission on appeal would be fairly good. On the other hand, one is here contemplating an outright sale not a conditional contract, and the possibility of losing every penny paid over the
D existing use value.

I think that is about as far as I am going to go. However, my attention has been drawn to section 17(1) of the 1971 Act and the fact that it is not too late to apply for a certificate with the consent either of the other party or of the tribunal. Mr Marder has not asked for leave, and I do not think I can give a leave which is not asked for; but I gather that in any event the acquiring authority would consent to an application for a certificate if one were made. So perhaps I had better say no more about that. And that I think is about as far as I can go in my decision. In the circumstances I adjourn this case *sine die* with liberty to
E apply.

SOUTH COAST FURNISHING CO LTD v FAREHAM BOROUGH COUNCIL; SAME AND HYMAN GREEN v SAME; HYMAN GREEN v SAME; CAPITAL & SOUTHERN PROPERTIES LTD v SAME

(REF/106-110/1975)

(E C STRATHON FRICS)

October 22 1976

Estates Gazette June 25 1977 and July 2 1977

F (1976) 242 EG 1057; (1976) 243 EG 53

Six adjacent, substantially vacant, back-land plots in Fareham, Hants, town centre—Included in CDA and now absorbed by town centre redevelopment—Held that planning permission could not have been reasonably expected for ad hoc individual shopping development in no-scheme world—Claimants' residual valuation rejected—Tribunal's review of residual method in "Clinker & Ash" case in "real world" of 1967 equally valid in "no-scheme" world of 1972—Valuation by reference to comparables accepted

G N Eyre QC and A Dinkin (instructed by Sherwin, Son & Roper, of Fareham) represented the claimants; J C Taylor G
(instructed by Fareham Borough Council) for the acquiring authority.

Giving his decision, MR STRATHON said: These five references concern the compensation payable by the Fareham Borough Council (in succession to Fareham Urban District Council) on the acquisition of six adjacent plots of land at the rear of the north side of West Street, Fareham, Hampshire. The owner of one plot is Hyman Green and the other four owners are companies in which Mr Green has controlling interests. The five ownerships were treated separately for the purposes of the compulsory purchase order and for the service of notices to treat. H

Graham Eyre QC, for the claimants, called E J Davies BSc FRICS, the senior partner of the firm of G L Hearn & Partners, of London. John Taylor, for the acquiring authority, called three witnesses: A G Parker MA MSc (Wales) MRTPI, the principal planning officer for development control of Fareham Borough Council; G J B Thomas ARICS, borough valuer to the Fareham Borough Council; and H E Couch FRICS, a partner in the firm of Hillier Parker May & Rowden, of London.

The references were heard together. It was agreed that for the purpose of valuation the six plots, comprising an aggregate of 3,616 sq yds, could be treated as if they were J
in one ownership. It was also agreed that for the purpose of valuation, the date of entry by the council, and thus the date of valuation, was November 10 1972. At that date the land was substantially vacant; it was agreed that the whole value of the subject property lay in the value of the site.

The subject land constitutes back-land immediately behind the northern frontage to West Street, lying to the west of High Street. The claimants' ownership includes 59 West Street, which is occupied as two shops and constitutes a physical link between the West Street frontage and the subject back-land. The claimants started to purchase their ownerships in 1956. K

I made my inspections after the hearing; the subject land and much of the terrain, the subject of evidence, had been absorbed by the town centre redevelopment. I included in my view properties at Eastleigh (Hants) and Bromley (Kent).

As the subject land was included in a comprehensive development area (hereafter to be referred to as the "CDA") the statutory background to the assessment of compensation, Mr Eyre said, is contained in sections 14, 16(4), (5) and (6) and 39(2), which require the assumptions to be made that planning permission would be granted for one of the range of uses which are specified as the purposes of the CDA, but only if planning permission might reasonably have been L
expected to be granted in what is now familiarly called the "no-scheme world"; that there were no CDA plans in existence; that no CDA development had taken place; and that the element of compulsory purchase must be ignored.

At this point I turn to the planning background, which may conveniently take the form of a précis of events in chronological order.

1951. A town centre development scheme for Fareham was first considered by the Fareham Urban District Council.

1955. The minister approved the first statutory town map for Fareham, which included no major redevelopment proposals for shopping in the centre of Fareham. M

1957. Hampshire County Council (the planning authority) submitted to the minister a proposed amendment to the development plan incorporating a comprehensive development area comprising an area of 20 acres, including the subject land. Its broad purpose was to secure this area for shopping.

1961. Following two inquiries, the minister approved amendment no 4 (1961) to the County of Hampshire Development Plan by adding:

(i) Fareham Comprehensive Development Area map together with its accompanying programme map.

(ii) designation map no 3 comprising approximately 9.2 acres of land at the rear of West Street and High Street within the CDA. The subject land was not included.

The following amendment additions to the approved written statement relating to Fareham town map were also made:

(a) The purposes of the CDA were stated to be "broadly as follows"

(i) to provide a town and civic centre for Fareham Urban District;

(ii) to secure development and redevelopment as a whole of land lying between West Street, High Street, Osborne Road and Westbury Road;

(iii) to make provision for local government offices and other public buildings and additional shopping and residences;

(iv) to provide additional car-parking accommodation for the central area of Fareham, to improve service access to existing shops and relieve traffic congestion on the trunk road (A27).

(b) CDA map. This area (20 acres) should be developed or redeveloped as a whole for the purposes of:

(a) dealing satisfactorily with conditions of bad layout;

(b) the creation of a civic centre;

(c) the provision of additional shops, residences and car parks;

(d) the provision of a rear access road to the shops fronting West Street including car parks near the access road.

(It was contemplated that West Street would remain a trunk road (A27), access to the hinterland to be provided by new streets from High Street).

Within weeks of the above approval, a review of the Fareham town map was submitted to the minister, who at once indicated that substantial modifications of the proposals would be required.

The urban district council appointed the firm of Shingler Risdon Associates, planning consultants, to advise on the form of redevelopment of the town centre.

1963. The claimants applied for planning permission (in respect of an area which included 3 acres outside their ownership) for development of the land as a shopping precinct, hotel, restaurant and car parking.

The claimants appealed against a deemed refusal of permission, although the planning authority issued a purported decision giving reasons.

1964. The minister's decision dated February 5 is reproduced hereunder:

A copy of the inspector's report is enclosed. His conclusions were that it would not be in the interests of the planning of Fareham town centre to permit the appeal proposal at present. Without prejudice to any later consideration of the revised central area scheme he regarded the complete segregation of vehicles and pedestrians as of paramount importance. The construction of an all-purpose road through the new centre would allow only limited segregation and would result in pedestrians crossing the new road to pass from one part of the centre to another. The link road would introduce more turning traffic in an already congested section of the trunk road but this might have to be accepted in any event to secure access to the service areas of the new shopping centre. The proposal could not be co-ordinated satisfactorily with the larger scheme and would be piecemeal development. In the absence of the link road, access to the site would only be gained by the three existing narrow ways (with the possible addition of a pedestrian arcade through the appellants' other property, 41 West Street) and in their existing form they would not provide suitable accesses. Moreover, for a satisfactory pattern of pedestrian movement and service roads, completely segregated, a broader plan than that embracing the appeal site would be necessary. While firm proposals by a developer should not be delayed for tentative planning schemes the inspector considered that the appointment of a general developer for the central area was reasonable assurance of early progress. On the other hand, in view of the opposition to the appeal proposal by other owners, the appellants' scheme itself could be no more than tentative. He saw no reason why the appellants should not reach terms with the appointed developer for a proper share in the overall scheme. The inspector's recommendation was that permission for the application should be refused.

The minister sees no reason to disagree with his inspector. He is aware that the authorities concerned are reviewing the approved scheme, and he notes that they are doing this pursuant to the advice he has given in Urban Planning Bulletin No 1.

In the recent joint circular on the Buchanan Report he and the Minister of Transport called for further study of the traffic and redevelopment problems of town centres. In these circumstances he would not think it right to permit substantial development which might prejudice a successful outcome to such study. For these reasons the minister agrees with the inspector's conclusions and accepts his recommendation. Accordingly he hereby dismisses the appeal.

1965. A town centre map—a non-statutory document—derived from a report by Shingler Risdon Associates was published, and in 1968, substantially in the same form, it was formally adopted by the urban district council.

1968. The minister approved the Fareham town map amendment map, which forms part of the County of Hampshire development plan. Submission of a review of the town map had been made to the minister in 1961.

The urban district council appointed the firm of Hillier Parker May & Rowden to prepare a developers' brief based on the town centre map. The first brief was produced in December 1970. It included the statement:

It is the council's intention to secure the immediate development for shopping purposes, of an area of some 4½ acres shown on the engineer and surveyor's plan entitled "Fareham CDA Phase I—Land for Commercial Development" inviting offers of ground rent and schemes in accordance with the terms set out in this brief for the first phase only.

The area covered by the brief included the subject land.

1969. The county council published "approved revisions" to the town centre map. In the introduction the county planning officer states:

The town centre map supplements the provisions of the Fareham town map amendment map, which was approved by the minister in 1968, and where any conflict exists between the maps, the town centre map, by reason of its greater detail and more recent preparation, represents the current policy of the local planning authority.

The town centre map is a diagrammatic plan designed as a simple statement of planning principles to guide development in the town centre and it does not attempt to show how individual properties or sites may be affected. This statement and the attached plan should be read in conjunction with the draft town centre map written statement prepared by the consultants in 1965."

1970. A further application by the claimants for planning permission to develop the subject land was refused. The claimants did not appeal.

1972. Following an inquiry in 1971, the minister approved a compulsory purchase order in respect of an area of 3.4 acres comprising the subject land and other land.

I will first set out the valuations (as amended during the hearing), which will show the extent of the differences between the witnesses as to value. I will later examine the circumstances and reasons which have caused the three surveyors to adopt their respective approaches to their valuations. It will be seen that Mr Davies has adopted the residual method of valuation. Mr Thomas bases his valuation on comparables. Mr Couch considers that the residual method of valuation is not appropriate but he puts forward calculations in criticism of that method; he formed his opinion of value by reference to the comparables cited by Mr Thomas. I have rearranged the residual valuations by excluding supporting detail, so that the differences may be more clearly identified.

A *Valuation on residual method by E J Davies BSc FRICS (for claimants)*

Capital value of completed development.

Estimated rental value per annum ...	£82,250	
Years' purchase at 6¼%	16	
		£1,316,000
Less development costs:		
Site assembly	£41,500	
Car parking	32,745	
Building costs	369,225	
Finance	45,529	
Letting fees	8,225	
B	497,224	
Developers' profit 10% of completed value	131,600	
	629,824	
	say	630,000
		£686,000

Apportionment:

Site value ...	100.00%
Interest thereon ...	20.25%
Fees and costs ...	3.00%
C	123.25%

Capital value of development site subject to following deduction:		£556,592
Less freehold value of 59 West Street, in the claimants' ownership but required for access		88,500
		£468,092
	say	468,000
Less		
D (a) Payments for extinguishment of rights of way	£2,765	
(b) Contribution towards construction by local authority of rear service road	10,000	
		12,765
Site value ...		£455,235

Summary of residual calculation by H E Couch FRICS (for acquiring authority), using Mr Davies' method of valuation.

Capital value of completed development

Estimated rental value per annum ...	£54,000	
E Years' purchase at 6¼%	16	
		£864,000
Less development costs	£598,047	
Developers' profit 20% of £864,000 ...	172,800	
		770,847
Residue ...		93,153
Less: On basis of decision of Lands Tribunal in *Stokes v Cambridge*, one third of £93,153		31,051
Net residue (site value) ...		62,102
F	say	£62,000

(*Note:* Certain fees and interest charges were not included by Mr Couch in the development costs.)

Valuation based on comparables by G J B Thomas ARICS (for acquiring authority)

Plot 17a	1,204 sq yds @ £18 per sq yd	£21,672	
	less to buy out pedestrian right of way in favour of 61 West Street (say) ...	150	
			£21,522
Plot 17b G	1,987 sq yds		
	360 sq yds (highway area)		
	1,627 sq yds @ £18 per sq yd ...	29,286	
	Nominal value of highway area (say)	100	
		29,386	
	less to buy out (a) pedestrian right of way in favour of 61 West Street (say) £150; (b) vehicular right of way in favour of 57 West Street (say) 2,500	2,650	
			26,736
Plot 17c H	232 sq yds @ £18 per sq yd	4,176	
	less to buy out vehicular right of way in favour of builder's yard (rear of 49 West Street)	3,500	
			676
Plot 17d	72 sq yds @ £9 per sq yd		648
Plot 17e	82 sq yds @ £18 per sq yd		1,476
Plot 19	39 sq yds @ £18 per sq yd		702
	Value of whole site ...		£51,760

Valuation by Mr Couch based on Mr Thomas's comparables	£60,000

J

Valuation by Mr Davies based on Mr Thomas's comparables

3,616 sq yds @ £30 per sq yd			£108,480
Less for extinguishment of rights of way:			
(a) 61 West Street	£300		
(b) 57 West Street	—		
(c) 49 West Street	2,000		
	2,300		
Interest on £2,300 for 2¼ years @ 9%	465		
		2,765	
K			£105,715

Mr Davies takes the view that in November 1972, conditions in a "no-scheme world" would have caused a prospective purchaser to assume that planning permission would be granted for the commercial development of the subject land in the form of a "schematic layout" which he, Mr Davies, had prepared (exhibit c1(6)). Having made that assumption, it followed, Mr Davies said, that only the residual method of valuation could properly measure the value of the development potential in the site. Mr Thomas and Mr Couch L denied that such planning assumption would have been made; their view was that purchases of comparable plots of land by the council fully reflected the levels of "hope value" which purchasers would be prepared to bid.

I turn to the opinions of the witnesses as to what would have happened in the "no-scheme world."

I will first take Mr Davies' evidence. The mood of the market in 1972 was "bullish"; the boom in property development and redevelopment reached its highest level in early 1973; risks were discounted by prospective purchasers to M whom finance was readily available and there was a lack of sites for development. By July 1973 prices were declining, and the collapse that occurred was inevitable. The present development of Fareham conforms not with the original CDA but with the town centre map, which clearly envisages the development of individual sites. In the Strategic Plan for the South-East, south Hampshire is identified as one of seven major growth areas, and in the Hampshire Consultation Draft Structure Plan (1972), Fareham was importantly mentioned. The claimants were included in a list of 13 developers to whom the first brief, prepared by Hillier Parker May & Rowden, was sent.

Samuel Properties were eventually selected as developers of phase I; development started in 1974, outline planning permission having been granted in September 1971, followed by detailed planning permission in June 1973, for 142,085 sq ft of shopping floor space.

At the time of the claimants' planning appeal in 1963, West Street (A27) was treated as a traffic road, but in 1972 it was known that it was to be converted to a pedestrian street. Mr Davies had taken into account (a) the landownership pattern at November 1972, including the extensive adjacent urban district council ownership, other than the areas acquired by the council under the compulsory purchase order, (b) the known estimated increase of population in the Fareham catchment area to 133,000 by 1981, (c) the known requirement of 252,000 sq ft of shopping space, (d) measures to relieve traffic congestion in West Street which could be expected to be proposed, (e) the known requirement for additional off-street car parking for the shopping area. If compulsory powers to purchase would not have been available, the council would have had to determine its policy for the town centre.

It was Mr Davies' opinion that the council would have sought to achieve their fundamental objectives by encouraging individual private developers to develop and redevelop and to make the best use of their ownership. Such objectives would include encouragement to provide shopping space in modern units of appropriate size, the segregation of vehicles and pedestrians, the provision of rear servicing for as many shops as possible, a pedestrian shopping mall system, improvements to the road system so as to relieve traffic congestion in West Street, the proper location of car parks and new and additional public buildings in the town centre. The definition of such a framework would be the responsibility of the council; its implementation would require access points at the rear of the north side of West Street and direction by the council as to the level at which main servicing would be carried out; having regard to the levels of the land, it is clearly predictable that access would be through back land in the ownership of the council; the depth of shopping development and the general line of a pedestrian mall would also be prescribed by the council.

In the short term the council would wish to secure additional shopping floor space as soon as possible; the ministry, in the decision on the claimants' planning appeal in 1964, had made known its attitude that the grant of permission in respect of firm proposals by a developer should not be delayed for some future tentative scheme. In 1972 the purchase of the subject land together with 59 West Street, would have attracted strong competition from developers and major retailers, who would also be aware of the council's significant ownership of adjacent back land.

Purchaser Influenced by Three Factors

An intending purchaser would prepare a schematic layout to establish a form of development that would have a reasonable chance of securing planning permission. When making his bid he would be influenced by three factors which would not require the prior grant of a planning permission, namely (1) the purchaser would establish his position in Fareham to enlarge the site by later purchases, (2) the purchaser would establish a position for negotiating with the council a form of participation in a larger scheme of development, and (3) he would anticipate that planning permission would be forthcoming for individual development of the site in accordance with the schematic layout c1(6), which provides for 34,000 sq ft of shopping floor space, pedestrian malls, modern shop units including a supermarket, rear servicing facilities, which, for future developments, could be extended to the east or west, and an east-west mall on a median line to permit future extension; overall, such a layout would provide an important catalyst for back-land development.

Another schematic layout was produced by Mr Davies to demonstrate how the layout c1(6) development could be incorporated in the development of adjacent back land in accordance with the criteria which would prevail in the "no-scheme world"; a further layout was designed to show how in the short and medium term, the access for rear servicing of shops would lie across land in the ownership of the council.

It was on this appreciation of the conditions which would have prevailed in the "no-scheme world" that Mr Davies had concluded that the market value of the subject land could be assessed only by the residual method of valuation.

Mr Parker for the acquiring authority agreed that in a "no-scheme world" proposals for major shopping development in Fareham, contained in the town centre map of 1969 and in the draft structure plan for south Hampshire of September 1972, would have been known at November 1972. The area north of West Street would have been the most appropriate for a substantial increase of shopping space. The principle of the provision of shopping space envisaged by layout c1(6) would have been acceptable in accordance with the approved town map (1968) use of land in the central area; also acceptable would have been the principle of first-floor-level servicing approached by a rear access. But the planning authority would have been seeking a much larger increase of shopping space, which could only be attained by coupling layout c1(6) with other development in the central area.

It was Mr Parker's view that planning permission would have been refused and that it would have been expressed as follows: "the site forms part of an area which should be developed on a comprehensive basis and the proposals of c1(6) represent a form of piecemeal development which would not only be undesirable but would prejudice the proper future development of the area as a whole."

Scheme not up to Acceptable Planning Standards

As to the detail of c1(6), there was no provision for a route from the rear service ramp to an access from a public highway, nor did it lead to an area with an acceptable surface for safe and convenient movement of traffic; a part of the ramp was on land not in the ownership of the claimants. Mr Parker concluded that it would not have been possible to provide in the "no-scheme world" an acceptable rear access for the individual c1(6) scheme; access would have been possible only from West Street, which would have interfered with the free flow of traffic on a trunk road and this would have constituted a further ground for the refusal of planning permission. Congestion and difficulty of manoeuvring on the parking area, and inadequacy of the width of the shopping malls were other details of the scheme which fell short of acceptable planning standards.

Mr Parker accepted that in principle the subject land could be used as part of a larger development scheme; many of the essential planning criteria were met by the c1(6) layout, but in the "no-scheme world" the local authority and the planning authority would need to be assured that it formed part of a larger scheme which would come to fruition; such an assurance could not be given. In criticism of Mr Davies' shopping layout of a "possible town centre—Fareham," Mr Parker said it required the assumption to be made that the land would become freely available and that rights over land would be extinguished, but this could only follow a long period of negotiation with the council (the owner of a large part of the shopping area) and 15 other owners.

Mr Thomas, the borough valuer of Fareham, had relied on the advice from the planning department of the council, including Mr Parker, that the subject land would be incapable of independent development. He had assumed that it would be included in a more comprehensive development at some future time.

Mr Couch was in agreement with Mr Davies as to the council's objectives in a "no-scheme world"; he agreed also the factors which Mr Davies had taken into account relating to the landownership pattern, the estimated increase in population, the known requirements for shopping space, traffic congestion and requirements for off-street car parking. But the council would not have sanctioned small schemes if they were likely to prejudice the principal aim of securing major traders and large increase of shopping space. Mr Couch thought that Mr Davies had given a fair illustration of the criteria which should be taken into account for the redevelopment of the town centre, but a potential developer would also have noted: (a) access to the site from West Street was restricted to Cawtes Place, only 11 ft 8 in wide, (b) a rear access from Osborne Road and/or High Street would be needed in view of the council's known wish to convert West Street to a pedestrian way, (c) the site was burdened with rights of way, (d) a stopping-up order for the public highway, Cawtes Place, would be required, (e) the area of the subject site—0.75 acre—was small for an individual development.

Mr Crouch had concluded that in the knowledge of these considerable constraints no developer would have contemplated an individual development of the site but development would have been considered in conjunction with adjacent land. Mr Couch considered the layout c1(6) a poor *ad hoc* scheme, designed to fit the site; it had the characteristics of a cul-de-sac; of the five shop units, two were of poor quality in this particular scheme; the restricted size of the proposed supermarket would not interest the major firms, and the lesser-known supermarket traders would have been concerned about layout, car parking and the practicability of future expansion.

I will now consider whether planning permission might reasonably have been expected to be granted in the "no-scheme world" in respect of the subject land in November 1972 in accordance with Mr Davies' layout c1(6). There is much common ground. The Shingler Risdon town centre map in its 1968-69 form would have been the instrument of planning control in an area which was to experience a large population growth. Fareham could be seen to occupy such a significant role in shopping that its status as a district centre would become closer to that of a sub-region; it was a role that could only be realised by substantial development and redevelopment in the centre. The national-multiple traders, restricted to existing premises on the West Street frontage, were poorly represented throughout the period 1957-72; such pressure had inevitably carried the unwelcome lineal extension of shopping in a westerly direction. The land which possessed the potential to meet demand lay in the back land at the rear of West Street, and, in any concept of development, links from the rear to the frontage would be an essential feature. The development infrastructure of the region and the local authority's commitments to highway improvements were known factors in November 1972. The property market was particularly buoyant and there was no reason to assume that it would not continue to be so.

These matters of common ground were, as Mr Eyre said, of equal significance both in the "scheme world" and in the "no-scheme world." But, Mr Eyre said, there could be a danger of being influenced by the real facts obtaining in Fareham and wholly attributable to the "scheme world"; for example, the comparison of any notional scheme with the actual town centre development that has happened; in more general terms, comparison should not be made with a scheme that could be achieved only in the "scheme world."

It was also common ground that the council would have been anxious to take advantage of the favourable climate for shopping space development in 1972, and as the dominant owner of the back-land potential, it would have had the opportunity to make its own contribution to attain its long-held objectives.

Witnesses in "Some Difficulty"

The witnesses, as Mr Eyre said, labour under some difficulty, since their empirical experience stems from their operations in the "scheme world." Nevertheless, firm opinions were given: Mr Davies was confident that the general pattern of redevelopment would have been sufficiently predictable to enable the planning authority to conclude that development of layout c1(6) would not hazard the overall objectives, including the provision of eventual total additional shopping space of 250,000 sq ft. I think the authority would have been more cautious in its consideration of the c1(6) application. Mr Parker and Mr Couch agreed that the pattern of ownership of the back land might not have remained static during the previous decade in the "no-scheme world," but it was a matter for speculation as to what changes in that pattern might have occurred. Such a possibility did not affect their opinions that the objectives of the planning authority would be jeopardised by a grant of permission for an *ad hoc* individual development of scheme c1(6). I accept these opinions.

Mr Davies rightly said that an appeal would lie from a refusal of permission to the Secretary of State for the Environment. On this matter Mr Eyre, referring to the minister's letter of February 5 1964, dismissing the claimants' appeal at that time, submitted that the Secretary of State would have to envisage that the scheme c1(6) application, if judged by the high standards achieved in the "scheme world," conforms with the instrument of control and the owner has waited for nine years before making this further application; these matters, said Mr Eyre, were very compelling indeed and this is the climate in which the decision would be taken. On the evidence adduced I have come to the conclusion that the appeal would be dismissed and I find that at November 1972 planning permission could not have been reasonably expected to be granted to carry out the c1(6) development.

Method of Valuation

This finding leads me to the next phase of my decision, namely to determine the method of valuation which in the circumstances of this case should be followed to arrive at the market value of the subject site. The alternatives canvassed in this case are (a) the use of the residual method of valuation (per Mr Davies) or (b) by reference to comparable transactions (per Mr Thomas and Mr Couch).

Rejecting alternative (a) as an acceptable approach, Mr Taylor said he relied on the decision of the Lands Tribunal in *Clinker & Ash Ltd v Southern Gas Board* (1967) 203 EG 735, where the characteristics of the method and of its use were the subject of a detailed review by the member R C Walmsley, who said:

> As I have mentioned, the expert evidence on behalf of the vendors was given by Mr Walton and Mr Vallis. Each produced a valuation and both valuations were prepared on the residual method. Now the tribunal has frequently rejected such valuations as having too many uncertain elements, and has almost invariably done so when some simpler method (such as the use of comparables) is available. Miss Viner was well aware of this, and indeed appeared almost reluctant to put forward the valuations of Mr Walton and Mr Vallis at all. In the present case, however, it will be seen later that I intend to accept the end-figure in each of these valuations at its face value. As this is an apparent departure from the normal practice of the tribunal, it is proper that I should explain the reasons which prompt this course.

A In the form in which it is normally presented to the tribunal, the residual method for the valuation of a development site shows a site value which is thrown up as the difference figure between the estimated value of a completed development of the subject land and the estimated cost of carrying out that development. The figures of completed value and development cost are usually both much greater than the difference between them, ie they are both much greater than the site value which is being sought. In the make-up of both the completed value and the development cost there are a number of variables; the appropriate figure to be adopted for each of these variables will depend on the viewpoint as well as on the knowledge and experience of the valuer; the choice from which each such figure may be made is a fairly wide B one, varying from what may be termed "conservative" to what may be termed "full"; and within this range whatever figure may be adopted is "correct," in the sense that it can be substantiated. From the viewpoint of a valuer who is retained by an intending vendor and who has therefore a responsibility to ensure that his client obtains not less than the full value of his land, there is a natural tendency to adopt somewhat full figures for the variables which together make up the completed value and/or to adopt somewhat conservative figures for the variables which together make up the development cost. Conversely, from the viewpoint of a valuer who is retained by an intending purchaser and who has therefore a responsibility to ensure that his client does not pay more than the full value of the land, there is a C natural tendency to adopt somewhat conservative figures for the variables making up the completed value and/or somewhat full figures for the variables making up the development cost. At this point of divergence, however, between the two valuers, the discipline of open-market conditions intervenes, imposing external sanctions which are highly effective. The external sanction facing the valuer for the intending vendor is that, if his choice of figures for the variables should throw up too great a difference between completed value and development cost, his client may well fail to find a purchaser at all because the calculated site value is above actual open-market value. The external sanction facing the valuer for the intending purchaser is that, if his choice of figures for the variables should throw up too small a D difference between completed value and development cost, his client may well fail to buy the land at all because the calculated site value is below actual open-market value. It is a striking and unusual feature of a residual valuation that the validity of a site value arrived at by this method is dependent not so much on the accurate estimation of completed value and development cost as on the achievement of a right balancing difference between these two. The achievement of this balance calls for delicate judgment but in open-market conditions the fact that the residual method is (on the evidence) the one commonly or even usually used for the valuation of development sites shows that it is potentially a precision valuation instrument. If there are two equally proficient valuers acting respectively for a willing vendor E and a willing purchaser they would thus expect to agree on a price for the site in question, it being irrelevant for this purpose that one valuer may have arrived at the agreed site value by using figures for completed value and development cost differing substantially from those used by his colleagues.

When a residual valuation is prepared for arbitration purposes, however, the conditions are very different; the valuation is then a calculation made *in vacuo;* and although there may be a *deemed* open market there are no external sanctions acting as an incentive to the achievement of the delicate balance which I have described, because there is in effect a captive purchaser and a captive vendor; thus there is no risk on the vendor's part of losing a sale by reason of the price advised by his expert being F too high, nor is there any risk on the purchaser's part of missing a buy because the price advised by *his* expert is too low. Possibly as a side-effect of this absence of any external constraint, the natural tendency of the vendor's (or claimant's) valuer to adopt full figures when calculating developed value and conservative figures when calculating development cost almost invariably results (in the experience of the tribunal) in his putting forward an undependably high opinion of site value. Similarly the natural tendency of the purchaser's (or authority's) valuer to adopt conservative figures when calculating developed value and full figures when calculating development cost almost always results in his putting forward an undependably low opinion of site value, and on occasion it may even throw up a minus site value. Having observed on so many occasions the working out of these tendencies in terms of widely conflicting "valuations," the deep G impression on the minds of the tribunal is that under arbitration conditions ". . . once valuers are let loose upon residual valuations, however honest the valuers and reasoned their arguments, they can prove almost anything" (*First Garden City Ltd v The Letchworth Garden City Corporation* (1966) 200 EG 123, 225, 365, 457, 507, 603, 714, 783). It is against this background and for this reason that the tribunal has reluctantly found itself compelled to reject the residual method when put forward as opinion evidence, unless there is no simpler method of valuation available.

The above are general comments, but they may be illustrated by the experts' valuations in the present case. Mr Walton and Mr Vallis had both prepared their valuations under the discipline H of open-market conditions and prior to any dispute arising, the former on behalf of intending vendors as the basis of an auction reserve price of £30,000, and the latter quite independently on behalf of a possible purchaser as the basis of an intended auction bid of at least £30,000.

In my judgment, that review, which concerned the "real world" of 1967, remains equally valid in the circumstances that would have prevailed in the "no-scheme world" of 1972.

The Residual Method J

I turn to what the valuers in the instant case had to say about the residual method. Layout c1(6) was the scheme on which Mr Davies relied to establish the market value of the subject land by the use of the residual method. It was his view that an extant planning permission to carry out the c1(6) scheme as an individual development would not necessarily be required by a prospective purchaser who, when making his bid, would be prepared to make the assumption that planning permission would be forthcoming. But my finding that planning permission for the layout c1(6) could not reasonably have been expected to be forthcoming leads K also to the conclusion that a valuation based on that scheme is of no assistance in establishing the value of the subject site. It is of interest that Mr Davies had accepted in cross-examination that in the four illustrations of his use of the residual method in practice, planning permission for development had been granted in all cases. (Gravesend (2), Eastleigh and Bournemouth). The experience of Mr Couch had been that the residual method was used in 1972 where the schemes were realistic and capable of certain implementation. It was an important tool in the hand of a developer to arrive at the values of land, but in doubtful or hypothetical situations— valuation of the subject land would be in this category—the L use of the residual method was not appropriate. A negative response came from Mr Thomas, who had no practical experience of the use of the residual method. As part of his philosophy, he took the extreme view that the method was so unreliable that it should not be used in any circumstances. Of Mr Couch's five categories of potential purchasers of land for development, I accept that a purchaser of the subject site would probably come from the band of small development companies; the purposes of buying would be to establish a position in Fareham with the aim of enlarging the site and/or to establish a position from which to negotiate with the local authority for participation in a scheme M of development.

In my judgment, on the evidence adduced, the use of the residual method of valuation to arrive at the value of the subject land is not acceptable. It is therefore unnecessary to consider the details of Mr Davies' valuation or the widely differing calculations produced by Mr Couch.

The alternative valuation approach, on which Mr Thomas relies (with Mr Couch in support), is based on 11 comparables all of which are freehold acquisitions by the Fareham Borough Council; all are within the CDA; nine of the 11 acquisitions are included with the subject land in the same

A compulsory purchase order; two of the purchases were not included in the compulsory purchase order, namely, reference nos 8 and 9, owned respectively by the Hampshire County Council and the Post Office.

Extracts from a schedule of details of the 11 comparables produced by Mr Thomas are given in the following table; the entries have been arranged in chronological order of the relevant dates of assessment.

Comparables (by G J B Thomas ARICS)

B

Plan Reference	*Vendor*	*Date of assessment*	*Site area*	*Price paid*	*Price per sq yd*	
			sq yds	£	£	
2	Exors. of W E Cole deceased	29.4.71	690	4,500	6.52	
7	Mr Etherington	12.5.71	1,667	11,350	6.81	
4	Capt Pullen	14.7.71	320	2,000	6.25	Part of larger interest.
10	Brutton & Co	24.9.71	721	10,000	13.87	
6	F J Bussey	11.10.71	42	400	9.50	
1	Halfords Ltd	10.10.71	3,646	30,000	8.22	
11	Exors of Abraham deceased	9.11.71	1,353	10,750	7.90	
5	Boots Pension Fund	23.10.72	90	1,300	14.44	Proportion of total price paid £3,800.
3	Barclays Bank Ltd	19.3.73	574	7,100	12.37	Proportion of total price paid £7,450. Effective proportion of total area of 626 sq yds.
9	Post Office	7.3.73	4,356	100,000	22.96	
8	Hampshire County Council.	1.9.73	3,579	80,000	22.59	

C

The vendors of the nine comparables within the compulsory purchase order were professionally advised; the sale prices of nos 8 and 9 were determinations by the district D valuer, whose opinion had been sought by the parties.

Mr Thomas relied on the 11 settlements to show the trend of increase in values over a period of two and a half years; he regarded some as more relevant than others and considered nos 8 and 9, at £22.59 and £22.96 per sq yd, as marking the upper limit of value. Mr Thomas had finally selected reference 1 as providing the best evidence from which to derive the value of the subject site. He used this comparable.

Settlement price per sq yd	£8.22
E Adjust for increase in price from date of settlement October 10 1971 to date of valuation of the subject site November 10 1972 Multiplier ...	2
	£16.44
The relevant land has access, so round up to ...	£18.00 per sq yd

Mr Thomas thereupon valued five of the plots of the subject land at £18 per sq yd, but the sixth plot, reference 17d, at £9 per sq yd; Mr Thomas thought that plot 17d, a small one of 72 sq yds at the eastern boundary, would not F be included with any other development within a reasonable time. His valuation, £51,760, represents "hope value" as part of a larger development at some future time.

Mr Couch concluded that there were no grounds to justify a value for the subject site outside the range of the prices of the comparables. He also selected reference 1 as particularly comparable. He had estimated the "existing use value" of the subject site at £25,000—Mr Thomas's figure was £8,000. In making a final judgment as to how much would be paid for "hope value" in excess of £25,000, Mr Couch thought that a developer having regard to the characteristics of the site might have paid somewhat more than Mr Thomas's £51,760. Mr Couch said he arrived at a figure in the region G of £60,000.

Mr Davies found no assistance from the comparables. But he had examined each transaction and he had taken the view that if comparison were to be made with reference nos 8, 9 and 10, he would arrive at a figure of £30 per sq yd.

Mr Taylor said that the Lands Tribunal has regarded with some reluctance the making of comparisons with settlements of compensation; this was recognised in the decision of *Xerri and Shanks v Enfield London BC* (1966) 18 P & CR 117 where the learned President, Sir Michael Rowe QC, said that the danger of the setting of a precedent by the first settlement was a very real one. The president added "but the tribunal H has also said that if settlements of this kind are to be disregarded, there must be evidence clear and positive to prove that they were on too low a basis."

There are no comparables of private sales. The vendors of nine of the comparables were professionally advised, and I was told that the district valuer acted in the capacity of an arbitrator to determine the prices to be paid for the plots owned by the Post Office and by the Hampshire County Council. On a review of all the comparables I accept Mr Davies' opinion that these two transactions (£22.96 and £22.59 per sq yd) are the most helpful. It is clear, however, J that Mr Thomas had regarded these settlements as having been assessed in conditions of the "scheme world," and following cross-examination of Mr Davies, the following written statement was submitted by Mr Taylor:

"I would agree with Mr Davis (*sic*) that the valuation basis for the acquisition of property not within the CPO by one authority from another is on a direct comparison with the no-scheme-world acquisitions. It must be recognised, however, that the vendor authority are in a powerful negotiating position and it is inevitable that, because of the urgency of the scheme-world situation, a settlement is influenced by that urgency. The vendor authority have to be persuaded to sell and to give possession, and the reality is that the scheme is a well-known fact. I do not wish K to suggest that such an influence has a substantial effect on settlement, but the influence of the scheme-world is nevertheless there."

Tribunal's Decision

In my judgment these two transactions should not be ignored; I have accepted Mr Davies' evidence that if the comparables are to be used as a guide, references 8 and 9 are relevant and I consider the value of the subject site should be derived from comparison with them. I heard the opinions of the three valuers about the comparative characteristics and L qualities of the two comparables and the subject site and I have come to the conclusion that the value to be ascribed to the subject site free of encumbrances should be £23 per sq yd. A deduction must be made for the extinguishment of rights of way; having regard to the evidence on this item I assess this deduction at £3,000. A summary of the details of my determination of compensation at £80,168 is therefore:

3,616 sq yds @ £23 per sq yd	£83,168
Less payment for extinguishment of rights of way ...	3,00
	£80,168 M

The acquiring authority will pay £80,168 compensation to the claimants in respect of the five references to this tribunal. In addition the acquiring authority will pay to the claimants their proper legal costs from the dates of the notices to treat to the date of the references to the Lands Tribunal, and a surveyor's fee on scale 5 of the scales of charges of the Royal Institution of Chartered Surveyors.

I was informed by counsel that although it had been agreed that for the purpose of valuation the subject land could be regarded as being in one ownership, the parties would require,

A for reasons of title to property, a single sum (if so awarded) to be apportioned between each parcel of land, the subject of claim. I have determined the value of the land as if it were held in one ownership. I consider that the single sum £80,168 should be apportioned between the several plots in the proportions that the respective areas bear to the area of the whole site, which the parties have agreed is 3,616 sq yds. The amounts of compensation to be paid in respect of the individual parcels will therefore be:

B

Reference no.	*Area sq yds*	*Apportioned compensation* £
17A	1,204	26,693
17B	1,987	44,052
17C	232	5,144
17D	72	1,596
17E	82	1,818
19	39	865
	3616	£80,168

Having opened the sealed offer of compensation lodged by the acquiring authority, I find that the amount thereof exceeds the amount of my award; accordingly the acquiring authority will pay the costs of the claimants of these references up to February 20 1976, the date of the sealed offer, and the claimants will pay the costs of the acquiring authority as from that date. All such costs, if not agreed, to be taxed by the registrar of the Lands Tribunal on the High Court scale. C

D

LANDS TRIBUNAL

LEASEHOLD REFORM

A **H A PATTEN v WENROSE INVESTMENTS LTD; N M LACKLISON v A A DAVIS (LIQUIDATOR OF FINSBURY DEVELOPMENTS LTD); G E THOMAS v SAME (LR/15, 28, 29/1976)**

(R C WALMSLEY FRICS)

November 17 1976

Estates Gazette February 5 1977

(1976) 241 EG 396

**Three terraced late-Victorian houses in Clapton, East London, each adapted and occupied as two residences—Question of correct entirety value—Tribunal accepts con-
B cept of house properly converted into two maisonettes—Likelihood that increase in market value due to conversion "might well be comparatively small"**

Lynn Dennis (instructed by Hutchins & Co) appeared for Mrs Lacklison and (instructed by Clinton Davis & Co) for Mr Patten and Mr Thomas; M Driscoll (instructed by Woodroffe-Robins & Co) for the landlords.

Giving his decision, MR WALMSLEY said: These three references were, at the request of the parties, heard together. Each reference is for the determination under the Leasehold Reform Act 1967, as amended, of the price which should be
C paid by the resident lessee for the freehold interest of the house in question. The three subject houses are close to each other in Hackney, near to where Lea Bridge Road runs into (Upper and Lower) Clapton Road. Their postal address is Clapton E5. All three are terraced Victorian properties, built at the turn of the century, and are typical of a great many other houses in the same locality.

At the hearing Miss Lynn Dennis appeared for the tenants and called evidence from Philip Tobin ARICS, a partner in the firm of Strettons. Michael Driscoll appeared for the landlords and called evidence from R A Creed FSVA, manager in the firm of Davies & Davies.

D I will take each reference separately.

LR/15/1976. 4 Cotesbach Road

The lessee of 4 Cotesbach Road is Henry Arthur Patten, whose interest is for the residue of a 99-year term from June 24 1899, and at a ground rent of £6 per annum. The relevant date for the assessment of the price is January 21 1976, being the date on which the lessee served notice on the landlords, Wenrose Investments Ltd, of his desire to have the freehold. The lease had then about 22½ years to run.

The respective valuations followed the familiar standing-house approach. For comparison purposes they may be con-
E veniently summarised as follows: (see top of next column).

It will be seen that there are only two matters in dispute. One is the appropriate risk rate. The other is the entirety value.

As to the risk rate, Mr Tobin uses 8 per cent and Mr Creed 7 per cent. The tribunal's decisions over the years have embodied risk rates, at the three stages of the calculation, varying in the main between 6 per cent and 8 per cent, the actual figures depending in each case on the evidence presented. In the present case I adopt 8 per cent as used by

	P Tobin for the tenant			R A Creed for the landlords			F
Present ground rent		£6			£6		
YP	23 yrs @ 8%	10.371	£62	22½ yrs @ 7%	11.167	£67	
Reversion to S 15 rent Assumed entirety value		£10,750			£17,250		
Proportion attributable to site		30%			30%		
Site value		£ 3,225			£ 5,175		
Decapitalised at		8%			7%		
Modern ground rent..		£ 258			£ 362		
YP in perpetuity	@ 8% def 23 yrs	2.128	£549	@ 7% def 22½ yrs	3.119	£1,130	
			£611			£1,197	G

Mr Tobin: interest rates by early 1976 were running at a materially higher level than in previous years. I am using the same risk rate at all three stages because that is what both valuers did.

The second matter in dispute, the entirety value, is less easily disposed of, because it involves first deciding between two different valuation concepts. To do this entails some further description of the circumstances.

No 4 Cotesbach Road is assessed as a dwelling-house at a H
rateable value of £274. The accommodation comprises six main rooms, thus: *First floor*—front room (1), rear room (2) used as kitchen; *Half-landing*—back-addition room (3), bathroom, separate wc; *Hall floor*—Front room (4), rear room (5); *Garden level*—back-addition room (6), kitchenette; *Outside*—wc.

The tenant's notice of January 21 1976 stated that since July 1969 his son (and family) had occupied the first-floor rooms, and since August 1972 had also occupied the ground-floor front room. The house is thus occupied as two residences, albeit without a sub-tenancy being created. There is a shared use of the bathroom and the wc. J

On these facts Mr Tobin, for the tenant, adopts the valuation concept of the house as it stands, deemed to be in good condition and modernised. His opinion is (and this is agreed by Mr Creed) that 4 Cotesbach Road is worth no more with one of its rooms adapted for use as a second kitchen than if it had not thus been made suitable for use as two residences.

The valuation concept adopted by Mr Creed, for the landlords, is quite different, namely that the subject house is deemed not only to be in good condition and modernised but also to have been properly converted into two maisonettes.

In considering which of these two concepts is the more K
helpful it must be borne in mind that the estimation of an "entirety value" is but the first stage of a valuation exercise directed solely to the finding of a notional ground rent described in section 15 as being:

> "a ground rent in the sense that it shall represent the letting value of the site (without including anything for the value of buildings on the site) for the uses to which the house and premises have been put since the commencement of the existing tenancy, other than uses which by the terms of the new tenancy are not permitted or are permitted only with the landlord's consent."

A I should interpolate here that I was not shown a copy of the original lease of 4 Cotesbach Road, and I was given no information as to its terms. In the absence of evidence to the contrary I must assume that the uses to which the house and premises are presently being put are uses permitted under the terms both of the existing tenancy and of the new tenancy described in section 15.

Returning now to the question of the concept: the customary way of deriving a section 15 site value by the standing house approach is to start with an "entirety value" which represents the value of the property in good condition *and fully developing the potential of its site*. The advantage
B of this valuation concept is that it enables an external check to be made by reference to the relationship which—as has been described to the tribunal in a number of earlier cases—has been found to subsist between the value of a site bought for residential development and the value of that development when completed. The percentages adopted in the context of a standing house calculation are linked with the implicit assumption that they are applied to an entirety value of this nature. This being so, I find the more helpful concept in the present case to be that adopted by Mr Creed. But in accepting Mr Creed's valuation concept it does not follow that I
C am accepting his valuation figure. That figure, of £17,250, was derived from a single comparable, 22 Newick Road. This comparable is a terraced house of the same period and in the same locality, and it was converted about two years ago into three self-contained flats or maisonettes. Mr Creed's evidence in respect of 22 Newick Road was that:

(a) the semi-basement flat was sold in January 1976 for £5,800.
(b) the hall-floor flat is currently under offer at £7,750.
(c) the first-floor flat is occupied by the owner, who is asking a price of £9,500.

D What Mr Creed does is to apply the two latter figures to the subject house, and to deem that part of 4 Cotesbach Road occupied by Mr Patten to be worth £7,750, and that part which is occupied by Mr Patten's son to be worth £9,500.

With respect, I find Mr Creed's figure of £17,250 arrived at in this way to be undependable. Mr Tobin for his part, when invited by myself to give his opinion of value related to the landlord's concept, suggested £10,000, which figure is lower than the value Mr Tobin had put the house at without conversion. I find this opinion unconvincing, because Mr Tobin, in cross-examination earlier, had agreed that if 4
E Cotesbach Road were to be deemed to have been properly converted into maisonettes, then its value would be greater than its value as a house. Nevertheless, on Mr Tobin's evidence, I accept the likelihood that the increase in market value consequent on a proper conversion might well be comparatively small. Mr Tobin told me that converted maisonettes in Hackney were difficult to sell because mortgages were not readily obtainable; and I note that although there are apparently many houses in this locality in multi-occupation, to the knowledge of the landlords' valuer only one house has been structurally converted, namely 22 Newick Road.

As to the value of 4 Cotesbach Road as a house, Mr Tobin's
F figure (as has been seen) is £10,750. Mr Creed regarded £10,750 as rather low and suggested £11,500. Mr Tobin, however, had cited a number of open market sales in support of this figure, which I regard as being nearer the mark. I find the value as a house to be £11,000, and I am deeming the entirety value for the purpose of calculating a section 15 rent to be £13,500.

It follows that the price payable is found to be £780, made up thus:

Present ground rent	£6	
YP for 22½ years @ 8%	10.29	£ 62
Reversion to section 15 rent		
Site value 30% of £13,500, £4,050		
Modern ground rent 8% of site value...	£324	
YP in perp def 22½ yrs @ 8%	2.214	£717
		£779
	Say	£780

G I now turn to the other two references. The circumstances relating to these, and the evidence about them, may be assumed to be the same as for 4 Cotesbach Road, unless otherwise mentioned.

H

LR/28/1976. 66 Lea Bridge Road

The lessee of the second house, 66 Lea Bridge Road, is Mrs Nellie Matilda Lacklison; her current lease is for 99 years from June 24 1902 at a ground rent of £6.75 per annum; the relevant date is March 3 1976 at which time the lease had about 25¼ years to run; and the landlord is A A Davis (liquidator of Finsbury Developments Ltd).

The rateable value of this house is £254 and its accommodation comprises six main rooms, thus: *First floor*—front room (1), rear room (2); *Half-landing*—back-addition room (3) used as kitchen, bathroom, separate wc; *Hall floor*—front room (4), rear room (5); *Garden level*—back-addition room J (6) leading to kitchenette; *Outside*—wc.

The tenant's notice of March 3 1976 stated that from 1969 until October 1974, and from December 1975 to date, the first floor had not been in her own occupation. The position is that Mrs Lacklison lives in the lower part of the house and there is subletting of the remainder. There is a shared use of the bathroom and the wc.

Mr Tobin's figure of entirety value was £11,000 as a house and £9,700 as converted into maisonettes. Mr Creed's figures, for comparison, were £11,750 and £17,600. The price proposed by Mr Tobin was £550, and by Mr Creed £1,031.

K On the same approach as for 4 Cotesbach Road I find the value of 66 Lea Bridge Road to be £11,250 as a house. For the purpose of calculating a section 15 rent I am deeming its entirety value to be £14,000. It follows that the price payable is found to be £670, made up thus:

Present ground rent	£6.75	
YP for 25¼ years @ 8%	10.72	£ 72
Reversion to section 15 rent		
Site value 30% of £14,000, £4,200		
Modern ground rent 8% of site value ...	£336	
YP in perp def 25¼ years @ 8%...........	1.78	£598
		£670

L

LR/29/1976. 4 Thornby Road

The lessee of the third house, 4 Thornby Road, is George Erasmus Thomas; his current lease is for 99 years from June 24 1899 at a ground rent of £6 per annum; the relevant date is January 30 1976 at which time the lease had about 22½ years to run; and the landlord is A A Davis (liquidator of Finsbury Developments Ltd).

The rateable value of this house is £302 and its accommodation comprises eight main rooms, thus: *Top floor*—rear M room (1); *Half-landing*—back-addition room (2); *First floor*—front room (3), rear room (4); *Half-landing*—back-addition room used as kitchen, bathroom/wc; *Hall floor*—front room (5), rear room (6); *Garden level*—back-addition room (7); *Semi-basement*—rear scullery (8); *Outside*—wc.

The tenant's notice of January 30 1976 stated that the first floor and the room at the top of the house (four rooms) were let to his sister-in-law. There is a shared use of the bathroom/wc and of the semi-basement.

Mr Tobin's figure of entirety value was £11,000 as a house and £10,000 as converted into maisonettes. Mr Creed's

A figures, for comparison, were £12,000 and £20,000. The price proposed by Mr Tobin was £646, and by Mr Creed £1,377.

On the same approach as for 4 Cotesbach Road I find the value of 4 Thornby Road to be £11,500 as a house. For the purpose of calculating a section 15 rent I am deeming its entirety value to be £14,500. It follows that the price payable is found to be £835, made up thus:

Present ground rent	£6	
YP for 22½ years @ 8%	10.29	£ 62
Reversion to section 15 rent		
Site value 30% of £14,500, £4,350		
B Modern ground rent 8% of site value ...	£348	
YP in perp def 22½ yrs @ 8%...........	2.214	£770
		£832
	Say	£835

Having read the above decision I opened three sealed offers which had been lodged by the tenants, and I received arguments as to costs. My decision on costs in the three references is as follows:

4 Cotesbach Road. The amount of the sealed offer was less than the amount of my award. The tenant will pay to the landlords their costs of this reference.

C *66 Lea Bridge Road.* The amount of the sealed offer exceeded the amount of my award. The tenant will pay the landlord his costs of this reference up to the date of the sealed offer and the landlord will pay the costs of the tenant as from the date of the sealed offer.

4 Thornby Road. The amount of the sealed offer was less than the amount of my award. The tenant will pay to the landlord his costs of this reference.

The costs in all three references, unless agreed, are to be taxed by the Registrar of the Lands Tribunal on scale 4 of the County Court Scales of Costs with certificate for expert D witness fit to qualify.

GAIDOWSKI v GONVILLE AND CAIUS COLLEGE
(LR/16/1976)

(V G WELLINGS QC)

March 7 1977

Estates Gazette April 9 1977

(1977) 242 EG 135

E **House in Cambridge—Question whether landlord college entitled to require insertion in conveyance of covenants reserving development rights under section 29 of 1967 Act—College's policy to use properties in area of subject house for student accommodation—No jurisdiction in tribunal to entertain question of validity of Secretary of State for Education's consent to covenants—Objects of covenants agreed by parties—Held that covenants should be in qualified and not absolute form—Vendors' consent not to be unreasonably withheld—Object of covenants "ought to be, as far as possible, to protect the tenant"**

D M W Barnes (instructed by Vinters, of Cambridge) F appeared for the tenant; N Hague (instructed by Francis & Co, of Cambridge) for the landlords.

Giving his decision, MR WELLINGS said: This is a reference by the tenant for the determination: (1) of the price payable for a house and premises, 6 Harvey Road, Cambridge, under section 9 of the Leasehold Reform Act 1967; (2) of the question whether the landlord college is entitled to require covenants to be inserted in the conveyance of the house and premises in accordance with section 29 of the Act of 1967; and (3) if so, of the question what provisions ought to be contained in the conveyance in accordance with that section.

At the hearing the price was agreed between the parties in G the sum of £5,740. They further agreed that the price would be unaffected by the decision of the tribunal on the other questions.

Section 29 provides:

(1) Where a tenant of a house and premises acquires the freehold under this Part of this Act, the landlord being a local authority, there shall, if so required by the local authority, be included in the conveyance under section 8 above such covenants on the part of the tenant restricting the carrying out of development or clearing of land as are necessary to reserve the land for possible development by the authority.

(2) . . . H

(3) Where a covenant is entered into to give effect to subsection (1) . . . above it shall be expressed to be so entered into, and Part I of Schedule 4 to this Act shall have effect with respect to the operation and enforcement of any covenant so entered into.

(4) Where a tenant of a house and premises acquires the freehold . . . , the landlord being a local authority, and afterwards the local authority or any other person acquires compulsorily any interest in the property, then for the purpose of assessing compensation in accordance with the Land Compensation Act 1961 no account shall be taken of any increase in the value of that interest which is attributable to the carrying out of development in contravention of a covenant entered into to give effect to subsection (1) . . . above, or to any prospect of carrying out J any such development. . . .

College Treated as a Local Authority

For this purpose a university body as defined in section 28(5)(c) of the Act (which definition includes a college of a university) is treated as though it were a local authority; see section 29(5). However, a university body is not entitled to require a covenant to be entered into under section 29 unless they have first obtained the consent of the Secretary of State: see section 29(5). K

Part I of Schedule 4 to the Act provides that a covenant entered into in accordance with section 29 is not to be enforceable by any means other than those provided by paragraphs 2 and 3 thereof (see paragraph 1). Paragraphs 2 and 3 contain a kind of forfeiture of the freehold interest which a university body is entitled to invoke when a breach of a relevant covenant imposed under section 29 occurs. After notice giving particulars of the breach or breaches has been served on the purchaser and he has been given an opportunity to serve a counternotice denying or explaining the breaches and to claim relief (as it were from forfeiture), then, if no relief be applied for or granted L (by the court), the university body is given power to execute a vesting declaration vesting the property in itself. After a vesting declaration has been made matters proceed as if there were a compulsory purchase order.

Where covenants under section 29 are contained in the conveyance and no breach of them takes place but the property or any part of it is required for development for the purposes (other than investment purposes) of the university body for whose benefit the covenants were imposed, or a related university body, the Secretary of State for Education and Science is given power at the cost and on behalf of the university body for which it is required to acquire the M property or that part of it by compulsory purchase: see Schedule 4, part II.

The Facts

At the hearing the facts were not in dispute. No evidence was called before me. On June 18 1975 Gonville and Caius College applied to the Secretary of State for Education and Science for his consent under section 29. Enclosed with the letter was a plan showing the location of 6 Harvey Road in relation to other properties belonging to the college in the area. The letter stated, as was the fact, that the tenant

A occupied the property under a lease dated March 8 1883 granting a term of 99 years from December 20 1881 at a yearly rent of £30. It mentioned the rateable value of the property and the tenant's notice dated January 12 1973 claiming the freehold of 6 Harvey Road. The letter stated that the grounds on which the college made the application were identical with those set out in its letter dated December 16 1974 to the Secretary of State in connection with its property 3 Harvey Road, Cambridge. The following paragraphs of the letter dated December 16 1974 are material:

> 6. The College has a settled objective that it should have in
> B its possession enough lodging houses or hostels to accommodate all undergraduates who cannot be housed in the College itself, and to provide accommodation for all graduates who wish it. It has been the College's policy for some time to use the properties in Harvey Road for College residential purposes. In keeping with this policy the College has declined to voluntarily sell the freeholds, and where the opportunity has presented itself, the leaseholds have been purchased. As a result the College now has the leasehold as well as the freehold interest in 1, 2, 4, 7 and 8 Harvey Road . . . 2, 4, 7 and 8 Harvey Road are already in use as accommodation for undergraduate and graduate students of the College. . . .
>
> 9. On the basis of the facts set out above, it is submitted that
> C it is very much in the interests of the College that the right for possible future development at Harvey Road be preserved and this is accordingly a proper case for the Secretary of State to give his consent to the inclusion of the appropriate covenants under section 29 of the Leasehold Reform Act 1967. . . .
>
> 11. Your attention is drawn to the fact that the College does not propose to develop this area for investment purposes.
>
> 12. We do not anticipate that planning permission will be required for the use of these properties intended by the College. Although permission would have to be obtained if any building works were envisaged.

Paragraph 12 of the letter of December 16 1974 was
D repeated in the letter dated June 18 1975.

On July 8 1975 by letter of that date the Secretary of State purported to give consent to the imposition of covenants in the conveyance by the college to the tenant. The letter stated:

> I am to say that the grounds of the application have been considered and that the Secretary of State, being satisfied that all the appropriate requirements of section 29 are met, hereby consents to the application by Gonville and Caius College, Cambridge, so that they may now require covenants to be entered into on the part of tenants restricting the carrying out of development or the clearing of land so as to reserve the land for possible develop-
> E ment of the university body as defined in sections 28(5)(c) and 29(6)(b) of the . . . Act.

Michael Barnes, on behalf of the tenant, said that the consent dated July 8 1975 was a nullity. It had been invalidated for one or both of the following reasons:

(1) In considering section 29, the Secretary of State had had regard to the possible use by the college of the property rather than to its development; he had misconstrued the meaning of the word "development," which is not defined in the Act, as including a material change in the use of the land, and had proceeded on a wrong basis of law;

F (2) the Secretary of State had had regard to matters which are irrelevant to the general purposes of this legislation, namely, to the need of the college to use the property to provide residential accommodation for its members.

Mr Barnes said that the question whether the Secretary of State had misdirected himself in the manner mentioned above is a question of law and a question which the tribunal must decide. He referred me to the decision of the House of Lords in *Padfield v Minister of Agriculture Fisheries and Food* [1968] AC 997 and to the decision of the Court of Appeal in *Associated Provincial Picture Houses v Wednesbury Corporation* [1948] 1 KB 223.

I agree with Nigel Hague, for the college, that the tribunal G has no jurisdiction to entertain the question whether the consent dated July 5 1975 be valid. It is sufficient that the consent, on its face, is valid. It is not for the tribunal to investigate the question whether the Secretary of State, in the exercise of his discretion, misdirected himself. The appropriate remedy, if there be one, lies in the court and not in the tribunal. I therefore proceed to consider the third question, namely whether any and if so what restrictions should be imposed in the conveyance to the tenant.

Covenants in Conveyance H

Mr Barnes conceded that if the consent be valid the covenants set forth in a draft which he submitted are necessary to reserve the property for possible development by the college. That concession would seem to have been rightly made because unless some covenants under that section be included in the conveyance there would seem to be no means by which the college could ensure that the property be made available for housing its undergraduates or graduates. Unless such covenants be imposed no power of compulsory purchase in aid of that purpose would seem to exist (unless, perhaps, the housing authority were to exercise its powers J under part V of the Housing Act 1957 and were then to transfer the property to the college).

Mr Hague also submitted a draft of the covenants for which the college asked.

I need not specify the covenants referred to in the two drafts because they were overtaken by an agreement between counsel at the hearing that the covenants to be imposed under section 29 should be as follows:

(1) A covenant forbidding the carrying out on the property of any development as defined in section 22 of the Town and Country Planning Act 1971, other than development included within paragraphs 2 to 5 inclusive of Class I K and Classes II and X of Schedule 1 to the Town and Country Planning General Development Order 1973; Mr Barnes also submits that development within paragraph one of Class I should not be prohibited;

(2) A covenant forbidding any demolition or removal from the property of any building or erection or any part or parts thereof and forbidding raising or lowering the level or letting down the surface of the land.

The parties are thus in agreement as to what the objects of the covenants should be, but the college thinks that they should be absolute and the tenant thinks that they should L be in qualified form. Thus the tenant wishes to add at the beginning of such covenant: "Not without the consent of the vendors first obtained (such consent not to be unreasonably withheld)." The college is not willing to agree to the insertion of those words because it is worried about the amount of compensation which it might have to pay in the event of compulsory purchase. If consent be given for particular development or be unreasonably withheld, the purchaser in carrying out the development will not be in breach with the consequence that section 29(4) of the Act will not apply and the compensation awarded will take account of the development carried out. Mr Hague said, however, that M the college was willing to go some way towards meeting the objections of the tenant. It was willing that the covenant should be subject to a proviso that the forfeiture procedure in part I of Schedule 4 should not operate in respect of any development:

(1) for which planning permission has been granted under paragraph 1 of Class I of Schedule 1 to the Town and Country Planning General Development Order 1973; or

(2) in respect of which the consent of the college could not reasonably be witheld (had the tenant been entitled to apply for such consent).

Tribunal's Decision

I do not think that that the purpose or main purpose of section 29 is to preserve the development value in the property for the benefit of the university body concerned. Provided that the covenants in question facilitate the reservation of the property for possible development by the university body, the object ought, then, in my opinion, to be, so far as possible, to protect the tenant. The covenants in question are intended to be of indefinite duration. The rule against perpetuities does not apply in respect of them and section 84 of the Law of Property Act 1925 does not apply either (see Schedule 4). There is further the point that if the covenants are absolute, then, while section 29(4) will preclude an enhancement of compensation attributable to development carried out in breach of covenant by the tenant, that subsection will not preclude an enhancement of compensation attributable to the use to which the university body proposes to put the premises: see section 15(1) of the Land Compensation Act 1961. I am thus not satisfied that absolute covenants are necessary. I am satisfied that qualified covenants are necessary. I am also of the opinion that the definition of "development" should not include paragraph 1 (the deemed permission for a 10 per cent extension of a dwelling-house). The form of the covenants will be that set forth in the annexure hereto. I so decide.

By consent I determine that the price payable for the freehold estate in the house and premises shall be £5,740. There will be no order as to costs.

Annexure

For the purpose of giving effect to section 29(1) of the Leasehold Reform Act 1967 the purchaser hereby covenants with the vendors as follows:

(1) Not at any time without the consent in writing of the vendors first obtained (such consent not to be unreasonably withheld) to carry out or permit to be carried out in, on, over or under the property hereby conveyed or any part thereof any development within the meaning of that expression in Section 22 of the Town and Country Planning Act 1971 including any development order within the meaning of Section 24 thereof PROVIDED ALWAYS that for the purposes of this covenant the expression "development" shall not include any of the classes of development described in Classes I, II and X of Schedule 1 to the Town and Country Planning General Development Order 1973.

(2) Not at any time without such consent as aforesaid first obtained (such consent not to be unreasonably withheld) to demolish or remove from the property hereby conveyed or permit to be demolished or removed therefrom any building, structure or erection standing thereon or on any part thereof nor to raise or let down the surface of the land thereof or any part thereof or permit the same to be raised or let down.

LANDS TRIBUNAL

RATING

A AVONDALE LAWN TENNIS CLUB v MURTON (VO)
(LVC/520/1976)
(W H REES FRICS)
November 26 1976

Estates Gazette January 1 1977
(1976) 241 EG 33

Tennis club near Fleet, North Hampshire—Club's contention that all tennis clubs should be assessed on same footing rejected—Tribunal sees no merit in making distinction between "social" and "changing" rooms in clubhouse—
B Sheffield & Hallamshire case followed—Deduction made for "somewhat isolated position"

J A C Farmer, chairman of the Avondale Lawn Tennis Club, appeared by leave of the tribunal for the appellant ratepayers; the respondent valuation officer in person.

Giving his decision, MR REES said: This is an appeal by the ratepayers against the decision of a local valuation court determining an assessment of rateable value £460 on the subject premises, a tennis club about two miles south of Fleet and half-way between Odiham and Aldershot. The matter originates with a proposal dated March 26 1975 by the respondent valuation officer
C to bring the property into assessment for the first time at rateable value £480. A note annexed to the decision reads:

> "The court expressed appreciation to the parties for the manner and content of their submissions.
> The chairman stated that on the face of it the court felt that the general level of valuation applied by the valuation officer to similar premises specified in the schedules he presented was correct, but had come to the conclusion that the changing-rooms should be on a lower basis than that applied to the main pavilion. Therefore the court reduced the rateable value to the figure set forth hereon, viz £460."

D The club has appealed to this tribunal on the grounds that:

> "the assessment is incorrect and excessive, and is not in accordance with the General Rate Act 1967, section 19 (3); ie what rent would a tennis club pay for the site as a whole which would include courts, changing and resting facilities, parking, and due to the standard and type of club—all the other facilities that are available at Avondale LTC, rather than as at present it has been assessed, in the main, on the type of premises that have been erected on part of that site."

Agreed Facts

There is no dispute as to the facts, and plans were agreed.
E The premises are off Ewshot Lane, which runs from Aldershot Road to a point about half a mile from road A287 (Odiham to Farnham). The club is about a mile north of A287, south-east of Crookham village and south-west of Church Crookham. Queen Elizabeth Barracks are on the opposite side of the lane, beyond open land. There is open land, too, on all other sides of the subject premises. It is fair to say that the club is outside the built-up area to the north. The premises comprise five all-weather hard-surfaced tennis courts, a clubhouse of 109.9m² and a car-parking area,
gravel-surfaced. Water and electricity are connected from F
the mains supplies and foul drainage is to a septic tank. The surplus land area has been grassed and in places banked to give some shelter from the wind: trees have been planted. The club purchased the land in June 1973, the courts were completed in October of that year and the clubhouse in June 1974. Its accommodation comprises: lounge, games room, bar, bottle-store, kitchen, and lavatory and changing accommodation.

Appellants' Case G

I gave leave for the chairman of the club to represent the club. He outlined its history since 1913, when it was formed, and its move from Albany Road, Fleet, to the subject site, the freehold of which is owned by the club. The total cost of building, including the land, was not far short of £50,000. He suggested an assessment of net annual value £310 made up as follows:

Five all-weather tennis courts at £30		£150
Pavilion		
Lounge/waiting Area 35.5m² at £2.00	£71	H
Games Room—containing a table-tennis table, and as table-tennis is not our primary sport we question whether the assessment on this area should exceed that of one tennis court, £30	£30	
Kitchen, bar, store and changing rooms 40.5 m² at £1.00	£41	
		£142
Parking etc, say		£ 17
		£309
	say	£310 J

He submitted copies of the documents put before the court below and of the assessments on a number of club premises including golf, hockey, cricket, football and rugby clubs in Fleet, Basingstoke and Camberley. He also submitted a copy of a questionnaire which the club had sent to 86 clubs throughout the country and the results obtained. These were all clubs having four to seven courts, except three local clubs with only two or three courts, and the questions related to the rating assessment on the club premises, the facilities
and number of members. Some clubs were as far away as K
Sheffield. He included the relevant population figures for the areas in which they are to be found. He gave details of the Avondale club's accounts for 1969 to 1976 and estimates up to March 1978 together with details of subscriptions and numbers of members in various categories. A set of club rules was submitted. The valuation officer had obtained details of the assessments on 12 of the clubs to whom circulars had been sent and the chairman included these details.

A

VO's Valuation

A S Murton ARICS, the valuation officer, spoke to valuations as follows:

Primary Valuation		
Tennis Courts		
5 all-weather hard-surfaced tennis courts		£250
Clubhouse		
Total effective floor area 109.9 m²	say	£300
		£550

Secondary Valuation	
B By direct comparison with assessments on other tennis clubs	
5 courts at £100	£500

He submitted analyses of his valuation of £500 and of the determination of £460 on the lines indicated by the local valuation court in their written note. He gave details of the assessments on 11 tennis clubs in southern England, one being as far away from the appeal hereditament as Southampton, but the majority were reasonably close, eg in Guildford. He also gave details of 15 premises referred to by the appellants, seven of various types of clubs which occupy C permanent buildings in the Fleet, Basingstoke and Camberley areas and four tennis clubs in the Basingstoke valuation area itself. He compared and contrasted many of these with the subject premises, but considered that the best comparable was the club at Camberley, which he said was more centrally situated but more exposed than the appeal hereditament.

The chairman of the appellant club considered that all tennis clubs throughout the country should be assessed the same, but he agreed that position, quality and size were factors to be taken into account. The valuation officer referred me to *Sheffield & Hallamshire Lawn Tennis Club Ltd v Elliott* (1966) 199 EG 187. He thought that the club D could, if it wished, make more money than shown in the accounts, but this was denied by the chairman, who considered that they could only pay a rent of £300 per annum, which he considered was fair and reasonable: they were fortunate in being freeholders.

I viewed the club premises and, externally, sufficient of the others referred to to illustrate the evidence. I adopt the approach in *Sheffield & Hallamshire*. This is summarised in that decision as follows:

> "The club is not a competitive profit-making concern and the accounts of its finances, designed to keep it efficient and solvent, cannot be taken as a direct guide to the determination E of the rent the hypothetical tenant could reasonably be expected to pay, and in the absence of rental evidence the accepted or undisputed assessments on the comparable hereditaments in the district must be regarded as the principal guide to the annual value of the subject property. I do not however ignore the accounts entirely, because they do offer some help in considering the burden of outgoings incurred by a club of this size and age."

I find much of the matter submitted to the local valuation court by the appellants to be of little assistance and the accounts to be only a little more helpful: many of the figures in the accounts relate to the club when it was established on its former site. I reject the view of the chairman F that all tennis clubs throughout the country should be assessed on the same footing: the factors affecting demand for such club premises and the supply of them no doubt vary in different parts of the country. I derive little help from the information relating to clubs many miles away. As to clubhouses, I see no merit in making any distinction between the "social" rooms on the one hand and "changing" rooms on the other unless the proportion of one to the other is abnormal, and no evidence was submitted to me that this is so here.

I agree with the valuation officer that the club premises at Camberley are the best comparison with regard to accommodation. The assessment on the Camberley club supports his G figure of £500 for the subject premises, which the chairman described as "one of the best in England," but that figure does not make allowance for its somewhat isolated position. Making a deduction for this factor, I arrive at a figure of £440 net annual value and rateable value for the subject premises. I disregard the additions made since March 26 1975: these include a practice court and floodlighting. The appeal is allowed to this extent and the respondent valuation officer will pay £27 towards the appellants' costs. The remarks of the local valuation court concerning the manner in which the parties presented their cases apply equally to the hearing before me. H

REASON v ATKINSON (VO)
(LVC/219/1976)

(J H EMLYN JONES FRICS)

October 28 1976

Estates Gazette January 29 1977

(1976) 241 EG 319

Caravan formed of two units joined together—One of 30 caravans of different sizes on site at Stondon, Beds—No firm foundation for VO's comparables evidence, restricted to other mobile homes—Tribunal attaches "very little weight" to evidence of site rents and capital costs—Ratepayer's comparables of prefabs held "the most cogent evidence"—Assessment reduced J

P J Brown appeared, by leave of the tribunal, for the appellant; the respondent valuation officer in person.

Giving his decision, MR EMLYN JONES said: This is an appeal by Mrs S M Reason, who is the occupier of a heredita- ment described as a mobile home, 22 Long Close, Station K Road, Stondon, in the county of Bedford. The mobile home is in fact a caravan standing on a pitch on a caravan site which contains in all some 30 other caravans of different sizes. The appeal caravan is a double unit which has been formed by the joining together of two separate caravans. The accommodation comprises a lounge with dining area, a kitchen, two bedrooms and a bathroom with wc. Main drainage, water and electricity services are connected, but there is no public gas supply. Outside and within the pitch is a small wooden shed. The caravan has a reduced covered area of 50 m².

At one time caravan sites comprising a number of mobile homes of this character were assessed as a single hereditament. L In 1966 the Court of Appeal in *Field Place Caravan Park Ltd v Harding (Valuation Officer)* [1966] 2 QB 484 decided that individual caravans on a caravan site should be separately assessed provided that there was a sufficient degree of permanence. When the new valuation lists came to be prepared, however, the caravan site at Stondon was assessed as a single unit and it was not until May 28 1975 that the valuation officer for the Bedford area made a proposal to assess separately the appeal hereditament at gross value £190, rateable value £132. In an appeal to the local valuation court this assessment was reduced to gross value £180, rateable value £123 and it is against this decision that Mrs Reason M now appeals to the Lands Tribunal.

Disabilities of Caravans

At the hearing Mrs Reason was represented, by leave of the tribunal, by P J Brown. Mr Brown set out a number of disabilities suffered by residential caravans in general and 22 Long Close in particular. He pointed out that their standard of construction was not high, since building regulations did not apply, and there were difficulties in insulation which led to extra heating costs and, under certain condi-

A tions, to heavy condensation. The water supply was susceptible to frost, the private road within the caravan site was not adopted and had minimal lighting, the site was covered to a high density, so there was little privacy, the floors tended to sag and this was particularly true in the case of a double unit. Mr Brown also referred to the internal height of the caravan of 7 ft, which he claimed was appreciably less than that encountered in more conventional types of dwelling, where it was frequently 8 ft and never less than 7 ft 6 in.

Mr Brown drew my attention to a number of other small residential units, some in Stondon and some within a radius of a few miles. One was a local authority prefabricated B bungalow, one a similar building on a private road, two semi-detached houses, one on a private road, a detached bungalow and a local authority terraced bungalow. The gross value for these comparable properties devalued at prices which varied between £2.62p per m^2 and £3.66p per m^2, which compared with a value applied by the valuation officer to the appeal hereditament of £3.60p per m^2. The assessment of the two local authority properties incorporated an allowance of 10 per cent, which the valuation officer apparently applied to all local authority housing.

In the light of this evidence, Mr Brown submitted that the valuation determined by the local valuation court could not C be supported and he asked for a reduction of 11 per cent in order to bring the assessments into line with the comparables and a further reduction of 10 per cent to take into account what he suggested was an element of "stigma" which applied to caravans just as much as it did to local authority housing.

The respondent valuation officer is G H Atkinson ARICS. Mr Atkinson made no attempt to deal with the evidence of the comparables put forward by Mr Brown. In his opinion mobile homes occupy a different sector of the market, which makes it difficult to form any direct comparison with traditional houses. In this latter category he D appeared also to include prefabricated bungalows. Mr Atkinson went on to claim that in the absence of direct rental evidence he looked to the assessment of other mobile homes and, in particular, the assessment of double units, that is to say, units manufactured in two parts and joined together on the site.

In support of the assessment as determined by the local valuation court, Mr Atkinson produced schedules giving details of the assessments of 57 other mobile homes on three separate caravan sites, one at Wilstead near Bedford, one at Clifton about 3 miles north of the appeal site, and one in Stondon. This evidence, however, did little to support Mr Atkinson's E basic valuation for mobile homes. It merely served to indicate the valuation officer's opinion of the comparable values to be attributed to mobile homes on the different sites.

300 Appeals Outstanding

I understood Mr Atkinson to seek to rely on these assessments on the grounds that they represented the established tone of the valuation list. I do not think, however, I can attach very much weight to this in view of the fact that some 300 appeals against the assessments of individual mobile homes are still outstanding in the area.

F In any case, having inspected all the sites I do not accept Mr Atkinson's opinion of their relative values. For example, he gave as his opinion that the value of the Wilstead site was 15 per cent below the values at the appeal site, comparing like with like. He justified this opinion by suggesting the appeal site was nearer to large centres of population and in evidence in chief said that it was 7 miles from Luton. He later agreed in cross-examination that it was in fact 13 miles from Luton, whereas the Wilstead site is about 5 miles from Bedford. In addition the Wilstead site is very nicely established, with growing trees and shrubs, and in my judgment is more attractive than the appeal site.

Mr Atkinson also tried to derive some support from G evidence of the site rent paid and the capital value of caravans, but again I can attach very little weight to this because of the many unknown factors—for example, the amount of rates included in the site rents, the nature and extent of services provided by the site owners, the actual cost of caravans and the extent to which the prices paid include non-rateable items such as furniture and fittings.

Having heard the evidence and carried out my inspections, I must say that I do not know on what foundation the valuation officer has built his pattern of values of mobile homes. In his evidence Mr Atkinson has done little to rebut the *prima facie* case made out by Mr Brown. I do not accept H that the occupation of mobile homes is of such a unique character that the rents paid, for example, for prefabricated bungalows are to be rejected as being of no help. In *Hanstock* (*Valuation Officer*) v *Robertson and Others* (1970) 16 RRC 337 I said:

> "I accept that the 'prefab' assessments provide the best starting point for the valuation of the appeal caravans, but I do not accept the interpretation put on those assessments by the appellant valuation officer."

The most cogent evidence put before me in the present case is that of Mr Brown concerning the prefabs at Arlesey and J Lower Stondon. The evidence of the valuation officer's comparables seems to me to have no firm foundation and in any case contains what I consider to be its own inconsistencies. The evidence of site rents and capital costs does nothing in my judgment to support a higher rate per m^2 for the appeal hereditament than for the bulk of Mr Brown's comparables.

In the light of the evidence taken as a whole I have come to the conclusion that Mrs Reason's appeal should be allowed and the assessment of the appeal hereditament determined at gross value £150, rateable value £98. An order will be issued to that effect. The valuation officer will pay £25 costs to the appellant. K

BURROUGHS MACHINES LTD v MOONEY (VO)

(LVC/484/1975)

(R C WALMSLEY FRICS)

December 22 1976

Estates Gazette March 19 1977

(1976) 241 EG 845

Showroom and offices in Baker Street, London—Whether ground floor to be valued overall or by zoning method— L VO says predecessor's zoning valuation "mistaken"—"Fluidity factor," giving internal changes of emphasis between showroom use and office use—Clear advantage of overall method for mix of two rental value elements—VO alleges comparable underassessed—Not tribunal's practice to allow VO to impugn his own list—Rejection of submission that decision made during hearing to initiate increase in assessment of comparable should enable tribunal to look behind actual assessment at rent paid—Ratepayer's appeal dismissed

A Anderson (instructed by Norton, Rose, Botterell & M Roche) appeared for the appellants; the respondent valuation officer in person.

Giving his decision, MR WALMSLEY said: This is an appeal by Burroughs Machines Ltd ("Burroughs") from a decision of the Greater London Central Valuation Court confirming, at gross value £23,350, rateable value £19,430, the assessment of "Showroom, offices and premises, basement and ground floor," 53-55 Baker Street, London W1, pursuant to an originating proposal lodged by the ratepayers on December 12 1973.

The appeal hereditament stands part of a very large development, 37-67 Baker Street ("the island block"), com-

A pleted in 1958 and covering the whole of a rectangular site bounded by Baker Street on the east, Blandford Street on the south, Rodmarton Mews on the west and Dorset Street on the north. This imposing development is in the form of a multi-storey structure of four tower blocks, each standing east to west, connected by a central spine running north to south. On the Baker Street frontage the design incorporated three shops or showrooms, with only one floor above ground level, as infillings between flanking towers. Nos 53-55 Baker Street may, for present purposes, be taken as comprising the central showroom of the three. The more southerly showroom is part of 37-45 Baker Street, B and the more northerly showroom is part of 57-67 Baker Street. (There is a smaller hereditament, 49-51 Baker Street, alongside the appeal premises, but its assessment is currently under appeal and I shall make no further reference to it.)

Inasmuch as *all* the properties mentioned in this decision are in Baker Street I propose hereafter simply to use street numbers, as for instance "nos 53-55," for the appeal hereditament.

The Island Block

Apart from the small hereditament nos 49-51, the island C block had, at the relevant date, only three occupiers, viz:

The Metal Box Co Ltd, who occupied one of the tower blocks, including the southerly showroom. Their full address was 37-45 Baker Street but, for convenience, I will refer to this hereditament as "no 37" (as was done at the hearing),

Burroughs, who occupy the central showroom no 53-55, and

Marks & Spencer Ltd, who occupy the other three blocks, including the northerly showroom. Their full address is 47 and 57-67 Baker Street but, for convenience, I will refer to this hereditament as "nos 47-67" (as was done at the D hearing).

The original planning permission for the island block was apparently for "offices with retail shops and office entrances on the ground floor." In 1959, when Burroughs took nos 53-55, they were granted a planning permission covering a use for "showroom and offices," with a condition attached that the shop window was to be used for display purposes only. Burroughs took the hereditament as clear space and they partitioned it to suit their own requirements. The purpose for which the hereditament is used was described by the ratepayers in a return made on September 1 1970 as "offices, demonstration and service centre" for the "marketing and E servicing of Burroughs business machines." The front part of the ground floor was at the relevant date partitioned to form a showroom, a sales office and a demonstration room; and behind these areas there was a larger demonstration room, a salesman's room, a manager's office, a tea room and toilets. The basement is used for servicing the advanced types of calculator demonstrated on the ground floor. An internal goods lift serves the two floors.

At the hearing Anthony Anderson appeared for the appellant ratepayers and called evidence from C H Bennett ARICS, an associate in the firm of Richard Ellis. The respondent valuation officer, A G Mooney ARICS, appeared F in person.

Mr Bennett's valuation was at gross value £18,900, made up thus:

Ground floor:			
Zone A (20 ft)	118.0 m²	@ £55	£6,490
Zone B (20 ft)	134.7 m²	@ £27.50	3,704
Zone C (remainder)	202.5 m²	@ £15	3,037
Basement:			
Workshops & stores	568.7 m²	@ £10	5,687
			£18,918
			Say £18,900

The valuation officer's valuation, defending the lower G court's decision at gross value £23,350, was made up thus:

Ground floor	455.2 m²	@ £40	£18,208
Basement	568.7 m²	@ £10	5,687
			£23,895
			Say £23,350

Although the hearing extended over three days, there is only one substantive issue to be resolved. It will be seen that the parties are agreed that the rental value of the basement, in gross value terms, may be taken as £5,687. The H substantive issue is whether the ground floor is more dependably to be valued, as Mr Bennett does, on the zoning method; or alternatively, as Mr Mooney does, at a value overall.

The Ratepayers' Case

The case for the ratepayers is that, for the purpose of the 1973 list, the assessment of nos 53-55 had been arrived at on the zoning method, as had the assessments of similar hereditaments; that Mr Mooney's predecessor, up to and including the local valuation court hearing, had himself relied on the zoning method; that the appeal property is J used for the purpose of demonstrating and selling business machines; that its ground floor is primarily a showroom; and that the floor space can indeed be equated to a large shop. Counsel invited me to accept that the following passage from *Ryde* (13th ed, at p 496) was directly in point:

"For many years before 1956 it was the practice to value large shops by a zoning method, taking the value of the zones nearest the street from the evidence of actual rents of similar areas in the smaller type of shop, and applying reducing rates to zones further away from the street. An allowance for quantity and for any disabilities was customarily made at the end of the valuation. In the 1956 valuation lists the valuation officers in K certain areas abandoned this well-established and tried method of valuation in favour of what the Lands Tribunal has described as 'the intuitive approach,' the main features of which were the adoption of a single overall rate per square foot for the sales space, and the refusal to make any allowance for quantity. This mode of proceeding had the undoubted advantages, from the valuation officer's point of view, of simplicity in application, and the impossibility of applying any real test to see whether the answer provided was right or wrong.

"A series of cases in the Lands Tribunal culminated in *Footman, Pretty & Co v Chandler* (*Valuation Officer*) (1960) 7 RRC 18, in which all the previous decisions were reviewed. Rejecting the 'intuitive approach' the tribunal (J P C Done) L summed up its conclusions thus:

'I cannot resist the conclusion that there are too many factors in the valuation of extensive trading premises to justify reliance upon a spot valuation unsupported by details, or upon any other valuation introducing methods which have not stood the test of time and experience. . . . I feel that where the zoning method is properly based upon local values which are established or argeed, it is susceptible of tests by precise comparison, and if applied by an experienced practitioner and confined to the type of hereditament for which it is suitable, more appropriate for submission to the tribunal. . . . Put simply, I consider that a zoning method of valuing shop premises provides reasons and results which can be examined and tested, that, properly and consistently M used, it is the best method of valuation which has yet been devised, and that it could be used with advantage in all cases for which it is suitable.'"

Valuation Officer's Contention

Mr Mooney, who has had 30 years' experience, told me that he had not long held his present appointment in the Inland Revenue; he had inherited this appeal from his predecessor at a date subsequent to the decision of the local valuation court (given on June 4 1975, after a hearing on May 1 1975). On studying the papers he had come to the conclusion that the zoning method, as adopted by the valuation officer hitherto, had been mistaken. Mr Mooney's contention was that a hereditament used for the purpose of a "showroom/office" must be valued for that use, and was not to be valued as if it were a "showroom" *simpliciter*. In support, the valuation officer referred me to the case of *London Transport Executive v Croydon London Borough Council and Phillips (VO)* (1974) 230 EG 987, and in particular to the following passage in the tribunal's decision:

> "We are satisfied that the practice which has stood for many years, from time immemorial in rating experience, of regarding the existing use as the only use to be valued is the right one.
>
> "In our view the law has been correctly set out in *Armour on Valuation for Rating*, 4th ed, at p 299, that so long as the proprietor makes a beneficial use of his property, the assessor has nothing to do with whether it might have been turned to other and more lucrative purposes. In our view that is the law in England, and we think it was rather understated in the Lands Tribunal case of *Fir Mill Ltd v Royton Urban District Council and Jones (Valuation Officer)* (1960) 175 EG 1029 at 1033:
>
> 'For all the foregoing reasons, subject to the qualifications we have indicated, we are of opinion that the doctrine *rebus sic stantibus* relates only to the physical characteristics of the hereditament.'
>
> "In our view the only exception to the rule that one does not take into account another mode of use is where no beneficial use is being made of the property. Whether that correctly comes under the doctrine of *rebus sic stantibus* or not seems to us not to matter. For our part, we are satisfied that one only looks to the actual use; where one has a shop one values as a shop for any purpose as a shop but not as a shop which may be used for an office, for example. While it is perfectly true that the Use Classes Order under the town planning Acts is a modern innovation compared with the age of rating law, it has stood, of course, now for many years and as a generalisation we think it is a useful guide—but we put it no higher—in deciding what other uses can be taken into account. Therefore one has a shop value of a shop, any kind of shop, or light industry or any other kind of light industry, and so on."

As I followed their evidence, both Mr Bennett and Mr Mooney regarded use for a "showroom" as being no different, in rental value terms, from use for a "shop"; indeed the two expressions seemed to be used interchangeably by the valuers. The valuers were equally in agreement on another matter, that—at any rate as regards premises in Baker Street—floor space which could be used for an office was more valuable than space which could be used only for a shop. It is also common ground that the permitted use of nos 53-55 enables the occupier to use the ground-floor space without any restriction as to the proportion allocated for office use.

As it happens, Burroughs have found that their own requirements are best suited by adapting a greater proportion of the ground-floor space for showroom use and a smaller proportion for office use, but another occupier might prefer to (and would be free to) reverse this emphasis. Envisaged as vacant and to let, it is not possible, therefore, to separate the floor space as between one part used for "showroom" and the other part used for "office." The significance of this fluidity factor, in a rental-value context, is that whereas it is generally accepted that floor space in a shop decreases progressively in rental value with a greater depth from the street frontage, no such characteristic is regarded as attaching to office space.

I accept the valuation officer's contention that the ground floor of nos 53-55 falls to be valued as a "showroom/office" and not as a "shop"; and, against the background which I have just outlined, I also agree with him that an assessment of rental value for this particular hereditament is more dependably arrived at by using an overall figure for the ground-floor space rather than by taking the first 20 ft of depth at a zone A figure and then halving back.

The question now to be resolved is whether Mr Mooney's figure of £40 psm for the ground floor is supported by evidence. For this purpose I propose to look at the assessments of other hereditaments rather than at the few rents which are available. Mr Bennett made the comment that "By 1976 rating surveyors tend to pay little attention to rents and considerable attention to assessments," and this seems to me right and proper with a list that has settled.

Along both sides of Baker Street the predominant user of ground-floor space is for shops or showrooms, and there are very few properties which incorporate an office use on the ground floor. A great deal of evidence was produced by both parties about a considerable number of hereditaments, but I find only three of these to be of assistance as direct comparables. These I shall be referring to in some detail shortly. A distillation of the remainder of the evidence, sufficiently accurate for my present purposes, shows that a general tone of assessment of shops and offices in this locality has been established thus:

For shops, in terms of a zone A of 20 ft depth:	
On the east side of Baker Street, in the block opposite to nos 53-55, between Dorset Street and Blandford Street	£75 psm
On the west side of Baker Street, north of Dorset Street	£65 psm
On the west side of Baker Street, south of Blandford Street	£75 psm
For offices, on either side of Baker Street (predominantly upper floors):	
If air-conditioned	£70 to £75 psm
If air-conditioned, but restricted in use to "showroom and ancillary offices"	£62.50 psm
If not air-conditioned	£57.50 psm
If not air-conditioned, and restricted in use to "showroom and ancillary offices not to exceed 15% of the total"	£42.50 psm

None of the above figures is directly relevant to the overall value of the ground floor of nos 53-55, but they all form part of the general pattern of assessment into which the correct assessment of nos 53-55 must fit. The inference I draw from this general pattern is that the appropriate overall figure for a "showroom/office" use should represent a mix of its two rental value elements, viz its office element and its shop element. As to the *office element*, the expected rental value would be of the order of £57.50 psm. As to the *shop element* the expected rental value would be—in terms of zone A—not £75 psm, as is found with the shops opposite, but some figure materially lower.

Reasons for Lower Figure

There are three reasons for the figure being lower than £75 psm: (1) Baker Street is a major through road, not particularly easy for pedestrians to cross, with all the traffic moving from north to south, so that the bus stops are all on the east side; (2) pedestrians making for, or coming from, the very busy Baker Street underground station to the north tend to walk along the east side because the underpass to Marylebone Road is on that side: and (3) the island block, in which the appeal hereditament is centrally situated, constitutes a comparatively "dead" frontage for shopping purposes.

From the above general evidence I turn to the particular. The three hereditaments which I find helpful as direct comparables are no 37, nos 47-67 and nos 19-25. I will take them in turn.

37 Baker Street

This hereditament, it will be recollected, is that part of the island block that was, at the relevant date, occupied by the Metal Box Co. It is currently empty. The assessment on this hereditament was dealt with by the local valuation court at the same time as the assessment on nos 53-55. A gross value of £543,000 was decided, this being the figure contended for by the ratepayers. That decision was not appealed. Attached to the lower court's decision was a computation (actually a copy of the valuation prepared for the appellant ratepayers), showing the make-up of the assessment thus:

Seventh floor: Dining room, canteen & offices	873.63 m²	@ £57.50	£50,233
Sixth floor: Offices	1064.00 m²	@ £57.50	61,180
First to fifth floors: Offices	6045.00 m²	@ £57.50	347,587
Ground floor: Showroom (zone A)	83.00 m²	@ £55.00	4,565
(zone B)	78.30 m²	@ £27.50	2,153
(zone C)	84.20 m²	@ £15.00	1,263
Ground floor: Reception, offices, etc	1029.80 m²	@ £57.50	59,213
Basement: Car spaces, 80 @ £150	—	—	12,000
Basement: Stores	493.00 m²	@ £10.00	4,930
			£543,124
		Say £543,000	

The above calculation was relied on heavily by Mr Bennett —indeed his figures of £55 for zone A, £27.50 for zone B and £15 for zone C had been derived straight from this decision. With respect, however, this use of the comparable is mistaken and would be justified only if nos 53-55 were used simply as a showroom. The inappropriateness of Mr Bennett's use of this comparable may be illustrated by noting that space in nos 53-55 to which he attributes a value of £15 psm is comparable with space in no 37 to which £57.50 psm was attributed by the local valuation court.

It is not difficult to make use of the lower court's decision for no 37, for comparison purposes, but in another way. Part of the ground floor of no 37 is a recognisable entity, similar in size and shape to the ground floor of nos 53-55; and an apportionment of the £543,000 assessment shows, for that comparable entity, a figure of £18,193 made up:

Showroom (zone A)	83.00 m²	@ £55.00	£4,565
(zone B)	78.30 m²	@ £27.50	2,153
(zone C)	84.20 m²	@ £15.00	1,263
	245.50		£7,981
Offices	177.60 m²	@ £57.50	10,212
	423.10 m²	= £43	£18,193

The overall figure of £43 psm above is directly comparable with Mr Mooney's £40 psm for nos 53-55 and provides solid support for that figure. It will be seen also that the other values above are closely in line with the values to which the general evidence points, as has been discussed earlier.

No 37 incidentally provides an interesting example of the potential fluidity of occupation within a showroom/office use. The areas used for "showroom" and "offices" within the 423.1 m² at the relevant date were as shown above, but these differed from the emphasis of occupation at an earlier date when no 37 had been referenced by the valuation officer. The valuation officer's recorded areas were:

Shop (zone A)	92.4 m²
(zone B)	88.5 m²
(zone C)	40.4 m²
	221.3 m²
Office-type uses	201.8 m²
	423.1 m²

The use of the zoning method for valuing "showroom/office" space has a clear defect—if the occupier should change the emphasis, eg by using less of the space for "office" and more for "showroom," an apparent reduction in rental value is thrown up by the zoning method of valuation, and yet the hereditament has remained unchanged. Thus in respect of no 37 there was no suggestion before me that the hereditament as referenced by the valuation officer had been any more valuable than when it was later referenced for the ratepayers, yet the logic of the local valuation court's decision would have put the assessment higher by over £1,500, because the showroom/office space, instead of coming out at £18,193, would have come out at £19,724 as follows:

Showroom (zone A)	92.4 m²	@ £55.00	£5,082
(zone B)	88.5 m²	@ £27.50	2,433
(zone C)	40.4 m²	@ £15.00	606
	221.3 m²		£8,121
Offices	201.8 m²	@ £57.50	11,603
	423.1 m²	= £46.50	£19,724

The overall method of valuation, on the other hand, has the clear advantage that any internal changes of emphasis, as between the showroom use and the office use, are of course absorbed.

47-67 Baker Street

This hereditament, it will be recollected, is that part of the island block occupied by Marks & Spencer. Its assessment was outstanding at the time when the local valuation court was dealing with no 37 and nos 53-55. In the light of the lower court's decision on no 37, the valuation officer discussed figures with the surveyors acting for Marks & Spencer. Agreement was reached that the very considerable area used for offices within this hereditament should be taken at £50 psm. Final agreement was later reached on a total assessment of £1,368,500. Mr Mooney told me that his own valuation had been:

Ground to seventh floors: Offices	26 402.6 m²	@ £50	£1,320,130
Ground floor: Showroom	437.1 m²	@ £35	15,298
Basement: Stores	1246.4 m²	@ £10	12,464
Basement: 102 car spaces @ £150			15,300
Yard: 63 car spaces @ £100			6,300
Generators			360
			£1,369,852

The final figure of £1,368,500, said the valuation officer, had been agreed by him at the end of the day as a prudent concession of £1,352 (or 0.01 per cent), which had obviated the uncertainty involved in coming to the tribunal. I find nothing in £35 psm (rounded down) for the showroom space in nos 47-67 to cast doubt on a figure of £40 psm for the ground floor of nos 53-55. Clearly the factor of "quantity" has come into play here, as is evident from the adoption of £50 psm for the office space instead of the more customary £57.50 psm. In any event the assessment of nos 47-67 was, on the evidence, derived from and has been based on, the assessment for no 37, and the latter assessment therefore carries the greater weight.

19-25 Baker Street

This hereditament is occupied as the Japan Trade Centre. Its assessment was entered in the 1973 list at gross value £27,500 and this figure still stands. Mr Bennett's analysis of the assessment differed from Mr Mooney's. After hearing the evidence, and having inspected, I prefer the analysis suggested by the valuation officer, which was as follows:

Ground floor:	532.9 m²	@ £40.45	£21,558
Add for return frontage to George Street 10%			2,156
			£23,714
Basement:	378 m²	@ £10	3,787
			£27,501
			Say £27,500

A The relevant figure here is £40.45 psm for the ground floor overall. Taken on its own, the above assessment for nos 19-25 does *not* support the assessment defended by the valuation officer for nos 53-55. The Japan Trade Centre, a high-quality property, was regarded by both Mr Mooney and Mr Bennett as being superior to the appeal hereditament.

The valuation officer, when presenting his evidence, made a comment: "I say that 19-25 is superior to 53-55, and I say that it is underassessed." At this point I intervened with an observation that it was not the practice of the tribunal to allow a valuation officer to impugn his own list; I pointed out that the assessment of nos 19-25 had apparently stood
B for over three years without challenge; and I indicated that the assessment must carry full evidential weight as part of the tone of the list. At a later stage of the hearing, after an adjournment, the valuation officer told me that the decision had now been made to initiate steps for an increase in the assessment of nos 19-25. He submitted that this circumstance should now enable the tribunal, for the purpose of the present appeal, not only to attach less than full weight to the actual assessment of £27,500 but also to look behind that assessment at the rent being paid, full details of which had been put in evidence. That submission I rejected, but I undertook to include in this decision both a note of the
C submission made and of my ruling on it.

My finding is that the evidence as a whole is sufficient to establish the valuation officer's defence of the assessment under appeal, despite the contra-indication provided by the assessment of nos 19-25.

The appeal is dismissed. The assessment is confirmed at gross value £23,350, rateable value £19,430.

The ratepayers will pay, towards the costs of the valuation officer, the sum of £30, this being the figure suggested by Mr Mooney.

D

SAUNDERS v MALBY (VO)
(LVC/199/1974)

(J D RUSSELL-DAVIS FRICS)

January 10 1977

Estates Gazette April 23 1977
(1977) 242 EG 299

Reconsideration, at direction of Court of Appeal, of obligation for repairs, as affecting 100-year-old Tunbridge Wells
E **house, in statutory definition of gross value—Tribunal holds that hypothetical landlord would have been prepared to expend money necessary to put house into reasonable repair—Inclusion of house in general improvement area would not have deterred landlord nor caused tenant to reduce offer of rent**

N Jordan, a solicitor of Inland Revenue, for the respondent valuation officer; no appearance by the appellant.

Giving his decision, MR RUSSELL-DAVIS said: On November 6 1974 the Lands Tribunal heard the appeal by Reginald William Saunders, the owner and occupier of the dwelling-house known as 2 Summer Cottages, Burdett Road, Rusthall,
F Tunbridge Wells, against the determination of the Central Kent Local Valuation Court, confirming the assessment in the 1973 valuation list of gross value £86, rateable value £55 (after April 1 1974 £51), the date of the originating proposal being May 3 1973. On November 27 1974 the tribunal upheld the valuation court and dismissed the appeal.

Mr Saunders appealed against this decision and the Court of Appeal on March 10 1976 ((1976) 239 EG 205) remitted the case to the Lands Tribunal for a reconsideration of the obligation for repairs, as it affects the appeal house, in the definition of gross value in section 19(6) of the General Rate Act 1967 in the light of the principles enunciated by the court with reference to the judgments in *Wexler v Playle (VO)* G [1960] 1 QB 217 and other cases heard before and after that case.

The appellant did not appear at this rehearing on December 6. I was satisfied that it had been properly appointed. It is significant that the appellant did not appear at the hearing before the Court of Appeal either. In view of the direction of the superior court that the case be reconsidered, I decided to proceed *ex parte* without applying Lands Tribunal Rule 53.

Nicholas Jordan called evidence from the respondent valuation officer, D W Malby ARICS, and evidence of fact from three other witnesses in regard to the state of repair and as H to the position in regard to the Rusthall Improvement Area as at the material date.

It is appropriate to set out the following passages from the judgment of the Court of Appeal which are particularly relevant to the points for reconsideration.

Per Lord Denning MR at p 207:

> "It seems to me that the tribunal has to consider whether [the want of repair] is such that it would be reasonable in all the circumstances to expect a hypothetical reasonable landlord to do the repairs. If the cost of doing the repairs would be out of all proportion to the value of the house, so much so that even a reasonable landlord would not do them all, then it must not be J assumed that he would do them. He would let the premises at a low rent. In those circumstances the low rent would be the basis on which to arrive at the rateable value."
>
> "[In the present case] it seems to me that the member of the Lands Tribunal fell into error. He only asked himself whether the defects were capable of remedy. He did not go on to inquire 'How much is it going to cost?'. If the expenditure was such that it would be out of all sense to do the repairs, then a hypothetical landlord would let it at a low rent. That fits in exactly with the statute, which says it has got to be in such a state of repair as to command that rent."
>
> "So far as the proposed redevelopment is concerned, or the proposed school, I think it may be a legitimate ground for a K lower assessment. That seems to follow from *Almond v Ash Brothers* [1969] 2 AC 366, when the House of Lords held that if it is likely that the house may be demolished within the near future, it will affect the hypothetical tenant's hypothetical rent."

Per Goff LJ at p 207:

> "In my judgment the hypothetical landlord is one who undertakes some liability to do repairs, and that imports an obligation to put the premises into such a state as accords with the ambit of his liability to do repairs. In my judgment, however, it is not an unqualified liability to do all repairs. The question is: What is the extent of the obligation which ought to be imputed to the hypothetical landlord? As Morris LJ put it in the case of *Wexler v Playle* at p 235, he should be required to do such repairs as L would make the premises reasonably habitable. Willmer LJ, at p 239, said that the hypothetical landlord's obligation was to remedy readily remediable defects; and, as Harman LJ put it at p 240, he should remedy repairable and temporary defects."

And at p 209:

> "Further, in the case in the House of Lords of *Dawkins v Ash Bros* [1969] 2 AC 366, . . . it is quite clear that the majority decided that the probability of the house being demolished by the requirement of the local authority in the near future was a factor which would have to be taken into account in considering what degree of repair it would be reasonable to consider the hypothetical landlord to be bound to adopt." M

Per Shaw LJ at p 209:

> "It seems to me that the repairs referred to in section 19(6) of the General Rate Act 1967 must be of such a nature that their execution by a hypothetical landlord can be contemplated as a reasonable and practical prospect having regard to all the circumstances."

Two questions have therefore to be answered:

(1) Can the house be repaired at economic cost, such that a reasonable landlord would contemplate carrying out, and

(2) what effect would inclusion within the Rusthall Improvement Area have on his mind in considering this obligation, or on that of the tenant in making his offer of rent?

State of Repair

Dealing firstly with the state of repair, Mr Malby confirmed and enlarged on the evidence which he had given at the first hearing. The house was 100-120 years old; the brickwork and pointing were fair for its age; an inexpensive overhaul would be required to the roof covering; there was no significant rising or penetrating damp; and the structure of the house was in very fair condition. Doors and windows needed an overhaul, and repainting, though due, was not urgent. The general structure and external condition were consistent with that of 3-6 Summer Cottages adjoining. Internally, the decorative condition was deplorable because the appellant had allowed it to become so. The sitting-room floor was "springy," but owing to an accumulation of rubbish a detailed examination had been impossible to make. Joinery was reasonable.

Mr Malby was of the opinion that a tenant would be prepared to take the house as it stood, but would reasonably expect the landlord to contribute the sum of £150 at 1973 level of prices towards the cost of repairs out of a total cost estimated to be about £225; thereafter the tenant would have been prepared to do the rest as part of a modest programme of improvements which he would himself carry out. If this expenditure of £225 were incurred the house would have an expectation of some 30-50 years of useful life. In the context of the judgment in the Court of Appeal he was convinced that a reasonable landlord would contemplate such an outlay.

This evidence was corroborated by D A Wild FRICS, Mr Malby's predecessor as valuation officer, who made the original inspection and appeared as respondent before the local valuation court.

B D Jeffery, a building surveyor and improvement grants officer for Tunbridge Wells council since 1970, described the application by the appellant for an improvement grant under the Housing Act 1968 and the negotiations which took place between 1969 and 1971. He had concluded that the building was structurally sound but suffered from lack of maintenance. He had been prepared to recommend a grant of £185 towards the cost of the installation of a bathroom and internal water closet, estimated at £394, provided that Mr Saunders carried out at his expense the cost of repairs, estimated at £154. The latter asked for 50 per cent of the cost of repairs, to which under the then legislation he had been unable to accede. There was further correspondence and an inspection, but the application was in the end withdrawn.

The significance of this evidence is that between 1969 and 1971 less than £200 was thought to be necessary in order to put the house in repair to give it reasonable life and to qualify for an improvement grant, which suggests that Mr Malby's opinion of the cost as at 1974 was probably not far wide of the mark. The evidence has not been controverted, and in the light of it and my own view in November 1974, I accept it fully. I conclude that an expenditure of some £225 at the material date to put the house into reasonable repair would not have been regarded by the hypothetical landlord as being out of all proportion, and that he would not have been deterred from undertaking it, in the belief that the house would then have a fairly long life; and that such repair would have been regarded reasonably as his obligation under the statutory definition of gross value.

Area Improvement Scheme

It has now to be considered whether such obligation would have been affected by the inclusion of the house within the Rusthall Improvement Area. W F Godwin BA Econ MRTPI, the forward planning officer for Tunbridge Wells Borough Council, said that originally in the late 1960s the area including Summer Cottages had been proposed for redevelopment not before 1978, but it had been decided to bring forward the proposals to 1971-72. At that stage Summer Cottages were excluded from the proposals and the residents so informed by letter from the town clerk on June 24 1970. Subsequently it was decided to make the area a general improvement area under the Housing Act 1969, and public meetings took place between May and November 1972. Owing to local government reorganisation a start was delayed and no scheme was implemented before a further meeting took place in 1975, although the commitment was still there. The scheme did not entail any disturbance to the occupation of no 2, and such was known at the date of the original proposal on May 3 1973.

I was referred to Ministry of Housing and Local Government Circular 65/69, which dealt with area improvement, and it is necessary only to quote a short passage from para 17:

> ". . . Area improvement means what it says; the area should be better to live in after improvement. Demand for property should not be adversely affected by proposals. It would be most unfortunate if at this or any other stage any form of blight were to be induced in these areas, and it is much to be hoped that from the beginning of their inquiries local authorities will make it absolutely clear that what is under consideration is a programme of action designed to raise the standards of comfort and amenity for the residents. It will follow from this that even if the first drafts of a possible scheme suggest some selective demolition, such proposals should be examined critically and dealt with in confidence until it is clear that they are justified. There may well be some unfit property in the area: any blight which falls on this is a consequence of normal slum clearance action and not the area improvement scheme. Ministers do not expect that area improvement schemes will provide for the clearance of fit property save in most exceptional cases."

This evidence was fuller than that given at the original hearing by this tribunal. It is much in point, and it clearly excludes the possibility of demolition or of any clearance scheme affecting the appeal house, such as would bring it within the ambit of the judgments in *Almond v Ash Brothers* or in *Dawkins v Ash Brothers* (*supra*). I find that this scheme would not have affected the mind of the hypothetical landlord that the necessary expenditure was reasonable in the circumstances; indeed, inclusion in a general improvement area would enhance the locality and could result in an increase in rental values generally. Nor do I consider that the hypothetical tenant would reduce his offer of rent on that account.

Accordingly, I have come to the conclusion that the decision of November 27 1974 should be confirmed, that the local valuation court be upheld and the appeal dismissed.

The original order of costs against the appellant is discharged. Mr Jordan did not ask for costs in respect of the rehearing and accordingly I make no order.

LANDS TRIBUNAL

LAW OF PROPERTY ACT 1925 — SECTION 84

A MANSFIELD DISTRICT COUNCIL'S APPLICATION

(LP/18/1975)

(SIR DOUGLAS FRANK QC, President)

November 10 1976

Estates Gazette January 22 1977

(1976) 241 EG 241

Mansfield cattle market—Application for discharge of restriction in 1876 conveyance by estate trustees to council's predecessors—Use of land restricted to purposes of markets and fairs and especially of cattle market—Council propose to develop site as leisure centre—Preliminary submission B that benefit annexed to lordship of manor and not to land rejected—"Other circumstances" in para (a) refers to physical circumstances and not to change in legal status—Proposal for leisure centre without providing alternative site for cattle market in conflict with development plan—Market "much used"—Public interest in favour of leisure centre does not necessarily outweigh market's convenience and importance—Tribunal's criticisms of council—Application fails

Alistair Dawson QC and Andrew Congden (instructed by the solicitor of Mansfield District Council) appeared for the applicants, and J Colyer QC (instructed by Baileys, Shaw C & Gillett) for the joint objectors, Welbeck Estates Ltd and the Rt Hon Alexandra Margaret Anne Cavendish Bentinck. No appearances by or on behalf of 62 other objectors.

Giving his decision, THE PRESIDENT said: This is an application under section 84 of the Law of Property Act 1925 for the discharge wholly of a restrictive covenant affecting the cattle market at Mansfield. By an indenture of July 18 1876, made between the trustees of the will of the Duke of Portland and the Mansfield Improvement Commissioners, there was granted to the commissioners all the fairs and markets usually held at Mansfield and all the tolls, D dues, fines, rents and profits receivable upon them, and there was conveyed the land the subject of this application now known as the Mansfield Cattle Market. The commissioners covenanted with the trustees and their heirs and assigns as owner or owners of the legal estate in the Manor of Mansfield to use the land for the purposes of markets and fairs generally and especially of a cattle market and in effect not to erect any buildings for purposes other than those. The applicants are the successors in title to the commissioners and the objectors to the trustees.

Copyholds Enfranchised

E The application is made under grounds (a), (aa) and (c) in subsection (1) of section 84. Mr Dawson for the applicants made a preliminary submission relevant to grounds (a) and (c). He pointed out that the covenant was expressed to be with the owner of the legal estate in the manor and that although the burden was annexed to the land conveyed, the benefit was annexed only to the manor. Accordingly he submitted that the benefit was annexed to the lordship of the manor and not to the land within its boundaries, or parts of it, owned by the trustees. He said that had it been intended to be annexed to the land it would have been F annexed to every part thereof as in the case of the burden. In any event the covenant says in terms that it is to run with the manor. Since 1876 copyholds have been enfranchised and land sold, so that to the whole the covenant can no longer be of benefit because the objectors no longer own the entirety of the land to which the benefit was annexed. Accordingly the interest of the objectors can only be paternal.

Mr Colyer for the objectors rejected this argument. He said that the 1876 concept of a manor is still very much alive, with its economic and social connotations. He argued that it was natural not to have specified each and every part G but to have reserved the benefit to "the manor." A manor, he said, is a bundle or aggregate of rights comprehending copyhold land and other land including land held by the lord in demesne, and the covenant was annexed to the lordship of the manor and the land in the manor. Accordingly, it runs with the land retained and with the lordship. He argued that the objectors' predecessors and the objectors held the market and the benefit of the covenant on trust for the community.

Moral or Paternal Obligations H

I have reached the conclusion that the purpose of this covenant is clear, namely, that it was for the benefit of each and every part of the land within the manor, that no part of the land has left the manor; all that has happened is that land has been taken out of copyhold tenure. I further think that the objectors are entitled to enforce this covenant whether or not any land retained by them would be injured by its discharge, not only because they have the benefit of the covenant but also because they are entitled to regard themselves as *quasi* trustees with a moral obligation towards their former tenants. Moreover, as was proved, the objectors own a considerable area, of which they have 3,500 J acres in hand and another 30 farms leased. It speaks for itself that the existence of the cattle market must have a potential benefit to the objectors *qua* landowners.

Mr. Dawson argued that a moral or paternal obligation is something which cannot be taken into account and referred to the case of *Gee v The National Trust* [1966] 1 WLR 170. However, in my view that case does not support him, for the *ratio decidendi* was that the modification sought would injure nobody. It was common practice in the last century for estates to be laid out by big landowners and the standard of development controlled and maintained by restrictive covenants; indeed that then was the only form of town K planning. Some of those estates still exist and the control is used for the benefit of each of the present owners; it is exercised mainly for the benefit of those owners, and perhaps the public interest at large, rather than for the benefit only of the successors to the covenantee. Provided that person has the benefit, as is conceded in this case, it seems to be unobjectionable that that benefit should contain or consist of the moral or paternalistic considerations to which I have referred. It further seems to me that this case is distinguishable in principle from the case I have just referred to.

I now turn to the submissions made under each of the paragraphs invoked under section 84 (1).

As to paragraph (a) it was pleaded in the application that " the use of Mansfield Cattle Market as a cattle market has declined. Market traders have diversified their activities to include motor auctions and other non-agricultural uses. The market buildings are dilapidated and incompatible with modern farming practice. The market is situated close to the centre of Mansfield and there is need for its use to be changed to one which will better serve the interest of the community at large." It seems to me that those considerations, even if proved, are irrelevant to paragraph (a) in so far as that paragraph is directed not to the land burdened but the land having the benefit of the covenant.

However, Mr Dawson relied not only on those considerations but on the enfranchisement of copyhold titles under the Law of Property Act 1922 which had to be completed at the latest by 1950. He said that accordingly the situation had radically changed in the last 100 years and that is " other circumstances " within the meaning of paragraph (a). However, for the reasons which I have already explained I regard the enfranchisement of copyholds, if not irrelevant, certainly as not rendering the covenants obsolete. Moreover, I think " other circumstances " refers to physical circumstances and not a change in legal status. Accordingly I hold that the application on this ground fails.

Coercion of Tribunal

It is convenient to deal with ground (c) at this stage because it must necessarily fail from what I have already said unless a further submission made by Mr Dawson is valid. He pointed out that the enforcement of the covenant cannot ensure that the land is used as a cattle market and he later led evidence to the effect that, win or lose this application, the applicants intend to close the market at the end of the year notwithstanding that it is making a profit. I think it is not to the credit of the council that they should take this point, for it can have no other purpose than to coerce me into making a decision which I might not otherwise make. I cannot think that if this application is refused the council will deprive a considerable number of farmers and butchers of the benefit of this market, which benefit in turn must to some extent accrue to the public of Mansfield. Moreover, the final decision does not necessarily rest with the council, because the Secretary of State for the Environment has certain powers; for example, he could in response to a memorial compel the council to retain the cattle market as such. Hence I find that the application fails on this ground.

Two Conflicting Public Interests

I now turn to paragraph (aa) and I am bound to say that this has given me great difficulty because it compels me to make a decision between two conflicting public interests. The reason why the council seek to discharge of the covenant is to enable them to develop the site as a leisure centre providing facilities for leisure and sporting activities during both daytime and evening for local inhabitants. Nobody denied that this was a commendable policy and that it would be a reasonable use of the land. Evidence was led to prove that there is a dearth of leisure and sports facilities in the town and that the cattle market site, being in the centre of the town and adjacent to existing baths and a park, would be a suitable site for the centre. Funds had been allocated by the applicants and the county council for phase 1 of the development, which is estimated to cost £660,000 and will be spread over three years.

Furthermore, it was argued for the applicants that the need for the cattle market had declined since 1876 because of cheap road transport, the availability of other outlets, and the establishment of large wholesalers. On the other hand, it was shown that the market is much used; thus in 1975 5,684 cattle were sold (which is almost exactly the same number as in 1970), 10,578 sheep and 4,207 pigs, which number would have been much higher but for swine vesicular disease. Evidence was given that the closure of the market would be a great inconvenience to farmers and butchers; they would have to travel greater distances to other markets; those markets would become overloaded. I have no doubt that the covenant does secure benefits of substantial value or advantage to farmers and others within the manor and certainly substantial potential advantages to the objectors themselves so far as their own land is concerned and that, as the objectors are the moral guardians of those advantages, the covenant does secure practical benefits of substantial value or advantage to the objectors.

Accordingly, I have to consider the much more difficult question of whether the covenant in impeding the leisure centre use is contrary to the public interest. In doing so I must have regard to the development plan and any declared or ascertainable pattern for the grant or refusal of planning permissions in the area, as well as the period and context in which the restriction was created or imposed and any other material considerations.

In the approved development plan, the site is allocated for business use. Attempts were made to persuade me that a leisure centre is a business use and although I reject that contention it seems to me that that allocation has little relevance today. The reason is that in 1967 there was submitted to the Secretary of State a review town map. A public inquiry was held in 1970, and in 1972 the Secretary of State indicated approval of the town map subject to modifications.

However, that approval was informal and had no statutory effect, the reason being that new forms of development plans were introduced by the Town and Country Planning Act 1968. Nevertheless that town map can be taken for all practical purposes to reflect the Secretary of State's latest opinion on the way development should be carried out or controlled in Mansfield. It is in my judgment an important document in this case. It shows the application site as " industrial," but I would not regard that in itself as particularly significant, because although a leisure centre cannot be said to be an industrial use it would not necessarily be inimical in an area so allocated. But the review town map also shows an alternative site for the cattle market at Oak Tree Lane on land to be acquired by the local authority.

Thus it seems plain that both the planning authority and the Secretary of State took the view that although there was a preferred use for the present cattle market site, nevertheless a cattle market should be retained either there or at another site. That I regard as a very important material circumstance relevant to the development plan and to the public interest. Yet on the other hand the council do not intend to provide another cattle market at Oak Tree Lane or elsewhere. I think that in the circumstances the proposal to use the land for a leisure centre, contrary to the development plan allocation and without providing an alternative site for a cattle market, is in conflict with the development plan. In my view that aspect of the matter should have been drawn to the attention of the Secretary of State, who I think should then have held a public inquiry and made the planning judgment whether or not to permit the leisure centre with the knowledge that there would then be no cattle market.

It is not for me to make that planning judgment, and in the absence of it the burden seems to be fairly and squarely upon the applicants to show that, notwithstanding the planning position, it would be contrary to the public interest to allow the covenant to stand in the way of the leisure centre.

A I think the applicants would at least have to show firstly that there is no suitable alternative site for the leisure centre and secondly that it was either wholly impracticable or uneconomic to continue the cattle market at the present site or on another site.

Council " Judges in their own Cause "

So far as planning permission is concerned, what they did was to submit an application to themselves as planning authority in July, and in early September to grant themselves permission: as they were judges in their own cause, I cannot regard that decision as being at all persuasive on B town planning grounds. In all the circumstances I come to the conclusion that the application fails under paragraph (aa).

I should add that in any event I would not have exercised C my discretion in favour of the applicants. In my view they should have considered the planning implications of this matter at a much earlier stage, drawn the attention of the Secretary of State to the provisions of the town map and asked him to hold an inquiry. I do not think that the public interest in favour of the leisure centre necessarily outweighs the convenience and importance of the cattle market. Further, even if the covenant were discharged, it would not necessarily follow that the site would be used as a leisure centre.

The applicants will pay the costs of the joint objectors represented at the hearing, such costs unless agreed to be D taxed by the Registrar of the Lands Tribunal on the High Court scale.